MW00610965

CASTING EQUIPMENT ENGINEERING GUIDE

Jagan Nath

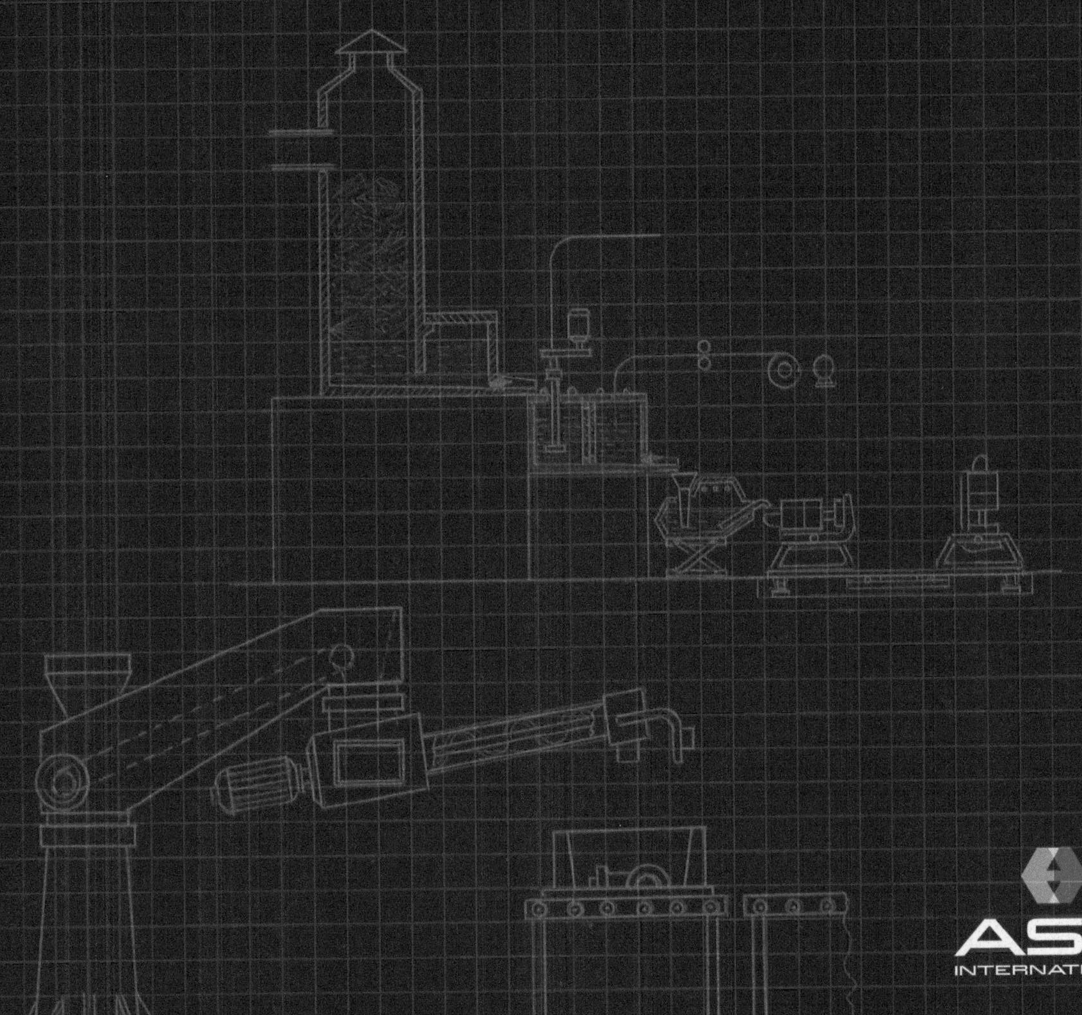

ASM INTERNATIONAL

ASM International®
Materials Park, Ohio 44073-0002
www.asminternational.org

Copyright © 2023
by
ASM International®
All rights reserved

No part of this book may be reproduced, stored in a retrieval system, or transmitted, in any form or by any means, electronic, mechanical, photocopying, recording, or otherwise, without the written permission of the copyright owner.

First printing, November 2023

Great care is taken in the compilation and production of this book, but it should be made clear that NO WARRANTIES, EXPRESS OR IMPLIED, INCLUDING, WITHOUT LIMITATION, WARRANTIES OF MERCHANTABILITY OR FITNESS FOR A PARTICULAR PURPOSE, ARE GIVEN IN CONNECTION WITH THIS PUBLICATION. Although this information is believed to be accurate by ASM, ASM cannot guarantee that favorable results will be obtained from the use of this publication alone. This publication is intended for use by persons having technical skill, at their sole discretion and risk. Since the conditions of product or material use are outside of ASM's control, ASM assumes no liability or obligation in connection with any use of this information. No claim of any kind, whether as to products or information in this publication, and whether or not based on negligence, shall be greater in amount than the purchase price of this product or publication in respect of which damages are claimed. THE REMEDY HEREBY PROVIDED SHALL BE THE EXCLUSIVE AND SOLE REMEDY OF BUYER, AND IN NO EVENT SHALL EITHER PARTY BE LIABLE FOR SPECIAL, INDIRECT OR CONSEQUENTIAL DAMAGES WHETHER OR NOT CAUSED BY OR RESULTING FROM THE NEGLIGENCE OF SUCH PARTY. As with any material, evaluation of the material under end-use conditions prior to specification is essential. Therefore, specific testing under actual conditions is recommended.

Nothing contained in this book shall be construed as a grant of any right of manufacture, sale, use, or reproduction, in connection with any method, process, apparatus, product, composition, or system, whether or not covered by letters patent, copyright, or trademark, and nothing contained in this book shall be construed as a defense against any alleged infringement of letters patent, copyright, or trademark, or as a defense against liability for such infringement.

Comments, criticisms, and suggestions are invited, and should be forwarded to ASM International.

Prepared under the direction of the ASM International Technical Book Committee (2022–2023), Richard Colwell, Chair.

ASM International staff who worked on this project include Scott Henry, Senior Content Engineer; Karen Marken, Senior Managing Editor; Madrid Tramble, Manager of Production; and Karesha Nutter, Production Coordinator.

About the Cover: Top image is a row of LPPM machines. Used by permission, LPM Group-Italy. Bottom image is an automated pouring unit in operation. Used by permission, Sinto America/Robert-Sinto.

Library of Congress Control Number: 2023941316

ISBN-13: 978-1-62708-445-1 (print)
ISBN: 978-1-62708-446-8 (pdf)
ISBN: 978-1-62708-447-5 (electronic)

SAN: 204-7586

ASM International®
Materials Park, OH 44073-0002
www.asminternational.org

Printed in the United States of America

To my wife Geetha

Copyright © 2023 ASM International®
All rights reserved
www.asminternational.org

Contents

Copyright © 2023 ASM International®
All rights reserved
www.asminternational.org

Preface

The casting process is one of the most efficient and unique cost-competitive methods of shaping a component. We have come a long way from the age-old foundries of the past which were noisy, dusty, dark, hot, and labor-intensive. Today's automated casting facilities are relatively quiet, less dusty, and bright. Modern machines and automation offer a high degree of productivity and consistency of product quality.

Labor shortages, difficulty of operator retention against competing industries, and increasing wages, have also compelled innovation in engineering and automation in machinery. The focus of mechanization has been to replace operators in hazardous and uncomfortable areas, reduce the risk of injury, and improve the overall environment.

Iron, steel, and aluminum casting manufacturers, as well as those pursuing manufacturing engineering of equipment for metal castings, benefit from a compilation of all aspects of preparation and manufacturing including sand and metal storage and handling, equipment for sand preparation, retrieval, conditioning, and reclamation, molding flask design options, melting and metal handling of iron steel and aluminum, sand molding, and core making equipment for iron and steel, and all aspects of material movement through the manufacturing processes of aluminum die casting, aluminum gravity permanent and semipermanent mold casting, low pressure permanent and semipermanent mold casting, and counter-pressure casting.

New casting facilities depend on consulting companies well-versed in the field of project planning and plant layouts. Many production facilities engaged in expansions of their current operations involve their staff and operators in designing the changes or expansions, and utilize their experience and expertise in their products. New market opportunities or changes in businesses also prompt new layouts, modern equipment, and improved automation.

The purpose of this book is to highlight all aspects of the entire line of equipment involved in the preparation and production of iron, steel, and aluminum castings, and their distinguishing features and innovations for higher productivity, higher dimensional capability, high integrity, and cost competitiveness. The book also offers guidelines for the selection of the right equipment, and for positioning them suitably, ensuring smooth product flow with high productivity.

This practical guide should be of special interest to all those involved in the casting manufacturing engineering field and those who are engaged in new equipment purchases, expansions, modifications, and automation of new machinery. Casting buyers and purchasing managers would also find this guide very useful.

Jagan Nath

Copyright © 2023 ASM International®
All rights reserved
www.asminternational.org

About the Author

As a mechanical, metallurgical, and materials engineer by profession, Jagan Nath has contributed to the advancement of the iron, steel, and aluminum casting industry. He became the vice president of advanced engineering of Amcast Automotive, a division of Amcast Industrial Corporation, in Michigan, and later, the vice president of engineering of General Aluminum Manufacturing Company's engineering division in Michigan.

The author has more than five decades of experience in iron, steel and aluminum castings manufacturing and engineering in a variety of industries.

Nath has contributed significantly to automotive vehicle mass reduction through many conversions to aluminum from iron and steel. He pioneered the successful application of the low-pressure process to safety-critical suspension components. His unique and pioneering ideas implemented in gravity and low-pressure permanent molding earned him the "Technology Leadership Award" from Amcast Industrial Corporation and the "Merton Fleming's Award" for innovation, from Advanced Casting Research Center in Massachusetts.

He has guided several teams of engineers in product-process development and problem solving of iron, steel, and aluminum castings in a variety of industries for over four decades. He has authored many papers that have been published and presented at ASM International's, American Foundry Society's, and SAE International's conferences and in many other forums. He is also the author of accompanying publications, *Aluminum Castings Engineering Guide* (ASM International, 2018) and *Iron and Steel Castings Engineering Guide* (ASM International, 2022).

Casting Equipment Engineering Guide
Jagan Nath
https://doi.org/10.31399/asm.tb.ceeg.t59370001

Copyright © 2023 ASM International®
All rights reserved
www.asminternational.org

CHAPTER 1

Casting Manufacturing Layout—Principles and Guidelines

A MANUFACTURING facility's location and the layout of the process equipment are two critical factors that impact product flow, productivity, operating efficiency, cost competitiveness, profitability, and the environment. These factors involve the selection of the most suitable site and the orientation of the plant; the infrastructure for incoming raw materials and finished goods; the availability of water, power, and labor; engineering of the right equipment for the manufacturing process and to assure product quality; automation for product consistency, in addition to addressing hazardous activities; and engineering for recyclability and regeneration. These vital elements all demand a structured approach, careful analysis, and professional project management to ensure the successful launch of a new manufacturing facility.

1.1 Planning Sand Cast Foundry Layouts

Planning and laying out casting facilities involve a number of vital factors, such as available infrastructure, selection of suitable sites, orientation of operations and process flow, markets and products, operating parameters, and targeted hourly output and annual capacity. Guiding principles and layout concepts with these factors in mind are presented in this chapter.

1.1.1 Planning Logistics — Infrastructure and Site Selection

The planning of a manufacturing facility begins with the selection of the site. Normally, three or four alternatives are explored and several factors are evaluated before the final choice is made. The main considerations for determining the best location include:

- Availability of resources — quality and volume of raw materials, power, and water —and nearest power substation
- Labor availability — minimum labor costs, proximity of township, and competing industries in the area
- Geological data — seismic data, safe load-bearing capacity of soil, underwater currents (if any)
- Direction of winds and velocities — ensuring that wind blows the effluent gases of the foundry away from the direction of the proximate township
- Humidity and temperature limits — affecting the ventilation, heating, and air conditioning costs
- Transportation infrastructure — highway and railroad proximity
- Topology of the site — influencing the construction of roads and railroad tracks
- Location of markets and freight costs
- Availability of land for future expansion
- Tax incentives offered by the local state or county to offset start-up costs

Figure 1.1 is a schematic illustrating some of these considerations; it shows the location of the casting facility relative to the township or residential areas and also the location of furnaces, wind direction, and the proximity of the site to railroads and highways.

A comparison matrix is compiled, listing a handful of options, and the priorities are identified. The priorities are assigned numbers as weighted averages. The management or consultant will make the final selection of the site based on the sum of the grades estimated for each option, multiplied by the weighted averages assigned for each priority.

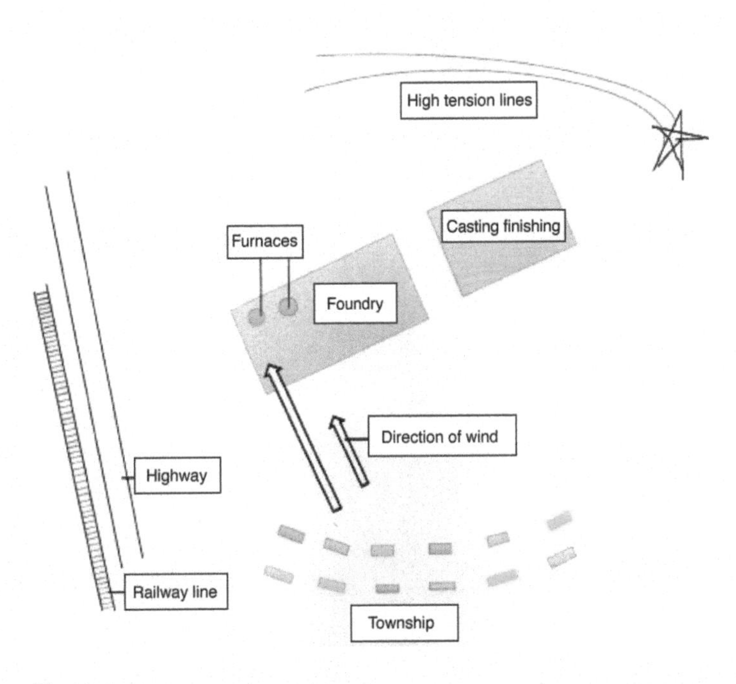

Fig. 1.1 Site selection considerations

1.1.2 Basic Elements of a Casting Facility

Most ferrous castings (iron and steel) are poured into sand molds. Castings produced in sand molds are called sand castings. (Thin-walled high-alloy steel castings for high temperature and corrosion resistance for aero-engines are poured into ceramic molds using the investment-casting process)

The sand casting process consists of melting metallic raw materials such as pig iron, steel, and scrap and pouring the resulting melt into sand molds, which are shaped into a finish casting configuration designed to perform a targeted function as a component in machinery or equipment. Metallic permanent molds are more common for casting nonferrous alloys such as aluminum. These are addressed in Chapter 8, "Aluminum Die Casting Equipment," Chapter 9, "Gravity Permanent and Semipermanent Mold Equipment," and Chapter 10, "Low Pressure Permanent and Semipermanent Molding, and Counter Pressure Casting," in this book. The casting process for converting raw materials into finished products is one of the most economical methods of shaping a product for use.

Figure 1.2 is a schematic showing the major elements of the sand-casting process. The refractoriness of silica sand, its abundant availability in nature, and its ability to be recycled are the main characteristics that make sand the most suitable and economical material for molds. In the green sand-molding process, sand is bonded with clay and water and compacted to achieve the strength and rigidity needed for retaining the shape of the mold while the metal is poured. Hollow castings are produced by inserting cores to form the inner cavity; the cores are produced out of sand, using a resin-based binder to give the strength and rigidity required for handling and setting the core in the mold.

After the poured casting has solidified, the mold and the casting are separated. The casting moves to the next steps of cleaning, degating, definning and grinding, before being shipped to the machine shop.

The mold sand is then recycled for reuse, and the core sand is mixed with it. Hot sand that has been knocked out of the molds passes through a cooler and returns to the sand hopper and the muller for blending with clay and water. The mulled sand is then ready for reuse in molding. The molding flasks used to contain the mold sand move from the knockout station back to the molding station. Gates, feeders, and any flash formed in the casting process are separated at the degating and definning stations and conveyed back to the melting station for reuse.

This description explains the basic elements of the process. There are several process steps and various machinery and controls that are used in actual practice. This book addresses these aspects in detail in the remaining chapters.

1.1.3 Market Segments and Strategies — Sizes and Volumes

The layout of the molding, melting, and cleaning equipment and the extent of their automation all depend upon the process steps, product sizes, and production volumes. Casting sizes and annual volumes are influenced by the market segment. The

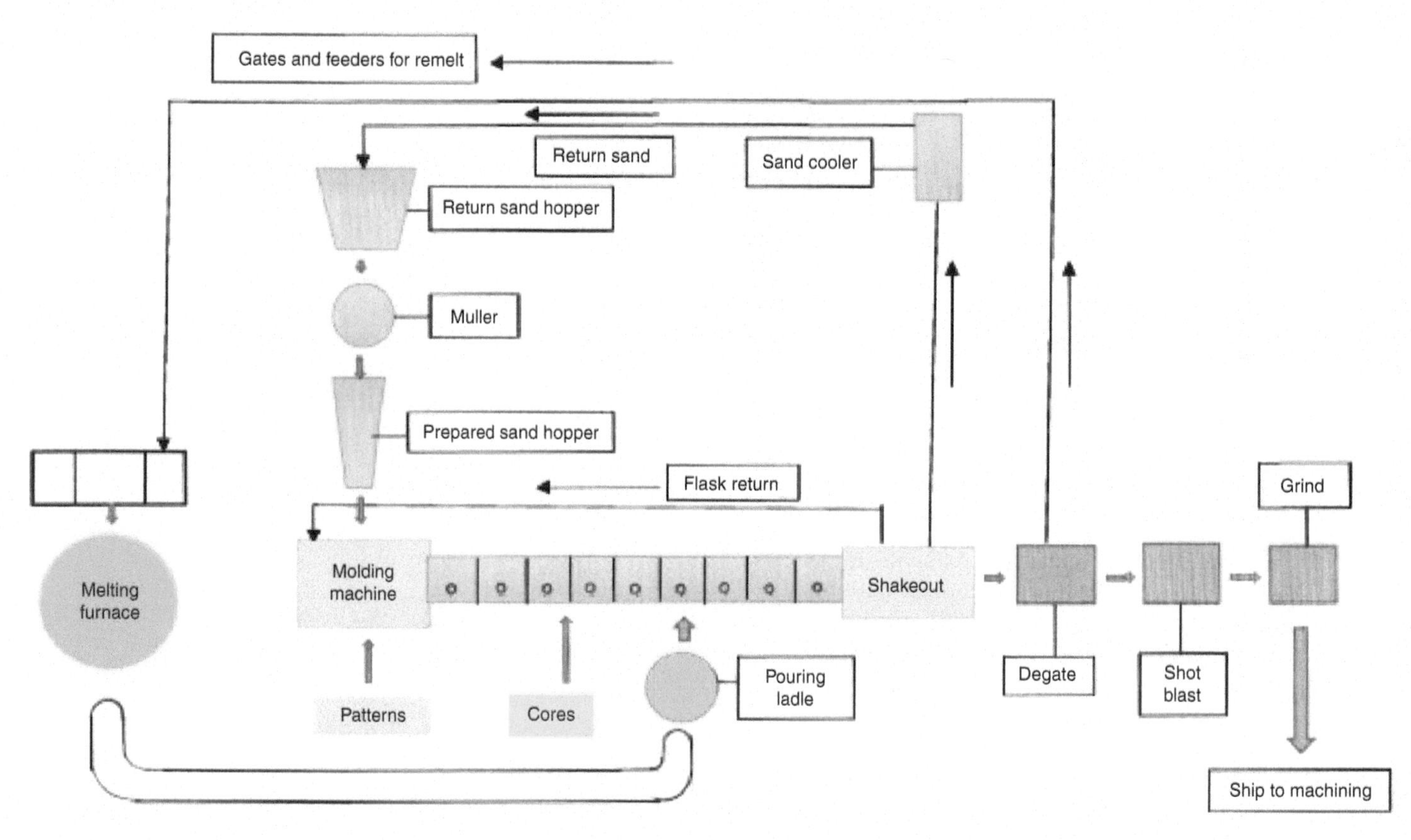

Fig. 1.2 Basic elements of a green sand-molding foundry

demand for consumer products such as pipe fittings, valves, pumps, and electrical products is high, and the casting sizes are small. The annual volumes of such items, which are known as mass production or large-volume jobs, number several hundred thousand castings per year or more.

This is followed by medium-volume jobs, such as those in the agricultural and transportation sectors. Castings for railroads, tractors, and luxury automobiles also belong in this category. These casting sizes are slightly larger, and their volumes run up to about 100,000 to 200,000 per year on average.

Heavy and large general engineering products for machine tools, steel plant equipment, and castings for fertilizer and chemical industries, earth-moving, and mining equipment belong to the third category. These are produced in volumes of a few hundred to a few thousand per year and are made to order. These types of products are designated as low- or small-volume jobs.

Targeted castings are grouped based on their weight, as shown in Table 1.1. The molding bays are classified as light, medium, heavy, very heavy, and extremely heavy bays, based on the number of molding bays in the layout.

Determination of the percentage of production in each weight class or the number of pieces in each class enables the construction of a casting weight and annual volume chart, also known as a Product-Quantity chart or P-Q chart. Figure 1.3 illustrates weight and volume relationships.

This chart can be used to divide the sand-casting facility into bays:

- Light bay—high automation (automated molding, core setting, pouring)
- Medium bay — medium automation with flexibility (match plate machine molding, manual core setting with or without fixtures, manual pouring with monorail)
- Heavy bay — low automation but with gantry and bridge cranes for mold handling, core setting, pouring, and knockout.

New manufacturing facilities are grouped into three categories, as indicated in Fig. 1.3:

- Group A — high-volume captive foundries
- Group B — medium-volume semicaptive foundries
- Group C — low-volume jobbing foundries

The casting sizes and the repetitive nature of the same or similar products of comparable weight or size will determine the extent of mechanization and capital outlay.

The large annual volumes of small castings in group A justify the significant capital investment that is required for high-production machines combined with a high degree of automation. Such high automation also improves the process capability and product consistency, in addition to the reduced labor cost that offsets the high capital costs. Foundries for group B products need to be engineered with reduced mechanization and flexible automation to handle different sized jobs. Large castings for group C are produced in smaller annual volumes. Such jobs are labor intensive, and skilled labor is required to produce them.

Some captive foundries have casting segments that fall in more than one category. The textile machinery industry is one example. In this situation, individual component castings in smaller sizes are needed in large quantities, while the frames of the textile machinery are large and heavy, and these are needed in smaller quantities. In such a case, one single foundry would need multiple bays with different degrees of automation and handling equipment.

Table 1.2 (Ref 1) lists several industries, cast metals, casting sizes, and output volumes. This table is provided for general orientation, as there may be exceptions depending on individual cases.

1.1.4 Operating Parameters

Market factors that influence capacity planning include:

- Assurance of continuity of the market for several years
- Scope and standardization of the product
- Need for diversification with fluctuation of market demand
- Variation of output needed — weekly, monthly, and yearly
- Warehousing or storage needed or freight availability
- Productivity increase anticipated in future
- Time to market — time lapse between the receipt of the order and the delivery of the product
- Need for a just-in-time system

The operating parameters that influence the equipment layout are:

- Process flow and the variations needed for different products
- Hours of work and the number of working days in a year

Table 1.1 Classification based on casting weights

Class	1	2	3	4	5
Classification of foundry or molding bay	Light	Medium	Heavy	Very heavy	Extremely heavy
Casting weight, kg					
Average	0–10	10–100	100–500	500–2000	2000–5000
Maximum	100	1000	5000	15,000	>15,000
Weight group	**Casting classification as per weight group, kg**				
1	0–1	0–5	0–10	0–50	0–100
2	1–2	5–10	10–30	50–100	100–500
3	2–5	10–30	30–50	100–250	500–2000
4	5–10	30–50	50–100	250–500	2000–5000
5	10–30	50–100	100–250	500–2000	5000–10,000
6	30–50	100–250	250–500	2000–5000	10,000–25,000
7	50–100	250–500	500–2000	5000–10,000	25,000–50,000
8	>100	>500	2000–5000	10,000–15,000	>50,000

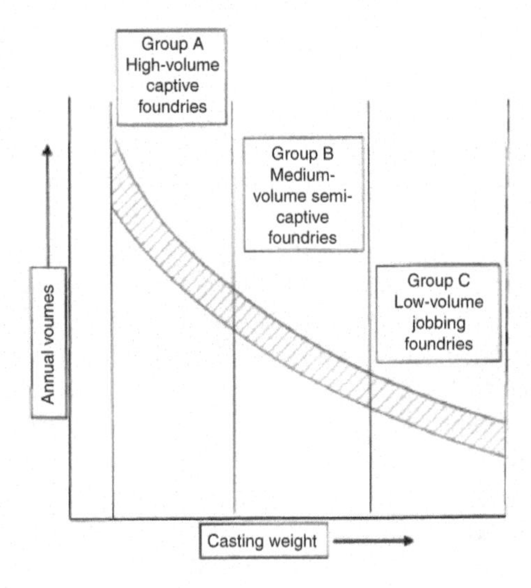

Fig. 1.3 Casting weight and annual volume relationship (PQ Chart)

Table 1.2 Casting industries, sizes, and volumes

Metal	Industry	Casting size	Annual volumes
Gray iron	Automotive	Small	Large
	Tractors	Small to medium	Medium
	Locomotives	Medium	Medium
	Heavy diesel engines, power generation, and marine engines	Large	Small
	Machine tools	Large	Small
	Steel plant equipment	Large	Small
	Turbines and power generation	Large	Small
	Boiler fittings	Small	Small
	Low pressure valves and fittings	Small	Large
	Rolling mill equipment	Large	Small
	Fertilizer and chemical industries	Medium	Small
	Paper and pulp machinery	Medium	Small
	Electrical machinery	Small to medium	Large
	Crane equipment	Medium	Small
	Presses and hammers	Large	Small
	Textile machinery	Small to medium	Large
	Agricultural machinery	Medium	Medium to large
Steel	Locomotives	Medium	Medium to large
	Turbines	Large	Small
	Steel plant equipment	Large	Small
	Rolling mill equipment	Large	Small
	Presses and hammers	Large	Small
	Cement mill machinery	Large	Small
	Sugar mill machinery	Medium	Small
	Excavators and earthmoving equipment	Medium	Small
	High pressure fittings	Small	Large
	Tractors	Medium	Medium to large
	Shipbuilding	Large	Small
Malleable iron	Automotive	Small	Large
	Valves and fittings	Small	Large
	Tractors	Medium	Medium to large
	Agricultural machinery	Medium	Medium
Ductile iron	Automotive	Small	Large
	Vans and SUVs	Medium	Medium
	Off-highway vehicles	Medium	Medium
Compacted graphite iron	Automotive	Medium	Medium
	Diesel engines	Medium	Medium

- Number of shifts working in a week
- Operating efficiency, or the percentage of uptime
- Activities scheduled for the nonmanufacturing shifts (third or night shift and weekends)

Highly mechanized and automated foundries of the group A type operate in three 8-hour shifts; the dependence on operators is minimal, and the high uptime helps to recover the significant capital investment. Foundries of the group B type generally operate in two or three 8-hour shifts or two 10-hour shifts, depending on the extent of mechanization and the skill set needed for operation. Group C foundries or low mechanized or heavy casting bays usually work in one or two shifts, depending upon the labor availability, skills needed, and the casting quality required.

1.1.4.1 Capacity Planning

Plant capacity is based on several factors:

- Cycle times and productivity of the machines chosen; the cycle times of the automated molding machines are the main guiding factors. The number of machines needed is based on the required output.
- Number of shifts of operation
- Number of working days in a year
- Estimated uptime of the equipment

Normally, the molding machine manufacturer will furnish details about cycle times and mold sizes. The size of the mold or flask should be chosen for the largest product in the group, to allow for space for mold sealing. Table 1.3 (Ref 1) provides guidance for selecting the flask size based on the pattern sizes and sealing areas. At this stage of planning, the parting plane and part orientation can only be estimated.

Machine size is determined by the flask size needed for the largest or thickest casting in the group. The number of

Table 1.3 Guidance for selecting flask size

Casting wall thickness, mm	Pattern height in cope or drag, mm (use higher value)	Sealing edge at gating A, mm	Other sealing edges B, mm
0–10	0–50	50	40
	50–100	50	50
	100–150	60	60
	150–200	60	60
10–15	0–50	50	40
	50–100	50	50
	100–150	60	60
	150–200	60	60
15–20	0–50	50	50
	50–100	60	60
	100–150	60	60
	150–200	70	70
	200–250	70	70
20–25	0–50	60	60
	50–100	60	60
	100–150	70	70
	150–200	70	70
	200–250	80	80
>25	0–50	70	70
	50–100	70	70
	100–150	80	80
	150–200	80	80
	200–250	90	90
	250–300	100	100

machines can be estimated based on the annual volumes of the class and weight group, machine cycle time rating, and estimated uptime. Bay width is selected to accommodate the number of machines and the standard size of the bridge crane, if needed. The length of the bay is estimated based on the length of the molding conveyor, which depends on the shakeout time.

1.2 Creation of a Plant Layout

Plant layout involves the arrangement of processing areas, machinery, and equipment for the efficient conversion of raw materials into finished products. Targeted objectives require that:

- All related activities are integrated
- Each process step moves the material forward
- Work flows smoothly, balancing the cycle times without any backflow of material
- The space is utilized effectively — horizontally and vertically (in all three dimensions)
- Provision is made for storage and handling of incoming material
- Provision is made for interim storage of cores and other similar needs close to the molding area for immediate use
- The arrangement of machines and their integration should be flexible to address alternatives in the case of breakdowns and to provide for equipment maintenance
- Design takes into consideration the safety of workers

A well-planned layout is essential for efficient material processing in order to maximize the throughput and minimize the cost of manufacturing. Planning for the receipt and storage of raw materials, balancing the production rates of the connected activities, and controlling the work in process are all critical to the profitability of the business and the growth of the enterprise. Engineering for delivery of raw materials, shipment of finished goods, and the disposal of waste products are important additional elements of layout planning.

1.2.1 Layout Concepts

The melting facility is the heart of the manufacturing process from both logistical and environmental viewpoints. Ferrous charge materials are very heavy, and they are commonly delivered by railroad or truck. The topology of the surrounding area needs to accommodate the rail track to facilitate delivery of the charge materials to the storage bins adjacent to the melt shop. The melt shop needs to be oriented to avoid the effluent from the stack, so that it can be blown away from the rest of the manufacturing plant, It is critical to capture the harmful effluents using filter bags and/or equipment that causes vortex motion.

The location of the furnaces permits flexibility in locating the molding bay if only one high-production molding machine is planned. The molding bay can be laid either parallel or perpendicular to the melt shop, as the metal delivery in this situation is automated, or at least a monorail loop is utilized. The molding shop occupies the largest area, and careful consideration must be given to the planning for efficient transport of the metal for pouring. The rest of the other operations — sand preparation plant, core room, casting cleaning shop, and others — are all nested around the molding area.

Figures 1.4 and 1.5 illustrate these concepts. Figure 1.4 shows the molding bay parallel to the melting bay, while Figure 1.5 shows the molding bay perpendicular to the melting bay. The rest of the support areas can be arranged suitably, depending on the area requirements and the footprints of the machines.

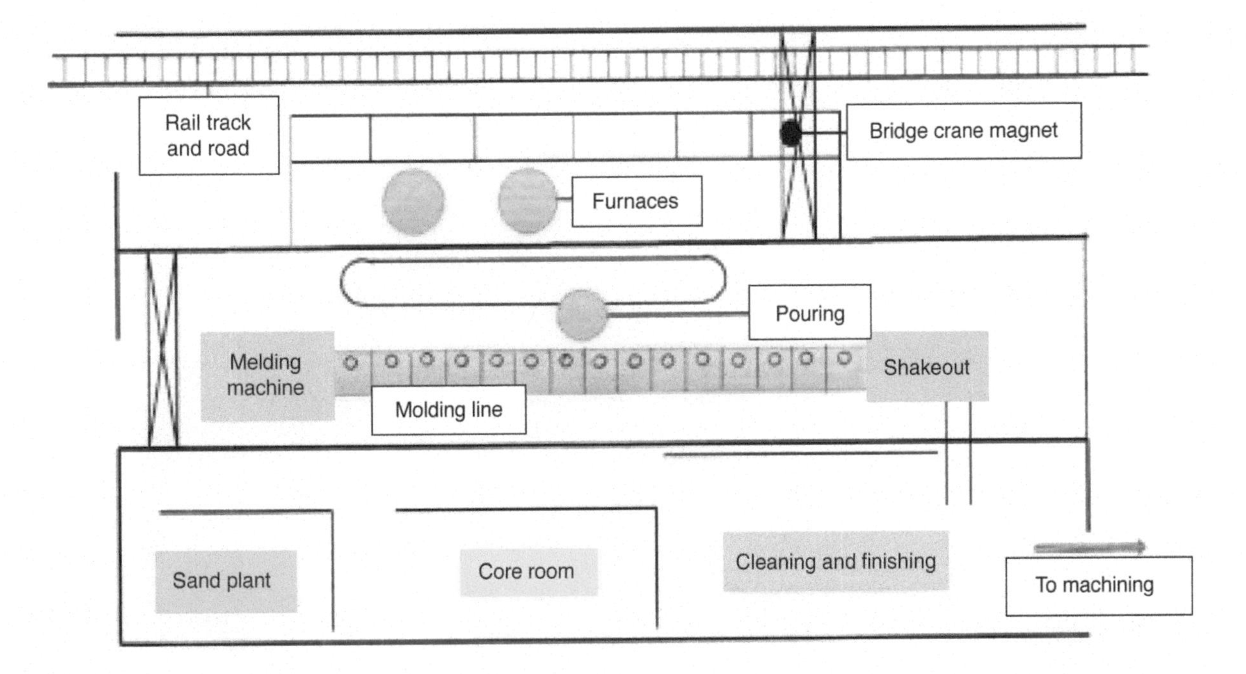

Fig. 1.4 Parallel bays layout concept

Many captive foundries are designed for multiple groups of castings of the various types (A, B, and C), which require different degrees of automation based on their annual output volumes. In such cases, the foundry is designed with multiple molding bays and centralized melting and sand preparation areas, core rooms, and cleaning rooms, and the handling of metal for pouring is engineered in different ways.

A layout with multiple parallel molding bays needs a cross-bay transport trolley to deliver the metal from the furnace to the individual bays, as illustrated in Fig. 1.6 (Ref 1). The high-production or high-volume bay may have a pouring monorail or an auto-pour furnace. A transfer car carries the ladle to the medium- and low-volume bays. The individual bridge crane picks up the ladle, and a pouring operator pours the castings.

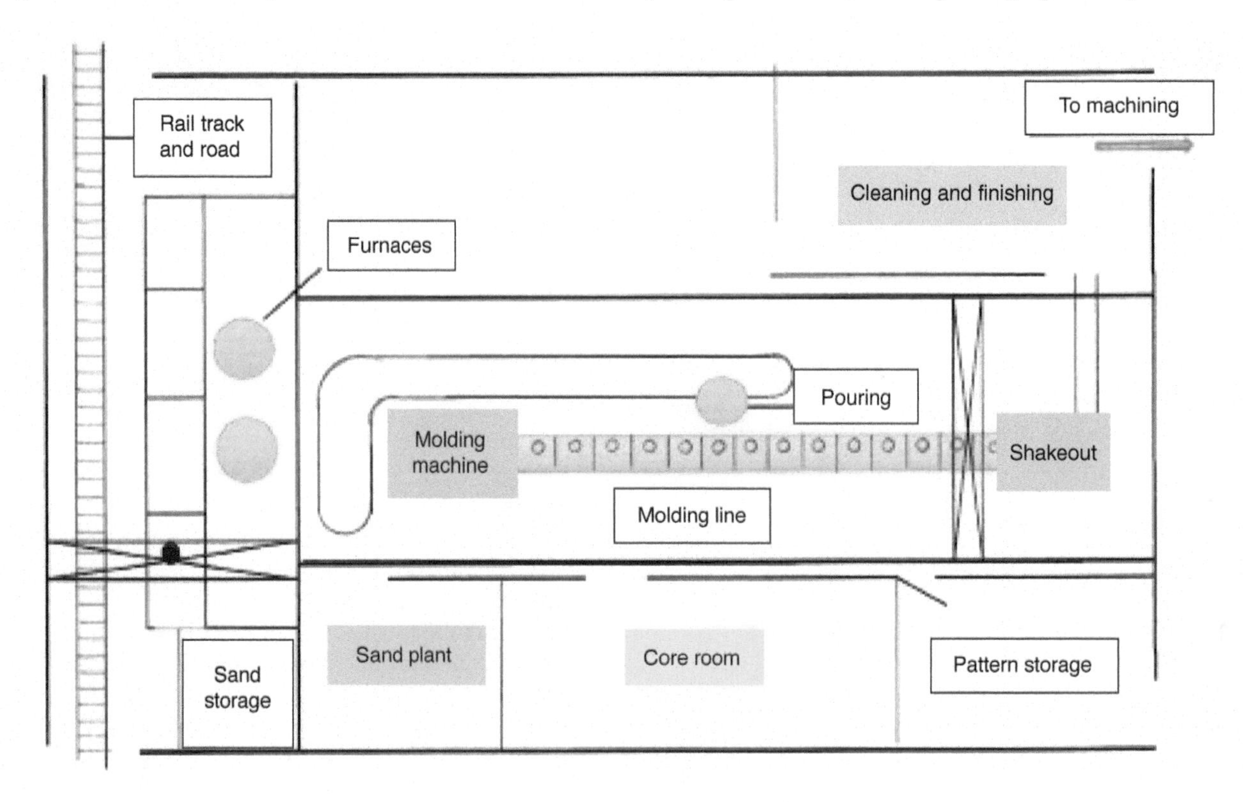

Fig. 1.5 Perpendicular bays layout concept

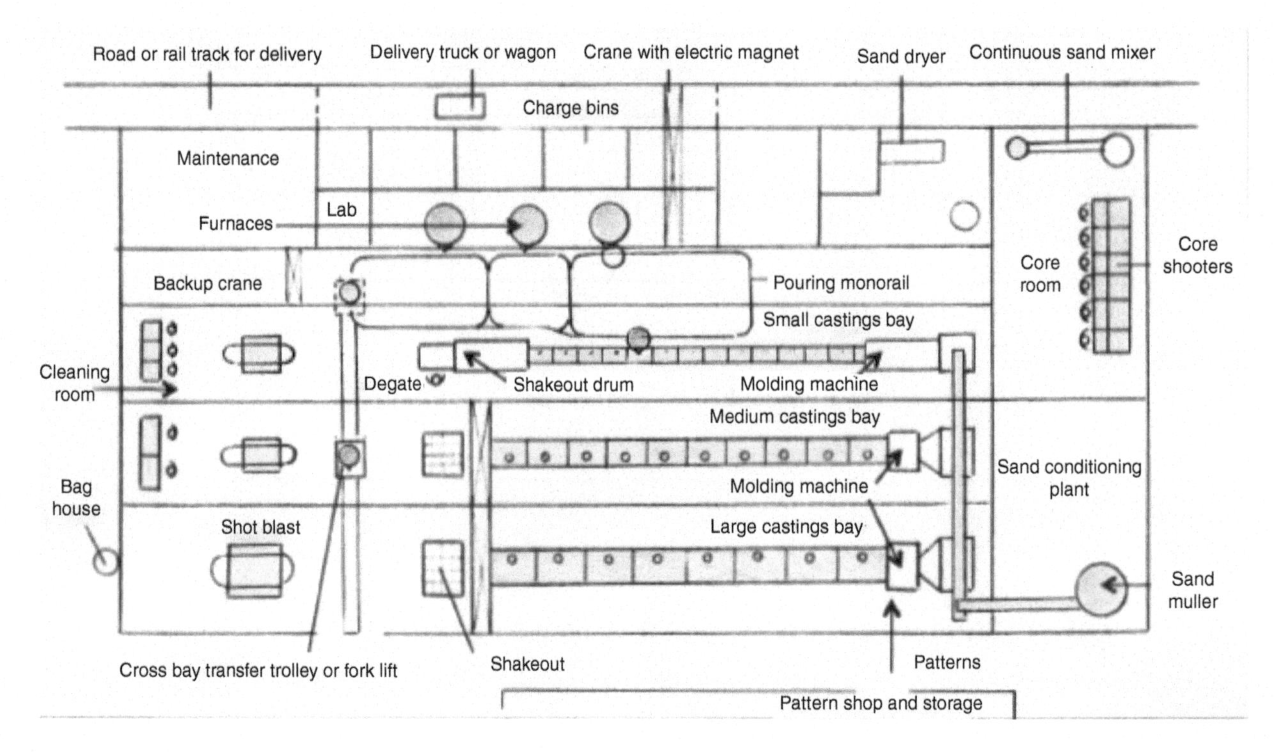

Fig. 1.6 Parallel layout of multiple bays

Some foundries use a forklift or a tow motor specially designed for metal handling for cross-bay metal transfer instead of a powered trolley.

The layout where the common melting bay is perpendicular to the multiple molding bays is engineered with an overlay of the crane access, as shown in Fig. 1.7 (Ref 1.) The metal is delivered to the medium-volume and low-volume bays by the overhead bridge crane of the melting bay, which sets it on the ladle station platform. The cranes in these individual bays pick up the ladle for manual pouring, or alternatively, a forklift can be used, as described in the layout for Fig.1.6.

1.2.2 Distribution of Areas

Table 1.4 (Ref 2) provides an idea of the distribution of production, storage, and other areas for a few typical industries. Other factors, particularly the extent of automation, will affect these values.

The provision of adequate areas for all production, storage, and auxiliary processes is critical. The layout of the equipment gives an idea of the production areas that are required, but some guidance is useful in benchmarking production, storage, and auxiliary areas.

1.2.3 Area Requirements for Different Alloys

Table 1.5 provides some guidance regarding areas in different types of casting facilities, to use as a guide or for benchmarking. This table accounts for parameters such as the alloy cast for both mechanized and jobbing foundries in iron, mechanized foundries of malleable iron, and jobbing steel casting facilities.

This table provides comparative estimates of areas and outputs for mechanized versus jobbing foundries. It also offers a comparison of the differences in areas and throughputs between jobbing gray iron foundries and jobbing steel casting facilities.

While the footprint of the selected machinery helps determine the layout plans, the storage area poses some challenges in ensuring that adequate provision is made. Table 1.6 offers some guidance for iron foundries. The areas for steel foundries can be determined by allowing a safety margin over the data found in Table 1.6.

Table 1.7 lists the outputs, molding, and cleaning room areas for gray iron and steel casting facilities in some typical industries with varying degrees of mechanization. This may help in benchmarking or in initial planning.

1.3 Guidelines for Developing a Layout

The development of a layout involves building up the different areas of the production process around the center of the activity, working out the details, and refining the arrangement to meet the equipment footprints and areas needed for material movement, while at all times ensuring the safety of operators.

General guidelines for developing a layout include:

- Generating a Product-Process Flow Diagram (Ref 1; see also the Appendix at the end of this chapter). This figure outlines the product movement as it traverses through the different process steps, including temporary storage, as required. It provides a roadmap for creating the layout.

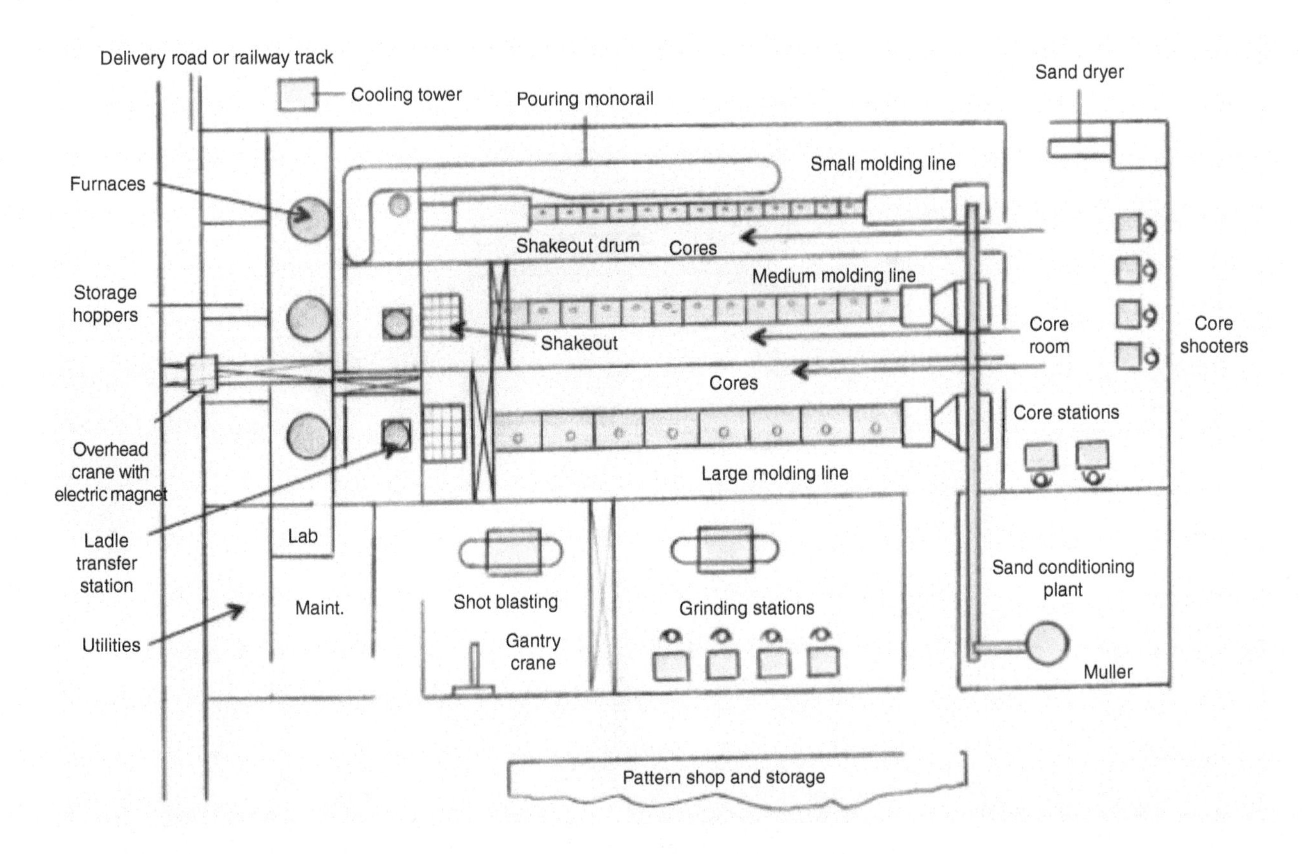

Fig. 1.7 Perpendicular layout of multiple bays

- Establishing the location of the melting shop, using the guidelines already provided. The molding area is fixed either to conform to Fig. 1.4 or 1.5 if only one high-production bay is envisioned. If more than one bay is needed, as per Table 1.1 or the P-Q chart (Fig. 1.3), the general layout may follow shown in Fig. 1.6 or 1.7. The logistics for receiving charge materials and distributing metal to the molding bays are priorities.
- Locating the sand conditioning plant adjacent or in close proximity to the molding area. Both the charge material transport and the incoming sand should share a common roadway or railway transport, if possible. This is not a high priority, however, because new sand is not received as often as charge materials are delivered.
- Locating new sand receipt and storage close to the core room, if sand drying is planned. A simple method for adding new sand to the molding sand to offset sand loss can be designed, if make-up additions are done once every two to three days.
- Locating the core room close to the molding shop; this will reduce handing damage to the cores or core assemblies. When cores are bought, adequate areas for receipt, assembly, and inspection need to be planned.
- Locating the stores for melting, sand conditioning, flask, and sprue bushing maintenance close to both the melting and the molding shop.
- Providing adequate length for the conveyors and the areas for casting in pit molds (where the drag is formed in a pit and the cope is formed in the flask), before the mold shakeout (The guidelines for casting cooling time and conveyor length calculation are provided later in this book.)
- Designing the type of shakeout equipment based on the alloy, casting configuration, and whether the flasks have crossbars or not. The hot return sand goes through a lump breaker, magnetic separator, and a hexagonal screen before entering a sand cooler. The hot sand return conveyor and the connected systems are engineered to be isolated from the foundry, for heat and environmental control.
- Locating the sand and metallurgical test laboratories; these activities are secondary, and therefore there is some flexibility allowable in their location.
- Superimposing the flowlines on the layout to ensure that there is no backflow of material and that adequate space is provided for the temporary storage of cores for immediate use in molds. The active and passive areas are confirmed.
- Holding brainstorming sessions with the individual departments for their input and evaluation. Changes should be made to meet their needs and to implement their recommendations.

Table 1.4 Distribution of areas for a few typical sand-cast iron foundries

Production, storage, and other areas, %	Type of market: Textile machinery	Agricultural machinery	Fittings and pumps	Electrical machinery	Machine tools	Others
Melting	3.8	4.6	3.9	4.7	4.1	3.1
Molding	47.5	41.0	36.4	40.8	38.1	39.7
Core room	3.6	3.6	5.5	8.1	9.9	5.5
Cleaning	11.0	10.0	7.6	9.7	12.3	11.0
A. Total production area	65.9	59.2	53.2	63.3	64.4	69.3
B. Total storage area	29.8	37.9	44.1	31.8	31.7	37.7
C. Other areas	4.3	2.9	2.5	4.9	3.9	3.0
Total of A + B + C	100	100	100	100	100	100

Table 1.5 Guidelines of areas for different cast metals

Type of foundry	Gray iron				Malleable iron		Steel	
Annual production of good castings, tons	5000–7000				10,000		8000	
	Mechanized		Jobbing		Mechanized		Jobbing	
	meter²/ton	%	meter²/ton	%	meter²/ton	%	meter²/ton	%
1. Melting	0.03	10	0.03	5	0.02	4	0.06	8
2. Molding and assembly	0.06	20	0.25	44	0.14	28	0.32	41
3. Pouring	0.04	13	...	...	0.04	13	...	...
4. Knockout	0.03	10	...	...	0.03	10	...	...
Total of 1 + 2 + 3 + 4	0.16	53	0.28	49	0.23	43	0.28	49
5. Drying ovens	...	...	0.04	7	...	...	0.06	8
6. Sand conditioning	0.03	10	0.03	5	0.02	4	0.02	2
Total of 1 + 2 + 3 + 4 + 5 + 6	0.19	63	0.35	61	0.25	47	0.46	59
7. Core room	0.05	17	0.11	19	0.04	8	0.06	8
8. Cleaning room plus Heat treatment	0.06	20	0.11	20	0.28	56	0.26	33
Total of production area 1 to 8	0.30	100	0.57	100	0.50	100	0.78	100
9. Storage charge plus sand	0.21	...	0.25	...	0.16	...	0.32	...
10. Other storage areas	...	...	0.15	...	...	...	0.11	...
Total storage area	0.21	...	0.40	...	0.16	...	0.43	...
11. Auxiliary sections	0.05	...	0.30	...	–	...	0.06	...
12. Cloak room plus offices	0.04	...	0.06	...	0.04	...	0.06	...
Total foundry areas	0.6	...	1.06	...	0.70	...	1.38	...
Molding shop utilization, t/m²/yr	6.25	...	3.57	...	6.25	...	2.64	...
Utilization including sand conditioning, t/m²/yr	5.30	...	2.86	...	5.55	...	2.17	...
Total overall utilization, t/m²/yr	1.77	...	1.00	...	1.53	...	0.79	...

Table 1.6 Storage area guidelines for gray iron foundries

	Annual output in tons						
Areas in square meters	5000	7500	10,000	12,500	15,000	20,000	25,000
Metal charges for daily needs	108	150	200	250	300	410	560
Fuels, fluxes, and refractories	144	216	252	336	360	480	570
Metal charge bunkers	180	216	252	288	288	288	432
Total area of charge materials	432	582	704	838	948	1178	1562
Molding materials	396	594	714	882	984	1296	1488
Sand storage for 4-month needs plus sand dyer area	216	216	252	252	288	288	432
Total area for molding materials	612	810	966	1134	1272	1584	1920
Total area for storage	1044	1392	1670	1972	2220	2762	3482
Total storage area with handling equipment	1296	1638	1764	2016	2268	2880	3600
Storage bunker length, meters	72	78	84	96	108	120	150
Storage bunker width, meters	18	21	21	21	21	24	24
Storage bunker height, meters	6	6	7	7	8	8	8
Height of crane track, meters	8	9	9	9	10	10	10
Number of cranes with lifting magnets and grab buckets	1	1	1	1	2	2	1

- Verifying the equipment footprints and access for maintenance. Headroom spaces for handling and replacement, safe clearances, and ventilation must be verified and confirmed.
- Examining the need and extent for future expansion and addressing any landlocked or topological bottlenecks. The future bays are drawn, and the expansion or modification of the metal delivery and prepared sand delivery to the molding machines, along with the return sand system and core delivery, should be thoroughly examined.
- Generating a 3D CAD model; this is very useful after the first 2D layout. It provides a very good visualization for clearances and interferences and provides opportunities for improvement. The ability to view in three dimensions is a great advantage in generating a workable layout once the basic layout and material flow and the required areas have been established.
- Using scaled models to assist in the presentation of the layout to management for their consideration and approval.

Table 1.7 Guidelines for casting output of different industries

Market segment	Annual output, tons (multiples of 1000)	Production type	Degree of mechanization	Output of molding area, t/m^2/yr	Cleaning room area as a % of molding area
Gray iron foundries					
Pumps and compressors	Up to 3	Series	Low	2.8–3.6	60–80
Fittings	10–12	Large series	Medium	6–8	30–40
Textile machinery	10–15	Large series	Medium	6–8	20–30
Tractors	80–100	Mass production	Hight	12–16	70–75
Heavy machine tools	10–12	Medium production	Medium	3.5–5	40–50
Steel plant equipment	15–20	Piece production	Medium	4–4.5	30–40
Steel foundries					
Tractors	25–40	Large series	Hight	7–9	100–120
Railroad locomotive	40–60	Series	Medium	5–6	100–120
Turbines	15–20	Small series	Medium	2–2.5	100–120
Forging presses	5–8	Piece production	Low	3.5–4.5	100–120
Steel plant equipment	20–25	Piece production	Medium	3–3.5	100–120

Appendix

Sample of a product-process flow diagram.

XYZ Company — **Product - Process Flow Diagram** — Doc. No PPFD No. 0001, Rev. Date : 01/05/2017

Customer: ABC Industries	Part Name : Rear Knuckle	Part No. 123456/LH
Date: 01/05/2017		Page 1 of 4

Step	Oper	Move	Insp	Store	Operation Description	PFMEA No	Key Product Characteristics	Oper. No	Key Control Characteristics
1a					Receive Tooling	1		1	
1b					Inspect Tooling	1	Designed Shape	1	Dimensions
1c					Assemble Tooling	1		1	
1d					Tooling Assembly Check	1		1	
2a					Alloy Coupon Analysis	2	Material Specification	2	
2b					Unload Alloys from truck	2		2	
2c					Verify & confirm Heat lot	2		2	Heat Lot No.
3a					Deliver alloy to melt shop	3	Material Specification	3	Alloy color code system
3b					Store in covered area	3	Material Specification	3	Alloy color code system
3c					Move alloy to melting	3	Material Specification	3	Alloy color code system
3d					Melt alloy	3		3	
3e					Transfer metal to holder through launder	3		3	
3f					**Flux & skim**	3	Melt Cleanliness	3	
3g					Alloy additions	3	Material specifications	3	
3h					Skim furnace	3		3	
3i					Inspect melt – Chemistry & Temperature	3	Material Specification & Process Specification	3	
3j					Hold melt in holder	3		3	
3k					Deliver metal to mold	3	Metal Volume Control	3	
4a					Pour castings	4		4	
4b					Solidify & cool castings	4		4	
4e					Store for trim	4		4	
5a					Move to trim & de-flashing	5		5	
5b					De-flash	5	Trim bead burr finish	5	Operator Insp. of burrs
5c					Move to de-gate station	5		5	
5d					Remove feeder		Feeder pad height control	5	Feeder pad height per design spec.
5e					Move castings to Radioscopic Inspection			5	

5f		Radioscopic Inspection		Internal soundness	5	Internal soundness specification
6a		**Load into baskets**	6		6	
6b		**Load into furnace**	6		6	
6c		**Heat treat -Aust.**	6		6	**Time and temperature**
6d		Quench	6		6	Quench time
6e		**Move to Tempering**	6		6	
6f		**Temper**	6		6	
6g		Verify heat treat cycle records	6		6	
6h		Test for Mech. properties	6	Mech. Props. TS, YS, Elongation, Hardness	6	Tensile properties & BHN
6i		Test for Fatigue Strength	6	Durability or Part Life	6	Fatigue results –Life and Failure mode
7a		Move to Dye -penetrant	7		7	
7b		Do Dye Penetrant Insp.	7	No Anomalies	7	Visual standard
7c		Move to packing	7		7	
8a		Packing castings in dunnage	8	Separate LH & RH castings in stacking	8	Automatic sorting through dunnage
8b		Hold for Audit	8		8	
8c		Final Audit and Release	8		8	Release Control
9		Ship castings to customer	9	Part count	9	Shipment tally

Process Symbols

◊	O	□	Δ
Operation	Move	Inspect	Store

REFERENCES

1. J. Nath, *Iron and Steel Castings Engineering Guide*, ASM International, 2021
2. J. Siroky, *How to Mechanize Foundries* (*Jak pracovat mechanizovanych slevaren*), Statni Nakladatelstvi Technicke Literatury (SNTL), 1965

Casting Equipment Engineering Guide
Jagan Nath
https://doi.org/10.31399/asm.tb.ceeg.t59370013

Copyright © 2023 ASM International®
All rights reserved
www.asminternational.org

CHAPTER 2

Sand and Metal Charge Storage and Handling

SAND AND METALLIC charge materials are two essential and heavy raw materials that are needed for molding and casting. Planning for efficient receipt, storage, and handling of these materials is a key element of manufacturing engineering. For every ton of casting shipped, nearly 1.4 to 1.6 tons of metal is melted, and 5.5 tons of sand is used for molding (Ref 1). Each of the raw materials has individual characteristics that require unique modes of transport, storage, and handling. Raw materials are procured in bulk and stored separately to ensure adequate supply to meet fluctuating production demands. The receiving system is automated to minimize labor, and the storage system is designed for first-in and first-out to prevent any deterioration. This chapter focuses on planning and provision for storage and handling of the raw materials needed for casting manufacturing.

Raw materials are procured according to specifications to ensure suitable process control and outgoing product quality. Suppliers are provided with specifications, and samples are tested for compliance before source approval. They are asked to provide certification of compliance before delivery to ensure that the wagons or trucks are not delayed on the unloading docks while awaiting test results and approvals. Random sampling is done at the foundry to validate the supplier certification. In some cases, the foundry is equipped with in-house testing, if the tests can be conducted quickly and the results confirmed before unloading the truck.

The major raw materials used for molding and casting are:

- Metallic charge materials — large bulk amounts
  a. Pig iron (for gray iron and ductile iron)
  b. Steel scrap (plates, bars, and railroad scrap)
  c. Briquetted steel turnings
  d. Briquettes or chips from cast iron machining
- Metallic charge materials — smaller packages
  a. Ferroalloys
  b. Inoculants
  c. Alloys
- Nonmetallic materials — large bulk amounts
  a. Silica sand
  b. Coke
  c. Limestone
  d. Olivine sand (steel foundries)
  e. Zircon sand (steel foundries)
  f. Chamotte (steel foundries)
  g. Refractories (furnace lining materials and tiles)
- Nonmetallic materials — smaller packages
  a. Bentonite
  b. Coal dust or sea coal
  c. Dextrin
  d. Core binders (resins and furan)
  e. Sodium silicate
  f. Mold coatings

In addition, foundries receive molding flasks, pattern plates, mold pallets, pins, and bushings periodically. The heavy items can be handled by cranes in the respective bays. The lighter items are handled by forklifts or tow motors, which have more flexibility.

2.1 Metallic Charge Materials

2.1.1 Large Bulk Amounts

An electromagnet suspended by cranes in the storage bay handles ferrous metallic charge materials. The charge materials are unloaded from the trucks and deposited in the concrete bunkers provided for each bulk category. The concrete bunkers are topped with heavy-duty wooden logs to prevent damage.

The bridge crane is operated from a cabin mounted on the bridge or remotely from an elevated position. Trucks carrying the material stop at a weighbridge to register the weight of the material.

The lifting magnet is equipped with a load cell to indicate the charge weight. This information is transmitted to a printer or a computer located next to the crane operator, and it is used to control the amount of charge that goes into each bucket. Large companies producing heavy iron castings equip themselves with a device known as a skull breaker, which consists of a steel ball that is lifted by an electromagnet and dropped on large castings to break them into smaller pieces suitable for charging. Bunkers are also provided close to the furnaces for the hourly replenishment of charge materials.

2.1.2 Smaller Packages

Ferroalloys, inoculants, and alloys are usually delivered on pallets. They are stacked in the allotted bunkers using forklifts.

Figure 2.1 illustrates the bunker layout for charge materials, coke, limestone, silica sand, bentonite, coal dust, and other materials. If the layout permits, (see Fig. 1.6 in Chapter 1, "Casting Manufacturing Layout—Principles and Guidelines" in this book), the bunkers for charge materials and sand are laid inline to use the same road and rail track for delivery. For the layout shown in Fig. 1.7 in Chapter 1, the rail track is used for the delivery of charge materials, while a separate access road is provided for sand and other materials. The schematic shows the layout for raw material transport, storage, and handling. It also shows the other connected elements for a study of the logistics involved. Ferrous charge materials are usually delivered by railroad. The bridge crane with an electromagnet unloads the materials in the respective bunkers allotted to the various materials. The amounts of charge materials for daily and hourly requirements are moved to bunkers closer to the induction furnaces (22). Automated charging is assisted by a vibratory conveyor with an upward swing chute. Different charging systems are available, depending on the layout. The furnace cooling heat exchanger and hydraulic pumps for furnace tilting are housed in area (7).

New sand is delivered mostly by road. Trucks unload the sand over a grating (10). The sand may be moist when it is delivered, and if so, it will need to be dried to use it for resin-bonded cores. A bucket elevator draws the sand from the hopper and delivers it to a sand dryer. Dried sand passes through a sand cooler (not shown). Cooled sand is delivered to the new dry sand hopper (13). The core sand mixer mixes the sand and delivers it to the core-making machines through the belt conveyor (18). Sand delivery to the molding sand hopper is engineered to skip the sand drying step if the moisture of the sand is at an acceptable level. Sand, clay binder (bentonite), and sea coal are mulled in the sand muller (15) and delivered to the hopper (19) above the molding machine and to all the molding bays through the belt conveyor (17). The molding machine (20) produces the molds and moves them onto the conveyor (21) for pouring.

The pouring ladles (24) supported by the pouring monorail (23) are used to pour the molds. The cooled molds pass through a shakeout unit where the castings are separated from the molds. The castings move to the cleaning room for shot-blasting, degating and deheading. The return sand from the shakeout unit passes through a magnetic separator, a lump crusher, and a sand cooler (not shown) before entering the return sand hopper (12). Cross-bay metal transport to the other molding bays is supported by the trolley (27).

Gates and feeders separated either at the shakeout unit or at the cleaning room are transferred to the charge storage and makeup bay using the trolley (29). Area (31) is the cleaning room, and area (32) is the core room. Area (33) is the next molding bay.

2.2 Raw Material Handling

2.2.1 Metallic Charge Materials — Large Bulk Amounts

The magnetic property of ferrous charge materials allows the relatively simple method of using electromagnets for unloading materials from railroad cars, storing it in bunkers, and delivering charge materials to the furnaces.

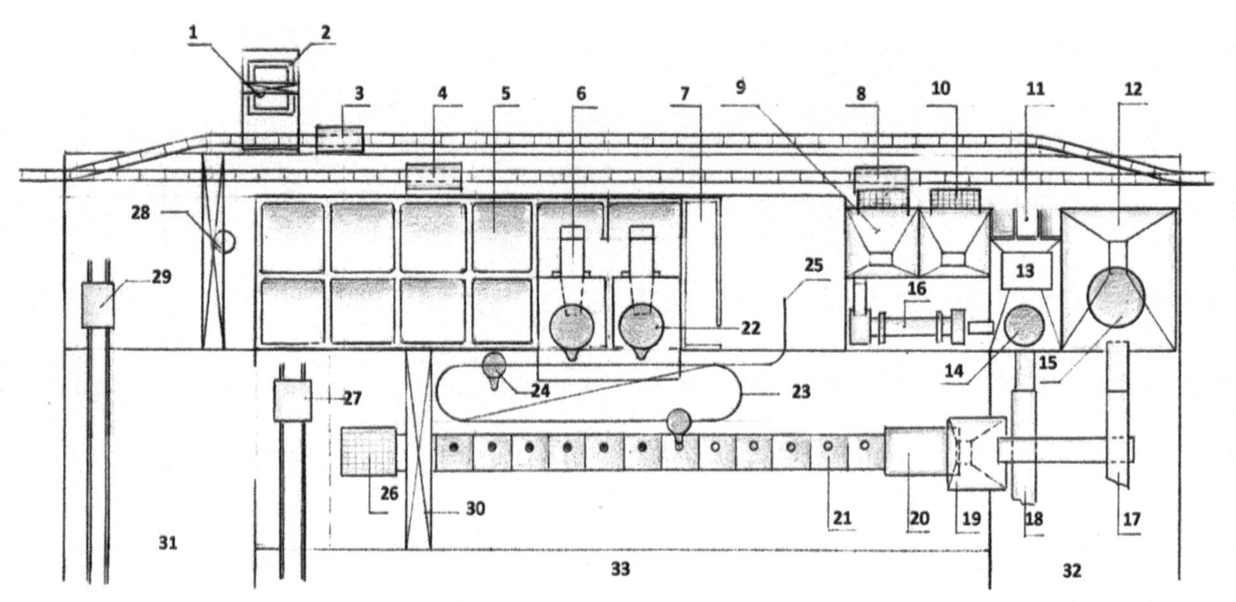

Note: 1. Skull cracker drop weight 2. Safety enclosure 3. Rail wagon hauling scrap for skull cracker 4. Rail wagon for hauling scrap 5. Metallic charge bunkers 6. Charging conveyor 7. Induction furnace room 8. Truck or rail car hauling new sand 9. New sand receiving hopper 10. New sand receiving grating 11. Bentonite, sea coal and dextrin storage 12. Return sand hopper 13. New dry sand hopper 14. Core sand mixer 15. Green sand muller 16. Sand dryer 17. Green sand conveyor to all bays 18. Core sand conveyor to all core-making machines 19. Sand hopper over molding machines 20. Molding machines 21. Molds produced 22. Induction furnaces 23. Pouring monorail 24. Pouring ladle 25. Ladle preparation area 26. Shakeout unit 27. Cross bay metal distribution trolley 28. Charge handling crane with electromagnet 29. Trolley for transfer of gates and feeders for remelting 30. Crane for handling transfer ladle 31. Cleaning room 32. Core room 33. Next molding bay

Fig. 2.1 Storage and handling bays for charge material, sand, and raw material

Electromagnets for these purposes are suspended from overhead cranes, as shown in Fig. 2.2. Features of electromagnets are:

- Designed for heavy-duty use under demanding environments with heat and dust
- A strong circular cast steel case to protect the coils inside
- Aluminum or copper coils around the soft iron core
- Offered in two versions: standard field or deep field
- Radio controlled for remote and wireless operation
- Battery backup for up to 20 minutes, preventing accidental unloading due to power failure
- Load cells for weighing with display and recording capability for charge makeup

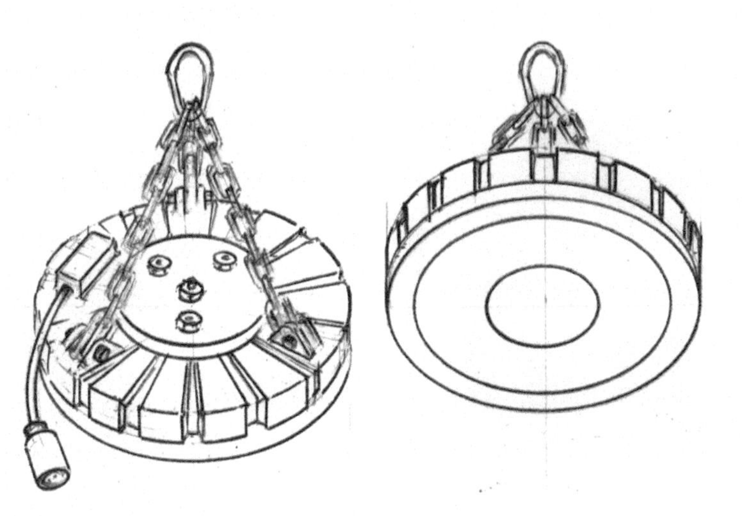

Fig. 2.2 Schematic of electromagnet for charge handling

- Lifting capacities depend upon the KW rating and the ferrous material handled (for example, ingots vs. plate steel or cast iron borings)
- Lifting magnets for production scrap handling – sizes range from 1200 to 1800 mm; lifting capacities are 1000 to 2500 kg. Larger sizes are available for scrap handling businesses. Industrial lifting electromagnets use 220 volts DC.

Electromagnets can handle pig iron, steel scrap, iron and steel briquettes, cast iron casting scrap, and cast iron machined chips. Ferroalloys and ferrosilicon (FeSi) inoculants are procured in granular form. They are loaded by forklift into small hoppers located close to the receiving ladle. A pneumatic door at the bottom of the hopper enables the granules to fall onto a rotating disc for good distribution. Some foundries connect the feed mechanism to a timer to control the amounts, especially if they are tapping into a larger ladle. Additions to the mid-sized ladles are done manually while tapping. Other alloys such as nickel and molybdenum are usually added in smaller quantities into the induction furnace to maintain close control.

2.2.2 Nonmetallic Materials — Large Amounts of Silica and Olivine Sands

Raw sand is relatively free-flowing unless it is very moist. Therefore, gravity is utilized in handling sand whenever the sand is able to flow downward. The trucks or wagons discharge the sand onto a grate, and the sand flows into a hopper located below ground level. The sand is carried through the system using conveyor belts and bucket elevators. Dry sand is handled using pneumatic systems. Figure 2.3 is an illustration of a delivery and handling system for sand.

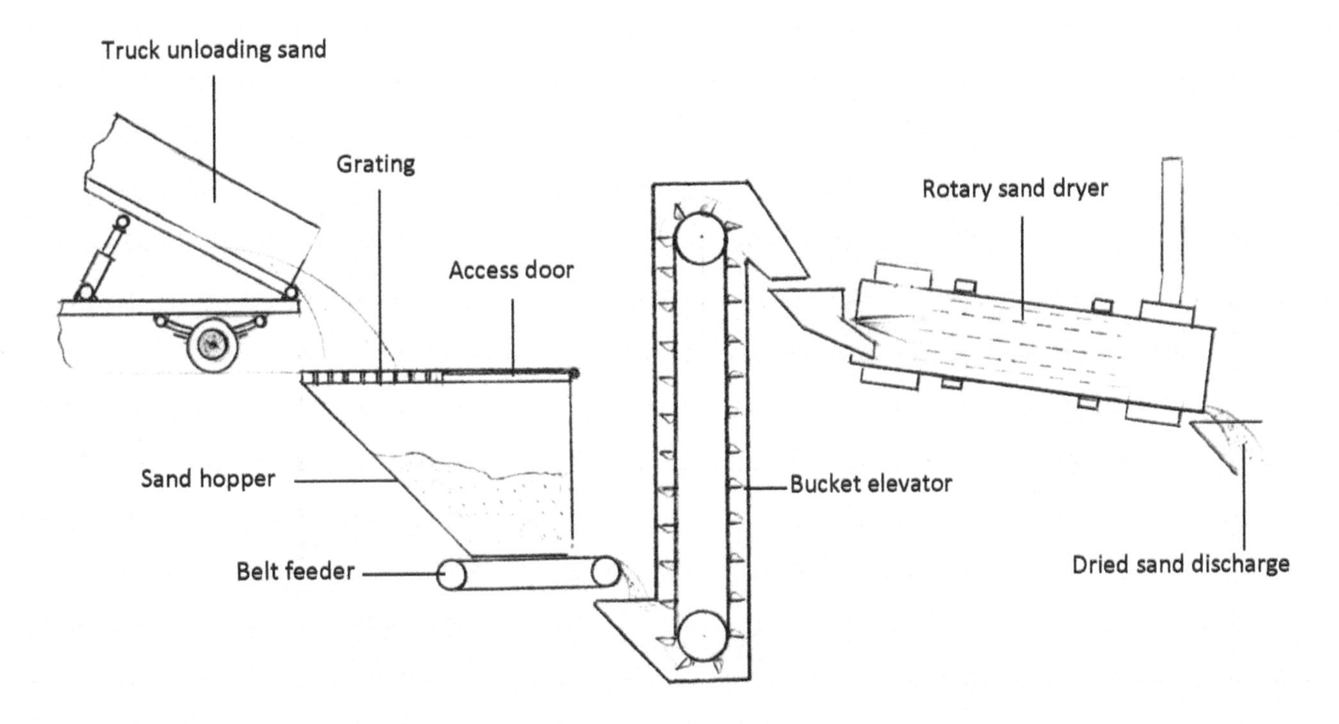

Fig. 2.3 System for handling sand delivery

2.2.3 Nonmetallic Materials — Coke and Limestone

Coke and limestone that are needed for melting in cupolas for gray iron castings are delivered by railroad or trucks. They are usually unloaded using a clamshell grab bucket, as shown in Fig. 2.4, which is suspended from an overhead crane in the charge storage bay and operated by the crane operator, either from the cabin or by remote control. They are either mechanically or hydraulically operated.

The grab bucket unloads coke into a hopper with a belt conveyor at the discharge end. The belt conveyor is timed to load the cupola charging bucket with the targeted amount. Clamshell grab buckets are very versatile and can be used for a variety of materials.

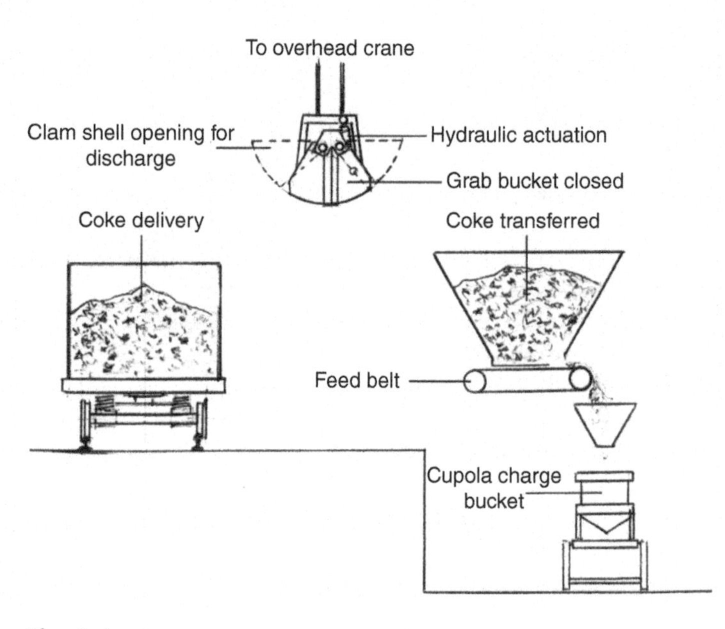

Fig. 2.4 Clam shell grab bucket handling coke

2.2.4 Nonmetallic Materials — Bentonite, Sea Coal, and Dextrin

Bentonite, sea coal, and dextrin are handled in powder form in packages on skids using forklifts. Improved control and automation have prompted the preparation of bentonite and sea coal slurry to replace the powder form.

2.2.5 Nonmetallic Materials — Resin Binders, Sodium Silicate, and Mold Coatings

Resin binders, sodium silicate, and mold coatings are obtained in liquid form in drums, and forklifts are fitted with an attachment for handing the drums.

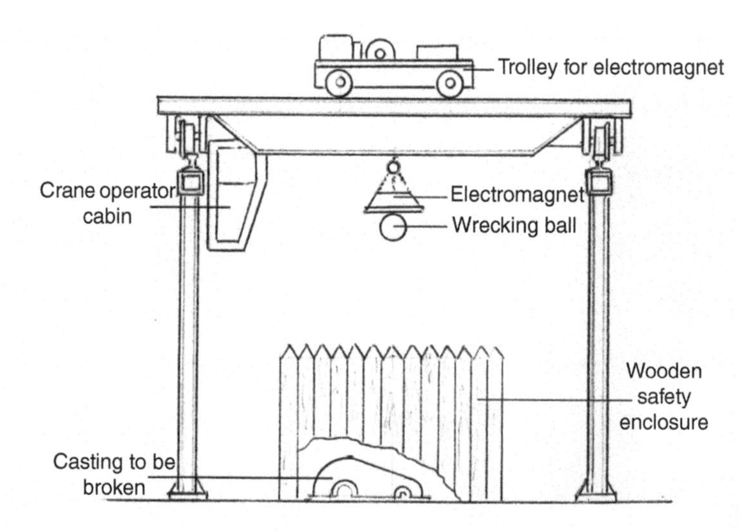

Fig. 2.5 Equipment for breaking castings

2.3 Scrap-Breaking Equipment

Medium and large foundries are equipped with skull breaking equipment, as mentioned earlier, to break scrapped castings into smaller pieces suitable for charging. Foundries manufacturing smaller castings do not need such scrap-breaking equipment.

A wrecking ball is dropped over the scrap castings, using an electromagnet or a winch. A derrick crane with a cactus grab is used, which has claws instead of clamshells for grabbing broken pieces. A sturdy wooden enclosure is built for operator safety, to allow the crane operator to maintain a safe height and distance. Video cameras mounted on the crane provide good visibility for positioning the wrecking ball for dropping, and the lifting magnet provides handling flexibility. Figure 2.5 illustrates the system using an electromagnet and steel wrecking ball.

2.4 Sand Drying, Cooling, and Storage

Figure 2.3 illustrates the raw sand delivery system. Sand is procured according to specifications of grain size distribution and moisture content. Raw sand has a moisture content of up to 8%, depending on the weather and the source. Sand is prepared for molding and making cores by drying, cooling, and sometimes screening.

The sand moisture level of 8% is too high for a molding sand mix. Sand needed for making cores is required to have low moisture content, below 0.5 to 1.0 % (depending on the binder system). The ambient temperature needs to be no more than 60 °C to prevent premature hardening in core boxes.

2.4.1 Types of Sand Dryers

Two types of dryers are used, and each has its advantages, limitations, and preferred applications. The two types are:

- Drum or rotary dryers (also known as rotary kiln dryers)
- Fluidized bed dryers (also known as vibration fluidized bed dryers)

Process- and product-specific criteria as well as operator skills influence the selection of one or the other type. Process-specific factors are continuous or intermittent operations and variation in moisture content. Product-specific factors are particle size and distribution and temperature sensitivity

2.4.1.1 Advantages of Drum or Rotary Dryers

- Insensitive to moisture variations of incoming sand
- Handles intermittent operation
- Accommodates feed rate and particle size variations
- Accommodates power interruptions
- Skilled labor not needed

2.4.1.2 Advantages of fluidized bed dryers

- Suitable for heating and cooling
- Lower power consumption due to efficient heat transfer and heat recovery by recirculation of cooled exhaust air
- Efficient for continuous operations
- Ideal for coarse-grained material of size around 4mm

2.4.2 Drum or Rotary Dryers and Coolers

Rotary drum dryers are more popular for sand drying operations for sand with higher moisture contents. Drum or rotary dryers consist of a rotating drum with lifting blades or vanes radially mounted inside. The blades or vanes lift the sand and allow it to fall, exposing the sand stream to hot gases (600 to 900 °C), which drives out the moisture. Thermally insensitive materials such as silica sand can be directly exposed to the flame generated by a burner at the farther end of the rotary drum.

The drums are inclined at a slope of 3 to 4 centimeters per meter to enable the sand to move downhill to the exit at the lower end. The shell and the lifting blades are made of heat- and abrasion-resistant steel to withstand high temperatures and the abrasion of sand. Larger diameter drums are made with two coaxial layers of steel. The drum is supported on two rings mounted as shown in Fig. 2.6, with the driving gear mounted in between. The exhaust gases pass through a bag house to catch any solid particles.

Hot sand needs to be cooled through a similar drum positioned to receive the hot sand from the sand dryer. The sand passes through the lifting blades, exposing the sand shower to air blowing upstream. Heat transfer occurs due to the counter-flow between sand downhill and air upstream. A bucket elevator or a pneumatic system is used to transport the sand to a new sand hopper. The new sand hopper supplies sand to the core room and to the return sand system.

Figure 2.7 (Ref 2 and 3) illustrates the concept of the rotary drum used for cooling. The sand is exposed to air as it traverses the staggered vanes mounted radially inside. The draft for cooling is created by the suction fan of the bag house.

Sand dryers and coolers are available for different capacities and with varied fuel choices. Manufacturers offer customized designs for different aggregates. The ratios of length to diameter vary, depending on the individual designs. Two shell designs are offered, one with a thick layer of carbon steel and the other with two layers, with the inner layer made of heat-resistant steel and the outer layer made of carbon steel of 6 to 8 mm thickness. Table 2.1 is an example of one mid-sized rotary dryer.

2.4.3 Fluidized Bed Dryers and Coolers

Fluidized bed dryers use hot air that bubbles through a bed of moist sand to dry it, and atmospheric air that bubbles through the bed of dried hot sand to cool it. Some appliances combine rotary drum dryers with fluidized bed coolers to deliver dried cool sand for molding and core making. Figure 2.8 is a schematic showing a fluidized drying and cooling installation. Some equipment manufacturers use vibration to move the bed of cooled sand forward toward the end where it is discharged (Ref 4).

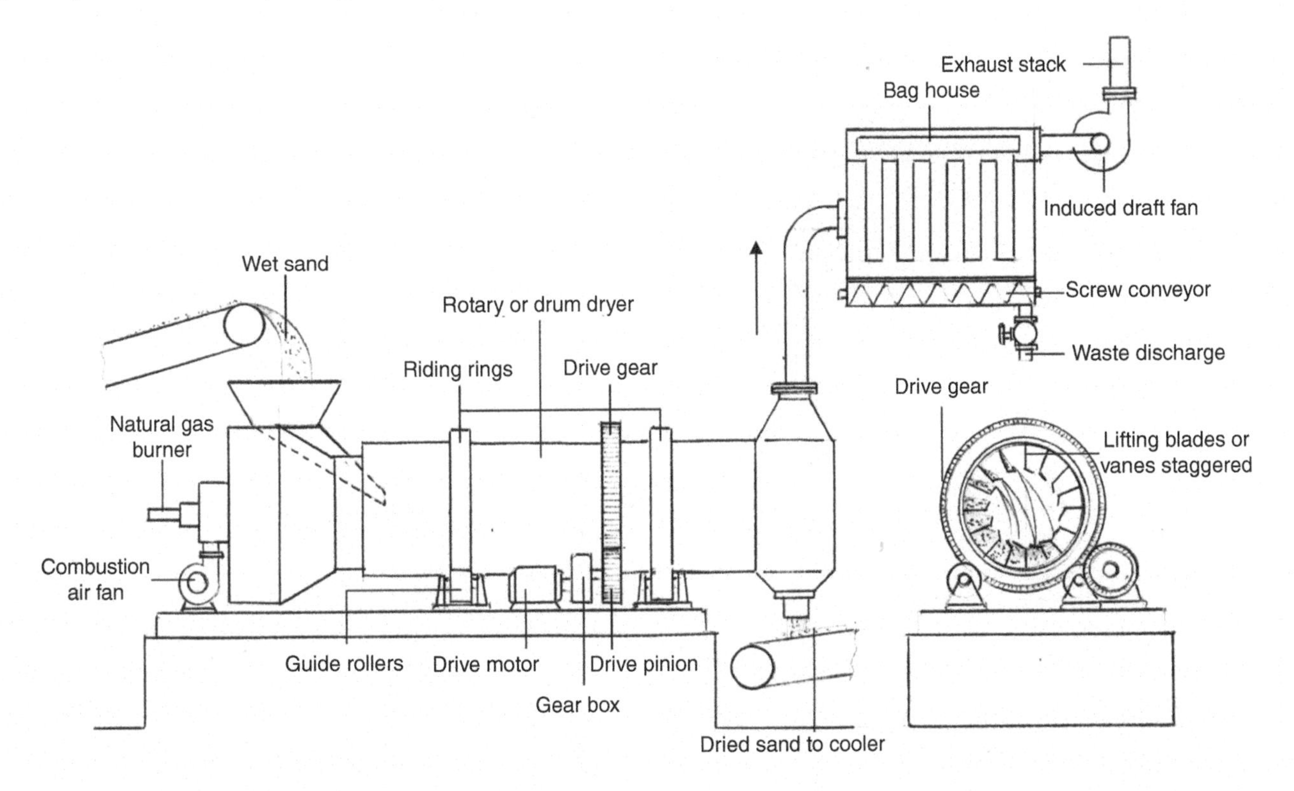

Fig. 2.6 Sand drying in a rotary or drum dryer

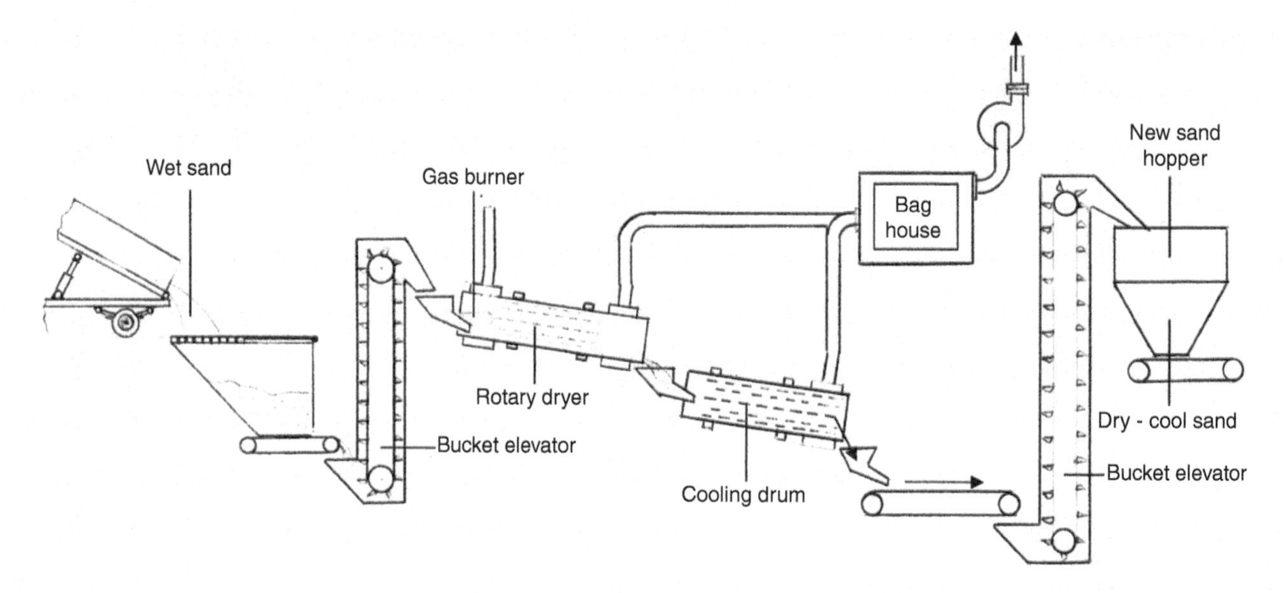

Fig. 2.7 System concept for sand drying and cooling

Table 2.1 Example of mid-sized rotary sand dryer

Parameter	Unit	Amount
Capacity	kg/hr	7500–10,000
Initial sand moisture	%	8–12
Final moisture	%	1.0
Drying temperature	°C	500–700
Rotary drum dimensions		
Diameter	mm	1500
Length	mm	9000
Slope	%	3
Motor power for rotation	kW	15
Rotation speed	rpm	2–6
Equipment weight	tons	17.0

2.5 Handling and Dosing of Bentonite and Sea Coal

Both bentonite clay and sea coal are fine powders, which renders the use of belt conveyors for dosing difficult and inaccurate. The bunkers or bins carrying the powders can bridge, causing the downward movement of the materials in the hoppers to be inconsistent. Screw conveyors are more dependable for dosing compared to belt conveyors. Some equipment manufacturers use a rotating disc below the hopper for dosing bentonite and sea coal.

Screw feeders are calibrated based on the weight transported per rotation of the screw feeder. Rotary encoders mounted on the screw shaft assist in calibration and control. Figure 2.9 illustrates the dosing of bentonite and sea coal into the muller using screw feeders at the bottom of the bins. These bins are either coated with polyurethane or lined with stainless steel to facilitate free downward flow. Some foundries use a constant ratio of bentonite to sea coal, buying a combination of bentonite and sea coal premixed and using the two twin bins alternately.

A few foundries use a bentonite and sea coal slurry mix to improve the blending of the sand. Bentonite slurries have been used in civil engineering construction projects such as filling around plies driven into the soil.

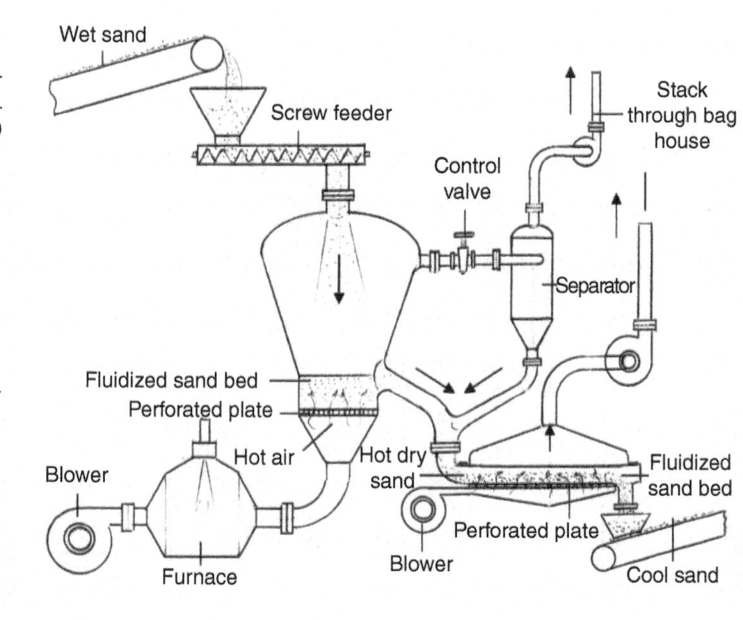

Fig. 2.8 Schematic for fluidized sand drying and cooling

Figure 2.10 (Ref 5) represents a concept of the preparation of the bentonite and coal dust (sea coal) slurry. The advantages of using slurry include:

- Reduction of mixing time by nearly 33% to achieve the same green strength
- Improved green strength for the current mixing time due to improved blending and elimination of the tendency toward lumping
- Improved consistency and accuracy of bentonite and coal dust additions compared to additions in powder form

The additions of bentonite and coal dust as a percentage of water range from 20 to 40%. The lower limit of 20% is dictated by the tendency for sedimentation. The upper limit of 40% is limited by the pumping equipment. Sodium silicate up to

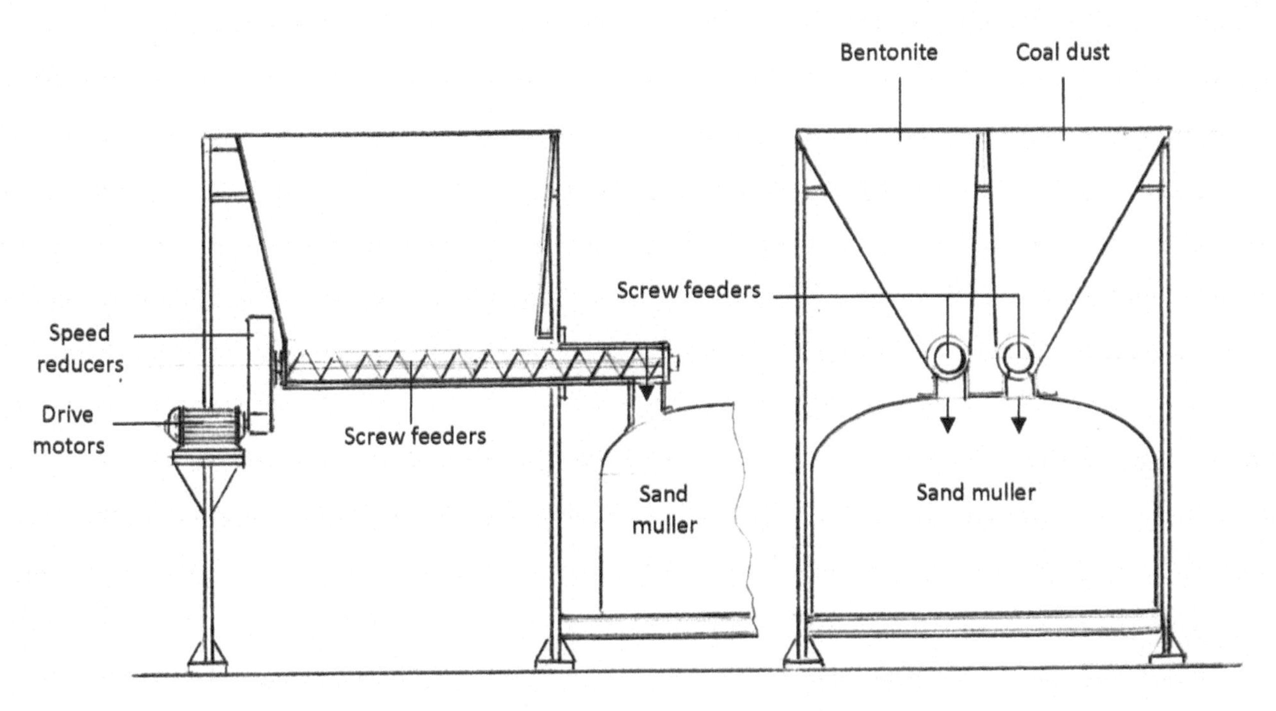

Fig. 2.9 Screw feeders for bentonite and coal dust

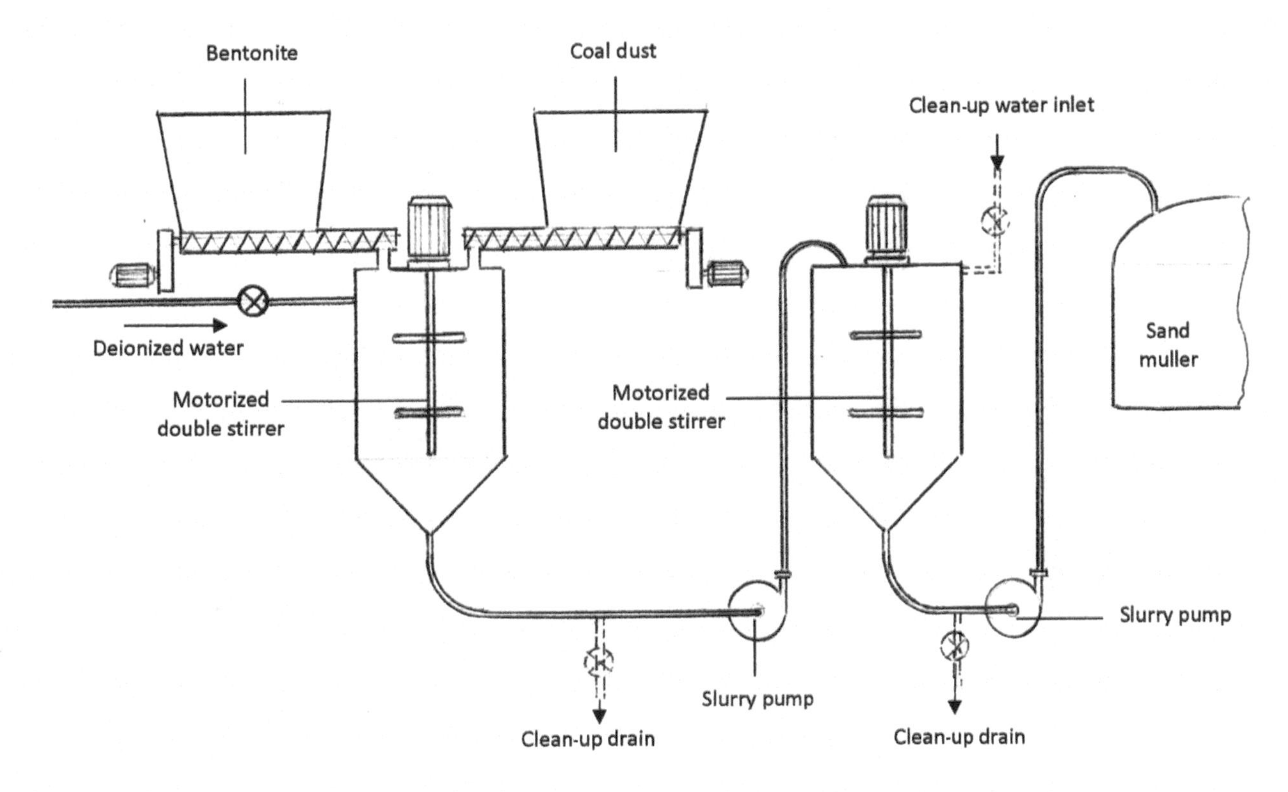

Fig. 2.10 System for addition of bentonite and coal dust slurry

0.25% of bentonite can be added to increase viscosity, if needed. The water added should be clean, with pH controlled at ≥ 7, with the temperature at 15 to 25 °C (Ref 5). The mixing time is around five minutes for each batch.

Table 2.2 provides some guidance on the composition of the slurry.

Table 2.2 Composition of bentonite and coal dust slurry

Type of addition	Bentonite, %	Coal dust, %	Sodium silicate, %	Water, %
Normal limit of bentonite	28	20	0.1	balance
Upper limit of bentonite	38	. . .	0.09	balance

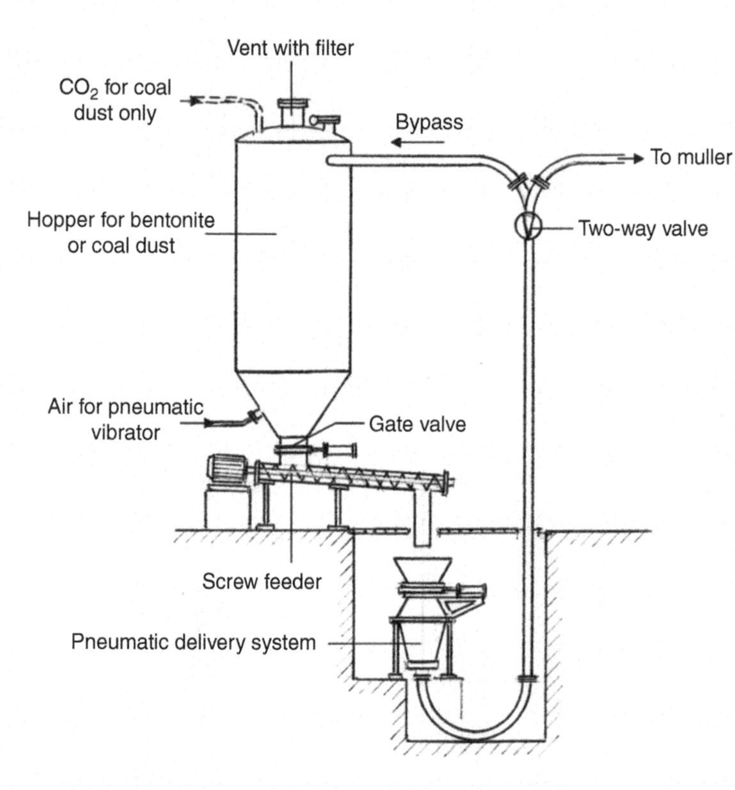

Fig. 2.11 Pneumatic system for delivery of bentonite and coal dust to the muller

Some foundries use a pneumatic system for conveying bentonite and coal dust to the muller. Figure 2.11 is a schematic illustrating such a system.

Because coal dust is a fire risk, CO_2 gas is injected into the hopper. Bentonite storage hoppers do not require this injection. A screw feeder conveys the bentonite (or coal dust) into a pneumatic delivery system. The delivery conduit is well rounded to allow free flow. A two-way valve in the delivery system enables occasional bypass back into the hopper. Bridging inside the hopper is addressed by locating a pneumatic vibrator at the truncated area.

REFERENCES

1. Energetics Inc., Energy and Environmental Profile of the U.S. Metal Casting Industry. Prepared for U.S. Department of Energy, Office of Industrial Technologies, Sept. 1999
2. FEECO International brochure 2021 at www.feeco.com
3. Vulcan Engineering brochure : Frac sand drying systems, 2021. www.vulcandryingsystems.com
4. Carrier Corporation brochure: 2021 www.carrier.com
5. J. Sirokich et al., "Zarizeni slevaren" (Foundry equipment), SNTL Statni Nakladatelstvi Technicke Literatury, 1968

Casting Equipment Engineering Guide
Jagan Nath
https://10.31399/asm.tb.ceeg.t59370021

Copyright © 2023 ASM International®
All rights reserved
www.asminternational.org

CHAPTER 3

Sand Conditioning Equipment

MOLD QUALITY is an extremely important factor with regards to casting quality. The quality of the mold impacts many aspects of the casting process, including the ability to reproduce the pattern shape, the tolerances that can be achieved, sand-related defects that can be avoided, and the mold dilation which affects feeding efficiency. The quality and consistency of the molding sand preparation (sand combined with clay and other additives), the effectiveness of the mulling or mixing of ingredients, the compactibility of the mix for mold packing, and the aeration of the sand mix for formability are all critical factors that contribute to casting quality. This chapter focuses on the elements and functions of sand conditioning equipment which are critical to achieving good quality castings.

One of the major cost advantages of sand molding is the ability to recycle the molding sand. Approximately 80 to 90% of the mold material used in sand molding is recycled. The molding sand is separated from the castings and reconditioned by adding a small amount of clay bond, water, and new sand (approximately 10%). Even the unburned bond and additives are separated by dust extraction systems for recycling. The ability to mechanize sand preparation and recycling with minimum human intervention adds to the advantages of the sand-casting process, making sand casting of iron and steel one of the most cost-effective methods for shaping products.

There are numbers of innovative types of molding equipment that have been developed for green sand conditioning and recycling (Ref 1). As the molding sand is recycled, sand cores used for green sand are mixed with the molding sands. A large percentage of core sand is embedded into the casting cavity even after the shakeout operation, and this sand is discarded during the decoring operation. Sand used for no-bake molding systems is reconditioned a little differently, to eliminate the burned and unburned resin used for bonding the sand. Reconditioning and recycling systems are addressed later in this chapter. Figure 3.1

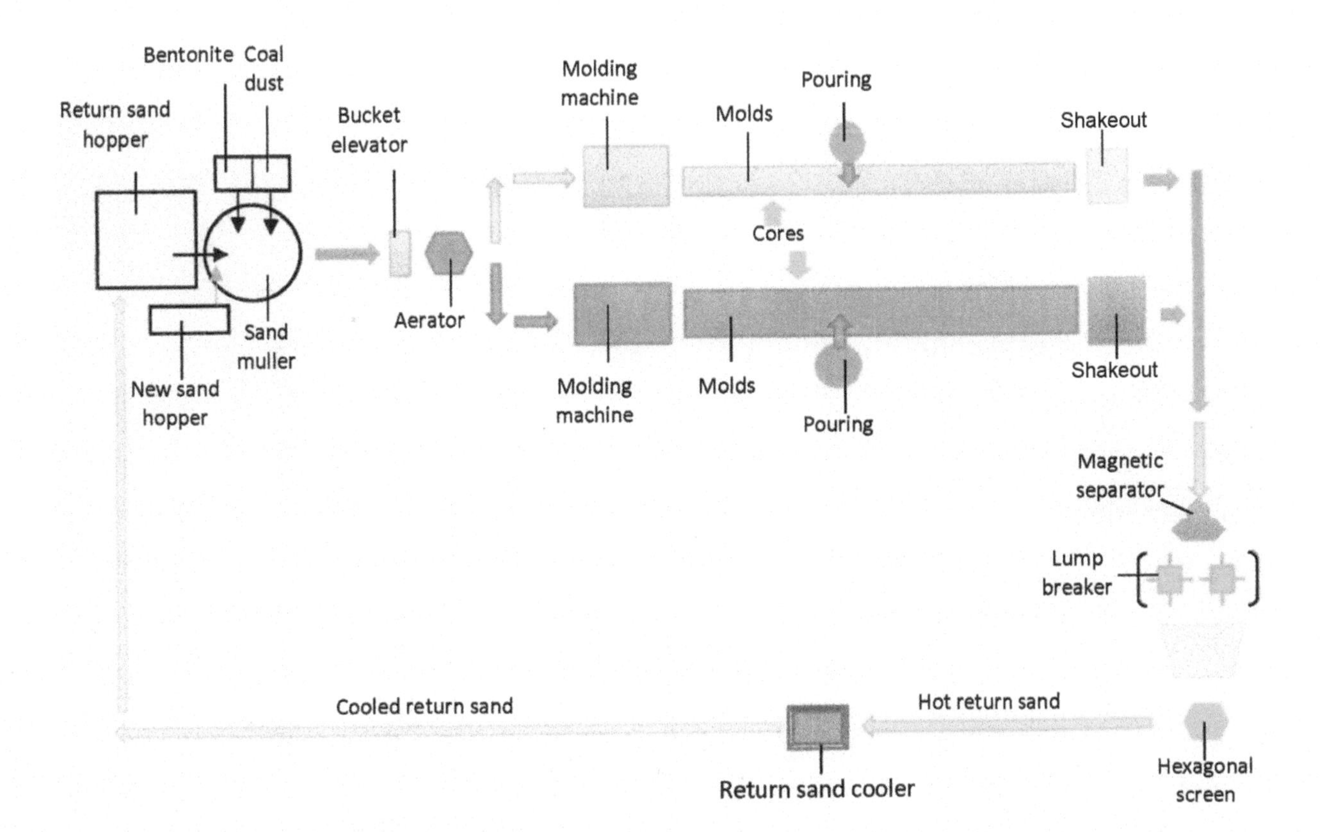

Fig. 3.1 Sand system schematic

presents a schematic showing the engineering of basic equipment for green sand preparation and recycling.

The return sand and a small percentage of new sand is added to the muller, followed by a targeted quantity of bentonite and coal dust. The amount of return sand is controlled to match the capacity of the muller. Water is then added, and the mixture is mulled for a designated time (or until it attains a preset compactibility value). The bentonite clay gels with water to develop the bond. The sand grains are bonded together with the clay, generating the compressive strength of the sand-clay-water mix. *Green strength* is the compressive strength of the mulled or prepared sand at room temperature.

The sand mix is discharged into a chute, and a short belt conveyor delivers the sand mix to a bucket elevator, which discharges the sand mix onto a belt conveyor with an over-the-belt aerator. The aerator breaks up the sand lumps, making the sand mix fluffy and compactible. The sand is then delivered to the hoppers over the molding machines using troughed belt conveyors and gates. Molds are made, cores are set, and then they are poured. The length of the mold conveyors is designed to allow for the appropriate cooling time for a safe knockout without distortion or damage to the castings. The molds are punched out or vibrated to separate the castings and the sand. The core sand partially mixes with molding sand, and the cores enclosed into the casting volume move to the cleaning room along with the castings.

The hot sand exiting the shakeout unit travels through an underground conveyor system, through a magnetic separator that catches any flash or broken gating pieces from the iron or steel castings. The sand then moves through a lump crusher, where sand lumps from the mold and the cores are crushed to finer pieces. The crushed sand then passes through a screen to eliminate any lumps.

Next, the hot sand passes through a sand cooler, where the sand is cooled to a preset temperature. The cooled sand moves through a series of belt conveyors and bucket elevators to be delivered to the return sand hopper. Additions of new sand to make up for the volume of sand that is lost during the casting process, are engineered for addition into the muller or the return sand hopper. This process describes how sand is recycled for reuse.

3.1 Sand Dosing into the Muller

The output of the muller should meet the sand requirements of the molding machines, with allowance made for spillage. Increasing the muller output by reducing the cycle time is not an option. The muller delivers the sand mix to the molding machines, and the muller capacity cannot be exceeded, because that would affect the quality of the sand mix and strain the motor of the muller. Therefore, good control of the amount of return sand delivered for each cycle is important. Dosing controls the amount of return sand that is delivered into the muller; this is achieved in one of two ways:

- Varying the speed of the feed belt
- Dosing the hopper with a load cell for weight control

3.1.1 Variable Feed Belt Speed

The feed belt underneath the return sand hopper that feeds into the muller can be engineered for variable speed to alter the amount of sand delivered to the muller. This system is simple, but the amount of sand delivered needs to be calibrated and verified periodically.

3.1.1.1 Calibration

To calibrate the amount of sand that is delivered, a diverter is used to bypass the sand at the bottom of the sand hopper. The feed belt is run for a targeted time, and the sand is collected and weighed. This collection is repeated for two increments of time by altering the belt speed. The collected sand is weighed and compared with the initial amount. The accuracy of the amount of sand being delivered can be established based on these measurements.

3.1.2 Dosing Hopper with Load Cell

A bottom-discharge hopper is engineered between the return sand hopper and the muller, as shown in Fig. 3.2. The load cell of the dosing hopper is programmed to control the feed belt until the targeted weight is achieved. The bottom doors of the dosing hopper open to discharge sand into the muller.

Dried new sand is delivered to the core sand mixer using pneumatic conveyors, belt conveyors, or screw feeders like the one illustrated in Fig. 2.9 in Chapter 2, "Sand and Metal Charge Storage and Handling" in this book. Pneumatic conveyors for dried new sand are popular because of their lower power consumption and maintenance costs.

Vibratory feeders are also used to deliver new sand from the hopper to the core sand mixer. The chute is angled at 5° for the

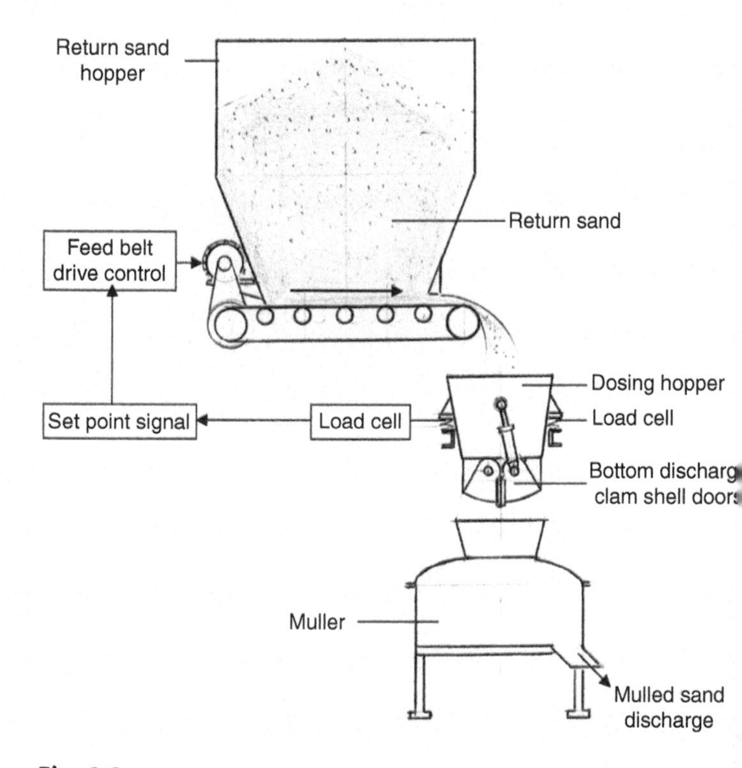

Fig. 3.2 Return sand dosing into the muller

flow of the sand. Level sensors in the mixer send a signal to the vibratory motor to turn off once the preset level has been reached. Some foundries use a timer control to regulate the feed, which is simpler but less accurate.

3.2 Sand Muller Types and Features

Mulling is the process of mixing sand, binder, water, and other ingredients under controlled pressure to generate the targeted strength and formability for quality molds. This process utilizes a gear-driven assembly of rollers for pressure and plows for blending; the rollers and plows are housed in a wear-resistant cylindrical container. The muller imparts a three-way action (Ref 2): pressing and squeezing; blending; and folding the aggregate toward the rollers for active mixing.

Sand mullers are classified as:

- Vertical roller-type — two rollers with a horizontal axis of rotation; each batch exits the muller before the new cycle starts
- Horizontal roller-type — three rollers at varied heights with a vertical axis of rotation capable of short cycle times, with or without cooling assistance
- Duplex muller — two vertical roller types combined for continuous operation, usually with cooling assistance
- Continuous screw-type muller — for continuous mulling operation

3.2.1 Vertical Roller-Type Muller

Figure 3.3(a) is a schematic illustration of the vertical roller-type muller.

Good quality mulling ensures that the sand grains are evenly coated with clay and other additives such as coal dust. Uniformity of coating is essential to ensure thermal stability and to prevent metal penetration defects during casting. The engineering design enables the two rollers to squeeze the sand and clay-water mix against the base for effective mulling, while the two plows (or scrapers), one inner and the other outer, continuously churn and blend to ensure that the sand is pushed toward and under the rollers. The rollers are designed to be off-center, and they rotate in opposite directions. A central shaft connected to a heavy duty reduction drive and motor, mounted at the bottom of the cylindrical base, drives the roller and plow assembly.

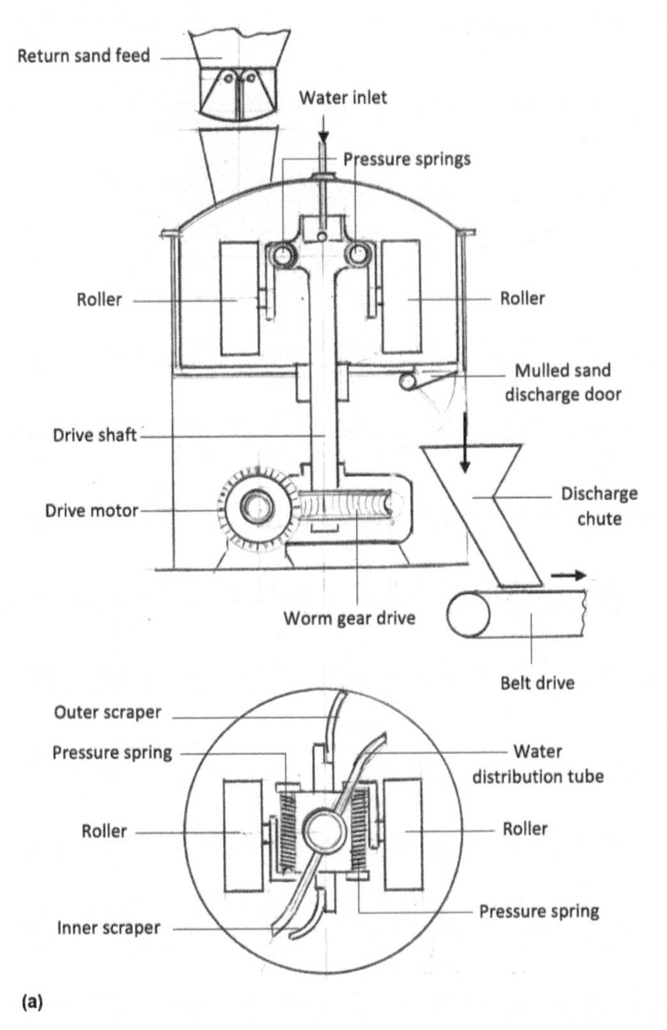

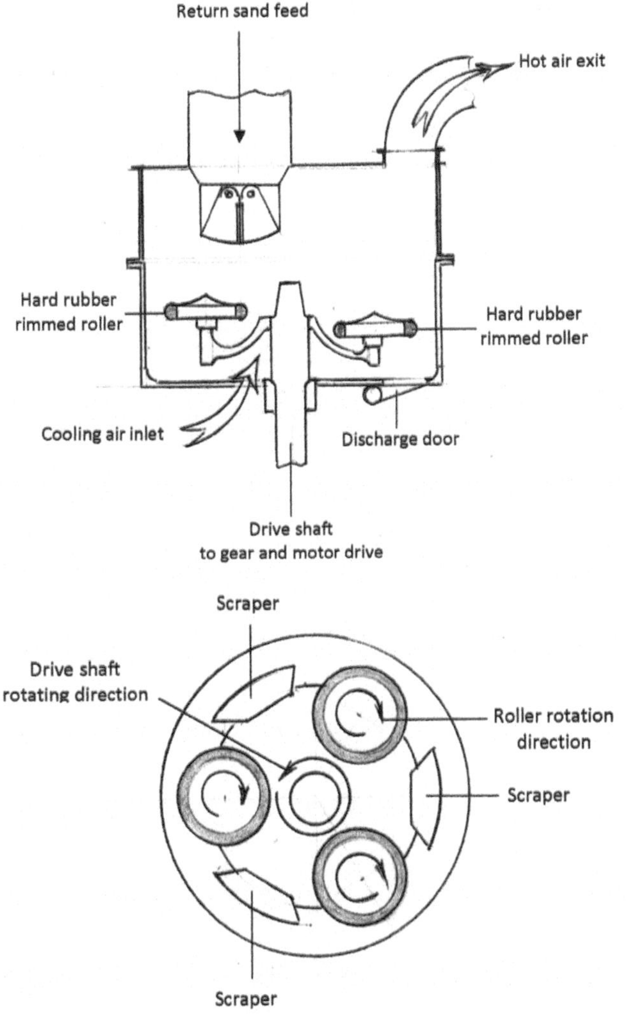

Fig. 3.3 Roller-type muller; (a) main features of a vertical and, (b) horizontal

A gap of 6 to 15 mm (0.24 to 0.59 in.), depending on the size of the muller, is initially maintained between the rollers and the base. The two rollers are mounted on spring-loaded arms, and as the sand volume increases, the rollers push back against the spring pressure, maintaining downward pressure. The wheel pressure can be adjusted by adjusting the spring pressure and the clearance between the roller and the base plate (Ref 3, 4).

Each wheel is mounted on two roller bearings, which are completely shielded and permanently lubricated (Ref 5). The plows are lined with tungsten carbide tips to withstand the abrasive wear of the sand. The cylindrical bottom is designed with austenitic manganese steel segment plates which have work-hardening characteristics to resist wear. These plates can be replaced if the wear is excessive over a long period. Water is injected through two tubes mounted radially from the central hollow shaft and a manifold, as illustrated. A pneumatically actuated discharge door is designed at the base for discharging the prepared sand mix. The operations are coordinated and controlled by a master programmable logic controller (PLC).

Mullers engineered for automated compactibility control have a sampling door for sand sample collection and a mechanism for automated ramming of the sand for a compactibility test measurement. The mulling time and water additions are controlled through a PLC, based on the test readout (Ref 6).

Table 3.1 (Ref 1) lists the various sizes and capacities of vertical roller-type mullers for small- to medium-sized foundries.

Table 3.2 lists (Ref 1) the vertical roller-type muller sizes for medium- to large-sized foundries.

3.2.2 Horizontal Roller-Type Muller

The horizontal roller muller is engineered for high productivity and is adaptable for sand cooling. Figure 3.3(b) is a schematic illustrating the main features of the horizontal roller-type muller.

The construction of this type of muller features three rollers rotating in a direction opposite to the main shaft and the roller carrier, as illustrated. The rollers are lined with hard rubber of 40 to 50 mm (1.57 to 1.97 in.) thickness, depending on the size of the rollers. The three scrapers positioned in between the rollers push the sand towards the rotating rollers, blending the sand, bentonite clay, and water. The central roller carrier is driven by a reduction gear drive and motor, like the horizontal roller-type muller.

Recycled sand is unloaded into the muller using a bottom discharge bin with clamshell doors. A bentonite-water slurry is injected into the sand mix. Cooling air enters the mixer chamber from the bottom, and hot air exits at the top, connected with a bag house or a dust collector. The air cools the sand mix exiting the muller. The cycle time ranges from 90 to 120 seconds, depending on the size of the muller.

Table 3.3 provides the sizes and capacities of different sizes of this type of muller. Larger sized mullers are also available (Ref 1).

Table 3.1 Vertical roller type muller sizes for small to medium sized foundries

Parameters	Units	Size 1	Size 2	Size 3
Nominal volume, based on density 1.5kg/dm^3	dm^3	150	330	600
Muller diameter	mm	1326	1830	2400
Main spindle speed	RPM	38	27	24
Mulling output	m^3/hr	2	3–6	5–12
Mulling output	t/hr	3	4.5–9	7.5–18
Mulling time	secs	4	4	4
Power rating	kW	7	22	29
Motor speed	RPM	735	975	975
Total weight	kg	1380	4500	7000

Source: Ref 1

Table 3.2 Vertical roller type muller sizes

Parameters	Units	Size 1	Size 2	Size 3	Size 4	Size 5	Size 6
Nominal volume	dm^3	125	355	710	1100	1500	2000
Batch capacity based on sand density of 1.5 kg/dm^3	kgs.	190	532	1065	1650	2250	3000
Muller nominal diameter	mm	1272	1650	2016	2250	2650	2805
Mulling output based on	m^3/hr	2.5	6	13	22	30	40
3 min cycle time	t/hr	3.5	10	21	33	45	60
Power rating	kW	7.5	20	35	55	90	120
Horse power	HP	10	27	47	73	120	160
Approximate total weight	kgs	1780	4090	6650	9650	12,050	15,545

Source: Ref 1

Table 3.3 Horizontal roller type muller sizes

Parameter	Units	Size 1 15 A	Size 2 30 A	Size 3 40 A	Size 4 50 A	Size 5 60 A	Size 6 70 A	Size 7 80 A
Nominal volume	dm^3	85	115	170	250	340	500	680
Batch weight (based on density 1.5kg/dm^3)	kg	127	172	255	375	510	750	1020
Muller diameter	mm	900	915	1090	1340	1575	1930	2285
Main spindle speed	RPM	115	115	95	95	88	76	62
Mulling output	m^3/hr	1.5–2.0	4.6–9.2	6.8–13.6	10–20	13.5–27	20–40	27–54
	t/hr	2.3–3.0	6.9–13.8	10.2–20.4	15–30	20.3–40.5	30–60	40.5–81
Main motor rating	kW	7.5	18.5	25.8	33.0	48.0	60.0	95.0
	HP	10.0	24.5	34.2	43.0	63.0	80.0	126.0
Exhaust fan motor rating	kW	0.7	0.7	1.5	2.2	2.2	3.7	5.5
Cooling air volume	m^3/hr	3060	3060	5860	9200	9200	13,400	19,400

Source: Ref 1

3.2.3 Duplex Muller

Two sets of vertical mullers are combined to work in tandem, setting up a mixing motion for a continuous operation of taking in recycled sand, binder, and water in one section and delivering the mixed sand to the other. Figure 3.4 is a schematic illustration of this concept. These mullers are also engineered for sand cooling by building a cooling air manifold all around and flushing with air to cool.

These types of mullers are chosen for large-scale continuous sand delivery. Cycle times range from 90 to 150 seconds, depending on the size of the operation and the desired degree of cooling. Return sand with temperatures as high as 49 °C (120.2 °F) is cooled, and the mulled sand is delivered with close moisture control and with high green strength. This cooling process is very effective, and typically, the mixed sand delivered is 10 °C above the ambient temperature (Ref 3).

3.2.4 DISAMATIC Sand Mixer

The sand mixer developed by DISAMATIC Inc. differs from other mullers. Figure 3.5 is a general schematic highlighting the major features of this equipment (Ref 6).

The main cylindrical housing is lined with a wear-resistant base and an outer shell. Return sand and bonding materials are fed into the mixer from overhead hoppers. Water is introduced to the mixture after a few seconds of premixing. The bridge motor works through a gear box; it turns a rotary arm that carries a rotor on one side and a wall scraper on the other side. It also turns the bottom scraper through a central column. The rotor is driven by a separate motor through a toothed belt drive. A door is provided for withdrawing samples for an automated compactibility measurement, and the addition of water is determined by these measurements through a controller.

The planetary mixer consists of a series of discs with radial spikes which are driven by the main drive shaft through a reduction drive. Two clamshell doors mounted off the bottom plate discharge the prepared sand into a hopper which delivers the prepared sand to a belt. These mixers form a part of the automated molding machine installation.

3.2.5 Continuous Screw-Type Muller

The continuous screw-type muller uses a horizontal drum-type outer container with a screw- or worm-type rotor that is continuously driven by a geared motor, as shown in Fig. 3.6 (Ref 1). The return sand is fed in at one end, along with the binder and water. The sand mix is then blended and discharged from the other end, as indicated in the figure.

The horizontal roller mixers offer the advantage of additional cooling and easier maintenance. This type of roller is being gradually replaced by duplex mixers, which offer more effective mulling and higher outputs. The screw-type auger concept is used for no-bake sand mixers, no-bake molding lines, and core sand mixers.

3.3 Automated Sand Conditioning Control

Automation of sand mixing control is essential to obtain consistent molds, which are necessary to ensure casting quality. In a practical casting environment where nearly 6 to 8 different patterns are molded per shift, the sand-to-metal ratio varies from pattern to pattern. The variability of sand to metal ratios affects the bond and sea coal that is burned. The varied sand-to-metal ratio also affects the casting cooling time and the mold knockout sand temperature.

Casting facilities equip themselves with sand coolers for the knockout sand, along with automated compactibility and green sand strength testing equipment synchronized with the muller operation. Air cooling is a popular feature in the mullers because it improves the consistency of the sand properties.

A master controller PLC is the nerve center for control of these critical variables. It is used to monitor and control the sand cooler, mulling time, bond, and water additions. Figure 3.7 is a

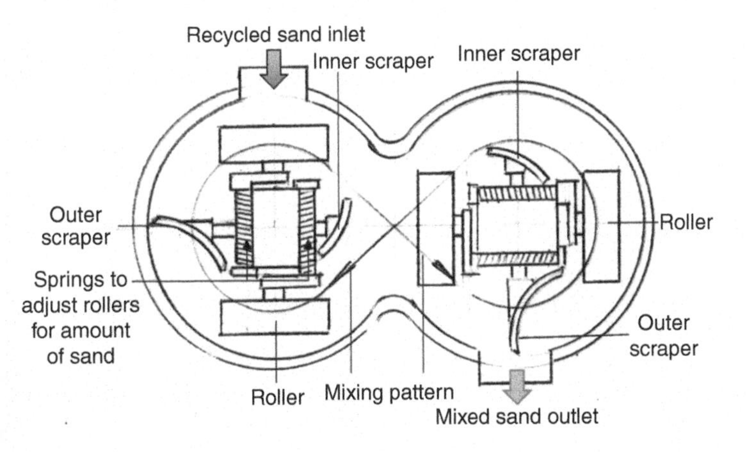

Fig. 3.4 Duplex muller schematic

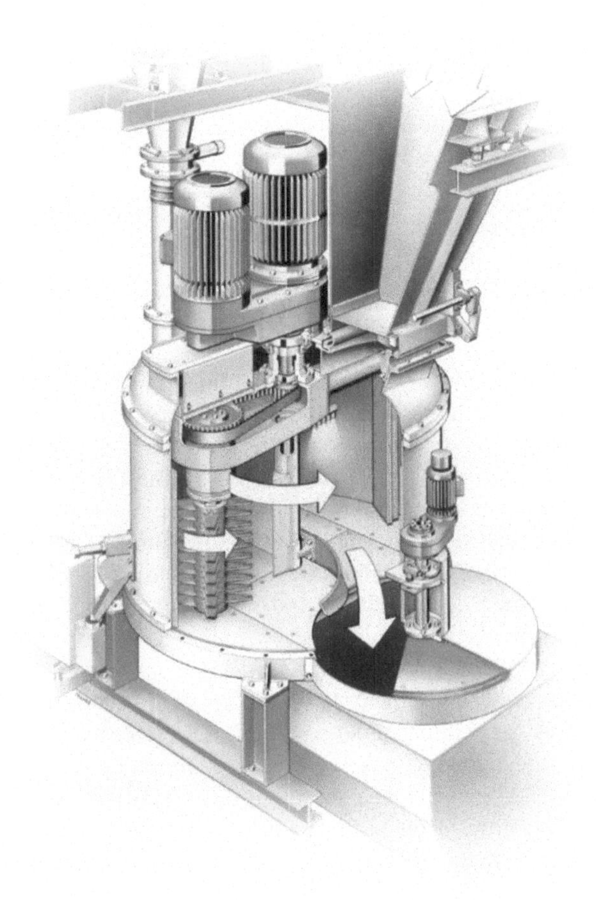

Fig. 3.5 Schematic of DISAMATIC sand mixer. Source: Ref 6, DISAMATIC

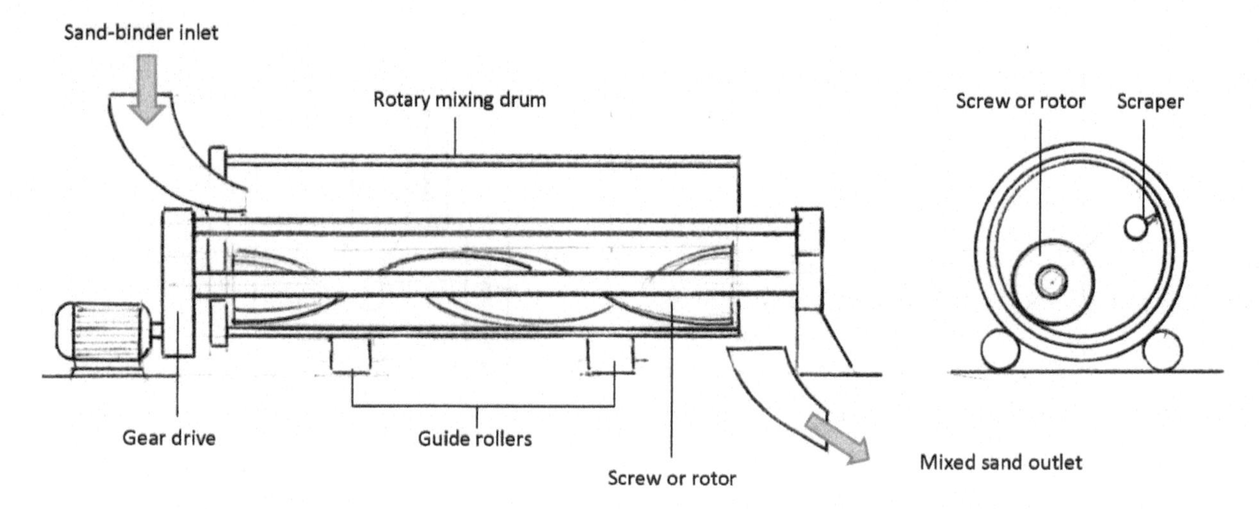

Fig. 3.6 Continuous muller concept

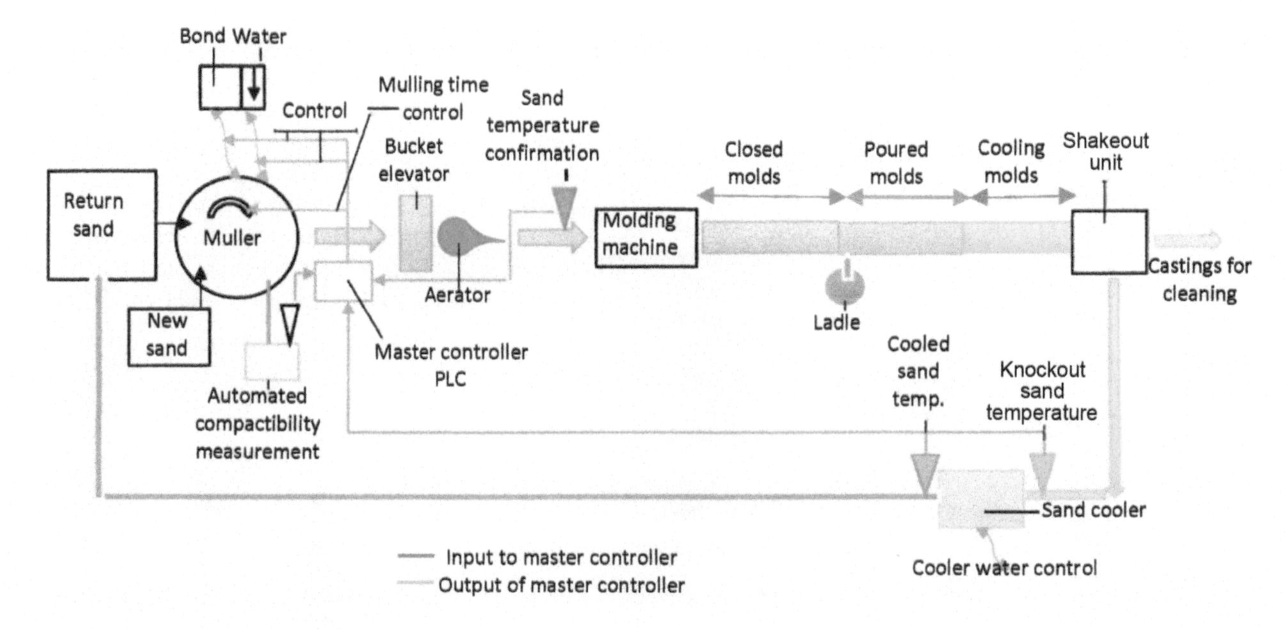

Fig. 3.7 Automatic sand conditioning system schematic

schematic showing the key elements of the system and the inputs and outputs that are connected to the master controller.

The master controller controls the amount of water added to the knockout sand (or shakeout sand), depending on the temperature of the knockout sand as it enters. The average of five readings is taken as a representative value. The temperature of the outgoing sand (average of five readings) is registered in the master controller. The automated compactibility and green compression strength testing apparatus takes a mixed sand sample and measures the compactibility as well as the green strength. Algorithms are established to make suitable adjustments to the amount of the bond and water, as well as the mulling time, based on the test results. Running averages are plotted to trace the trends for fine tuning.

The bucket elevator and the aerator also cool the sand mix. The temperature of the sand mix at the sand hopper above the molding machine is measured and fed back to the master controller for confirmation.

3.3.1 Time-Controlled and Energy-Controlled Mulling

Most of the muller installations are designed for mulling for a targeted time. This assumes that mulling efficiency stays the same. However, mulling efficiency tends to decrease as the plows or scrapers wear out, so some casting manufacturers track the effect of the plow wear by connecting a kW or power meter with a recorder.

The power consumed by the muller also changes during the cycle, as the bond activation increases. As the clay bond added is activated with ionic bonding, the viscosity of the clay increases, which requires more energy to coat the sand grains and develop green strength. The batch-to-batch variability of the bond activation can be reduced by tracking the energy consumed instead of the cycle time (Ref 7). The energy is measured as kW/sec. The advantages experienced in a trial involving three facilities were:

- Reduced energy consumption
- Reduced cycle time
- Decreased sand friability

3.4 Core Sand Mixers

Core sand mixers are intended to blend raw sand with any type of resin binder (or sodium silicate as an organic binder) for packing into the core box (or core-mold box) before compaction.

Core sand mixers do not have any rollers because there is no mulling action needed.

Figure 3.8 (Ref 8) illustrates a no-bake casting line using core assemblies instead of green sand molds. These assemblies are called *core-molds*. Core sand mixers are either batch-type or continuous. Typically, foundries needing very few cores or those with less automation may use a batch-type core sand mixer. No-bake molding lines such as the one illustrated normally use a continuous sand mixer.

Figure 3.9 is a schematic of a batch-type core sand mixer. This type of mixer can be engineered for automation. Two mixer blades mounted diametrically opposite to each other enable the mixing of sand and resin in a cylindrical container. Scrapers mounted on the arms of the mixer blades scrape the inside of the cylindrical surface, keeping it free from any sand buildup.

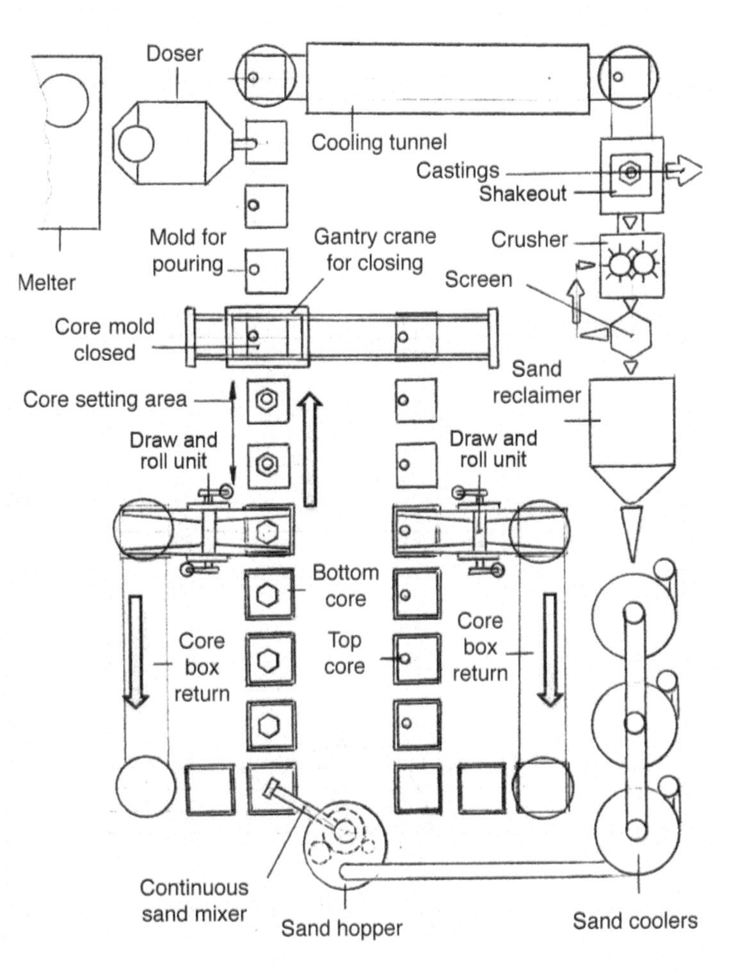

Fig. 3.8 No-bake casting line schematic. Source: Ref 8

A worm and worm wheel drive mounted at the base of the cylindrical container drives the blade assembly. A discharge door at the bottom of the cylindrical container allows the discharge of the mixed sand onto a distribution belt conveyor.

Engineering for automation involves providing for a pneumatic cylinder that actuates the discharge door in addition to dosing the sand and metering the resin addition. An operator may supervise to ensure the smooth running of the core sand mixer.

Table 3.4 (Ref 1) provides the throughput, cycle time, and kilowatts for a typical core sand mixer.

Automated no-bake casting lines such as the one illustrated in Fig. 3.8 use continuous sand mixers which mix and deliver the mixed sand just before packing and compacting the sand mix into the core boxes. An *activator* or *catalyst* added to the

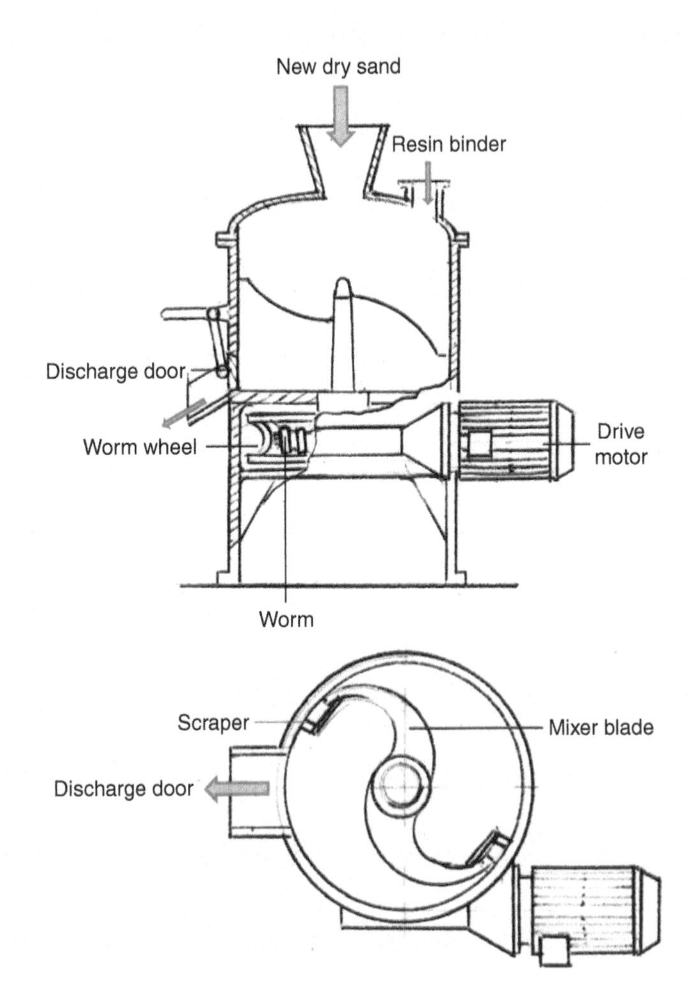

Fig. 3.9 Core sand mixer

Table 3.4 Core sand mixer parameters

Parameter	Unit	Amount
Nominal capacity of the mixer	kg	100
Diameter of the cylindrical container	mm	700
Rotational speed of the mixer blades	rpm	50
Output for a 2-minute cycle time	m^3/hr	1
Power requirement	kW	5.5
Equipment weight	kg	375

Source: Ref 1

sand-resin mix hardens the cores due to the resin-catalyst interaction. It is essential that the compaction is completed before the onset of hardening. Smaller no-bake mixers are engineered with one swing arm, and larger units are provided with two arms, permitting more flexibility and longer reach of the mixer head, to accommodate varied sizes of core boxes.

Figure 3.10 illustrates a smaller unit with one swing arm. The swing arm houses a screw-type (auger-type) mixer driven by a motor mounted at one end, as illustrated. Mixed sand is delivered into a slinger head, driven by another motor mounted on the swing arm. A slinger head consists of two or three blades mounted radially in a housing, to sling or throw the mixed sand into the core-mold box, for further compaction.

The swing handle has push-button controls, conveniently mounted. The resin and catalyst can be programmed to be pumped in to allow for more accurate metering. These types of mixers are very useful and efficient.

Larger molds of no-bake lines require increased ability for manipulation of the slinger head. Double-arm continuous mixers meet this need. Figure 3.11 illustrates such a concept.

The double-arm continuous mixer incorporates most of the features of the single-swing continuous mixer illustrated in Fig. 3.10. The larger arm houses a belt conveyor to convey the sand from the hopper to the auger mixer. These continuous mixers are engineered for multistation layouts, either with translation movements or turntables, for core box compaction, top surface leveling, stripping, and core box return to the compaction station. This book illustrates this application in Chapter 4, "Molding Flasks and Molding Machines," Fig. 4.20.

3.5 Sand Conveyors and Bucket Elevators

Foundry sand is transported in various forms:

- Wet sand from the quarry to the foundry
- Dried unbonded sand
- Bonded molding sand
- Hot sand returned from shakeout
- Cooled sand reclaimed from no-bake molding

Sand may be transported nearly horizontally, up a small incline, and vertically upward. Rubber belt conveyors are used for the horizontal and inclined transport of wet sand, bonded molding sand, and hot return sand. Bucket elevators are used for vertical transport.

Dried unbonded sand for use as molding sand, makeup sand, or reclaimed sand for no-bake lines can be transported using either belt conveyors or pneumatic transport.

3.5.1 Sand Conveyors

Sand conveyors are either flat or troughed, using the guide rollers to shape the conveyor. Trough angles may vary from 20 to 45°. Figure 3.12(a) and (b) illustrate these two types. The idler

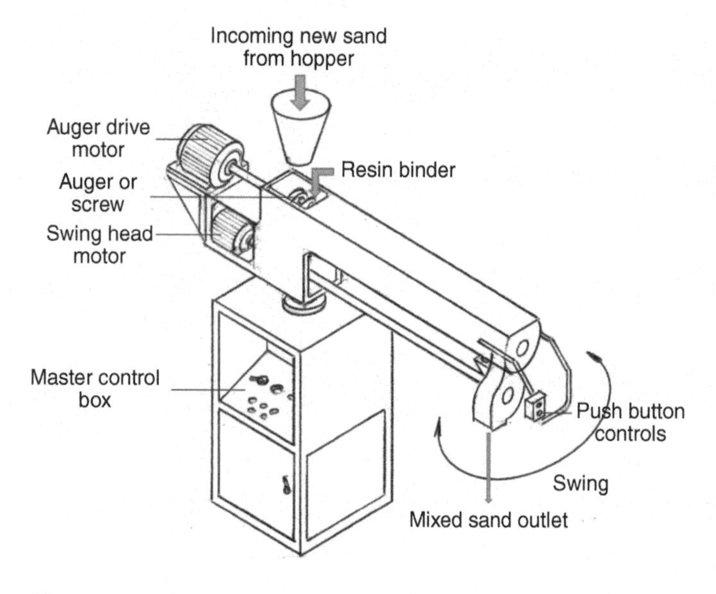

Fig. 3.10 Single-swing continuous sand mixer

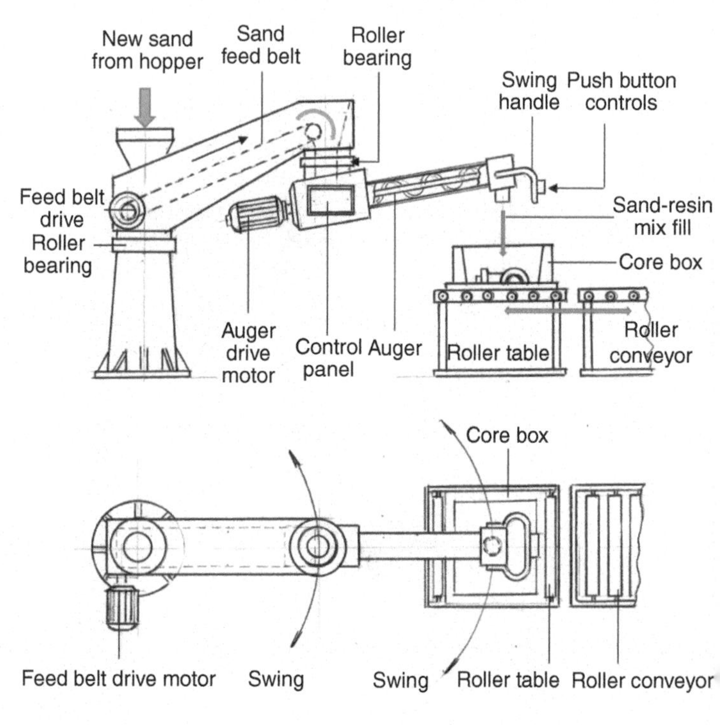

Fig. 3.11 Double-arm continuous sand mixer

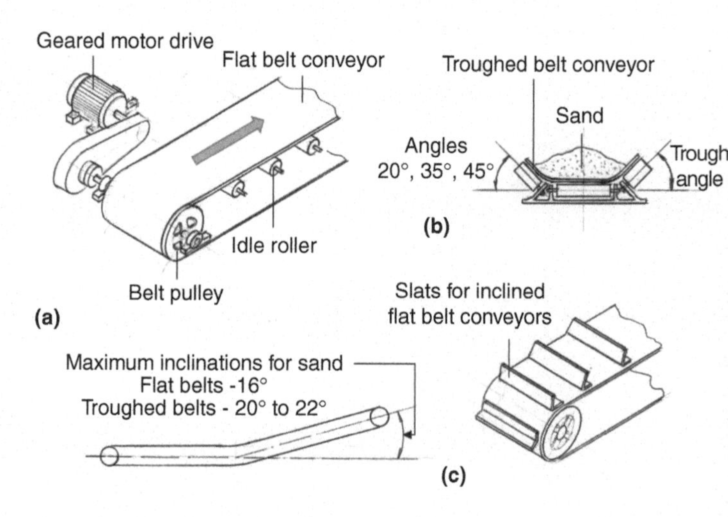

Fig. 3.12 Belt conveyors; (a) flat, (b) troughed, (c) flat with slats

rollers have sealed bearings for long life, minimizing downtimes. Rollers whose bearings get damaged are replaced over shutdowns. The maximum inclination desirable to ensure that the sand does not slip is typically about 16° for flat belts and about 20 to 22° for troughed conveyors. Some conveyors use slats mounted on the flat belts to improve the efficiency for inclined belt applications. Figure 3.12(c) illustrates a belt conveyor with slats. These are driven by geared electric motors.

Normally, conveyor belts are made from a rubber compound combined with one or more layers of fabric material, such as styrene-butadiene rubber and several layers of nylon fabric. Conveyors for the knockout are invariably made of steel, to be able to withstand the heat and wear and tear. The return sand conveyors may use layers of steel wires instead of fabric for improved heat resistance. Some use flame-retardant compounds such as antimony oxide in the manufacture of conveyor belting (Ref 9).

The transportation capacity of the belt depends upon:

- Belt width
- Belt speed
- Trough angle
- Material density

The maximum output of a belt conveyor is given by Eq 3.1 (Ref 10):

$$W = \frac{60 \cdot A \cdot d \cdot S}{2000} = 0.03 \cdot A \cdot d \cdot S \qquad \text{(Eq 3.1)}$$

where W is output of the belt conveyor in tons per hour, A is cross-sectional area of the load in square feet, d is density of the material in pounds per cubic foot, S is belt speed in feet per minute.

Belt speeds range widely from 100 to 800 feet per minute. The cross-sectional area of the load increases as the trough angle increases. Normal trough angles are 20°, 35°, and 45°. Tables are available for quick computation (Ref 10).

3.5.2 Bucket Elevators

Bucket elevators are useful to transport sand from a lower level up to mezzanine heights in foundries. These compact elevators are manufactured in various sizes and heights to suit the layout. Depending on the application, bucket elevators are driven by geared motors, and they sometimes use a sprocket and chain in the drive.

Steel buckets are produced in varied shapes, depending on the width of the belt, capacity needed, and the bulk density. The buckets are either riveted or bolted to the belt. Bucket elevators haul sand on one side. After delivering the sand, the other side carries the empty returning buckets. The driving gear train is designed for rotation of the bucket elevator in only one direction, usually clockwise. In the event of a power failure, the weight imbalance between the loaded and unloaded sides can force the load into a counterclockwise direction, which results in dumping of sand at the bottom of the loading side. This causes the buckets to tear off the belt, and the repair process results in a long downtime. Therefore, it is critical to design a nonreversal mechanism on the top pulley of the bucket elevator to prevent the rotation of the belt drive in a counterclockwise direction in the event of a power shutdown. The nonreversal mechanism is also called back-stop.

Different mechanisms of bucket elevators are the pawl and ratchet, locking ball, and the hold back clutch.

Figure 3.13 illustrates the main features of the bucket elevator. Figure 3.13(a) shows the pawl and ratchet nonreversal mechanism. The rachet is mounted on the shaft that carries the upper drive pulley. The rachet has angled teeth to allow clockwise rotation only, with the spring-loaded pawl gliding over it. In the event of power failure, the spring engages the pawl onto the rachet, preventing counterclockwise rotation.

Figure 3.13(b) illustrates a second type of nonreversal mechanism for bucket elevators in the event of a power failure. The inner housing is keyed to the bucket elevator shaft, and the outer housing mounted on the structure has a gradually diverging groove. With the clockwise rotation of the drive pulley, the balls ride up toward the larger area of the diverging groove. When there is a power shutdown, counterclockwise rotation of the drive pulley is prevented by the balls sliding down to the narrow end of the groove and locking the pulley against a backward rotation.

A third mechanism to prevent counterclockwise rotation is the *hold-back-clutch* system. The hold-back clutch is a compact housing mounted on the top belt pulley shaft. If there is an unexpected power shutdown, the clutch engages on the brake drum, preventing counterclockwise rotation.

The output of a bucket elevator can be calculated by Eq 3.2 (Ref 10):

$$W = \frac{60 \cdot V \cdot d \cdot S}{2000\, b} = \frac{0.03 \cdot V \cdot d \cdot S}{b} \qquad \text{(Eq 3.2)}$$

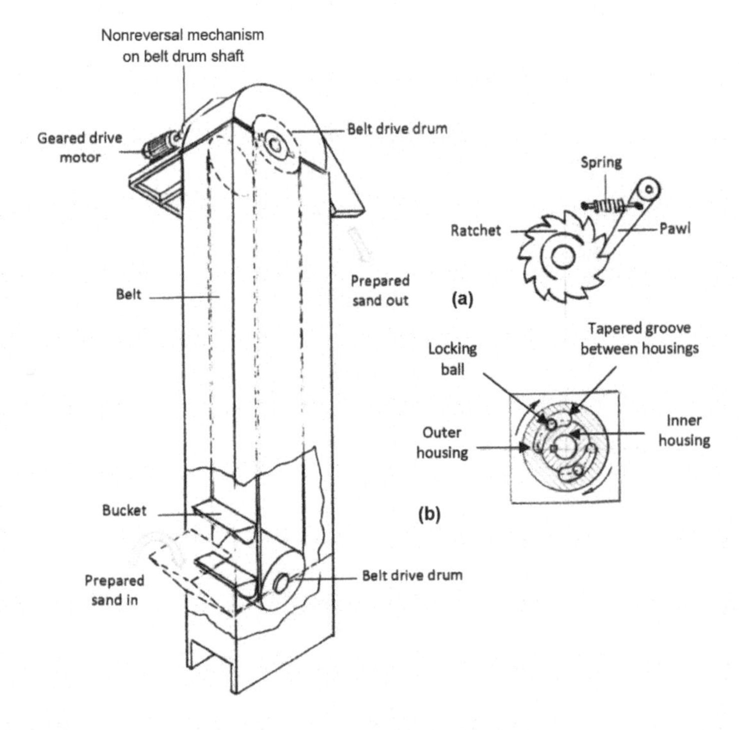

Fig. 3.13 Features of bucket elevators, (a) pawl and ratchet nonreversal mechanisms, (b) ball locking system for preventing anticlockwise rotation

where W is output of the bucket elevator in tons/hour, V is volumetric capacity of a bucket in cubic feet, d is density of the material in pounds per cubic feet, S is speed the bucket travels in feet per minute, and b is spacing between the buckets in feet.

3.6 Sand Aerators

Prepared sand coming out of a muller tends to lump, especially toward the end of each discharge, as the blades push out the last portions of the mulled sand. These compacted sand lumps do not flow readily and do not fill the space between the flask and the pattern properly without voids. The sand compaction will be nonuniform and the mold quality will be poor.

Aerators improve the compactibility of the sand, resulting in mold density that is more even. Aeration reduces the mold surface friability to below 10%, even with a lower compactibility of 30%, compared to typical values of friability of 35 to 38% (Ref 11).

Additionally, aerators reduce the blow pressure for sand filling. High-pressure molding is engineered to blow the sand into the molding chamber or flask frame, typically at about 0.5 MPa (72.52 psi). The aerators reduce the blow pressure down to 0.1 MPa (14.5 psi), lowering the energy consumption (Ref 11). Reduction of blow pressure reduces sand spillage and lowers the maintenance costs while improving housekeeping.

3.6.1 Types of Aerators

There are two types of aerators:

- Belt-type aerator — where prepared sand passes through a drum with combing pins
- Pin bushing-type aerator – where sand passes through two rings with pinned bushings rotating in opposite directions.

3.6.1.1 Belt-Type Aerator

The belt-type aerators have a combing drum that is housed in a steel casing and driven by a motor through V-belt drives. Figure 3.14(a) is a schematic representation of this type of aerator.

Prepared sand either from a belt conveyor or a bucket elevator enters the aerator chamber through a top funnel, below which the aerator drum is mounted. The aerator drum has a number of closely mounted spikes to break the sand lumps using an operation similar to *combing*. Such drums are called *combing drums*. As the sand impinges on the combing drum rotating at high speed, any lumps and cakes in the prepared sand are broken and thrown onto a steel curtain made of alloy steel chains. This chain curtain prevents the aerated sand from impinging on the shell of the housing and sticking to it. Aerated sand falls on to the distribution belt underneath, carrying it to the molding machine hoppers.

This type of aerator can be engineered to be placed in one of three ways:

- Over the belt (Fig. 3.14a)
- Between two belt conveyors (Fig. 3.14b)
- Downstream of a belt conveyor (Fig. 3.14c)

Table 3.5 lists the sizes of the two types (b) and (c) manufactured by one foundry equipment company (Ref 12).

3.6.1.2 Pin Bushing-Type Aerator

Effective aeration can also be achieved by passing prepared sand in between two rings carrying a series of hard rubber

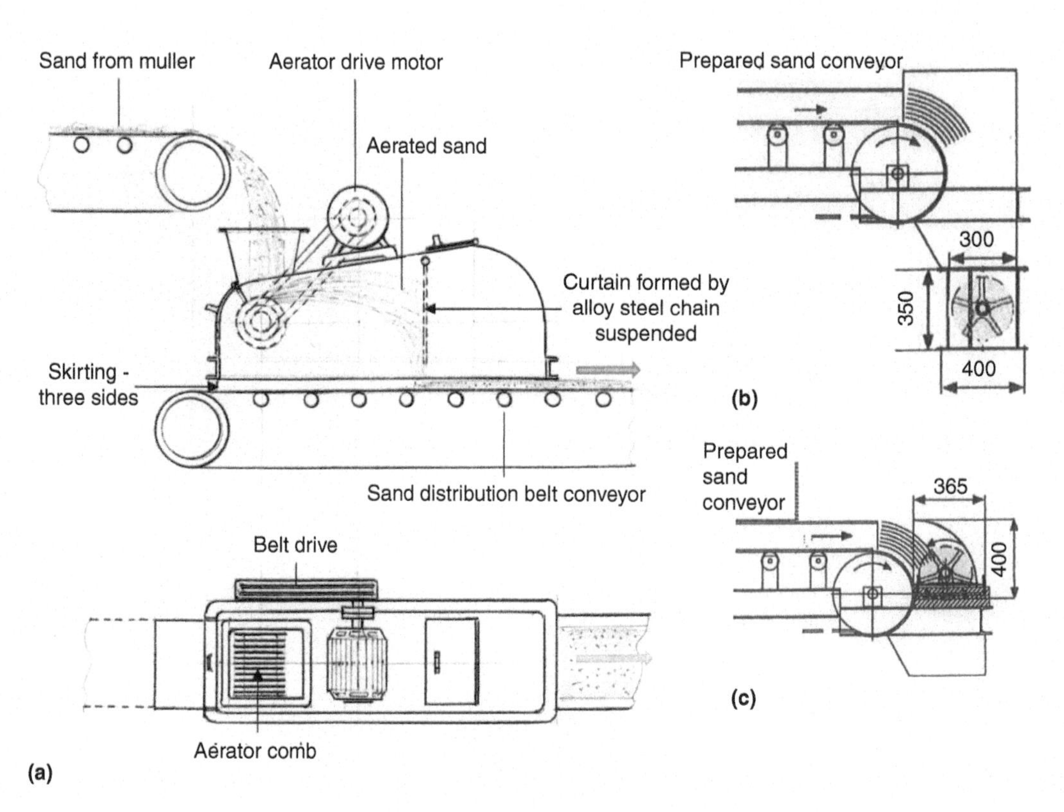

Fig. 3.14 Aerators (a) over-the-belt conveyor, (b) between belt conveyor, (c) downstream belt conveyor. Source: Ref 12, Eirich Machines Inc.

bushings that rotate in opposite directions. Figure 3.15 illustrates a pin bushing-type of aerator.

Two coaxial shafts carry two rings with pinned bushings, as illustrated. A drive motor drives the outer and inner rings in opposite directions using V-belt pulleys. Prepared sand from the muller enters the aerator chamber either through a belt conveyor or a bucket elevator. As sand passes through the passages between the two sets of rings rotating in opposite directions, it is aerated and discharged to the bottom onto a distribution belt. The hard rubber bushings are inspected during shutdowns and replaced once or twice a year.

Table 3.5 Sizes of Aerators

Type	Rating(a), m³/h	Motor, kW	Belt width, mm	Weight, approx., kg
Installation above belt conveyor				
FS1	25	3.0	400	350
FS2	40	3.0	500	350
FS3	75	5.5	650	400
FS5	100	5.5	800	500
FS7	160	7.5	1000	600
FS8	350	7.5	1200	700
FS9	450	1.5	1400	900
Installation downstream of belt conveyor				
FV1	25	3.0	400	100
FV2	40	3.0	500	100
FV3	75	3.0	650	125
FV5	100	3.0	800	150
FV7	160	3.0	1000	175
FV8	350	3.0	1200	200
FV9	450	3.0	1400	225
Installation between belt conveyors				
FZ1	25	3.0	400	100
FZ2	40	3.0	500	125
FZ3	75	3.0	650	150
FZ5	100	3.0	800	200
FZ7	160	3.0	1000	250
FZ8	350	3.0	1200	300
FZ9	450	3.0	1400	350

(a)Depending on belt speed and material level. Used with permission from Eirich Machines Inc. Source: Ref 12

3.7 Mold Conveyors—Types and Lengths for Cooling

Mold conveyors are used to move the compacted molds from the molding stations to the pouring stations, and onto the cooling section before shakeout. Conveyors are needed to return the empty molding flasks and the bottom pallets back to the molding stations for molding.

Flaskless molding lines and no-bake molding lines need a conveyor line to return the empty pallets to the drag-molding line.

3.7.1 Mold Conveyors

The heavy molds that are produced need to be moved to the pouring stations. Poured molds need to be moved to the cooling section and on to the shakeout stations. Different types of conveyors are used to transport the molds through molding, pouring, cooling, and shakeout.

The most common mold conveyors are:

- Roller conveyors
- Guide rail conveyors
- Multilevel mold conveyors
- Walking beam mold conveyors

The heights of these conveyors are engineered according to the molding machine table height and pouring system design. Some foundries provide a platform for the pourers to adjust the elevation for a comfortable pouring height, especially if the stacked height of the molds is high. High mold heights can impair the

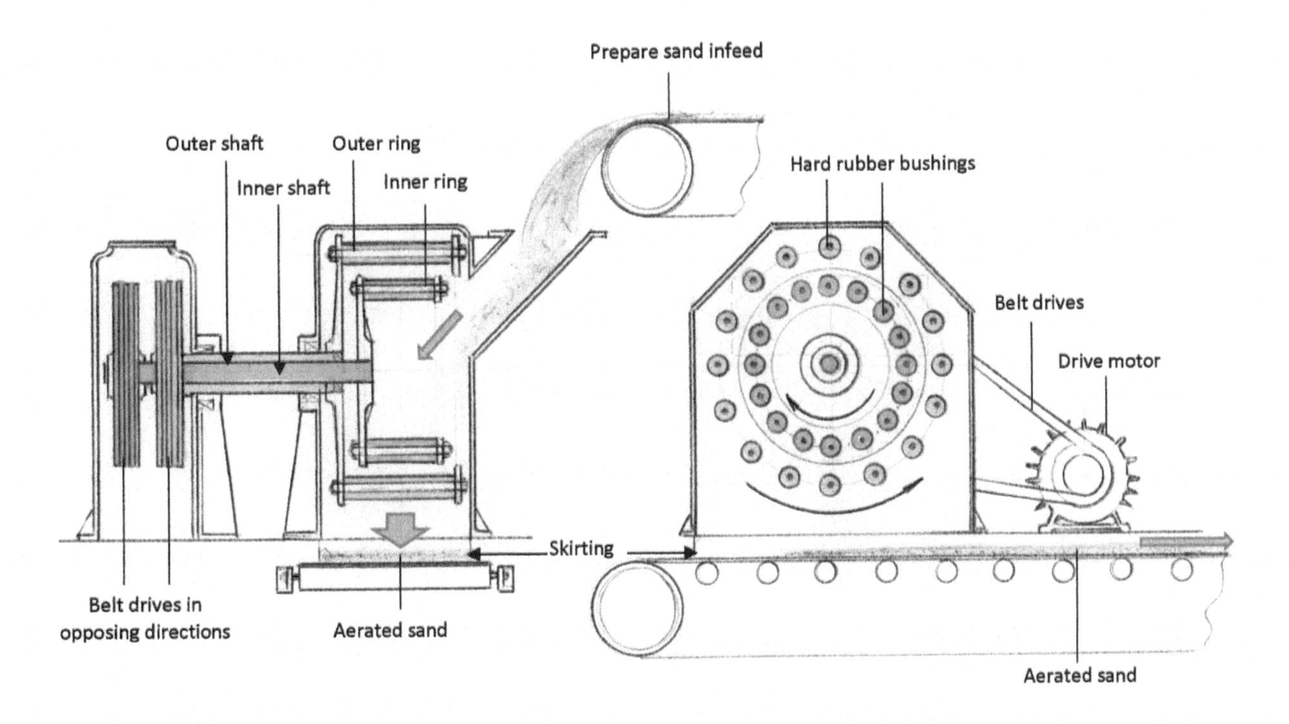

Fig. 3.15 Pin bushing-type aerator

view of the sprue and affect pouring accuracy. An elevated platform in pouring stations can also assist the design of the fume exhaust system.

3.7.1.1 Roller Conveyors

Roller conveyors are designed with chain drives to move the molds. Intermediate rollers are designed as idlers. Rollers are made from steel tubing with sealed bearings on either side. These rollers are periodically inspected and replaced as needed. The design of their mounting allows for easy replacement.

Figure 3.16(a) is a schematic of the general construction featuring the roller conveyor, the pallet, and the molds over the roller conveyor. Figure 3.16(b) is an illustration of a cross section of a roller, showing the roller bearings housed inside the steel tubing.

Whenever possible, gravity is leveraged for mold movement by providing a slight incline of the roller conveyor. All sequential process steps such as molding, closing, pouring, and shakeout should synchronize with compatible cycle times. Synchronization is achieved through regulation of the movement by a pneumatic stop which is actuated by proximity switches. The use of gravity for flask and pallet return is common.

Some layouts call for a positive drive to push the mold to the next station. In such cases, a motor drive is used. Figure 3.17 is a schematic of a roller conveyor drive. It features two rollers driven by sprocket and chain mechanisms from a central sprocket and chain drive. The motor and the gear box can be located suitably, depending on the layout.

Roller conveyors can be designed to move molds over a curve if the molding line layout requires a change in direction. In such a case, the rollers are made conically, with the outer edge roller diameters larger than the inner edge. A guard rail to prevent the mold from sliding outward is provided, as shown in dotted lines in Fig. 3.18(a). Figure 3.18(b) is a design for perpendicular movement of molds using a turntable. The cross-sectional view

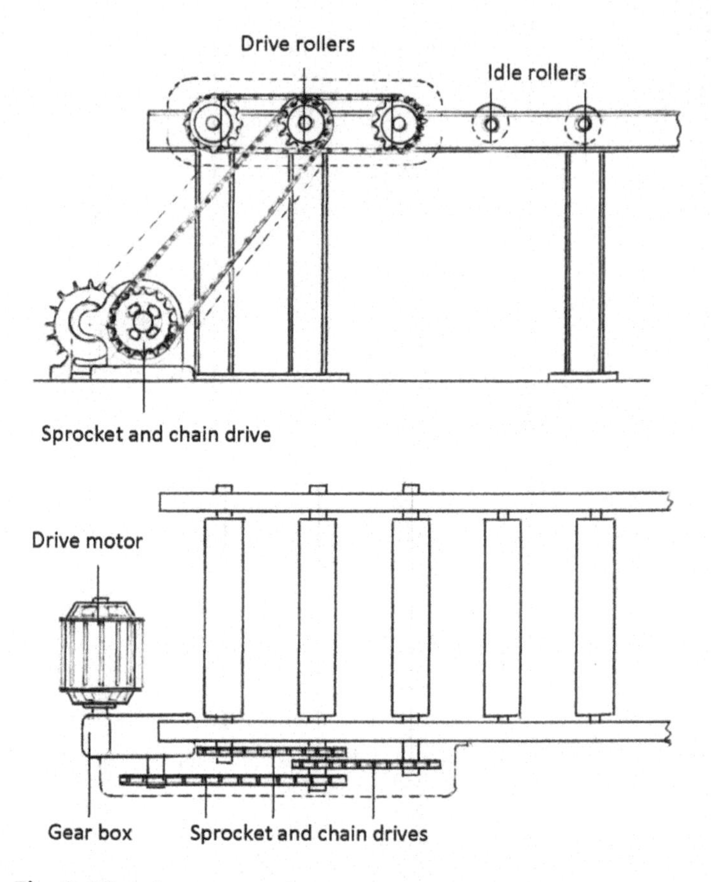

Fig. 3.17 Roller conveyor drive

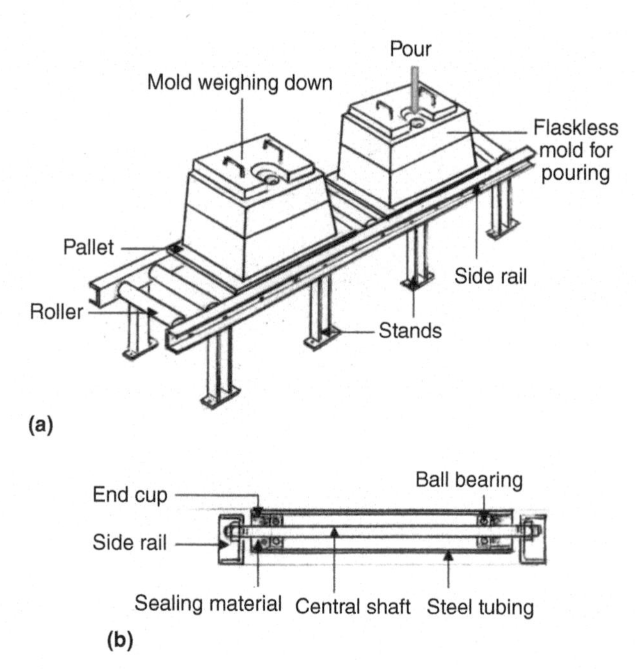

Fig. 3.16 (a) Roller conveyor for molds. (b) Roller cross-section details

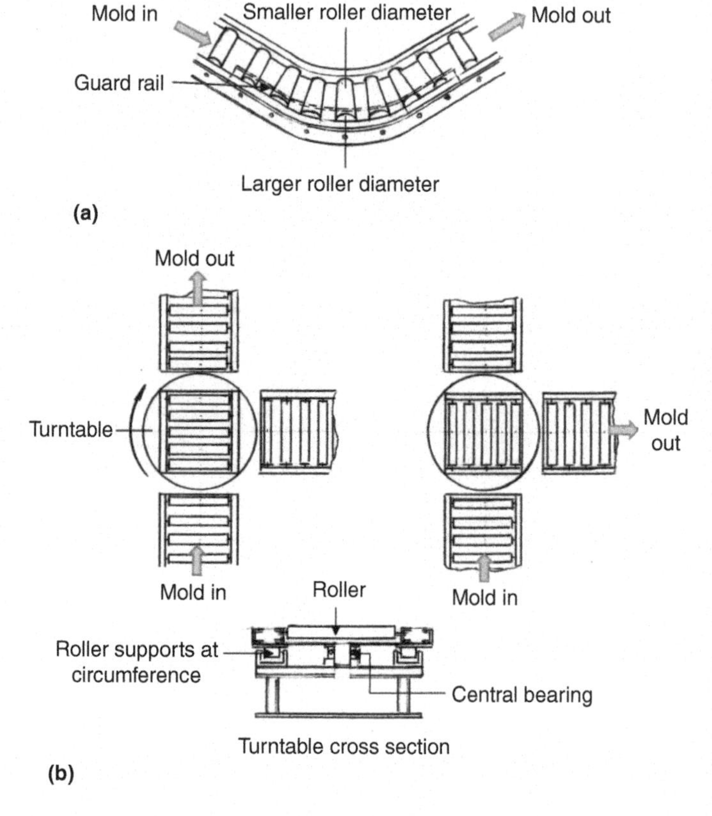

Fig. 3.18 Conveyors for right-angled movement, (a) curved mold line, (b) turntable mold line

at the bottom of the figure provides the details of the design. The turntable carries a roller frame which carries about 6 to 8 rollers. The spindle and the central ball bearing allow the swinging motion of the table. A set of supporting rollers at the circumference of the table frame carry the weight of the molds on the table. A 90° crank attached to a pneumatic cylinder underneath the turntable enables the molds to move in a perpendicular direction, as shown in the top views of Fig. 3.18(b).

3.7.1.2 Guide Rail System for Mold Movement

The guide rail system applies to medium and smaller flask sizes. The system uses a steel platen with grooved wheels that ride on guide rails. This conveyor system is less expensive than the roller conveyor, but its load-bearing capacity limits the maximum size of the flask that can be used. Figure 3.19 is a schematic showing the guide rail system for mold movement.

Figure 3.20(a) shows a bottom view of the platen with four casters riding on the guide rails. Figure 3.20(b) is a cross section of one caster, showing the details. Some of the designs feature a roller bearing instead of the two ball bearings shown in the figure. The design for the shaft also varies from a bolted stud to a stud secured by split-pins. Designs with a small flanged I-section, instead of a flat guide rail, are also feasible.

3.7.1.3 Multilevel Mold Conveyors

Long cooling durations for the castings before shakeout may require long conveyor lines, but adequate space for a long line may not be available. One foundry equipment manufacturer has engineered a unique solution using linear and circular conveyors at multilevels to suit the targeted cooling time (Ref 13). These linear and circular conveyors consist of roller conveyors that are engineered for two or three levels (depending on the targeted cooling time) and include automatic mold weighing and unloading before the molds reach the shakeout unit. See Fig. 4.30 in Chapter 4, "Molding Flasks and Molding Machines," in this book.

3.7.1.4 Walking Beam Mold Conveyor — Precision Mold Conveyor

For larger vertically parted molds, a precision mold conveyor (PMC) is used to move the molds through pouring, solidification, and the first part of the cooling zone. The walking beam conveyor is a mold-conveying system developed by DISA. The PMC is engineered to match vertical flaskless green sand molding machines (Ref 6). The molds are supported on two sets of support grates. The first support grate (hatched black) has four degrees of movement: up, down, forward, and backward. This is called the movable grate. The other support grate (white,) placed between the first support grate, has two degrees of movement: up and down. This is called the fixed grate. The grates advance the molds in synchronization with the molding machine as each mold is completed. Figure 3.21 (a) illustrates these operations schematically.

The working cycle of the PMC consists of six stages, as numbered in Fig. 3.21(a):

1. The movable grate carries the molds forward.
2. The fixed grate is lifted to the level of the movable grate.
3. The movable grate is lowered, and the mold now rests on the fixed grate.
4. The movable grate returns to its starting position.
5. The movable grate is lifted to support the molds.
6. The fixed grate is lowered, and the PMC is ready to begin another cycle. When the molding cycle has completed, the mold closes up for the next mold.

3.7.1.5 Sliding Mold Conveyor — Automatic Mold Conveyor

For smaller, vertically parted molds, an automatic mold conveyor (AMC) is used to move the molds through pouring, solidification, and the first part of the cooling zone. The operation of the conveyor is shown in Fig. 3.21(b).

The AMC operates synchronously with the molding machine. The working cycle of the AMC consists of four stages, as pictured in Fig. 3.21(b).

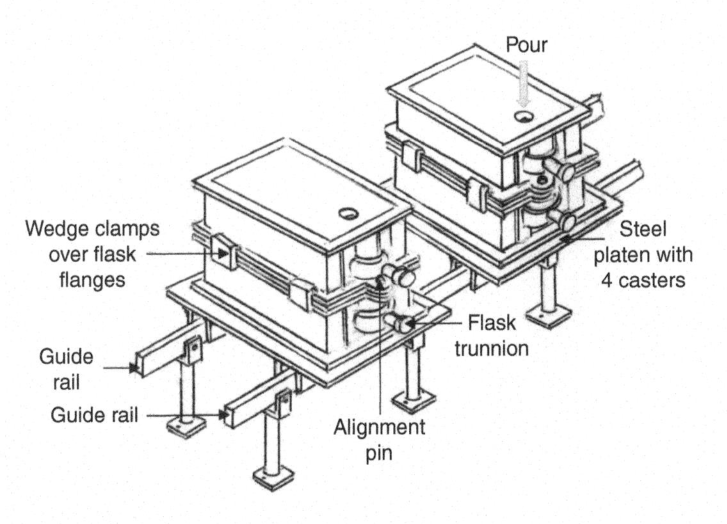

Fig. 3.19 Guide rail system for mold conveyors

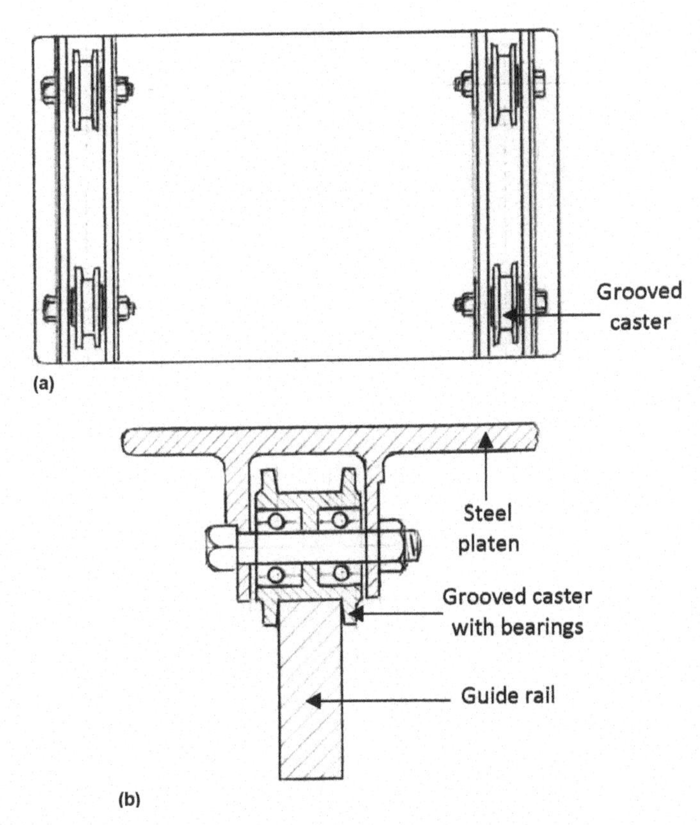

Fig. 3.20 (a) Platen underside showing casters. (b) Caster cross section

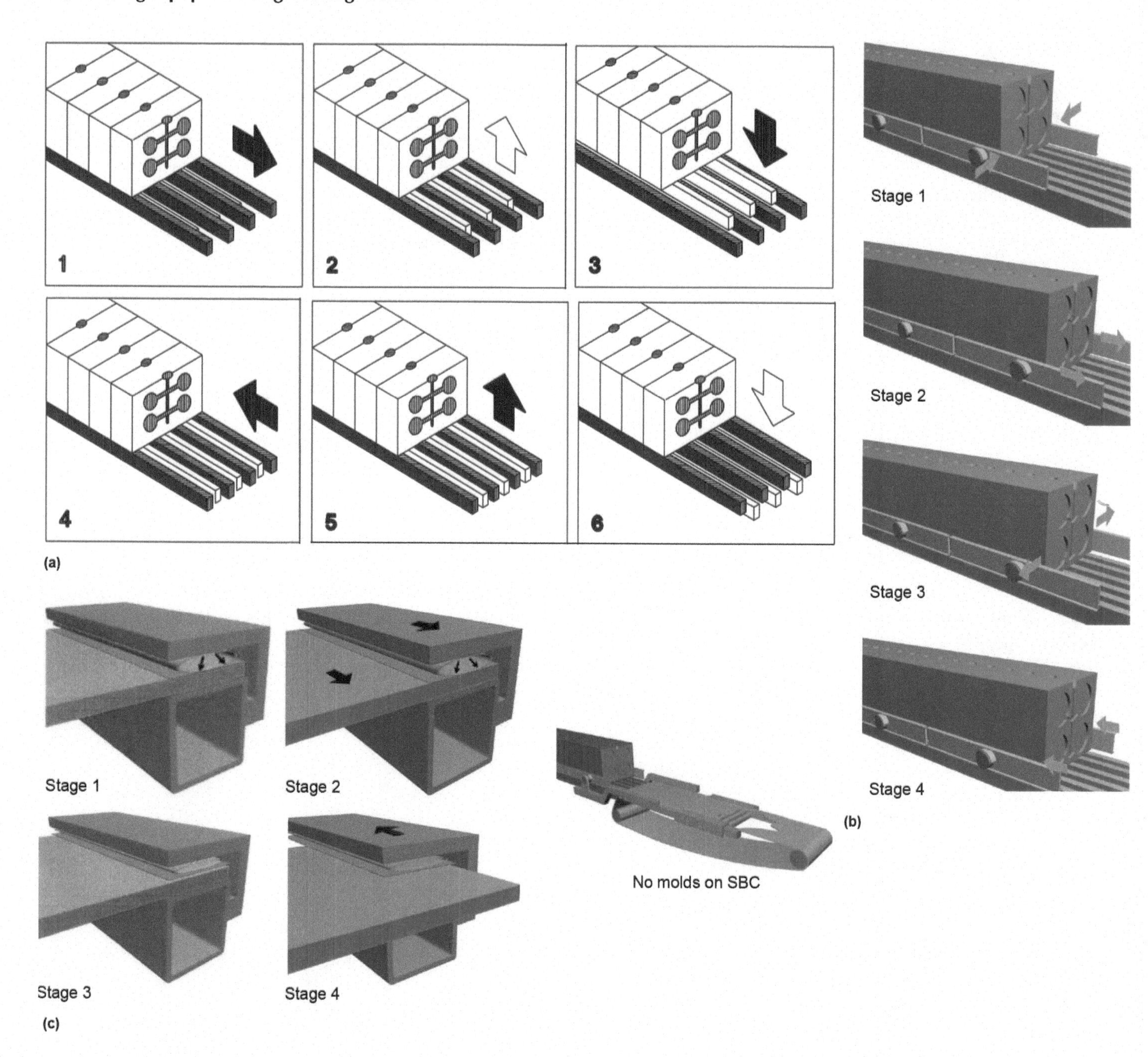

Fig. 3.21 DISAMATIC mold conveyor operating principles, (a) precision mold conveyor, PMC; (b) automatic mold conveyor, AMC; (c) synchronized-belt conveyor, SBC, without molds. Source: Ref 6, DISAMATIC.

1. In their starting positions, the thrust bars are as close to the molding machine as possible. This position is where they are activated and grip the molds.
2. While holding the molds, the thrust bars transport the mold string forward by the thickness of one mold. This movement is carried out synchronously with the movement of the pressure piston of the machine.
3. The thrust bars release the molds.
4. The thrust bars return to their starting positions as close as possible to the molding machine.

3.7.1.6 Belt Mold Conveyor — Synchronized Belt Conveyor

The synchronized belt conveyor (SBC) is an extension of the first part of the cooling line, AMC, or PMC. The SBC is mechanically connected to the transport movement of the AMC or PMC. The cooling of the casting from the solidification temperature down to the shakeout temperature takes place on the SBC.

The working cycle of the synchronized belt conveyor consists of four stages which operate synchronously with the AMC or PMC. The working cycle is illustrated in Fig. 3.21(c):

1. The clamping clutch is activated by compressed air which inflates the hoses. This allows the clutch to grip the entire length of the belt on both sides simultaneously.
2. The synchronized belt conveyor clamping clutch moves the belt forward by the thickness of one mold.
3. The hoses of the SBC clamping clutch are deflated and release the sides of the belt.

4. The clamping clutch returns to its starting position synchronously with the return stroke of the AMC or PMC.

3.7.2 Engineering for Cooling Time

Castings need to cool before the mold reaches the shakeout unit to avoid undesirable microstructure, distortion, damage, and cracking. The lengths of the conveyor should be engineered to allow for the targeted cooling time.

The engineering of a conveyor line is influenced by factors such as:

- The maximum weight, size, and wall thickness of the castings to be produced
- Desired maximum temperature of the casting at shake-out, without potential for undesirable microstructure, distortion, damage, or cracking
- Number of molds per hour produced on the molding machine
- Size of the pallet (for horizontally parted molding)
- Thickness of the mold (for vertically parted molding)
- Heat transfer rate of the mold medium (higher for green sand molding compared to no-bake molding sands)
- Ratio of the mold volume to the casting volume

The casting shake-out temperature has to be lower than its solidification temperature.

The casting solidification time is a function of the heat content based on the casting volume (V) and the casting surface area (S), transferring the casting heat content to the sand mold.

The ratio of the casting volume to the surface area, which is the main parameter influencing the solidification time, is called the modulus (Ref 14):

$$\text{Casting modulus } M = \frac{\text{casting volume}}{\text{casting surface area}} = \frac{V}{S}$$

The measuring unit used for the casting modulus is centimeters (cm).

Solidification time T is $\propto$ to $\{\frac{V}{S}\}^2$ or $\{M^2\}$

Solidification time is $T = K \times M^2$, where K is the constant of proportionality.

Figure 3.22 outlines the formulas for calculating the modulus of a few simple shapes. Casting shapes are usually very complex. Calculations are simplified by dividing the castings into several sections of simpler geometric shapes, as shown in Fig. 3.22.

The highest section modulus is selected for the computation of the cooling conveyor length.

Captive foundries, where fewer types of castings are produced in large volumes, can justify the investment in computer systems and operating personnel to compute casting modulus and simulate solidification times and cooling times.

Figure 1.3 in Chapter 1, "Casting Manufacturing Layout—Principles and Guidelines" in this book, illustrates the relationship between annual volumes and casting weights. Conveyors are needed for castings of groups A and B. Typical examples of captive industries are:

- Gray iron: automotive, tractors, textile machinery, fittings
- Malleable iron: fittings and valves
- Ductile iron: automotive, fittings
- Steel: high pressure fittings, valves

Taking into consideration the known and identified products makes it easier to fix the flask or mold size, and to compute the highest casting cooling modulus.

Figure 3.23 is a plot of the cooling time in minutes for different casting modulus values for green sand molded castings. To establish the cooling conveyor length, the highest value of the modulus is used. The maximum shakeout temperature that is permitted depends on the stresses generated in the casting due to the design, collapsibility of cores, microstructure requirements, and the casting strength at elevated temperatures.

The total heat content of the castings in the mold depends on the casting weight in the flask or mold. The rate of solidification and subsequent cooling are influenced by the wall thickness or the modulus of the individual casting in the mold.

	1	2	3	4	5
	Large plate L > 5 x t	Long Bar	Cube	Cylinder	Ring
Shape	t, L, L	b, a	c	r, h	b, a
Formula for Modulus	$M_C = \frac{t}{2}$	$M_C = \frac{a \times b}{2(a+b)}$	$M_C = \frac{c}{6}$	$M_C = \frac{r \times h}{2(r+h)}$	$M_C = \frac{a \times b}{2(a+b)}$

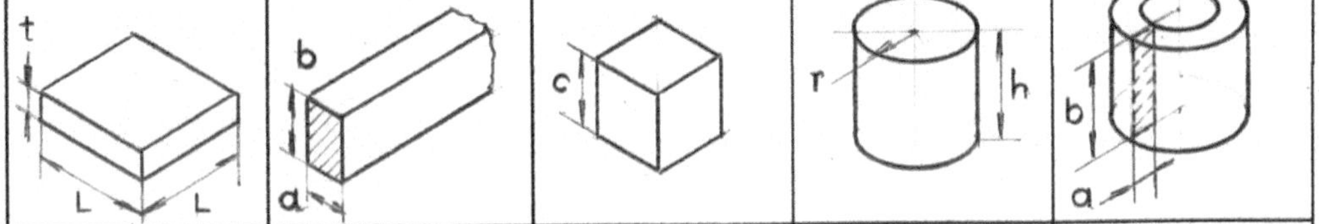

Fig. 3.22 Formulas for modulus calculation of simple shapes

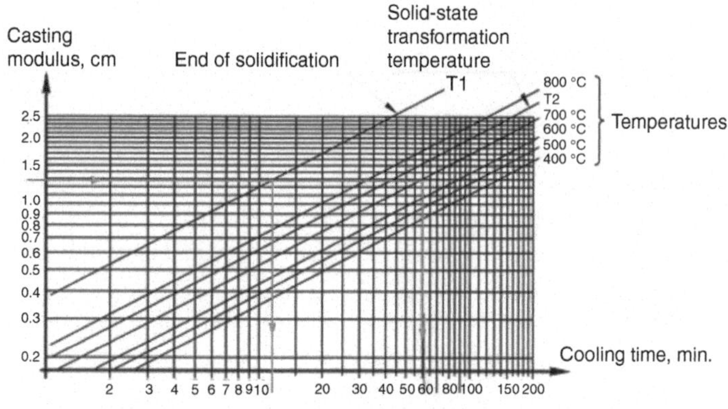

Maximum shakeout temperatures for gray iron castings

a. Relatively less complex castings with few cores, 700 °C (1292 °F)

b. Moderately complex castings with more cores, 600 °C (1112 °F)

c. Complex castings with many cores, 450 °C (842 °F)

Fig. 3.23 Casting modulus and shakeout time relationship. Source: Ref 6, DISAMATIC.

It is challenging for noncaptive and jobbing foundries to engineer the optimum length of the cooling conveyor. The design requires a few assumptions for a conservative estimate of the conveyor length because the exact sizes and shapes of the products are not always known.

To simplify the computations, the parameters assumed are:

- Maximum casting wall thickness
- Modulus assumed as half the maximum wall thickness
- Maximum shakeout temperature for gray iron castings (Fig. 3.23)
- Pallet size — 20% larger than the flask width (core mold width) in the direction of the conveyor length
- Cooling time for no-bake molds — 15% higher than green sand molds
- Number of molds produced per hour

3.7.2.1 Example Calculation of Conveyor Length for Gray Iron Foundries

Assumptions:

1. Maximum casting wall thickness is 20 mm (0.79 in.)
2. Shakeout temperature is 700 °C (1292 °F)
3. Flask width along conveyor length is 400 mm or 0.4 meters (15.75 in.)
4. Production rate is 60 molds per hour

Computation:

1. Casting modulus equals maximum wall thickness divided by two:

$$^{20}/_{2} = 10\,\text{mm or 1 cm}$$

2. Using Fig. 3.22, the cooling time is 36 minutes
3. At a production rate of 60 molds per hour, each minute the conveyor advances by 1 mold
4. With a flask width of 0.4 meters (15.75 in.), allowing for the pallet size of 20% over the flask size, each minute the pallet advances by 0.4 times 1.2, which is 0.48 meters (18.9 in.).
5. To allow for a cooling time of 36 minutes, the conveyor has to be 0.48 times 36, or 17.28 meters or 18 meters (59 ft.). The conveyor can be designed in two parallel sections, each of half the total length, if space is a constraint in the layout.
6. It is common practice to allow a margin of about 20% to accommodate for any minor errors in the assumptions, such as the modulus or the safe shakeout temperature.

In practice, it is possible that a small percentage of castings with a modulus higher than the assumed maximum for the conveyor length computation may have to be produced on the same conveyor line. In this case the heavier part is molded toward the end of the second shift if the foundry works in two shifts, or before a weekend shutdown, to allow for more cooling time. Alternatively, the line must run at a lower speed to allow for longer cooling time.

Figure 3.24 (Ref 6) is a nomogram for estimating the cooling conveyor length, generated for a vertical molding operation. Using the mold thickness as the pallet width, the same nomogram can be used to obtain cooling times for horizontally parted molds. The red lines in Fig. 3.24 indicate the method of using the nomogram for conveyor lengths (a) to complete the casting solidification, and (b) to achieve the targeted shakeout temperature. Example steps to obtain the length of the cooling conveyor for both cases are:

1. The required mold thickness is determined based on the casting features and dimensions. This is estimated according to the required mold height, sand-to-metal ratio, and machine limitations. The mold thickness (for example, 250 mm or 9.84 in. in Fig. 3.24) is by point A in Fig. 3.24.
2. The production speed (molds per hour) is estimated based on machine output limitations, such as the limitations of pouring rates, liquid metal available per hour, and mixed sand available per hour. Point B (for example, at 375 molds/hour in) is selected in Fig. 3.24. Point C is the intersection of A and B.
3. Line CD is drawn from point C.
4. Point E is the casting modulus (for example, 1.25 cm or 0.49 in.) as shown in Fig. 3.24. The intersection of E and T1 (the solid-state transformation temperature) is point F. A vertical line through F indicates the cooling time (in minutes) to reach the solidification temperature. The vertical line intersects line CD at G.
5. A horizontal line through point G intersects at point H (18 m or 59 ft.). This is the length of the AMC/PMC for the casting to reach the solidification temperature.
6. The targeted temperature for shakeout is chosen based on the modulus and complexity of the casting. The diagonal and parallel lines in the upper graph of Fig. 3.24 indicate the targeted temperature for the shakeout. Point I is the intersection of the horizontal line through E and the selected shakeout temperature of 700 °C (1292 °F).
7. A vertical line from point I indicates the cooling time needed for the casting to be at 700 °C (1292 °F) at the shakeout.
8. The vertical line through point I intersects the diagonal line CD at point D.

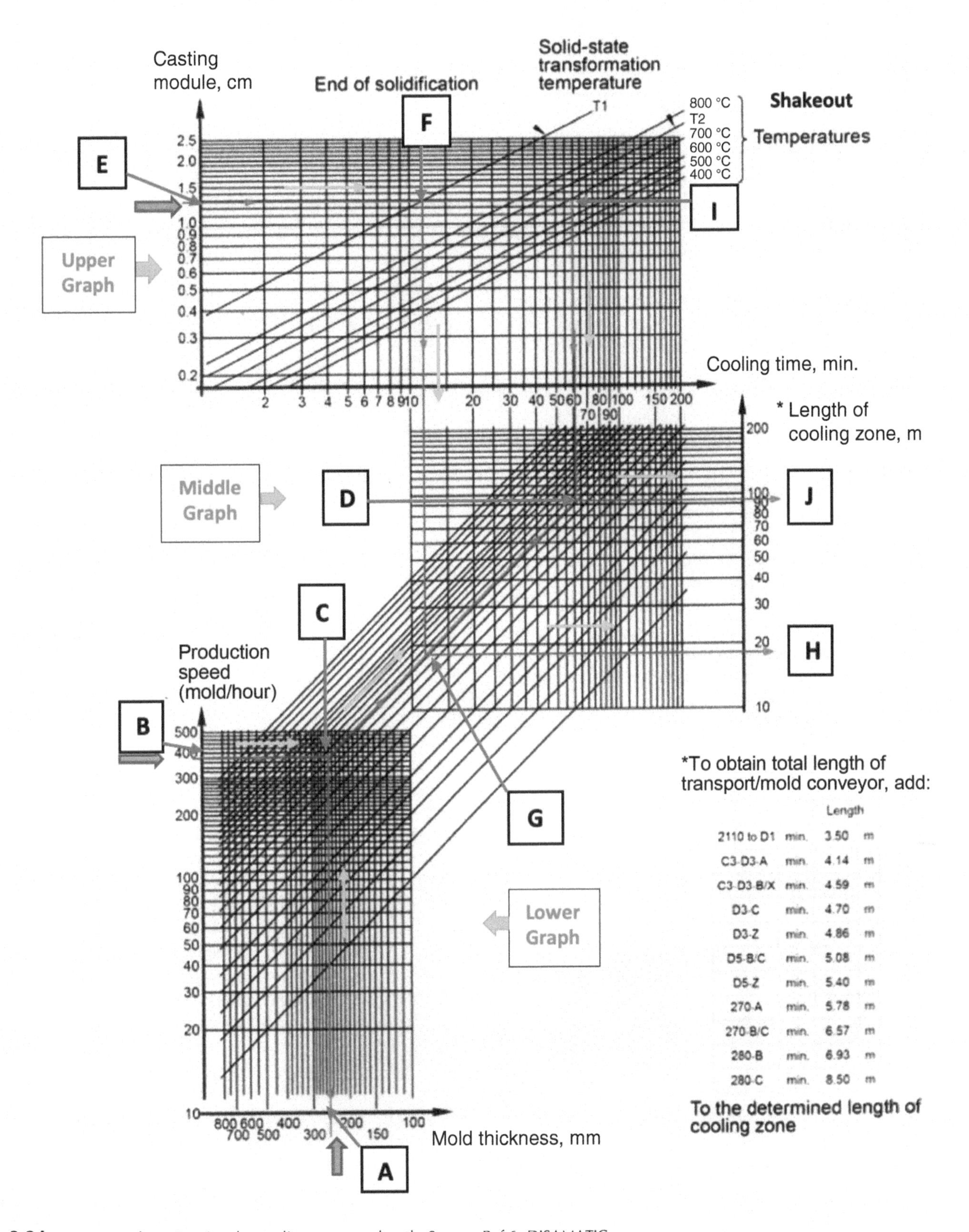

Fig. 3.24 Nomogram for estimating the cooling conveyor length. Source: Ref 6, DISAMATIC

9. A horizontal line through point D intersects at point J. This gives the length of the cooling zone (92 meters or 100 yards) for the shakeout temperature of 700 °C (1292 °F).
10. This gives the length of the AMC or PMC and respectively the total length of the AMC or PMC and the SBC.
11. Additionally, an allowance must be made for a safe distance between the vertical molding machine and the pouring position to allow a safety margin to the established length of AMC or PMC and SBC. This margin is needed to prevent mold leakage at the parting plane as a result of the pressure

from the liquid metal in the mold. The table in Fig. 3.24 offers guidance for this allowance.

3.7.2.2 Conveyor Length for Ductile Iron Castings

The microstructure of ductile irons is significantly influenced by the cooling rate. Faster cooling rates of irons of standard chemistry produce a higher percentage of pearlitic microstructure. But excessive cooling rates may cause higher internal stresses, resulting in distortion and cracking. Castings above 480 °C (950 °F) have lower strength that may result in damage and distortion during shakeout and handling. The constraint offered by cores against contraction is difficult to compute or simulate. Castings retained on the mold until the hottest area is below 315 °C (600 °F) do not normally experience internal stresses. Such castings display ferritic microstructure.

Conveyors are designed for longer cooling durations which produce ferritic grades with normal chemistry. Most of the ductile iron foundries produce multiple grades such as D 4512 (with higher ferrite content for structural applications such as steering knuckles) and D 5506 (with lower ferrite and higher pearlite content for applications such as brake calipers). The conversion of the melt from D 4512 to D 5506 is achieved by adding pearlite-promoting ferroalloys to the induction melts, keeping the shakeout time the same as those for D 4512 grades.

Conveyors for dedicated foundries with identified products can be designed to accommodate those specifications. Jobbing foundries without specific products need to target some parameters such as the maximum size and wall thickness for computing the length of the conveyors.

Figure 3.25 shows a relationship between the casting modulus and the shakeout time for both ferritic and pearlitic grades.

The assumptions of this relationship include:

- The castings are poured into green sand molds at about 1400 °C (2550 °F).
- Castings are cooled below 660 °C (1200 °F) by the time they separate from the mold.
- Castings poured in no-bake molds need slightly longer times before shakeout.
- The modulus of the maximum section thickness of the casting is the guiding factor to establish the shakeout time to achieve the targeted microstructure.
- Achieving ferritic structure requires longer cooling times before shakeout compared to those for obtaining pearlitic structure, as the figure indicates.

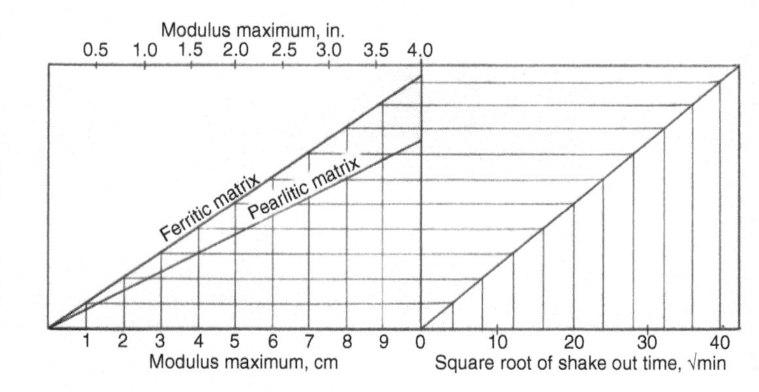

Fig. 3.25 Ductile iron castings shakeout time guide. Source: Ref 15

3.7.2.3 Conveyor Length for Steel Castings

Steel castings for valves and pipe fittings are produced in relatively large volumes using conveyors for molding. Castings for railroad and tractor applications are produced in medium volume on conveyors, using green sand for smaller sizes and no-bake for medium sizes.

Steel castings have more feed metal requirements to offset the volumetric shrinkage compared to other ferrous metals. Therefore, steel castings have relatively larger feeders compared to ductile and gray iron castings. The coefficient of linear contraction for steel is higher than that of ductile and gray iron. The larger linear contraction values of steel have the potential to induce higher internal stresses, especially if the contraction of the feeders and gates is impeded. Although steel castings do not generally have complex cores, care is taken to hollow the cores out to avoid cracks and hot tears.

Feeders solidify and cool slower than the casting segments they are feeding. The modulus of feeders is usually 20% higher than the casting modulus of the segment it is feeding. Conveyor lengths are designed conservatively for relatively diverse castings within the market segment. The shakeout temperatures are between 500 and 350 °C (932 and 662 °F) to induce minimum internal stresses (due to different wall thicknesses solidifying at different rates). Almost all steel casting grades are subjected to some type of heat treatment (normalizing and tempering or quenching and tempering) to achieve the targeted microstructure and mechanical properties.

Many jobbing steel foundries plan to mold in the first shift, pour in the second shift, and shake out in the third shift, a sequence which provides ample cooling time. But medium- and high-volume steel foundries need to engineer the conveyor lengths adequately. The feeder sizes which have a modulus higher than the casting sections solidify slowly, and these dictate the duration of the cooling time before the shakeout. For example, a feeder size of 12 cm (4.72 in.) diameter and 18 cm (7.09 in.) height is capable of feeding about 25 kg or 55.12 lb. (for 5% shrink volume) of the casting. If the casting weight is about 50 kg or 110.23 lb., two feeders are adequate. The corresponding casting modulus is 1.8 cm (0.71 in.).

Figure 3.26 offers guidelines for establishing the shakeout times. The graph can be used to establish the shakeout times when the specific products and feeding systems are known. When products are not specifically known, the guiding factor is the maximum feeder size, based on the shop floor experience.

When the feeder height to diameter ratio is 1.5, the feeder modulus is calculated as:

$$M_f = 0.2145\, D \tag{Eq 3.3}$$

where M_f is the feeder modulus, D is the feeder diameter, $H/D = 1.5$.

The feeder modulus $M_f = 1.2 \times M_c$, where M_c is the casting modulus. The length of the conveyor is calculated like the

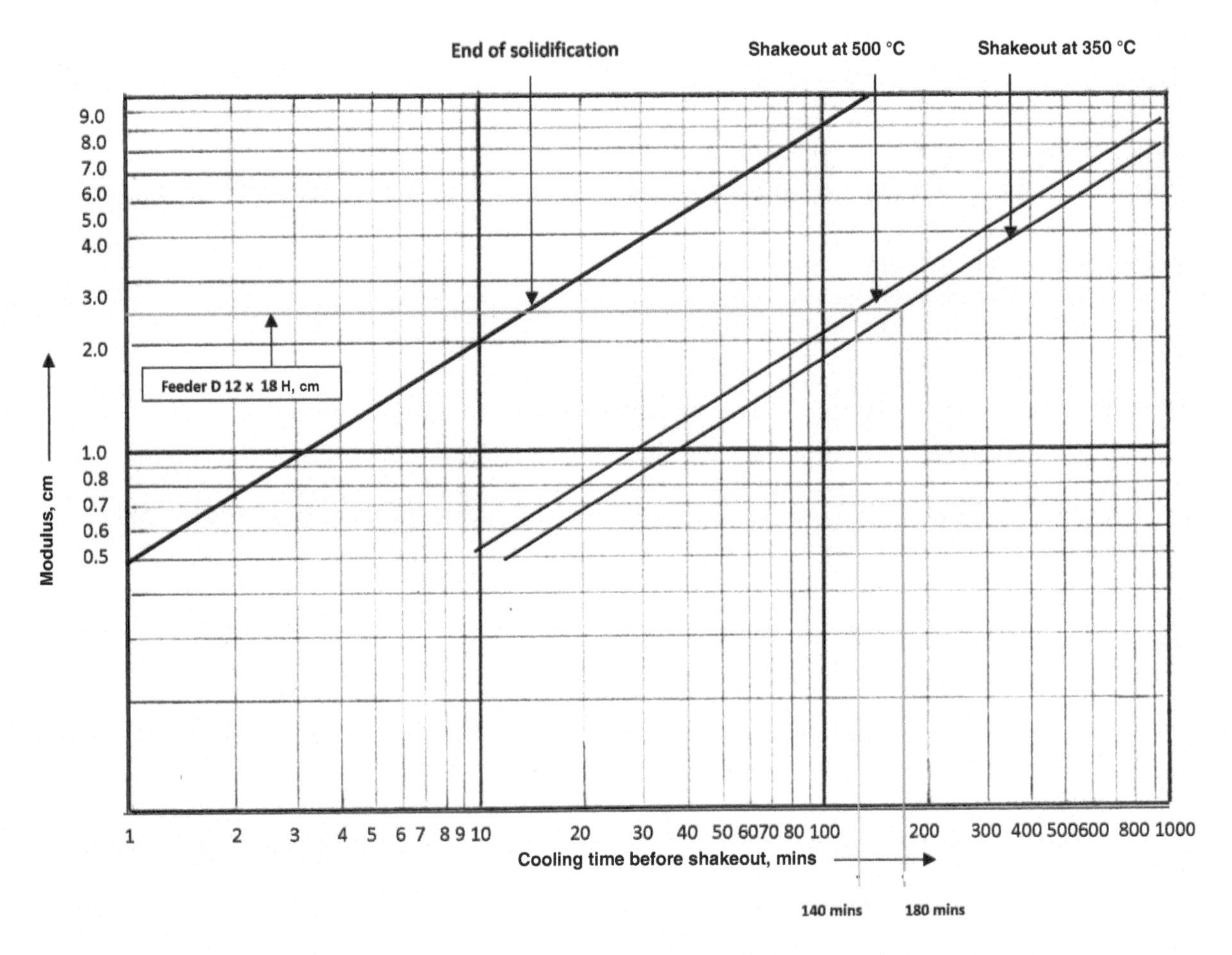

Fig. 3.26 Relation between modulus and shakeout time

example for gray iron in the section " Example Calculation of Conveyor Length for Gray Iron Foundries" in this chapter.

An example for assumptions based on designated product geometry is:

1. Overall casting wall thickness is 1.0 cm (0.39 in.)
2. Localized thick wall thickness is 2.0 cm (0.79 in.)
3. Casting modulus of localized section equals approximately half of the localized section (2.0/2 equals 1 cm, or 0.39 in.)
4. Flask size is 50 cm by 65 cm (19.68 by 29.53 in.), with 50 cm (19.68 in.) along the conveyor
5. Number of molds per hour is 30
6. Pallet size is 20% larger than the flask width

Calculations

1. Feeder modulus equals 1.2 times casting modulus of localized thick section (1.2 by 1 equals 1.2).
2. Pallet size equals 50 times 1.2 = 60 cm or 23.62 in. (pallet size assumed 20% larger than the flask size).
3. Cooling time of feeder for shakeout temperature of 350 °C or 662 °F is 50 minutes (Figure 3.26).
4. In 1 hour (60 mins) the molds move 30 times 60, which equals 1800 cm or 18 meters (19.69 yards).
5. In 50 minutes the molds move 18 times 50 divided by 60 equals 15 meters (16.4 yards).
6. The conveyor length would be 15 meters or 16.4 yards.

Unknown product geometry

1. In a jobbing environment, the number of molds per hour can be assumed to be fewer than 20 molds per hour.
2. Assume the maximum feeder diameter of 12cm (4.72 in.) in diameter
3. The feeder modulus M_f = 0.2145, D = 0.2145 x 12 = 2.574 cm (1 in.)
4. The shakeout time for shakeout temperature of 500 °C or 932 °F is 140 minutes (Figure 3.26).
5. In 1 hour (60 mins) the molds move 20 x 50 divided by 60 meters equals 16.6 meters (18.15 yd.)
6. The conveyor length would be 16.6 x 140 divided by 60 = 38.7 meters (42.32 yd.)
7. This length can be engineered with 2 or 3 parallel sections, totaling to 38.7 or 40 meters (42.32 or 44 yd.).

3.7.3 Flexibility in Conveyor Design

Pouring should keep pace with molding so that the molding pace is not interrupted. In practice, interruptions of the molding pace are inevitable due to pattern changes and any short-term maintenance that may be necessary. Similarly, there may be delays due to delivery from the distribution ladle or from deslagging (the skimming of slag over the melt surface in the ladle). A conveyor bypass should be provided where molds can be diverted and returned to the pouring station using proximate switches.

3.7.4 Extended Cooling Times

Mold conveyor lengths for mass produced castings, as illustrated in the section 3.7.2.1 in this chapter, may occupy extended lengths of the bays, which may not be practical.

The mold conveyors can be designed to include additional or spare conveyor lines so the molds can be diverted after pouring for additional cooling. Figure 3.27 illustrates this concept.

Figure 3.27(a) illustrates the provision of two optional conveyor tracks for additional cooling. The automated transfer cars can be quickly and easily programmed to divert the molds to the optional conveyor tracks. Finally, the molds are conveyed to the tracks that advance the molds to the shakeout unit or a rotary drum cooler. Such an arrangement represents a low capital investment, although the additional conveyors require more space.

Conveyor space can be optimized by using a patented system (Ref 13) where the conveyors are engineered to occupy the three-dimensional space more effectively. Figure 3.27(b) illustrates this concept.

Green sand molds produced by a high productive molding machine pass through a conveyor line with a gantry crane. Weights are placed over the top of molds to prevent the cope from lifting due to the force of buoyancy of the liquid metal. These are called *mold weights*. The gantry crane moves the mold weights from the cooled molds to the molds waiting to be poured. The poured molds move through to the end of the line, passing beneath the pouring line because the conveyors are designed with horizontally mounted guide drums. They then advance to the pouring level. A gantry crane moves the molds to the second row after moving the mold weight to the first line of molds successively. The molds continue to cool in the second parallel row and move down to the lower level at the end of the line. As the molds move to the end of the conveyor, they are pushed to a shakeout unit or a rotary cooling drum, where the castings are separated from the molds. The return sand is recycled, and the castings are moved to the cleaning room.

The bottom of Fig. 3.27(b) is a schematic of this mold movement. Molds that are cooled with the mold weights *on,* have their mold weights switched by another pick-and-place gantry crane, as indicated. The lines can be designed to lay the slip jackets simultaneously with the weights, or the jackets can be slipped on before the mold weights are positioned. Ref 13 illustrates layouts with three adjacent molds.

Another space-saving alternative is to use a rotary turntable configuration, as shown in Fig. 3.27(c) (Ref 13). Molds produced by the high production flaskless molding process pass through a gantry crane with a pick-and-place mechanism to switch the mold weights from the cooled molds to the molds awaiting pouring. Molds poured along the outer circle move around a circle, where the pick-and-place unit unloads the weights. The same crane elevates the molds from the outer circle to the inner circle.

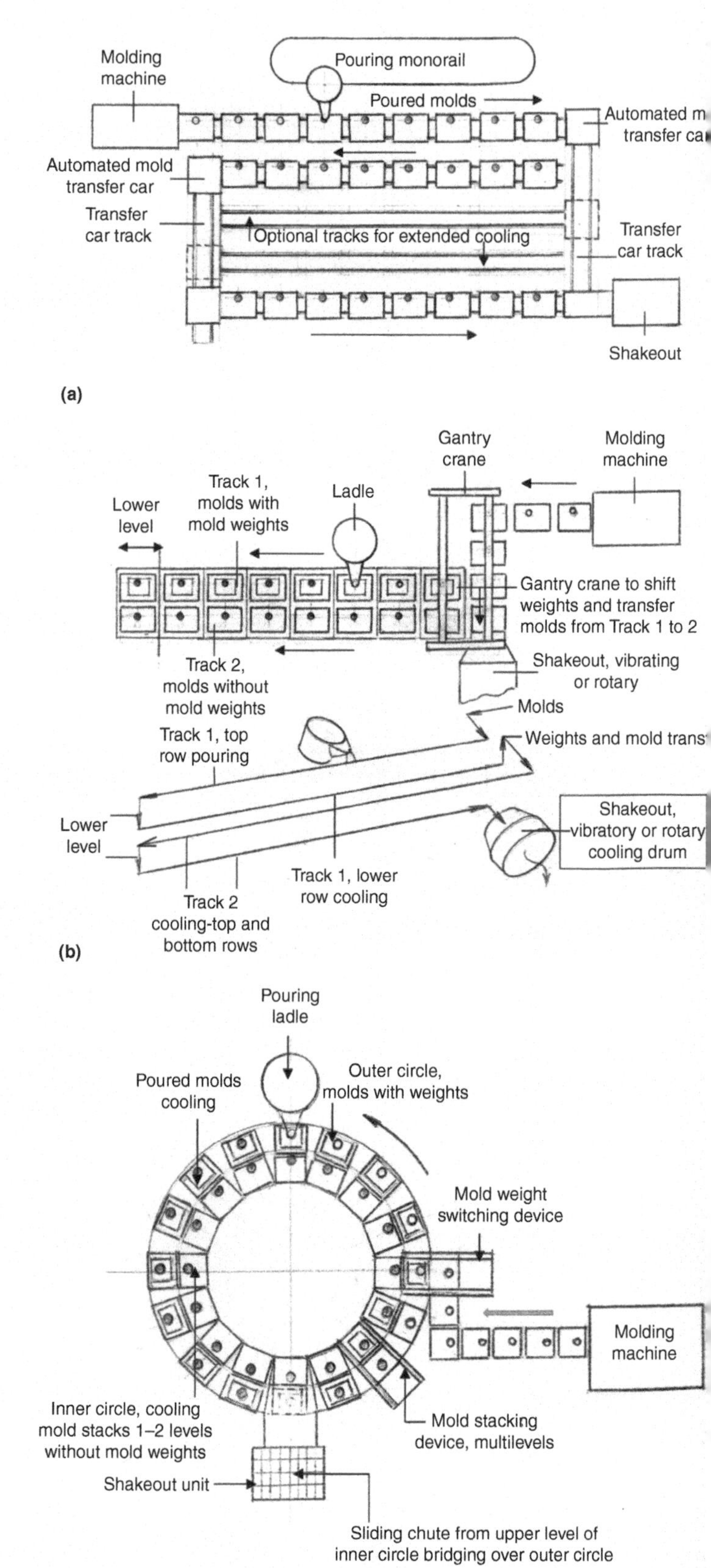

Fig. 3.27 Mold conveyors for cooling, (a) optional tracks for extended time, (b) multirow section for increased time, (c) turntable configuration for extended time

The inner circle carries shelves from one to three levels where the molds can be stacked by the crane, depending on the required cooling time.

The molds from the inner circle are designed to slide down to the shakeout unit or the rotary cooling drum, as illustrated.

3.8 Shakeout Units and Features

The shakeout operation separates the castings from the mold and disintegrates the cores. This process also achieves the separation of the gating and feeders from the castings without damaging them (with the exception of steel castings). Some shakeout units are designed to cool the sand by moisturizing it. Nearly 20% of the casting cost is consumed in shakeout, cleaning, and flash-grinding.

Smaller flasks with widths up to 600 to 800 mm (23.62 to 31.5 in.) do not have crossbars in the cope and drag. These molds produced in flasks without crossbars are punched out over a shakeout unit, and the flasks are returned to the molding line for reuse. Larger flasks with crossbars rest on the vibrating grid of the shakeout unit to loosen the sand, and the empty flasks are moved by the crane to the molding station. Green sand molds made without flasks (flaskless molding) and no-bake molds enter the shakeout unit directly.

The shakeout operation consists of vibrating the mold to loosen the sand around the core sand inside the castings. The technologies and features of different types of shakeout equipment vary, depending on the applications and objectives. Often, the shakeout units with grids are designed to discharge the castings on to a slowly vibrating trough or a table and a heat-resistant conveyor, where additional degating or deheading (feeder removal) is accomplished manually. When multiple types of jobs are processed simultaneously, castings are picked out manually with a hook to separate and load them into separate bins for the next operation of shot blasting. Hot sand falls from the grid or the grating onto an underground hopper.

Figure 3.28 (Ref 16, 17, 18) outlines the different types and features of shakeout equipment. The figure also indicates the scope of their applications.

3.8.1 Basic Brute and Variable Drives

The simpler units, the basic brute and variable drive types, feature one motor on each side of the box structure. The box structure, which houses a hopper at the bottom, is supported on springs to isolate the vibration from the supporting structure. The motors are designed with an eccentric weight to create the imbalance that causes vibration. The motors are also provided with heavy-duty bearings to withstand the vibratory forces.

These types are sometimes combined with a punchout unit, as illustrated in Fig. 3.29(a).

Mass or large production layouts with molding in flasks may use a punchout unit that keeps pace with the speed of molding. A trolley riding on a frame carries a pick-and-place unit for picking up the mold assembly, bringing it to the punchout head, and returning the empty flasks to the return conveyor.

Punched out molds dump the castings and the mold core sand on to the vibrating grid of the shakeout. Some shakeout units are engineered with a lift mechanism on one side to tilt the grid to advance the castings faster. Sand passes through the grid to an underground hopper on the return sand belt conveyor. Castings move to an adjacent vibrating table, where operators knock out runners and gates that have not been broken on the grid (typically for gray iron and ductile castings) and sort the castings into

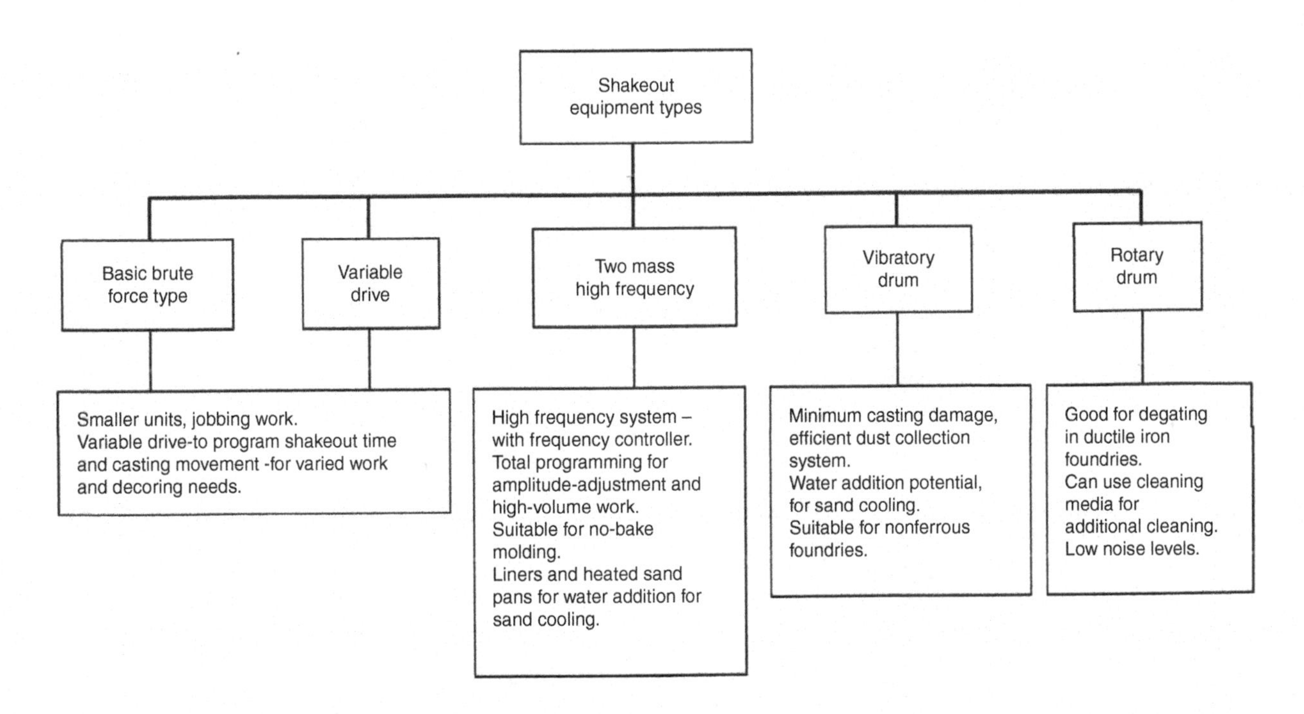

Fig. 3.28 Shakeout equipment classification

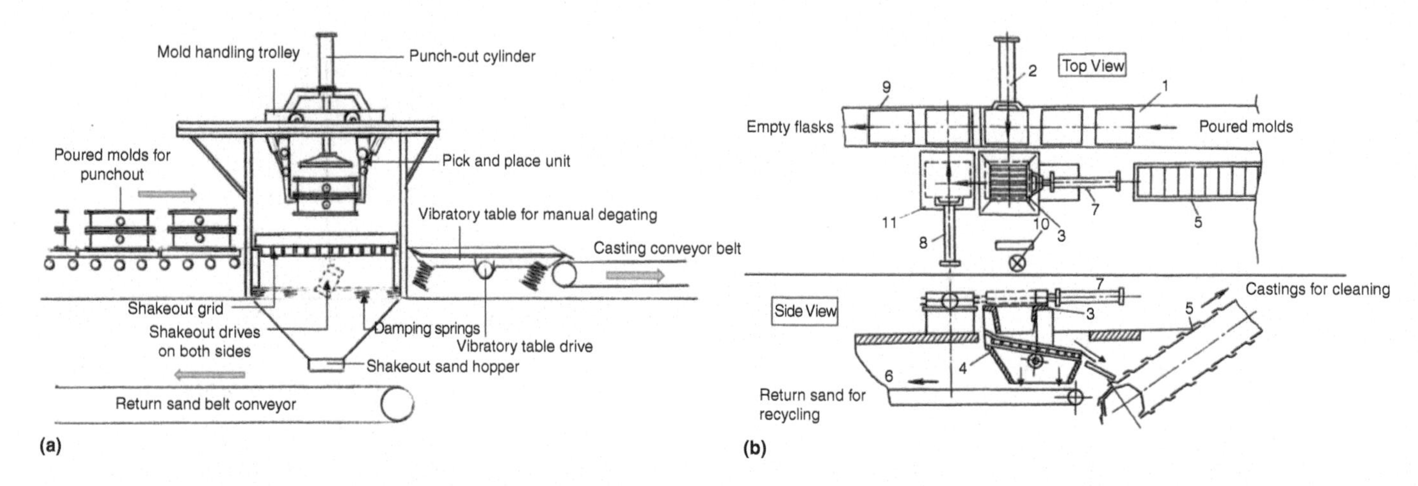

Fig. 3.29 Brute and variable drives, (a) mold punchout unit, (b) vibratory shakeout for molds in flasks. Notes: (1) Roller conveyor; (2) Pusher cylinder; (3) Flask vibrating grid; (4) Sand-castings vibrating grid; (5) Casting conveyor; (6) Return sand conveyor; (7) and (8) Pusher cylinders; (9) Empty flask return; (10) Operator control console; (11) Auxiliary table. Source: Ref 1

different hoppers, as needed. Sorting is also required when different parts are nested together on one match plate (or on a universal match plate that can accommodate two to four inserts with different patterns). If only one type of casting is run, the castings move to a conveyor that takes them to individual hoppers positioned at the end.

Molding in flasks allows fuller utilization of the pattern plate area while nesting multiple cavities, thus increasing output per flask. But the maintenance of the flasks, cleaning the flasks for sand buildup inside, and engineering the flask return to the molding station are additional costs.

Casting producers that mass produce small sized castings (for example, pipe fittings, textile machinery, refrigeration and air conditioning parts, or engine accessories) are increasingly opting for flaskless molding, which also allows the poured molds to be dumped onto the shakeout units.

Open-type vibratory shakeout units are manufactured in various sizes to suit a variety of flask and mold sizes. The simpler units are less expensive. Large flasks are suspended over the shakeouts by overhead cranes, with part of the flask resting on the vibrating grid. As the mold sand disintegrates, the flasks are lifted by the crane and moved away. The crane unloads the flasks and then returns to pick up the castings on the grid to decore them. The open units radiate a lot of heat and create a huge volume of dust. These types are not conducive to dust extraction enclosures.

Figure 3.29(b) (Ref 1) illustrates a layout where a vibratory unit shakes the flasks for the sand and the castings to drop into another vibratory grid where the castings and sand are separated.

Molds poured into the flasks arrive at a push-out station where the flasks are pushed onto a vibrator. The flasks supported on a frame loosen the sand and the castings. They drop onto another vibratory grid where sand and castings are separated. The castings retained on the grid slide onto a conveyor to be transported for cleaning. Return sand passes through the grid, and a belt conveyor transports the sand for reconditioning for reuse.

3.8.2 Two-Mass High-Frequency Type (Ref 16, 17)

The two-mass high-frequency shakeouts offer several advantages over the brute and variable drive types. Figure 3.30 is a schematic of a two-mass high-frequency type, with motors mounted at the top. Some equipment manufacturers engineer the drive motors to be mounted at the side toward the base, and some designs feature the drive motors mounted at the top. Top mounting allows for an enclosure for dust extraction and easy access for drive maintenance.

The two-mass high-frequency type is fitted with two motors on each side or on top, with opposing directions of rotation to create vibration. Both the frequency and amplitude are adjustable to meet the needs of different jobs. This type is ideally suited for large series production and for water addition to control the sand temperature. When the motors are mounted at the sides, the orientation of both the motors can be adjusted for movement along the grid or for increased duration of shaking or decoring, depending on the configuration of the part. Some manufacturers design the grids for changing the inclination towards the exit to control the residence time over the grid.

3.8.3 Vibratory Drum-Type

The vibratory drum is a hybrid that ensures minimum casting damage. It also facilitates a good connection to the dust collection system. Water can be added for sand temperature control. The vibratory types are efficient, especially for decoring. Intricately cored jobs and boxy types of designs are well suited to this type of shakeout unit.

3.8.4 Rotary Drum-Type

One of the limitations of the vibratory shakeout units is that a significant amount of sand adheres to the gating system and is lost from the recycled sand volume. These sand-covered gating systems also contaminate the induction melts and reduce furnace lining life and melting efficiency. The heavy-duty bearings inherent in these types are also expensive to replace.

Rotary drum-types offer advantages that suit several types of small and medium-sized castings. The main advantages are:

- Reduction and elimination of the need for shot-blasting
- Lower capital expenditures.
- The concept and design are well suited for integration of rotary drums with flaskless green sand molding systems.
- Eliminates the need for human intervention to remove gating and to unclog jams over the grid, improving the working environment significantly
- Eliminates the need for human intervention for degating or deheading, because even chunky ductile iron castings are degated in the rotary drum
- Return sand is cooled in the drum as it circulates with the castings, resulting in cooler castings at the exit.
- Rotary drum types are versatile for handling a variety of alloys — gray iron, ductile iron, steel, and bronze castings.

Figure 3.31 illustrates a cross section of a rotary drum-type cleaning unit (Ref 18).

Flaskless molds are tipped by pneumatic or hydraulic cylinders into the rotary drum. Castings and sand are separated as the drum rotates. The spiral ridges of the liners help to provide the necessary attrition. When grinding material is loaded along with castings, the surfaces of both castings and gating are cleaned. The sand and castings are separated at the discharge end. The liners are produced from austenitic manganese steel, which work hardens with use.

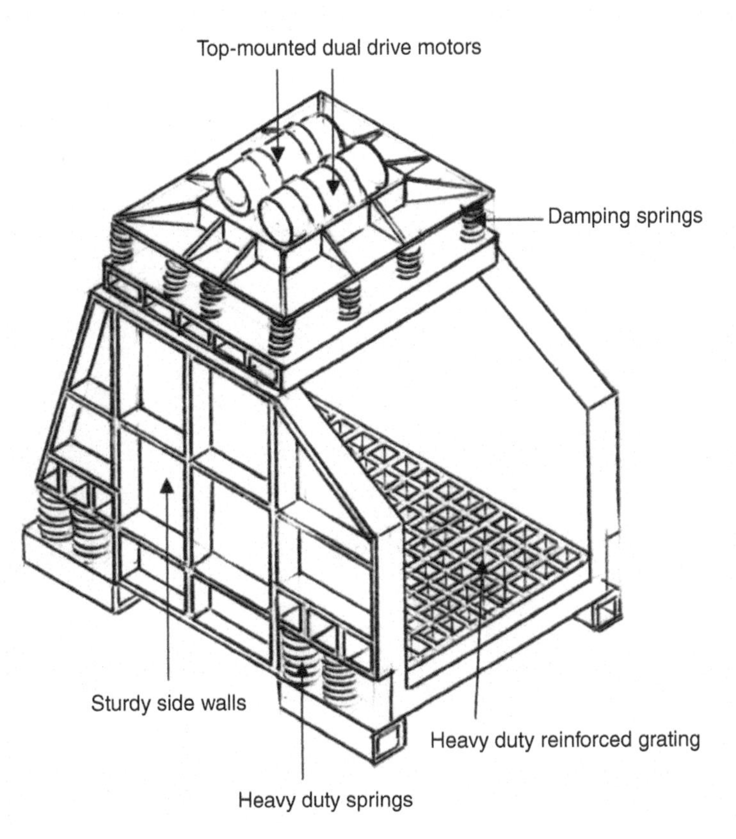

Fig. 3.30 Two mass high-frequency shakeout unit

3.8.5 Hydroblasting

Very large castings in gray iron and steel are cleaned by hydroblasting. The hydroblasting process uses high pressure water jets to remove any sand that sticks to the casting. Since large castings are produced in low volume, it is more economical to position these castings in booths or enclosures where the operators stand outside the booth and operate these jets. An overhead crane operated by the hydroblast operator orients the casting to facilitate the cleaning. Table 3.6 lists products and casting sizes for hydroblasting.

3.8.6 Summary

There is a wide range of shakeout equipment available. Table 3.7 lists suggestions to casting manufacturers.

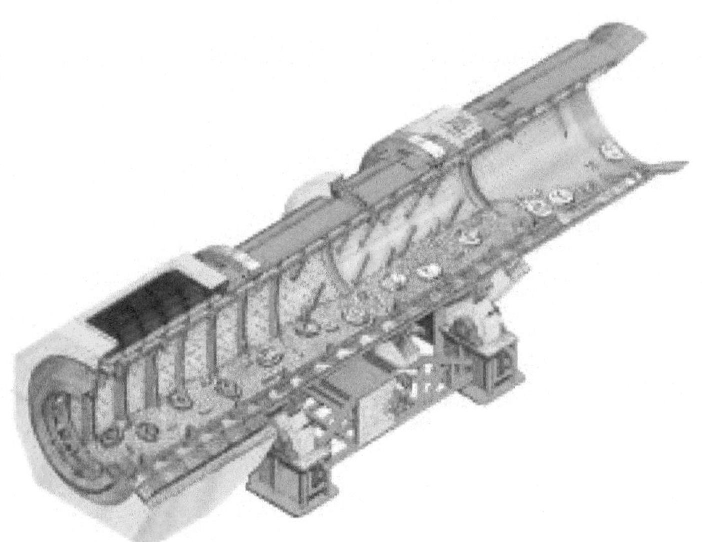

Fig. 3.31 Cross-sectional view of a rotary drum-type casting cleaning. Source: Ref 18, Didion International Inc.

Table 3.6 Typical industry products and casting sizes for hydroblasting

Material	Products	Casting size	Annual volumes
Gray iron	Heavy diesel engines for locomotives, power generation and marine applications	Large	Small
	Machine tools	Large	Small
	Steel plant equipment	Large	Small
	Turbines and power generation	Large	Small
	Rolling mill products	Large	Small
	Paper and pulp machinery	Large	Small
	Presses and hammers	Large	Small
Steel	Turbines and power generation	Large	Small
	Steel plant equipment	Large	Small
	Rolling mill equipment	Large	Small
	Presses and hammers	Large	Small
	Cement mill machinery	Large	Small
	Sugar mill machinery	Large	Small
	Excavators and earthmoving	Medium	Small
	Ship building	Large	Small

Table 3.7 Typical castings and suggested type of shakeout equipment

Metal	Industry	Typical components	Casting sizes	Annual volumes	Suggested choice(a)
Gray iron	Automotive	Disc brake rotors, drums, turbocharger housings	Small	Large	5 / 4 / 3
	Steel plant equipment	Ingot molds Stove grates	Medium	Medium	2 / 3
Malleable iron	Threaded fittings	Gas and water line fittings	Small	Large	4
	Electric transmission	Pole line hardware	Small	Large	4
Compacted graphite iron	Steel plant equipment	Pig molds	Medium	Medium	3
	Engine components	Diesel blocks and heads	Large	Medium	3
	Locomotive	Disc brake rotors	Medium	Medium	3
Ductile iron	Automotive	Steering knuckles, brake calipers, gears and pinions, differential carriers	Small	Large	5 / 4 4 / 3
	Fluid handling	Pump bodies, valve bodies	Small	Large	5 / 4
	Highway transport	Axle housings	Medium	Medium	3
Austempered ductile iron	Automotive	Engine mounts, heavy duty knuckles, hypoid gears	Small	Medium	3
	Construction	Crawler track shoes	Medium	Medium	3 / 2
Steel castings	Railroad	Draft sill, draft gear bolster, side frame	Small to medium	Medium	
	Highway transport	Axle housings knuckles, control arms	Medium	Medium	3
	Mining	Shovels, digger teeth	Medium to small	Medium	3
	Ore handling	Mantles, wear plates	Medium	Medium	3
	Fluid handling	Pump bodies, valve bodies	Small	Large	5 / 4

(a) Types of techniques, 1. Basic brute force, 2. Variable drive, 3. Two mass frequency, 4. Vibratory drum, 5. Rotary drum

3.9 Magnetic Separators

A magnetic separator is one of the most critical elements of a return sand system. It catches metallic tramp elements such as flash, pieces of gating, separated feeders, and embedded external chills from the return sand, protecting the lump crushers.

The contaminants in the return sand from the shakeout units vary depending on the alloy cast. Gray iron foundries often see flash and gates mixing with sand. Ductile iron foundries may have some separated feeders in the return sand. Steel foundries use chills more extensively than gray iron and ductile foundries. These chills are mixed with the return sand. The size of lumps in a green sand system are likely to be smaller than those in no-bake systems.

It is essential that these metallic contaminants are separated from the sand before the recycled sand enters the lump breakers and core crushers, to avoid any damage to the lump breakers. A magnetic separator is an essential and critical component in the process of recycling the return sand.

There are three basic types of magnetic separators:

- Magnetic drum
- Magnetic pulley segment
- Belt-type magnetic separator

3.9.1 Magnetic Drum

The return sand belt is fitted with a magnetic drum at a transfer point, as illustrated in Fig. 3.32(a). Electromagnetic coils are embedded into a soft iron drum, with power connected through a commutator segment system.

Tramp elements such as flash, broken gating pieces, and separated feeders and chills embedded in the mold are attracted by the

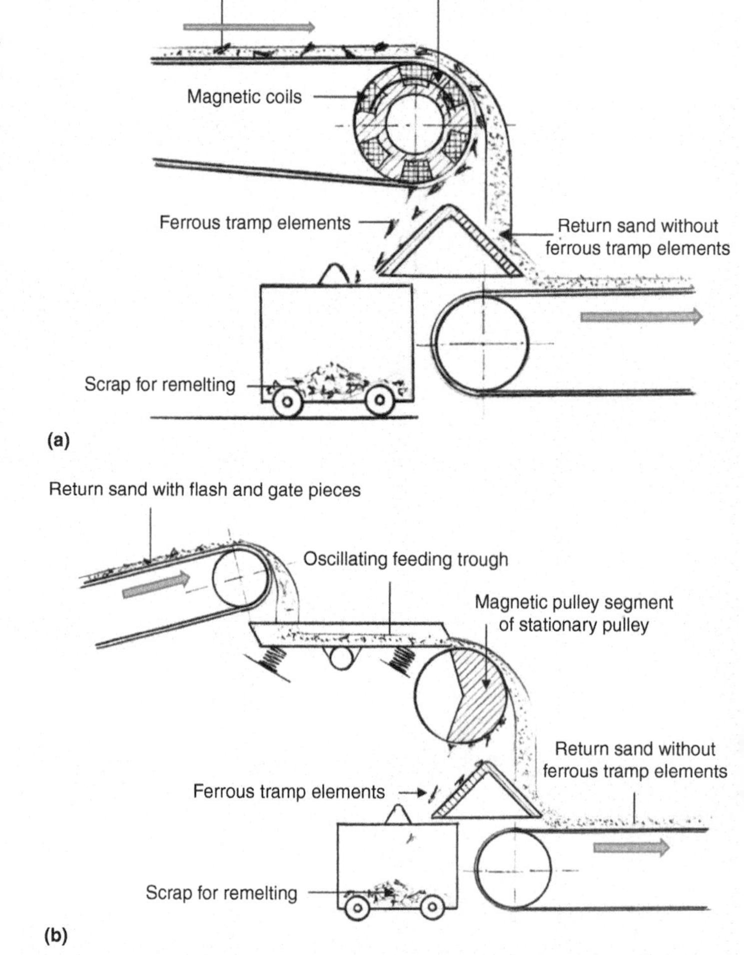

Fig. 3.32 Types of magnetic separators, (a) drum, (b) pulley segment

magnets grabbing onto the drum until the belt leaves the magnetic drum, thus separating the sand from the ferrous tramp elements. The sand leaves the belt first, and the tramp elements held by the rotating magnetic drum exit as the belt leaves the drum.

These magnetic drums come in different diameters and widths. The drums rotate at approximately 60 to 70 rotations per minute. The drum diameters range from 300 to 500 mm (11.81 to 19.69 in.). The drum lengths vary from 500 to 800 mm (19.69 to 31.5 in.) to suit different belt widths. Widths of 500 to 600 mm (19.69 to 23.62 in.) are common. The wattage ranges from about 4 to 8 kw. These units are also enclosed in a chamber and connected to the bag house for dust extraction.

3.9.2 Magnetic Pulley Segment

The magnetic pulley segment is stationary, unlike the magnetic drum, which rotates. The return sand passes over the magnetic pulley segment, where it is separated from the tramp elements as these ferrous materials cling to the magnet. Figure 3.32(b) is a schematic illustrating this type. This type is effective in sorting out the flash that is generated in gray iron foundries. Ductile iron foundries where there is a potential for feeders to be broken in shakeout units may find the other types more suitable.

An oscillation feeder takes the return sand from a belt conveyor and feeds it over a stationary magnetic pulley with a magnet segment spread over an angle of nearly 210 degrees. Sand falls over the pulley, and the flash is carried over more than 180 degrees to separate it, as illustrated. The diameter of the magnet pulley segment ranges from about 300 to 500 mm (11.81 to 19.69 in.), with widths ranging from 400 to 800 mm (15.75 to 31.5 in.). The wattage of the magnet ranges from about 0.50 to 0.75 kw.

3.9.3 Belt-Type Magnetic Separator

The belt-type separator offers the advantage of moving the tramp elements away from the return sand belt after they have been attracted by the magnet. It utilizes another short belt drive running over the return sand belt, as illustrated in Fig. 3.33. Figure 3.33 (a) features a magnetic separator over the return sand conveyor belt. Figure 3.33 (b) shows an installation over the transfer point between two conveyors.

This type is quite popular in many ductile iron and steel foundries, where there are large amounts of tramp elements such as feeders and chills . These separators are engineered with higher magnetic power, from 0.8 to 1.5 kw. Automated belt tensioners are provided to keep the space between the sand conveyor and the magnetic separator constant over time, irrespective of the tramp element load on the belt. These belt-type magnetic separators are suitable for larger belt widths of 600 to 1000 mm (23.62 to 39.37 in.). The belt widths of the pick-up magnet drive may range from 650 to 800 mm (29.59 to 31.5 in.).

3.10 Lump Breakers and Core Crushers

Sand exiting from the shakeout units may be lumped together both in green sand and no-bake systems. The cores may not be completely disintegrated in the shakeout units. Recycling the return sand or reclamation of the no-bake sand system requires lump breakers and core crushers in order to be effective.

It is important to install the magnetic separators ahead of the lump breakers to avoid damage. Some foundries invest in more than one magnetic separator to ensure that no metallic pieces enter the lump breakers.

Lump breakers vary in design, based on the sand systems. Green sand systems lump a little less than no-bake sand systems. The size

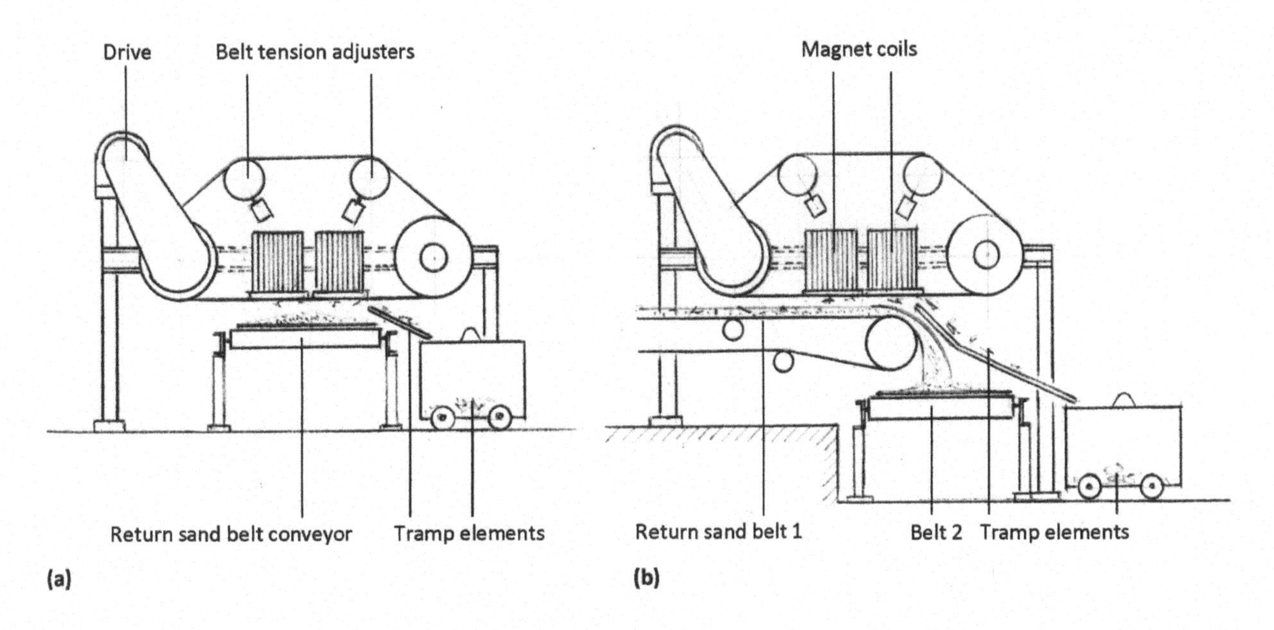

Fig. 3.33 Belt-type magnetic separators, (a) over belt, (b) over transfer belts

and number of cores influence the choice of system. Lump breakers can be classified as:

- Over-screen-type lump breakers
- Roller-type breakers
- Rotary breakers
- Attrition mills
- Hybrid units

3.10.1 Over-Screen-Type Lump Breakers

Lump breakers over the screen are relatively simple and are chosen by foundries where the parts do not have many cores. There are two types of designs:

- Arresting plate-type
- Pounding rollers-type

The arresting plate increases the residence time of small lumps over the vibratory screen, improving the crushing of the lumps. The mechanism is simple, and the investment is minimal. Figure 3.34 illustrates this concept.

Foundries producing more cored parts may need a set of roller bars that ride on the screen to pound lumps gently. These are effective but not expensive. They may slightly impair the life of the screen over prolonged times. Figure 3.35 illustrates this installation.

3.10.2 Roller-Type Lump Breakers

Roller-type lump breakers are suitable for no-bake molds where the sizes of lumps are larger than those from green sand molds. The equipment consists of two rollers rotating in opposite directions, one fixed and the other spring-loaded, to apply pressure on the lumps that pass between them from a hopper. Figure 3.36 illustrates the roller-type lump breaker.

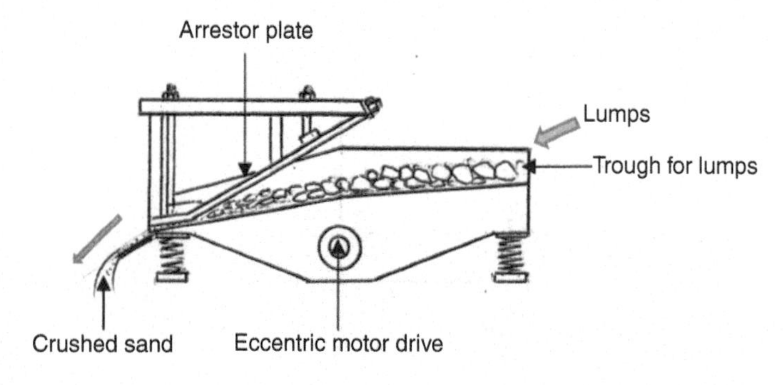

Fig. 3.34 Arrester plate-type lump breaker

3.10.3 Rotary-Type Lump Breakers

Rotary-type lump breakers offer an economical and environmentally efficient solution for breaking mold and core lumps with less noise and dust. Wear-resistant attrition spikes are fitted inside a rotating drum, and they disintegrate the lumps as the lumps advance through the spiral wear plate elements. Perforations in the lining collect the sand, which moves from there into the scrubbing section. These rotary lump breakers serve the needs of sand reclamation. This type of lump breaker is typically used in flaskless molding installations and no-bake molding lines of heavily cored parts.

Figure 3.37(a) is a cutaway section of the rotary-type lump breaker (Ref 18). Figure 3.37((b) illustrates the details of the attrition spikes inside.

3.10.4 Attrition Mills

No-bake cores using resin-bonded sand mixes and sodium silicate-bonded, CO_2-cured cores are good applications for attrition mill lump breakers. Designs vary from shearing blades to spiked pins that break down the lumps. The shearing spikes are mounted on twin shafts driven by geared motors. Smaller sizes use one geared motor to drive both the shafts, while larger units are designed with a geared drive for each shaft.

Figure 3.38 is a schematic of this concept. The units are designed with an oscillating feeder that feeds the lumps into the hopper. The attrition blades shear the lumps with the fine aggregate. The crushed sand exits at the bottom. Units are also designed with sieves and classifiers at the discharge end.

3.10.5 Hybrid Units

Hybrid units are popular with larger installations, where they carry out multiple functions of magnetic separation, lump

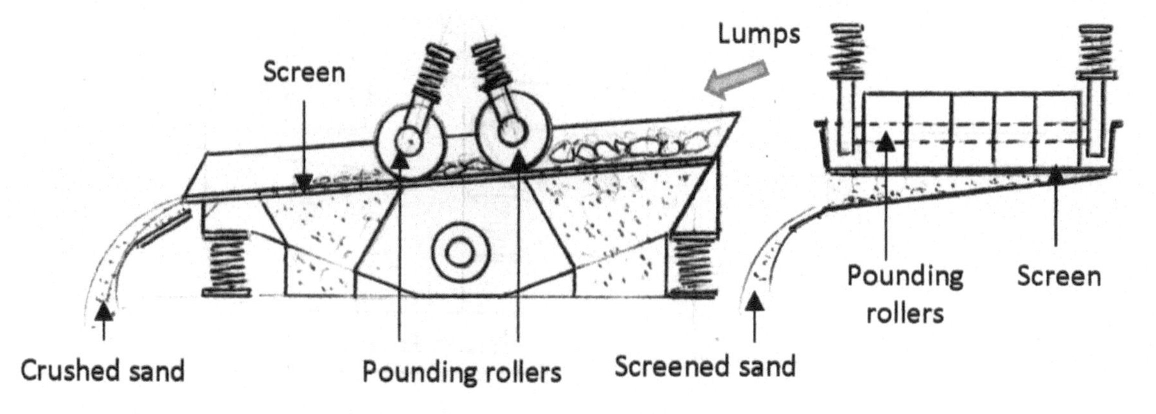

Fig. 3.35 Pounding roller bars in over-screen-type lump breaker

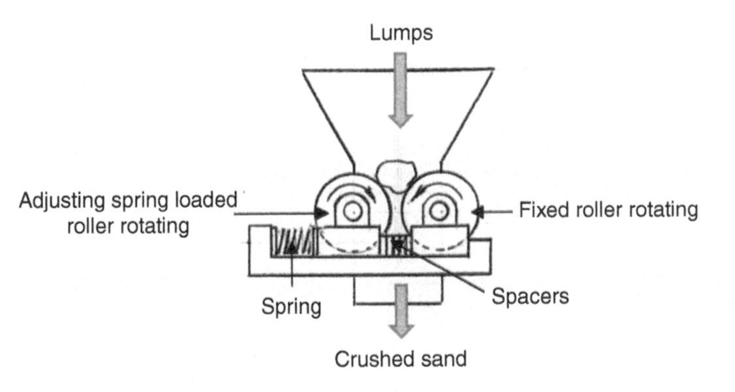

Fig. 3.36 Roller-type lump breaker

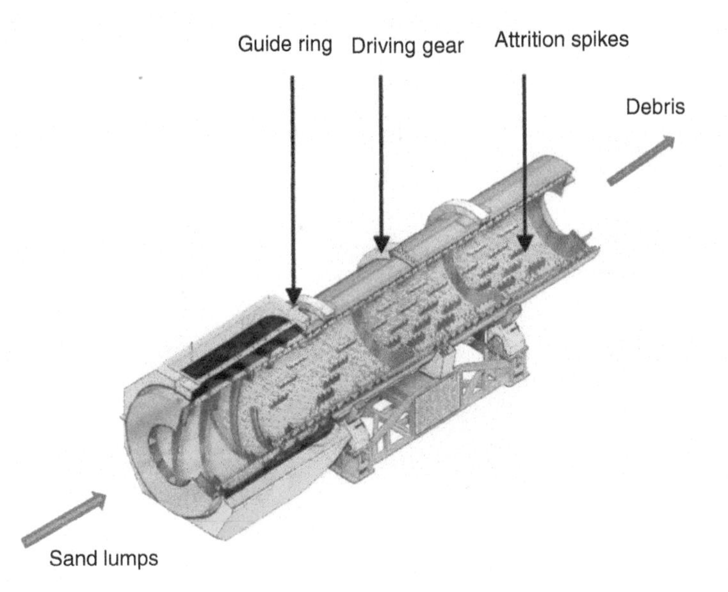

(a)

Attrition spikes

Perforations for sand collection

(b)

Fig. 3.37 Rotary-type lump breaker, (a) cut away section, (b) attrition spikes inside drum. Source: Ref 18. Reprinted with permission from Didion International Inc.

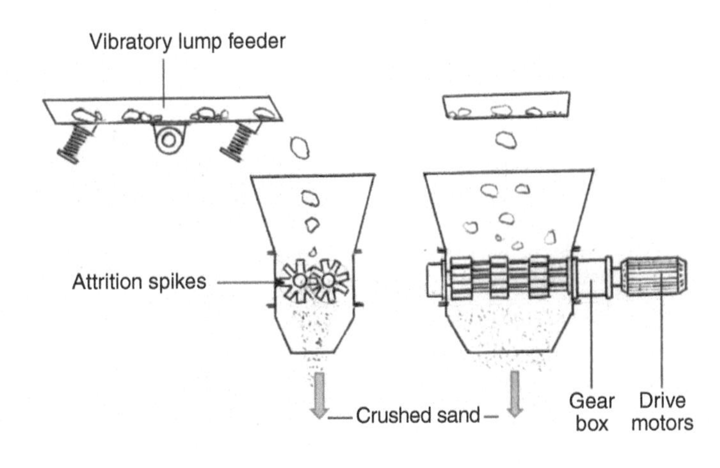

Fig. 3.38 Attrition lump breaker concept

disintegration, and sand classification for reclamation. Both green sand molding lines and no-bake molding lines find these hybrid units useful.

High throughput and productivity are achieved by engineering an automated discharge door for removal of the debris without stopping the machine (Ref 16). The tendency of sand to stick due to the moisture in green sand foundries is eliminated by heating the lower deck to avoid moisture condensation.

3.11 Screens

Sand discharged from the lump breaker is sieved to ensure that core butts that did not get crushed in the lump breaker are separated by a screen. The screen also catches metallic elements such as flash, gate stubs, nails, or any tramp elements or debris that may have escaped the magnetic separator.

Screens are classified into three groups:

- Hexagonal screens
- Rotary screens
- Vibratory screens

Rotary and vibratory screens have been more widely used in recent years. Figures 3.39 to 3.41 show schematic representations of these three types, with some minor variations.

3.11.1 Hexagonal screens

Hexagonal screens are used by small and jobbing foundries because they offer a low-cost solution to screening. In recent years, foundries have opted more often for rotary and vibratory screens. Figure 3.39 is a schematic illustration of the hexagonal screen system. Some installations design for a slight inclination for the material flow if core butts and other debris are minimal. Another option is to tilt the axis of the screen assembly over a higher angle for emptying the screen as the debris builds up. Apertures of 4 by 4 to 5 by 5 mm (0.16 by 0.16 to 0.2 by 0.2 in.) are common. These screens are enclosed in a casing and connected to the bag house for effective dust control.

3.11.2 Rotary Drums

Rotary drums offer an option for a continuous screening operation with no or minimum manual intervention. The maintenance costs are low, and the efficiency is high (Ref 18). These systems use approximately 100% of the screening area.

The rotary drums include two coaxial drums. The inner drum has perforations for sieving, and the outer drum is for sand collection. The inner drum is supported on ring guides for rotation through a ring gear. Guides which provide up to a fraction of the drum length help to move the material. A slight inclination of the drums is beneficial to enhance material movement. Figure 3.40 is a schematic of this type of equipment.

Screened sand collected in the outer drum exits through a window at the lower end onto a belt conveyor. The debris collected in the inner ring is designed to be dumped into a bin which is periodically emptied.

3.11.3 Vibratory screens

Vibratory screens have two drives that work in opposite directions; they are popular with larger casting facilities. Screens with a single motor are used in mid-sized foundries. The two drive designs (Ref 17) offer advantages such as:

- Significant energy savings (close to 80% compared to a conventional design)
- Improved load responsiveness — the sieving efficiency increases with the increase of load

Purchase options include:

- Quick change screen designs
- Special corrosion-resistant steel screens
- Provision for multiple decks

Figure 3.41 is a schematic of the designs with single or dual motor drives and a schematic of a unit in production.

3.12 Sand Coolers

The temperature of the return sand influences the sand mix properties such as compactibility and green compressive strength. The return sand temperature also affects mold qualities such as friability and sand erosion. Most modern foundries design sand systems with sand coolers, moisture sensors, and conductivity sensors for close control.

Molding sand is heated as the mold extracts the heat from the solidifying casting. The heat extraction by the sand mold from the casting continues during the casting cooling up until the shakeout operation. The temperature to which the sand is heated depends primarily on:

- The sand-to-metal ratio: the ratio of the volume of sand in the mold to the volume of the metal cast. This ratio may be as low as 5:1 or as high as 12:1, depending on the part cast. The lower the ratio, the higher the return sand temperature.
- Contact residence time: the time elapsed between the pouring and the shakeout
- The moisture in the sand mix

The foundry ambient temperature plays a role, but seasonal temperatures do not result in hourly or daily fluctuations.

Secondary factors such as the conductivity of sand and the type of sand play a role. But in a foundry with a chosen sand system such as green sand or no-bake and either silica or olivine sands, these do not vary from part to part or from time to time.

After the shakeout, the return sand cools down slowly as it is exposed to the ambient temperature in the belt sand conveyors and bucket elevators. Typically, the target sand temperature

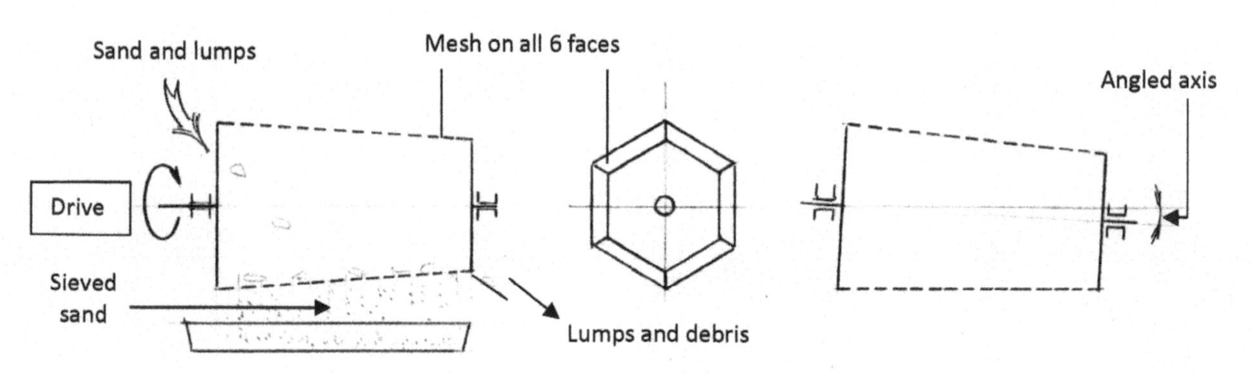

Fig. 3.39 Hexagonal screen

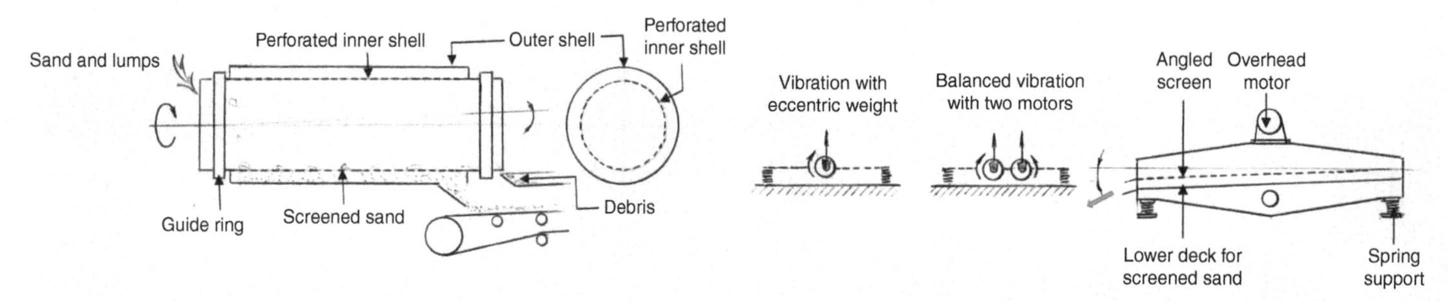

Fig. 3.40 Rotary screen concept

Fig. 3.41 Vibrating screen concept

should not be higher than 10 °C above the ambient temperature in the foundry. In the winter months, the ambient temperature in foundries decreases significantly.

Many foundries design return sand cooling equipment for automated sand temperature control. Table 3.8 (Ref 19) lists the effect of sand temperature on mulling efficiency and properties of prepared (mixed) sand.

The sand temperature decreases as the sand is transported on belt conveyors and bucket elevators. The transfer of sand from one belt conveyor to the other results in some cooling. Table 3.9 lists the average temperature changes in such cases (Ref 1, 3).

Water added in the muller helps to cool the sand, but the cycle time may be prolonged to generate the targeted green strength and compactibility. Foundries may opt for specially designed mullers with provisions for cooling, or they may provide for a sand cooler in the return sand system.

3.12.1 Sand Cooling Processes

The processes used for return sand cooling are:

- Spraying water onto the hot return sand that is on the conveyor belt. Part of the water turns into steam, which extracts heat from the sand
- Heat extraction by removing stratification on the belt between the top layer (higher moisture) and the bottom layer (lower moisture), through homogenization by aerators
- Evaporative cooling by air, extracting heat from sand by increased surface exposure, mechanically or by fluidizing

Table 3.8 Effect of return sand temperature on sand quality and mulling effectiveness

Temperature range °C (°F)	Effect on mixed sand quality	Effect on mulling
38 to 49 (100 to 120)	Consistent properties	Good mulling
49 to 60 (120 to 140)	Reduced green strength	Increased mulling time required
60 to 71 (140 to 160)	Poor strength and friable sand	Inadequate water available for clay
>71(> 160)	Unacceptable sand mix strength	Quick water evaporation, Inability to activate clay bond

Source: Ref 19

- Heat exchange between hot sand showering over water running in tubes, using a heat exchanger
- Flushing the sand mix with air in the muller as water is added

3.12.2 Sand Cooling Types

The different types of sand coolers are classified as:

- Sand cooling by water addition
- Blending by aeration
- Evaporative cooling
  a. Rotary drums
  b. Fluidization with or without vibration
- Heat exchanger-type sand cooler
- Sand cooling in mullers

Table 3.10 provides an overview of the types of sand coolers.

3.12.3 Stages of Sand Cooling

Figure 3.42 is a schematic flow diagram of the different elements of the return sand system and temperature management. The diagram shows how the different temperature corrective actions are combined. The three stages of sand cooling are identified in the figure.

Sand from the shakeout may be over 100 °C or 212 °F for sand-to-metal ratios of under 10:1 (Ref 18). The magnetic separator picks up any ferrous metallic tramp elements. The sand passes through a lump breaker, where mold and core lumps are disintegrated, helping to cool the sand. A screen isolates any tramp elements that the magnetic separator has missed.

Table 3.9 Sand temperature changes due to sand handling and mixing

Transportation	Decrease of temperature, °C
Belt conveyor, length about 10 meters	2
Sand transfer from one belt conveyor to the other	3
Bucket elevator, height 10 meters	5
Cooling in specially designed mullers	Below 49 °C or 10 °C above ambient in foundry

Table 3.10 Sand cooler types

Type of cooling				
1	2	3	4	5
Water spray	Aeration	Evaporative cooling	Heat exchanger	Cooling in muller
Water spray on return sand belt for sand temperatures above 100 °C Moisture controlled to about 3 %	Over-the-belt conveyors reducing temperature from 100 to 85 °C Some equipment manufacturers use water spray in aerators	Exposure of hot sand to high volume, low pressure air by using either: 1. Rotary drum (Ref 20) or 2. Fludization (Ref 21) with or without vibration (Ref 20) Some equipment manufacturers use water spray in return sand belt followed by aeration and evaporative cooling by rotary drum or fluidization	Reclaimed no-bake or core sand passes across a bank of water-cooled tubes for precise temperature control within 1 °C Curing of resin bonded sands is temperature-sensitive, and these systems are more popular for resin-bonded and no-bake applications	This cooling feature is offered both in horizontal roller types as well as in duplex mullers. Air enters the muller from the bottom during the preconditioning stage, resulting in intimate contact of sand and water. Figure 3.3(b) illustrates the concept for the horizontal roller type. In duplex mullers air is blown from the bottom as the counter rotation of the duplex mixing provides for intimate contact of air, water and sand for effective cooling

A temperature sensor measures the sand temperature and sends a signal to a programmable logic controller (PLC), which directs the amount of water to be added to the hot sand. The sand temperature decreases by about 15 to 20 °C in stage 1 cooling.

The sand passes through an aerator where the sand temperature is homogenized. (Some equipment manufacturers add the water in the aerator instead of the belt.) This is stage 2 cooling.

The sand enters the sand cooler (either rotary or fluidized bed type) for stage 3 cooling. The interaction between the air and the hot sand by either sifting in a rotary drum or fluidizing in a vibratory unit cools the sand substantially to around 45 to 50 °C (113 to 122 °F). A temperature sensor at the exit end of the cooler interfaces with the PLC, directing the amount of water to be added in the muller. A moisture sensor can also be used.

Figure 3.43 illustrates the three stages of cooling. It illustrates the water spray system (stage 1), homogenization with an aerator (stage 2), and a fluidized bed system (stage 3).

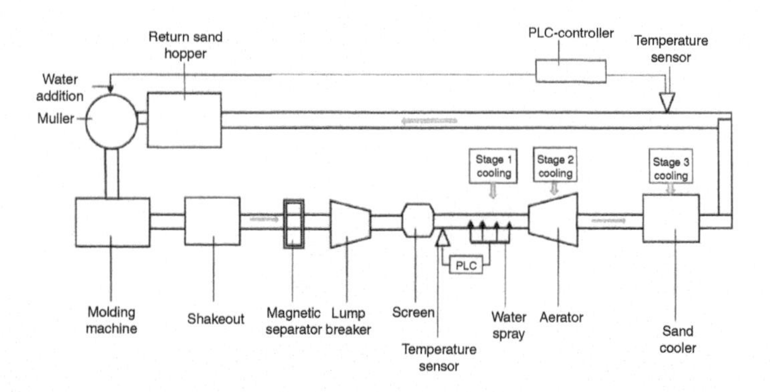

Fig. 3.42 Return sand cooling system schematic

Figure 3.44 shows a rotary cooling drum for stage 3. The rotary drum cooling offers several advantages, such as low energy consumption and maintenance costs and a quieter environment.

Figure 3.45(a) is a cut section of the rotary drum showing the bands of rings inside that turn the sand grains and expose them to the cooling air.

3.12.4 Heat Exchangers for Sand Cooling

The systems with water-cooled heat exchangers are popular for no-bake sand systems, especially those using thermal sand reclaimers (Ref 22). The equipment consists of a bank of cooling tubes assembled in columns in a rectangular box-like container. Water circulates through these tubes, using headers for the inlet and outlet.

Figure 3.45(b) shows a cooler based on the heat exchanger concept. Cold water enters through the bottom header and picks up heat from the descending hot sand. The hot water exiting through the top header is connected to a cooling tower. A PLC regulates the water inlet temperature and the cooling tower fan

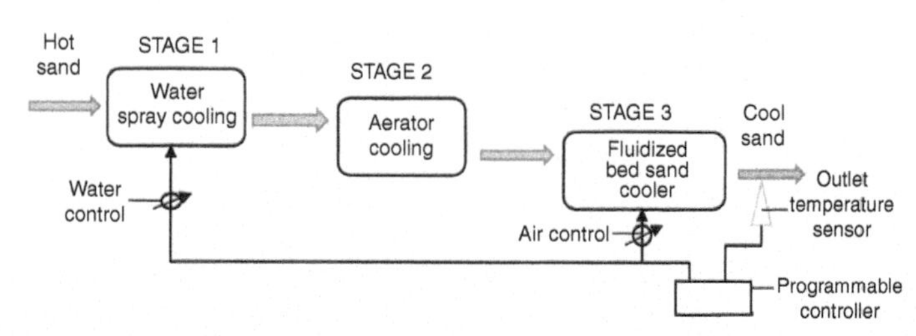

Fig. 3.43 Three stages of sand cooling with fluidized bed cooler for stage 3

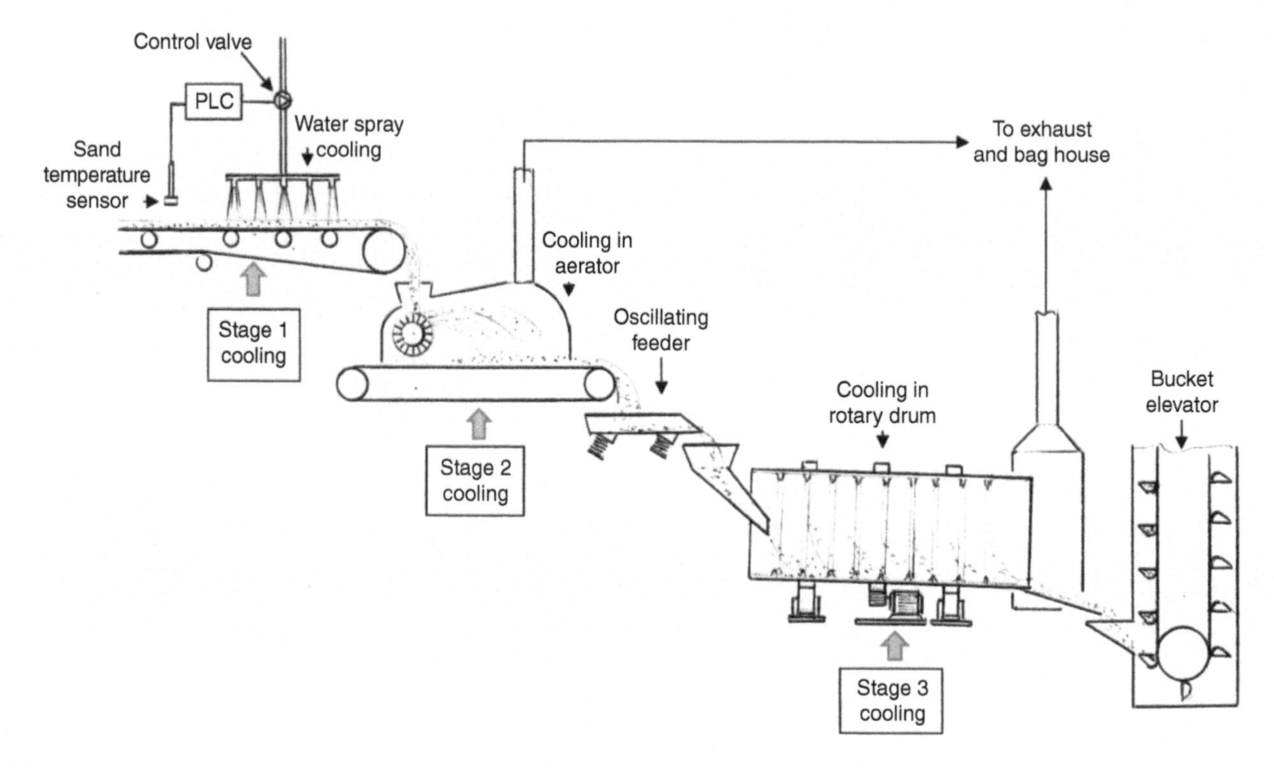

Fig. 3.44 Schematic of three-stage sand cooling with a rotary drum

parameters. Cooled sand exiting at the bottom is transported through a belt conveyor onto a bucket elevator and from there to a storage hopper.

3.12.5 Sand Cooling in Mullers and Mixers

Sand cooling is offered in horizontal roller-type mullers and duplex mixers. In both cases, the cooling air enters from the bottom.

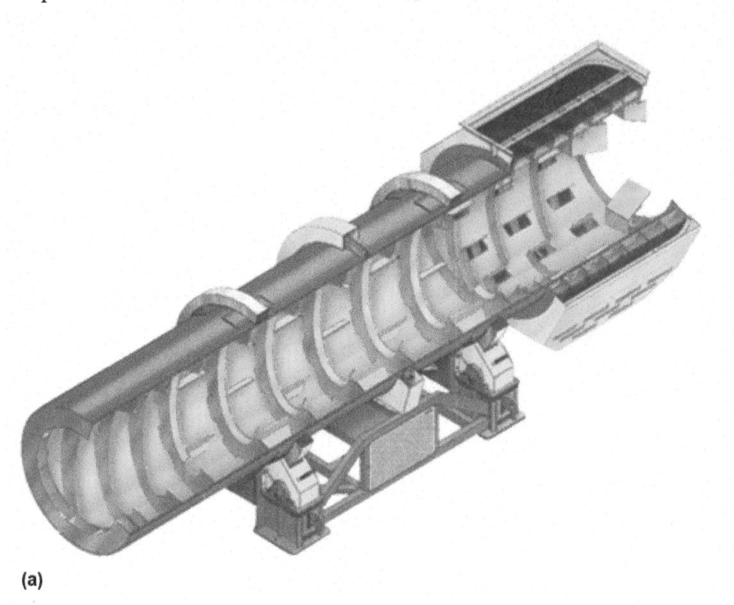

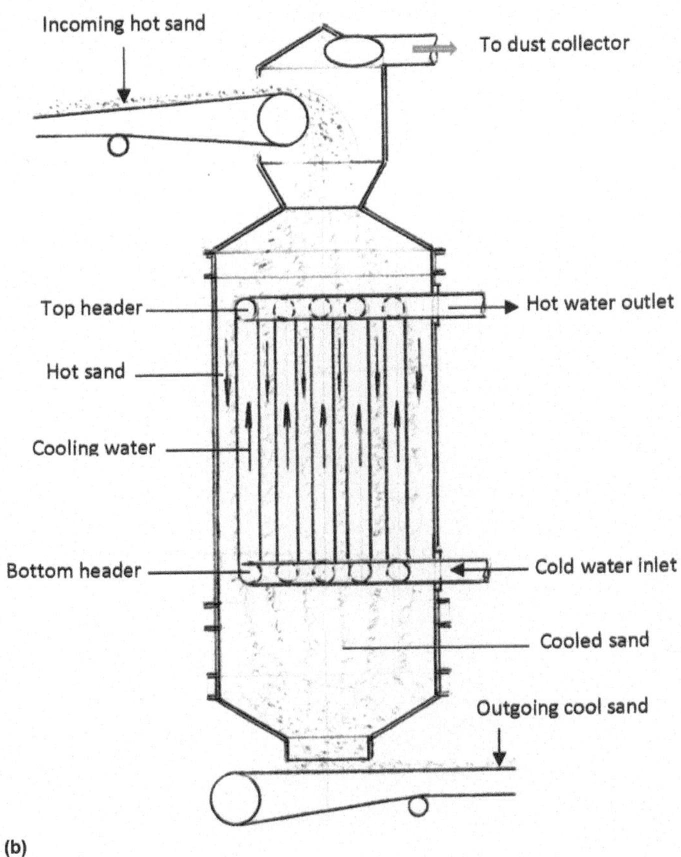

Fig. 3.45 Sand cooling units. (a) Cut section of a rotary drum cooler. Reprinted with permission from Ref 18. (b) Sand cooling using a heat exchanger. Source: Ref 20

The amount of water to be added is calculated based on the temperature of the exhaust air and the thermal conductivity of the sand mix. The air blown from the bottom enables contact between the water and the sand, bringing the sand temperature down. Ducting for the clay bond addition has a valve that is operated by the controller. This ensures that the air is shut off at the time the valve opens for bond addition to the sand mix. The opening of the discharge door is operated by the controller, based on the amperage the motor is drawing.

Figure 3.46 is a schematic illustrating this concept (Ref 3). The figure illustrates a duplex muller with a cooling blower that blows air from the bottom through the sand. The counter-rotating plows move the sand around, resulting in exposure to the cooling air and thorough mixing of the water with sand.

3.13 Sand Reclamation Systems

The economics of recycling sand and adherence to environmental regulations for sand disposal require foundries to explore and implement reclamation units. The amount of clay and coal dust that are burned during the casting process depends on the sand-to-metal ratio in the mold. The burned binder is removed by the dust collector. The amount of the unburned binder depends upon the extent of coring in the parts produced. Heavily cored parts such as engine cylinder blocks and cylinder heads have large amounts of core sand that is mixed with the recycled sand. Reclamation is important in such heavily cored captive operations. Reclaimers are classified as outlined in Fig. 3.47 (Ref 21, 23).

The suitability of the main categories for different binder systems and the capital and operating costs are listed in Table 3.11

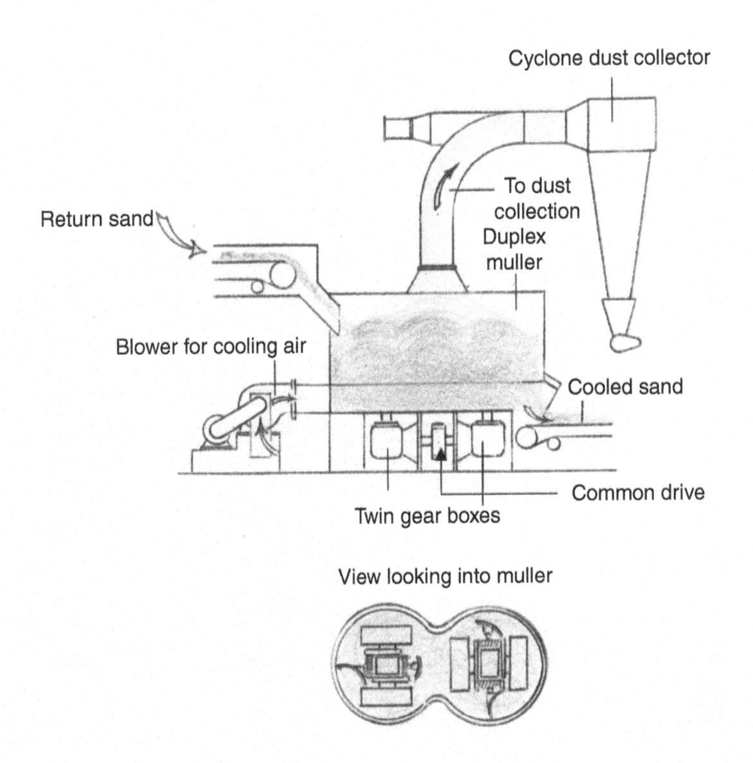

Fig. 3.46 Sand cooling in duplex muller. Source: Ref 3

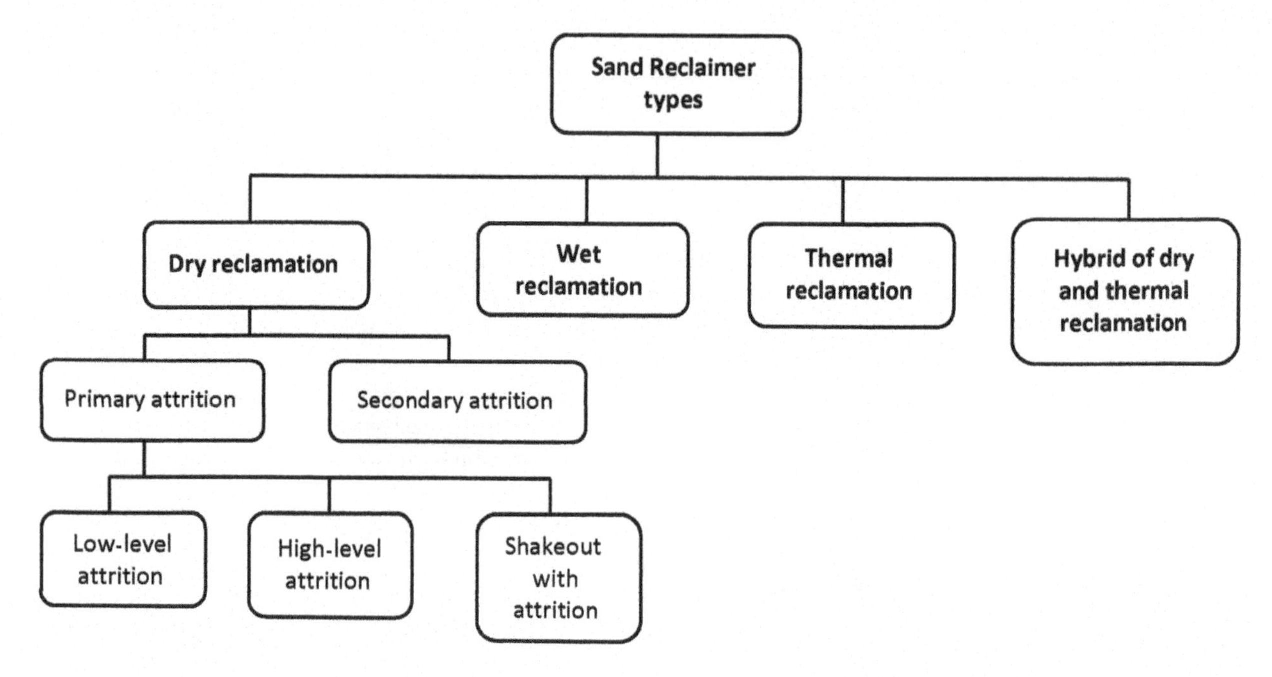

Fig. 3.47 Types of sand reclaimers

Table 3.11 Reclamation system types, suitability and costs

Reclamation type	Application for binder system type	Capital costs	Operating costs
Dry reclamation	Furan, phenolic urethane	$10 K per ton per hour	$2–7 per ton
Wet reclamation	Inorganic binders like sodium silicate binders and alumina phosphate	$40 to 50 K per ton per hour	$5–10 per ton
Thermal reclamation	Alkaline phenolics	$1 million for 5 ton/hour unit	$10–15 per ton
Dry + Thermal - Hybrid	Clay bonded sand systems and all organic no-bake binders	>Thermal reclamation units	>Thermal reclamation units

(Ref 20), for comparison. Both the capital and operating expenses of the dry reclamation process are minimal. Many equipment manufacturers have improved the attrition process to increase its ability to separate the unburned clay or resins. Figure 3.47 highlights the modifications to the attrition processes under the heading "Dry reclamation."

Inorganic binder systems such as sodium silicate or alumina phosphate can only be reclaimed using the wet reclamation process. The capital cost and the operating costs are less than those for thermal reclamation. Thermal reclamation is the most expensive, but the quality of the reclaimed sand is superior to that of sand reclaimed from the other systems. The hybrid system combines dry and the thermal reclamation and delivers the best sand quality. The reclaimed sand has more rounded grains, which requires less binder when it is mixed.

3.13.1 Impact of Reclamation

Sand reclamation influences both the economy and environment. Highlights of the influence of reclamation are:

- Reduction in new sand purchases
- Drastic reduction of used sand disposal costs
- Rounded sand grains, resulting in reduced binder content
- Improved quality due to reduced resin content
- Increased sand mixture strength due to rounded grains
- Improved surface finish of the castings

3.13.2 Dry Reclamation

Dry reclamation requires a sand moisture content of less than 1 to 1.5% for effective processing through attrition. The preparatory equipment for lump breakage is the jaw crusher or the attrition mill, illustrated in Fig. 3.38.

Primary attrition equipment uses vibratory shakeout units followed by impingement in an attrition cell. Banks of these individual cells are assembled for improved performance. Pneumatic scrubbers (Ref 21) can also be used.

The vibratory units of the primary attrition consist of a heavy duty shakeout grid and a secondary steel screen with 6 mm (0.24 in.) diameter perforations (Ref 21). Following this are a stainless screen of 1.6 mm (0.06 in.) square perforations and a final 1.6 mm (0.06 in.) square aperture screen that is used for sand classification. The sand enters a belt conveyor and is sent to an attrition-impingement cell, as illustrated in Fig. 3.48.

Figure 3.48 is a schematic illustrating the attrition principle and bank of dry reclamation units. The equipment consists of a bank of vertical tubes with a *venturi* at the bottom and a plate at the top. Air blown from the bottom of the tube picks up the return sand fed from a belt conveyor. The sand is drawn up to impinge on the plate. The impact of sand underneath the plate results in the separation of the fine particles of burned resin, unburned resin (or clay), and fines. The cluster of these attrition cells is connected to a bag house which collects the resin (or clay). The sections are stacked in such a way that the sand from

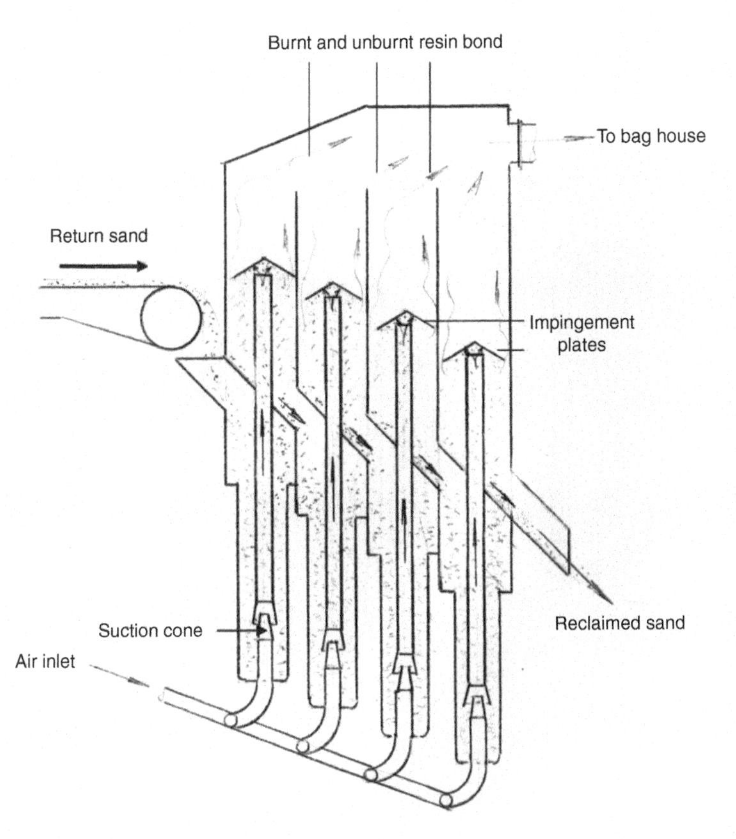

Fig. 3.48 Primary attrition cell

the first cell flows in to the second cell for repetition of the impingement action. The stack of multiple sections results in cleaner sand without a buildup of resin and fines. The reclaimed sand flows out of the last cell onto a belt conveyor for storage and then into a mixer for recycling.

Secondary attrition is used for further binder removal. Such units are preferred for no-bake sand systems based on alkaline phenolic resins. The equipment consists of a rotary drum that to propels the sand by spinning at a velocity high enough for binder separation without damaging the sand grains. The speed of rotation is adjustable to suit different sand grains and binder removal rates.

There are two types of secondary attritions, soft and hard. The soft system is suitable for furan resins, where the centrifugal force of the drum is adequate for binder removal. Alkaline phenolics require hard secondary attrition, where a pair of squeeze rollers are used to obtain higher attrition force for more thorough binder removal.

Figure 3.49 is a schematic of a rotary lump breaker used to crush the sand lumps before dry or wet reclamation. Jaw crushers can also be used. The main component of the lump breaker is the shearing arbor that carries carbide-tipped anvils on its periphery. The sand lumps are fed from an oscillating feeder onto the shearing arbor, and the catching levers prevent the lumps from moving up the rotating arbor without being crushed. Pulverized sand from the crushed lumps descends onto a belt conveyor and is carried to the next stage in the process.

Figure 3.50 illustrates the arrangement of the equipment for dry reclamation with secondary attrition. Sand lumps from the

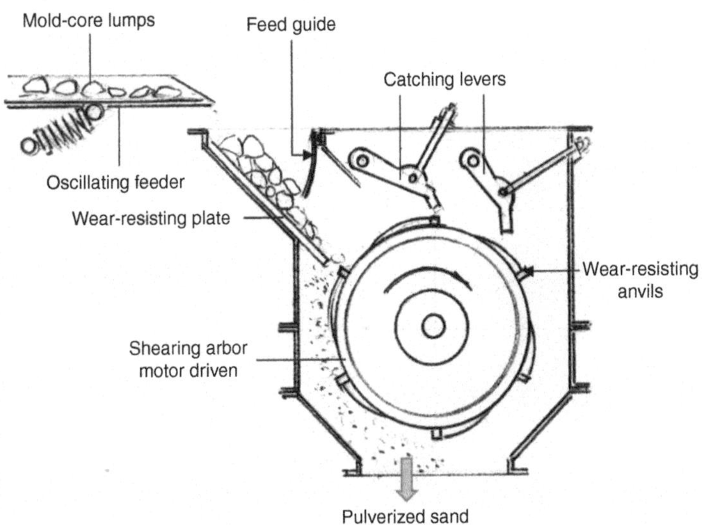

Fig. 3.49 Rotary lump breaker

hopper are fed into the lump breaker by an oscillating feeder. Pulverized sand from the lump crusher is carried by a belt conveyor. The sand passes through a magnetic separator, where any fine tramp metallic materials are isolated.

Pulverized sand feeds into a bucket elevator. Another oscillating feeder feeds the sand onto a vibratory screen which catches any smaller lumps. The sand passes through a dry reclaimer or a primary attrition unit where most of the burned and unburned resin bond is separated from the surface of the sand grains. A cyclone separator catches small lumps. The connected bag house collects fine particles of size less than 0.1 mm (0.004 in.).

Sand from the dry reclaimer passes through a secondary attrition unit. This unit consists of a ceramic-lined rotating drum with rollers inside to separate any bond adhering to the sand grains. The sand passes through a classifier to separate the sand grains according to the mesh size.

The classifier consists of a rotating drum in an enclosure. Sand fed over the top surface of the drum is thrown to different distances due to the centrifugal force of spinning. The distance over which these are thrown depends on the grain size. Troughs mounted to receive the respective sand grains automatically classify the sand grains according to their size.

3.13.3 Wet Reclamation

Sodium silicate-bonded sands cannot be reclaimed by either the dry reclamation or thermal reclamation process. The high temperatures of ferrous castings dehydrate and vitrify the coating on the sand grains. The vitrified coating is much harder than the coating at the time of bonding during molding or core making (Ref 20). Low sand-to-metal ratios and thick casting sections raise the temperature of the sand grains to degrade the sodium silicate-sand bond, which makes dry reclamation difficult. Wet reclamation requires a higher capital investment and is more expensive to operate compared to dry reclamation, but it is the best option for reclaiming sodium silicate-bonded sand mixes.

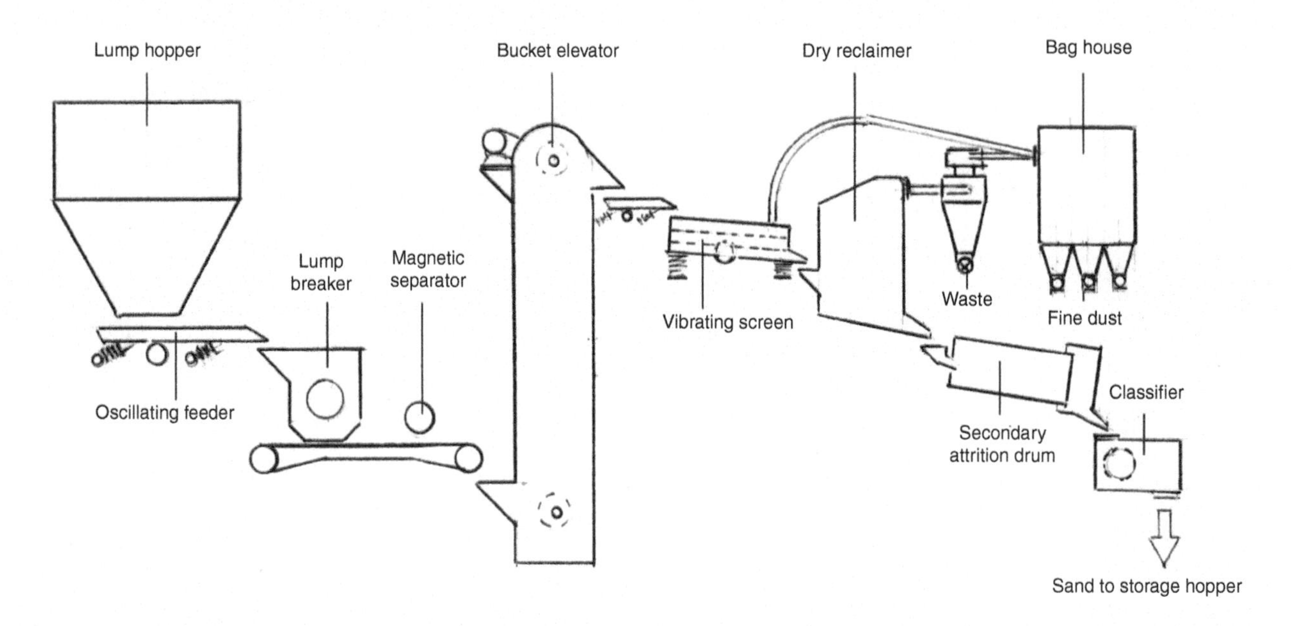

Fig. 3.50 Dry reclamation system concept

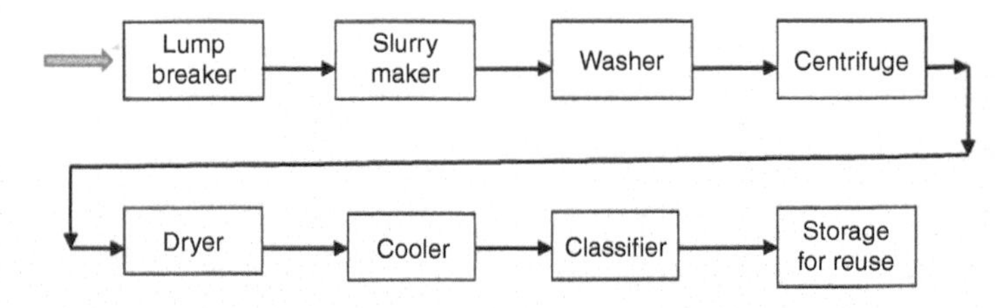

Fig. 3.51 Wet reclamation overview

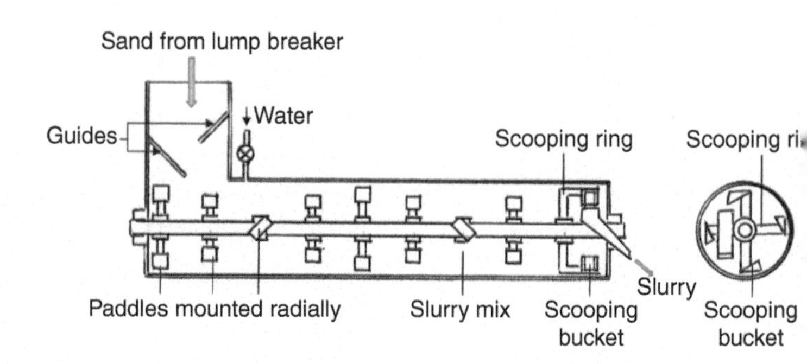

Fig. 3.52 Slurry maker schematic

The percentage of sand that can be reused is limited by the soda buildup in the sand, which alters the silica-soda ratio. This accelerates the cure rate, due to the higher rate of gelation. A buildup of soda over 0.6 percent decreases bench life, reducing the working time. Many foundries limit the percentage of reclaimed sand to about 50% to allow for an acceptable working time. In addition, the additives for improving shakeout decrease the storage life of the cores, limiting the amount of reclaimed sand available for reuse (Ref 20).

Aluminum phosphate binders offer the advantage of improved shakeout as well as reclamation potential. Aluminum phosphate in a dilute solution of phosphoric acid in water along with magnesium oxide and zinc compounds acts as a binder. It is cured using either ammonia or amine.

Figure 3.51 provides an overview of the different equipment needed for wet reclamation.

The cell consists of a lump breaker that receives the lumps from the shakeout process and crushes them to pulverize. The pulverized sand enters a slurry maker where water mixes with sand to produce a slurry with about 20 to 30% water. A washer washes the binder, and the slurry enters a centrifuge where it is spun to expel most of the water. The wet sand is fed into a dryer, where the moisture is reduced to the target value. The hot sand passes through one of the types of sand coolers. If needed, a classifier is used to sort the sand into the desired grain sizes. The sand is now pneumatically conveyed to silos for use.

Figure 3.52 (Ref 1) is a schematic of one type of slurry maker. It consists of a cylindrical chamber with a series of agitators mounted on a shaft. A scooping ring (carrying scooping buckets) at the end of the tube picks up the slurry and sends it to the washer station.

The types of washers range from horizontal washers to vertical washers, vertical counter-flow washers, and hydrocyclones. Each separates the water that is carrying the binder from the sand. Figure 3.53 shows a hydrocyclone for slurry-water separation.

The hydrocyclone (or cyclone) is a device that is used to separate sand particles suspended in water (carrying dissolved sodium silicate) based on the centripetal or centrifugal forces created within the vortex. It is a conical vessel with an inlet for the slurry and two outlets, one for the sand slurry and the other for water. The slurry enters the hydrocyclone tangential to the cylindrical section at a pressure of 1 to 2 atmospheres. The centrifugal force continuously changes due to the conical shape, which pushes the light factions upwards. The water at the top carries the sodium silicate that is separated from the sand. A bank of hydrocyclones can be configured for more effective removal of sodium silicate. They are run in combination with the intermediate slurry-making to bring the percentage of water to 20 to 30% in the slurry. Slurry pumps can handle sand-to-water ratios as low as 1:10.

Hydrocyclones are easy to install and simple to use. They may experience a significant amount of wear, especially at the discharge end. Some are lined with ceramic refractory. They can be designed with rubber devices for automatic or periodic discharge of the slurry.

After confirming that most of the sodium silicate has been removed from the sand, moisture must be removed before the sand can be reused. Centrifuges (Ref 1) are used to lower the moisture to between 5 and 8%. Sand dryers (see section 2.4 in Chapter 2, "Sand and Metal Charge Storage and Handling" in this book) reduce the moisture to between 1 and 1.5%. Figure 3.54 is a schematic of a centrifuge.

The centrifuge consists of a perforated drum with a diameter ranging from 400 to 800 mm (15.75 to 31.5 in.), which rotates at 900 to 1200 rotations per minute. The slots are 0.3 to 0.4 mm (0.012 to 0.016 in.) wide. The perforated drum is enclosed in a case with a central inlet for the intake of the slurry and outlets for the sand and water. A pressure plate that moves back and forth directs the slurry to the drum. The slurry pumped through an elbow enters through a conical funnel to the inside of the rotating drum, pushing the water through the perforations. The wet sand exits through a bottom outlet passage. Water collecting outside the drum exits through ports. The outputs of these units range from 3 to 6 tons per hour.

Wet sand is directed to a rotary sand dryer, which dries the sand to a moisture level of about 1%. Hot sand enters a cooler, as described in section 3.11. A classifier sorts the sand according to grain sizes and is distributed over four sieves. The sand is pneumatically conveyed into a silo or a hopper for reuse.

3.13.4 Thermal reclamation

Thermal reclamation delivers sand of the best quality, ensuring that there is no burned or unburned resin binder, and sand grains are more rounded than the starting new sand. Both the capital and operational costs are higher than the costs for dry and wet reclamations, as seen in Table 3.11, but no-bake systems based on alkaline phenolic resins (Ref 24, 25) benefit from thermal reclamation.

Sand is heated to a temperature range of 750 to 800 °C (1382 to 1472 °F), at which temperature 100% of the binder is burned. The sizes of thermal reclaimers range from 250 kg/hour to 12 tons per hour. Sand systems based on an alkaline phenolic system require an inhibitor to be premixed with the sand to prevent the fusing of the sand grains together before blowing or compaction. The remnants of inhibitors that contaminate the sand also are burned at these thermal reclamation temperatures.

Thermal reclamation can also be used for reclamation of green sand to supply sand back to the core room, but these units need a pre- and post-mechanical scrubbing (Ref 25, 26), making recycling more expensive.

The thermal reclamation units use mostly gas or electricity for heating, and they are energy-intensive. A heat recovery system reduces fuel consumption, and furnaces with a fluidized bed are common. Thermal reclamation in the furnace is followed by a cooler and classifier to bring the sand temperature down to 25 °C (77 °F) for reuse.

Figure 3.55 (Ref 24) illustrates a thermal reclamation cell with a fluidized bed furnace, equipped with a furnace pre-heat

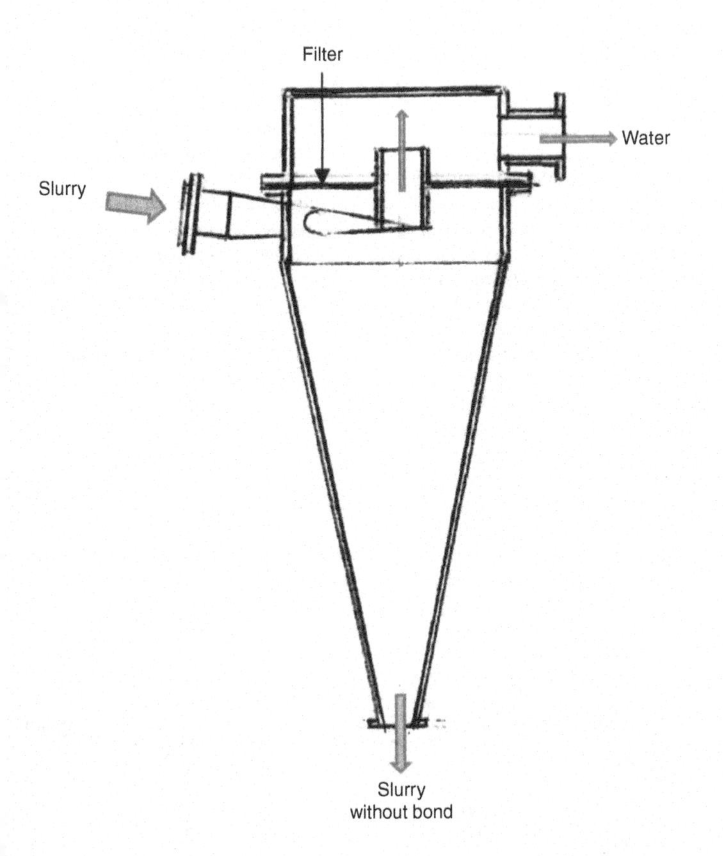

Fig. 3.53 Hydrocyclone washer

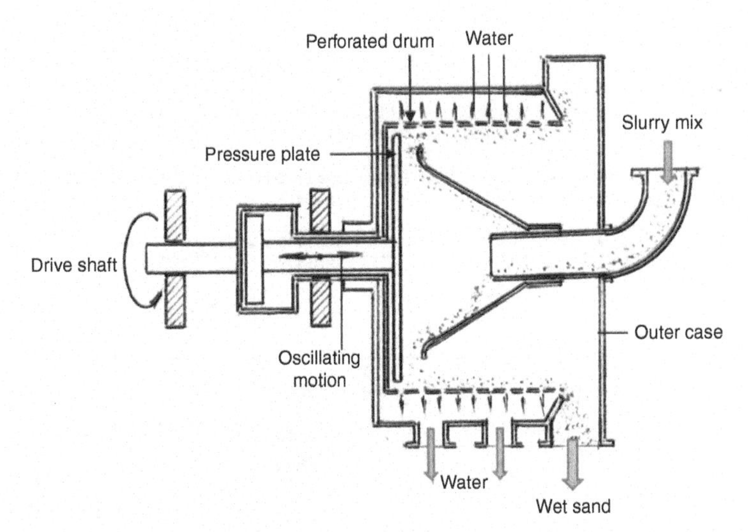

Fig. 3.54 Centrifuge for water extraction

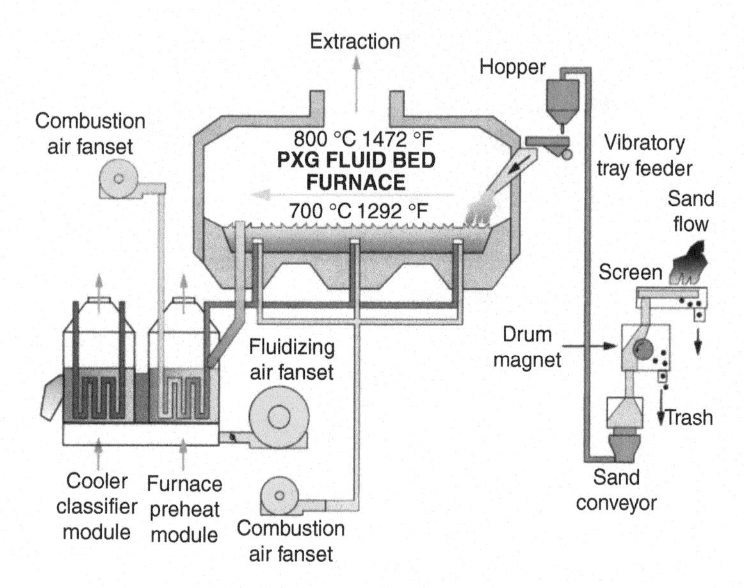

Fig. 3.55 Thermal reclamation schematic. Source: Ref 24, Tinker Omega Inc.

module. The figure features the sand passing through a magnetic separator (drum magnet) and pneumatically conveyed to a hopper. A vibratory feeder feeds the sand to a furnace designed with a fluidized bed. A fan set provides air for combustion. The layout shows the preheat furnace using the furnace exhaust heat. Reclaimed sand passes through a cooler classifier module.

Figure 3.56 (Ref 24) displays the fluidized furnace, the preheater, and the cooler.

3.14 Prepared Sand Hoppers and Automated Level Control

Prepared or conditioned sand from the muller is transported by belt conveyors and bucket elevators (depending on the relative elevations). The sand passes through aerators before being stored for the molding machines. The hoppers are mounted just above the molding machines and are replenished when the levels decrease to a preset low value.

Continuous working of the high production molding machines requires a continuous and uninterrupted supply of sand. The hourly sand storage bins need to be properly engineered for smooth flow. The engineering requirements for continuous sand delivery are:

- Proper shaping of the hoppers to prevent bridging
- Lining or coating inside hoppers to avoid sand sticking to the sides.
- Design for vibration to remove any jamming
- Automated level measurement to trigger sand delivery to the respective bin

Sand is discharged at the end of the bins, either by clamshell doors or by belts that run below the rectangular hoppers. Figure 3.57 indicates the nonsymmetric shapes of hoppers or bins, as well as the belts below the larger hoppers.

Fig. 3.56 Fluidized furnace with the preheater and cooler. Source: Ref 24. Reprinted with permission from Tinker Omega Inc.

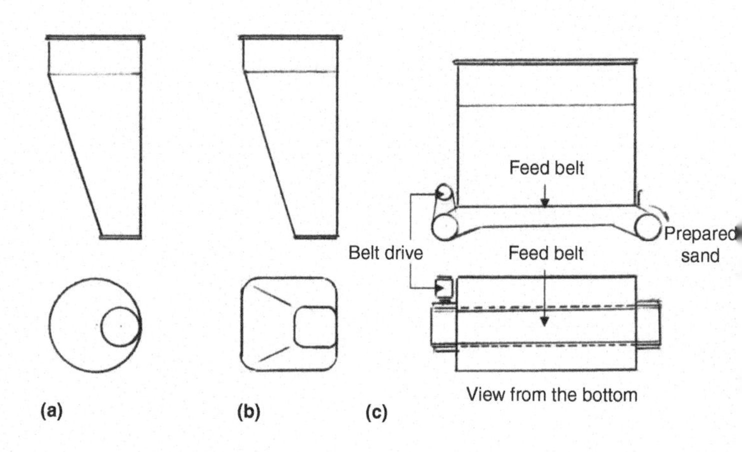

Fig. 3.57 Prepared sand hopper shapes, (a, b) nonsymmetric, (c) rectangular with feed belt

3.14.1 Bin or Hopper Shapes for Reduced Bridging

The nonsymmetric shapes reduce the potential for bridging. Both conical and square shapes can be designed without symmetry, as indicated in Fig. 3.57(a) and (b). Bridging is the buildup of an arch of sand that prevents the smooth flow of sand downwards. The lack of symmetry discourages the buildup of an arch, allowing unimpeded downward sand flow.

The rectangular hoppers in Fig. 3.57(c) include a belt at the bottom and have less tendency toward bridging. But even these may need auxiliary vibration to remove bridging periodically.

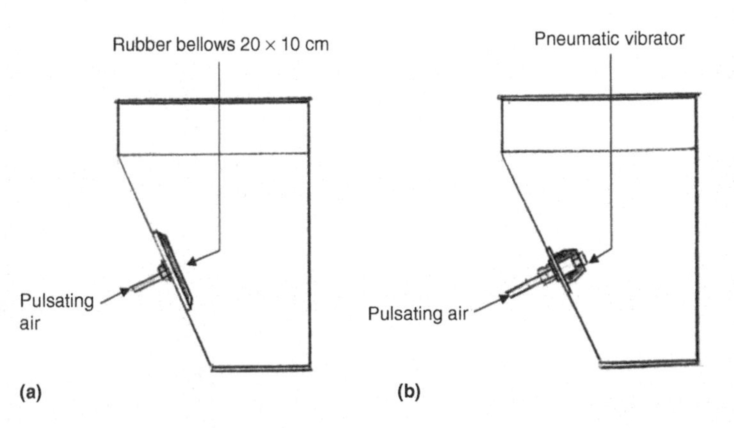

Fig. 3.58 Vibrator designs, (a) rubber bellows 20 by 10 cm, (b) pneumatic for hoppers

3.14.2 Bin or Hopper Linings or Coatings

Moisture in the sand has the potential to corrode the interiors of the hoppers over time, resulting in buildup and interruptions of flow. One economical method to avoid this is to coat the inside surface with polyurethane. The hoppers may be recoated every year.

Stainless steel linings can prevent rust, although they are relatively expensive. Hoppers can also be lined with rubber belting. These liners work well with the addition of vibrators.

3.14.3 Design for Vibrators to Remove Jamming

Pneumatic bellows or vibrators are successfully used to remove bridge formation inside the hoppers. Figure 3.58 illustrates bin vibrators. The bellows shown in Fig. 3.58(a) are less noisy, while the vibrators in Fig. 3.58(b) tend to be noisier.

3.14.4 Sand Level Measurement in Hoppers

The level of sand is continuously measured by sensors to trigger automatic replenishment of the sand in the hoppers. These sensors use a variety of technologies for height sensing.

REFERENCES

1. J. Sirokich, Foundry Equipment (Zarizeni Slevaren), SNTL, 1968
2. K. C. Bala, Design, Analysis and Testing of Sand Muller for Foundry Applications, *AU Journal of Technology*, Vol 8 (No. 3), Jan 2005, p 153–157
3. Brochure of Simpson Technologies, www.simpsontechnologies.com
4. H.C. Verakis, Flammability in the Mining Sector, *Advances in Productive, Safe and Responsible Coal Mining*, 2019, www.researchgate.com
5. Brochure from MIFCO Foundry and Heat Treatment Furnaces and Accessories, www.mifco.com
6. DISAMATIC Inc., www.disagroup.com
7. Rose Torielli et al., Mull to Energy Strategies for Controlling Green Sand Systems, *Modern Casting*, Oct 2020
8. J. Nath, *Aluminum Castings Engineering Guide*, ASM International, 2018
9. D. Day and N. Benjamin, *Construction Equipment Guide*, John Wiley Publications, 1991
10. Pooley Inc., Supplier and Distributor of Conveyors and Belts, www.pooleyinc.com
11. S. Ramrattan, Foundry Management and Technology Publication, Jan. 15, 2009
12. Eirich Machines Inc. brochure, Molding Sand Aerators, www.eirichusa.com
13. Hunter Foundry Machinery Corporation Brochure, www.hunterfiundry.com
14. R. Wlodawer, *Directional Solidification of Steel Castings*, Pergamon Press, 1966
15. J. D. Mullins, Publication No. 100 Sorel Metal, Rio Tinto Iron and Titanium Inc.
16. General Kinematics brochure, www.generalkinematics.com
17. Carrier Corporation brochure, www.carriercorporation.com
18. Didion International brochure, www.didioninternational.com
19. M. J. Mroczek et al., Effect of Hot Sand and Its Cure by Use of a Sand Cooler, A Case Study, Paper 11-059, American Foundry Society Proceedings, 2011
20. P. Carey, Fundamentals of No-bake Sand Reclamation, ASK Chemicals, Dublin, OH, www.askchemicals.com
21. C. Wilding, Sand Reclamation of Chemically Bonded Sand Systems, Omega Foundry Machinery Ltd.
22. Solex Thermal Sciences Inc. Foundry Management and Technology, *Sept* 2017, www.solexthermal.com
23. Omega-Sinto Foundry Machinery brochure, www.omegasinto.com
24. Tinker Omega Inc., Advantages of PXG Thermal Reclamation, www.tinkerusa.com
25. Harrison Castings Inc., Thermal Sand Reclamation, www.harrisoncastings.com
26. BHM Thermal Sand Reclaimers, EC & S Inc. Birmingham, AL, www.escands.com

Casting Equipment Engineering Guide
Jagan Nath
https://10.31399/asm.tb.ceeg.t59370059

Copyright © 2023 ASM International®
All rights reserved
www.asminternational.org

CHAPTER 4

Molding Flasks and Molding Machines

THE MOLDING MACHINE is the heart of a casting facility. Molding flasks and other supplementary equipment are essential for molding complex shapes at competitive production rates and costs. This chapter addresses the design aspects of molding flasks and accessories, the features and handling accessories of molding machines, core making machines and innovations for productivity and quality, and automated core-setting aids.

4.1 Molding Flask Designs

Smaller castings made in captive foundries are produced in green sand, especially if the castings are not intricately cored. Medium-sized castings that have multiple cores are produced in no-bake lines. Both can be produced using flaskless molding. Molding in flasks accommodates more patterns than is possible using flaskless molding, because flaskless molding requires a wider mold-sealing area. Castings with intricate and complex cores are often produced by molding in flasks.

Molding flasks are capital-intensive, and a plan for their periodic replacement needs to be included in budgeting. The molding flask requires a carefully engineered design to ensure high strength and rigidity so that the molds will maintain their shape during the pouring and shakeout operations. Small flasks for hand molding are made of aluminum, and medium sizes are made of cast iron or fabricated steel. Large flask sizes are made by assembling cast steel sides. Trunnions for handling are usually produced out of cast steel and bolted to the sides. Fabricated steel flasks usually have welded trunnions (Ref 1).

4.1.1 Materials for Flasks

A general guideline is presented here regarding the materials used for flasks, although individual foundries may vary in their approaches. Cast aluminum (Grade 356-T6 or 319-T5) is used for small flasks for hand molding (Ref 2). Medium-sized flasks are made of gray iron, grade 20 or 25, or of fabricated mild steel. Larger flasks are produced using cast carbon steel sides (ASTM A 27, A 27M -16, or A 216 -WCA) and bolted together.

Molding flasks play an important role in holding the molding sand and cores together during molding, mold stripping, reversals, core setting, closing, and shakeout. The design of the flasks is critical to meet the following requirements:

- Rigidity—limited deflection without mold damage during lifting and reversal
- Strength—ability to withstand forces of clamping and weighing down against the buoyancy of the liquid metal
- Accuracy—close matching between two halves
- Ruggedness—ability to withstand forces during shakeout without damage
- Longevity—acceptable lifespan to justify the capital investment
- Affordability—optimum design for the least cost

4.1.1.1 Rigidity

The sizes of the flasks are chosen to provide for sufficient sand thickness around the pattern to allow for compaction and to accommodate the gates, runners, and feeders. In a captive foundry, where the sizes of the castings are predetermined, it is possible to optimize the sizes of the flasks. In a jobbing foundry, where the casting sizes can be diverse, the flask design and dimensional standards need to be established. When a new casting is produced, the most suitable size is chosen from among the existing standard flask sizes.

Flasks are designed in regular shapes—rectangular, square, or circular—to contain the packed sand around the pattern. Excessive deflection of the flasks during molding may buckle and break the mold. Flasks need to have adequate stiffness to limit deflection. Flanges are provided at the top and bottom to improve vertical, lateral, and torsional stiffness. Lateral stiffness is critical when the flasks are rotated.

Figure 4.1 illustrates the basic elements of the flask. Specially shaped flasks are designed to suit uniquely shaped pattern

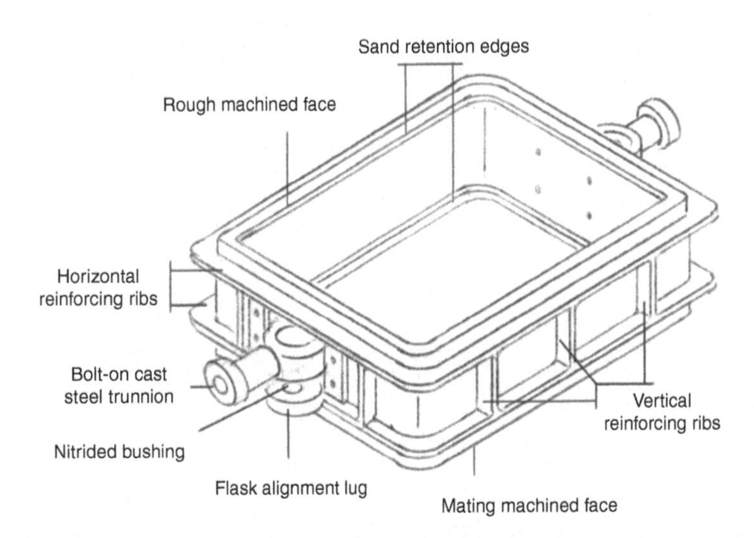

Fig. 4.1 Basic elements of a molding flask

profiles. A *match plate* is where the pattern profiles of each of the halves are mounted on two faces of the same plate. The top half is called *the cope* and the bottom is called *the drag*. A *pattern plate* is where each part of the pattern is mounted on two different pattern plates.

Molding flasks consist of a box-like structure, reinforced with ribs for vertical and lateral stiffness. Two trunnions at opposite ends allow lifting and rotating. These can be made of cast or fabricated steel. The trunnions are bolted to the flask sidewalls. Larger flasks have four trunnions, two on each face, for lifting and rotating.

A small ledge on the inner periphery helps to hold the packed sand inside the mold. Some flasks are designed without inner sand retention ledges to minimize the tendency for sand to stick. The mating flasks are aligned using a pin and bushing system for accurate matching.

The length and width of the flasks are determined by the way the patterns are parted and to provide space for the sprue, runners, gates, feeders, and sand packing. The height of the flasks is chosen according to the pattern height and to avoid interference with the crossbars. Crossbars are either bolted or welded at the top of the cope flask and the bottom of the drag flask to improve lateral stiffness. The height of crossbars is about 20 to 30% of the height of the flasks.

Table 4.1 (Ref 1) provides general guidelines for selecting the flask size. When the part geometry is known and the parting plane is established, it is easier to establish the flask size. Table 4.1 is based on the assumption that the weight of the casting bears an approximate relationship to the casting size.

4.1.1.2 Strength

A general guideline for strength is offered here, although individual foundries may deviate from these guidelines. Small flasks for hand molding—up to 450 by 500 mm (17.7 by 19.7 in.)—are made of cast aluminum (grade 356-T6 or 319-T5). Medium flasks of 2000 by 2500 mm (78.7 by 98.4 in.) are made in gray iron of grade 20 or 25 or from fabricated mild steel. Larger flasks are made using cast carbon steel sides (ASTM A27 A , A 27M -16, A 216 -WCA, or A 216 -WCB) and bolted together.

Table 4.1 Molding flask sizing guidelines

	Minimum distance between features, mm					
	A	B	C	D	E	F
Casting weight, kg or ton, t	Top of pattern and top of mold	Bottom of pattern and bottom of mold	Pattern and side of mold	Runner and side of mold	Distance between two patterns in mold	Pattern and runner
0 – 5	40	40	30	60	40	40
5 – 10	50	50	40	60	50	40
10 – 25	60	60	40	70	50	40
25 – 50	70	70	50	80	60	50
50 – 100	90	90	50	80	70	60
100 – 250	100	100	60	80	100	70
250 – 500	120	120	70	80	150	80
500 kg – 1 t	150	150	90	90	180	120
1 – 2 t	200	200	100	100	200	150
2 – 3 t	250	250	125	125	250	200
3 – 4 t	275	275	150	150	300	225
4 – 5 t	300	300	175	175	350	250
5 – 10 t	350	350	200	200	400	300
Above 10 t	400	400	250	250	400	350

The flasks made of gray iron are the least expensive and are dimensionally stable. But they are liable to break due to rough handling in the shakeout operation. Fabricated steel flasks are more expensive than gray cast iron flasks, but the machined mating faces of the flasks also may be damaged by rough handling in the shakeout units. One-piece cast steel flasks are expensive to manufacture, but they are more dimensionally stable compared to fabricated steel flasks. Individually cast steel side walls bolted together are more economical and are generally chosen for larger flask sizes.

Table 4.2 (Ref 1) provides guidelines for designing cast iron flasks. The smaller sizes, up to 500 by 500 mm (20 by 20 in.),

Table 4.2 Cast iron and steel molding flask side cross sections

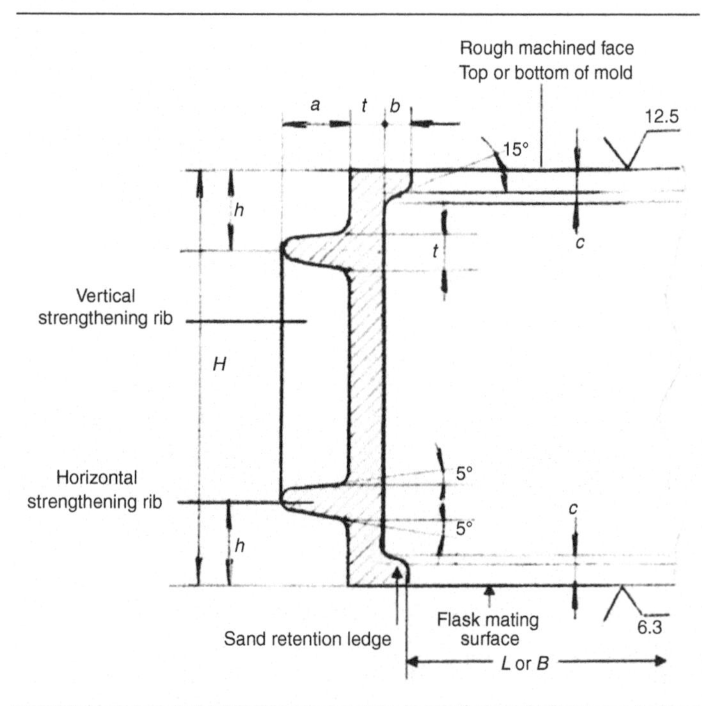

Average size $\frac{(L+B)}{2}$ or Dia. D mm	t mm	a mm	b mm	c mm	h mm
250	8	18	8	9	20
300	8	18	8	9	20
350	9	20	8	9	20
400	10	22	10	11	22
450	11	24	10	11	22
500	12	26	10	11	22
600	14	30	10	11	22
700	16	32	12	14	24
800	17	36	12	14	24
900	18	38	14	16	26
1000	19	40	14	16	26
1100	20	42	16	18	28
1250	22	45	16	18	28
1400	24	48	18	20	30
1600	26	52	18	20	32
1800	28	56	20	22	34
2000	30	60	20	22	36
2250	32	65	22	25	38
2500	34	70	22	25	40
2750	36	75	25	28	42
3000	40	80	25	28	45

The height H is chosen as per Table 4.1. The number of horizontal ribs is taken from Figure 4.3

may be cast in aluminum alloy if they are intended for hand molding. The sizes are valid for aluminum as well as cast iron. The suggested dimensions need to be multiplied by 0.8 to design one-piece cast steel flasks.

Table 4.3 provides guidelines for establishing the number of horizontal ribs, based on the height of the flasks.

Table 4.4 offers guidelines for determining the number of vertical reinforcing ribs on the side length L, and Table 4.5 offers guidelines for the number of vertical ribs on the side length B.

Molding flasks require trunnions for lifting, reversing, closing, during molding and handling in the shakeout unit. Most of the small and mid -sized flasks have two trunnions, one on each end. Large flasks, greater than 3000 mm (118 in.), have two trunnions on each end face for manipulation and handling by overhead cranes during molding and shakeout.

The trunnions are either bolted on, cast in, or welded onto steel flasks. Their sizes depend upon their load-bearing capacity. Table 4.6 guides the selection of trunnion diameters based on the mean dimensions of:

$$\frac{L+B}{2}$$

Table 4.7 (Ref 1) provides the overall dimensions of four sizes of bolted-on cast steel trunnions. These are produced in A 216 WCB -NT grade steel, with austempered ductile iron Grade 1 (Ref 3) as alternative material.

Fabricated steel flasks are made from channel sections or flats and plates. The trunnions can be designed as bolt-on or welded-on. Figure 4.2 illustrates an example of a fabricated flask.

Flasks can also be assembled by bolting on separately cast sides, with or without corner pieces. Such flasks may have cast iron or cast steel side walls. These flasks can be ribbed for sturdiness. These are not preferred for molding operations where the flasks need to be rotated.

Figure 4.3 shows an example of a sturdy side wall of an assembled flask made of cast iron.

4.1.1.3 Accuracy

There are two aspects of accuracy needed in all varieties of flasks.

- Accuracy of alignment of cope and drag flasks affecting the mold shift
- Accuracy of perpendicularity between the guide pin axis and the mating plane. (This ensures the stripping of the mold from the pattern, without damaging the mold.)

Table 4.3 Guidelines for determining the number of horizontal ribs

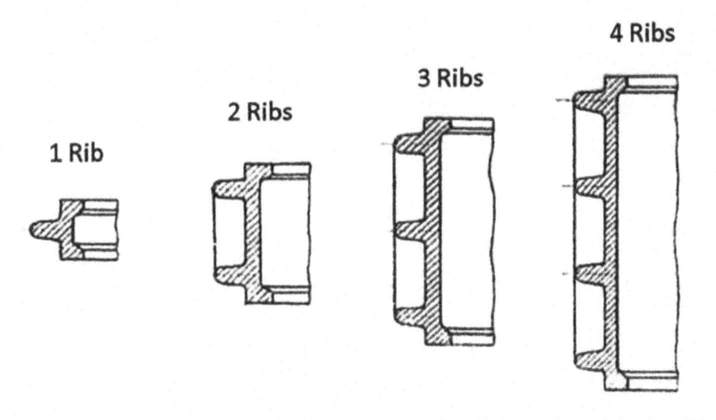

	Flask height, mm																	
$\frac{L+B}{2}$ or diam., D, mm	40	60	80	100	120	140	160	180	200	250	300	350	400	450	500	600	700	800
250	1	1	1	2	2	2	2											
300	1	1	1	2	2	2	2	2										
350	1	1	1	2	2	2	2	2	2									
400		1	1	2	2	2	2	2	2	3	3							
450			1	2	2	2	2	2	2	3	3							
500				2	2	2	2	2	2	2	3	3						
600				2	2	2	2	2	2	2	3	3						
700					2	2	2	2	2	2	3	3	3					
800					2	2	2	2	2	2	3	3	3					
900						2	2	2	2	2	2	3	3	3				
1000						2	2	2	2	2	2	3	3	3				
1100							2	2	2	2	2	3	3	3	4			
1250							2	2	2	2	2	3	3	3	4			
1400								2	2	2	2	2	3	3	3	4		
1600								2	2	2	2	2	3	3	3	4		
1800									2	2	2	2	3	3	3	4	4	
2000									2	2	2	2	3	3	3	4	4	
2250									2	2	2	2	2	3	3	3	4	
2500										2	2	2	2	3	3	3	4	4
2750										2	2	2	2	2	3	3	4	4
3000										2	2	2	2	2	3	3	4	4

Table 4.4 Guidelines for vertical ribs on the side walls *L*

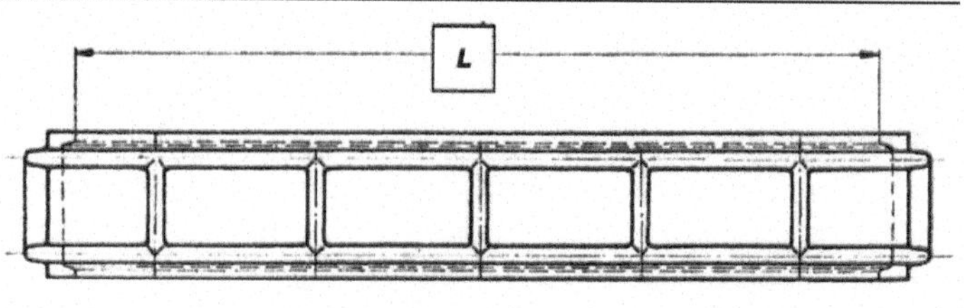

L, mm	Number of vertical reinforcing ribs 2	3	4	5	6	7	8	9
250	x							
300	x							
350		x						
400		x						
450		x						
500		x						
600		x						
700			x					
800			x					
900			x					
1000			x					
1100			x					
1250				x				
1400				x				
1600				x				
1800				x				
2000				x				
2250					x			
2500					x			
2750					x			
3000					x			
3500						x		
4000						x		
4500							x	
5000								x

Table 4.5 Side wall widths and number of vertical ribs

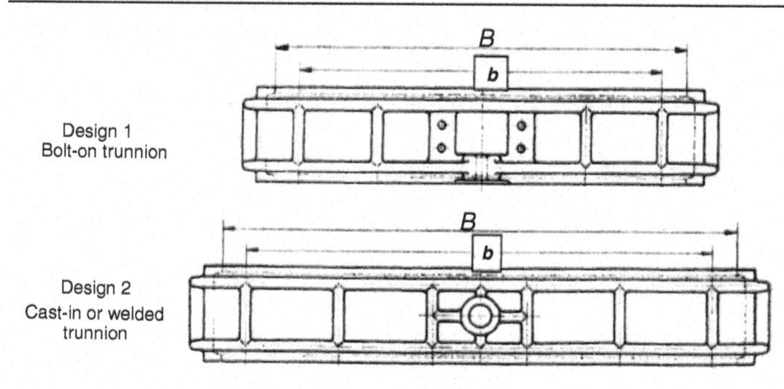

Side wall width B, mm	Distance between end ribs b, mm	Number of ribs 2	4	5	6
350	250	x			
400	290	x			
450	300	x			
500	320	x			
600	350	x			
700	400	x			
800	440	x			
900	800		x		
1000	880		x		
1100	960		x		
1250	1060		x		
1400	1170		x		
1600	1320		x		
1800	1500			x	
2000	1660			x	
2250	1800				x
2500	2150				x
2750	2400				x
3000	2600				x

Table 4.6 Guidelines for trunnion diameter selection

Mean dimension, mm $\frac{(L+B)}{2}$ or diameter	Trunnion diameter, mm	Mean dimension, mm $\frac{(L+B)}{2}$ or diameter	Trunnion diameter, mm
450	40	1100	
500		1250	
600	50	1400	
700		1600	80
800		1800	
900		2000	
1000	63	2250	

4.1.1.3.1 Alignment Accuracy

The accuracy of alignment between the flasks is assured by:

- Controlling the center distance between the alignment holes or guiding holes
- Accuracy of the drilled holes in the flasks for bush-fitting
- Accuracy of the guide pins mounted on the pattern plate
- Monitoring the wear of the guide pins and bushings

Table 4.8 (Ref 1) provides guidance for establishing the size of the alignment or guide pin holes, their center distances, and the diameter of the circular and flat bushings.

If both alignment holes are circular, there are more chances of the cope locking onto the pins, even if there is only a minor tilting of the cope during closing. By making one of the holes flattened, and providing clearance along the non-flattened faces, the tendency to lock is minimized.

Alignment accuracy also requires that the guiding pins mounted on the pattern plates should be provided with close tolerances, and that the bushings are also manufactured and maintained to close tolerances. The guiding and closing pins are hardened and chrome plated. The bushings are *nitrided* to withstand wear.

Table 4.9 illustrates the design and major dimensions of the guide pins. Table 4.10 details the surface finish and tolerances specified for the bushings. Figure 4.4 illustrates the dimensions of the closing pins used for closing the cope over the drag.

Table 4.7 Cast steel A 216 WCB-NT trunnion design

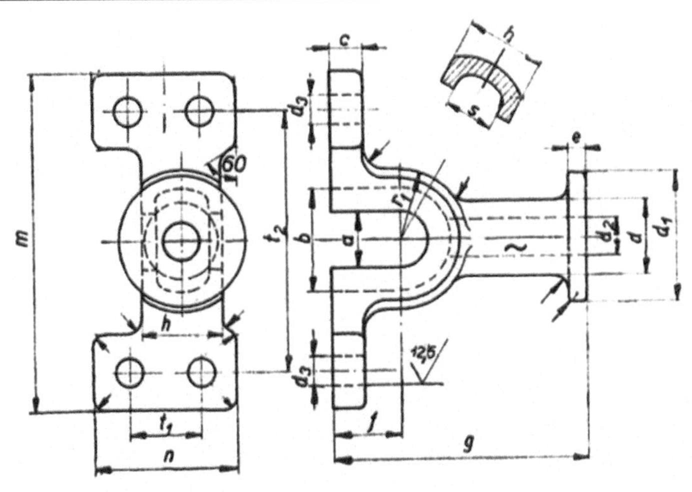

Trunnion diam., mm	Capacity, kg	a	b	c	d_1	d_2	d_3	e	f	g	h	m	n	r_1	t_1	t_2	s
40	800	32	63	16	63	20	14	10	34	125	45	170	85	40	45	125	28
50	1250	40	80	22	80	25	18	12	45	160	56	210	105	50	56	160	36
63	2000	50	100	28	100	32	23	16	56	200	70	260	135	63	70	200	45
80	3150	63	125	36	125	40	2x7	20	70	250	90	330	170	80	90	250	56

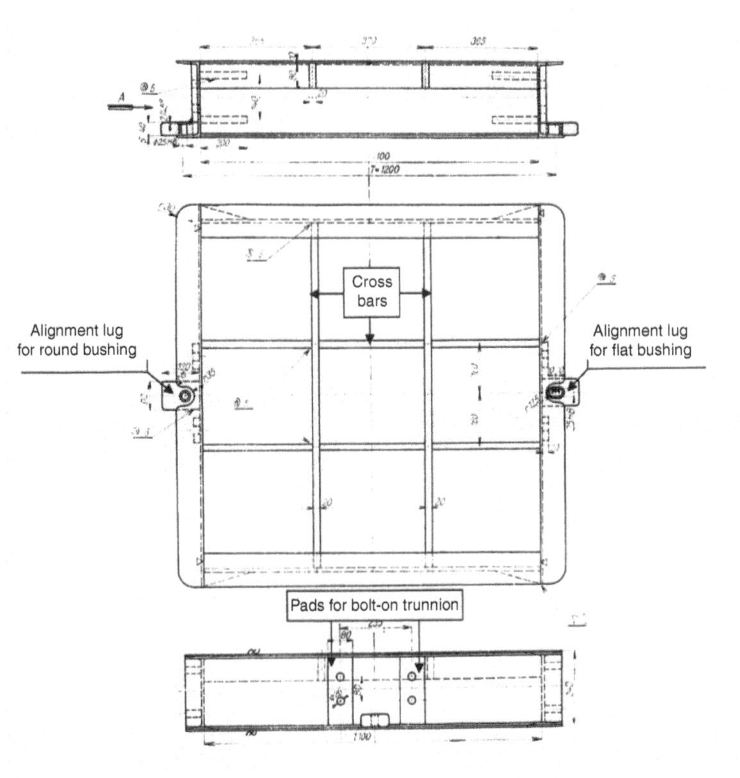

Fig. 4.2 Fabricated steel flask

4.1.1.3.2 Perpendicularity of Guide-Pin Axes and Mating Surfaces

The perpendicularity of the guide-pin axes to the plane of the cope-drag mating surface is important because it affects the ease of stripping the pattern from the mold or the mold from the pattern. During the manufacturing of the flasks, the mating surfaces of the flasks are machine-finished, and the opposite parallel surface is rough machined. The holes are jig-drilled into the two side lugs, and bushings are pressed in.

While the flasks are being reconditioned to repair a damaged mating face, the machining fixture uses the two guide pin holes for reference, and the mating faces are ground. Dents in steel flasks are welded before resurfacing.

4.2 Mold Pallets

The drag (or bottom mold) is placed over a pallet, after stripping and reversing, with the profiled molded face up. The cores are placed in the drag mold, and the cope (or the top mold half) is closed over the drag. The pallets move the closed mold over a conveyor, as illustrated in Fig. 3.20 and 3.21 in Chapter 3, "Sand Conditioning Equipment" in this book, to the pouring station and onto the shakeout grid. The molds are lifted off of the pallets before mold-punching or placing them on the vibratory grid for the shakeout operation. The pallets are designed to move to the return-mold conveyor and back to the molding stations.

Pallets are usually cast in gray cast iron or fabricated out of steel and are ribbed to resist any buckling under the load of the mold. Figure 4.5 illustrates a ribbed gray cast iron pallet. Pallets are likely to have metal spills on them. The top surfaces of pallets need to be cleaned on the return mold conveyor before the next mold is set. Some casting manufacturers automate the cleaning of pallets.

No-bake lines engineered for automated handling can use gray cast iron pallets for setting the core-molds, assembly, and pouring. Aluminum pallets offer the advantage of light weight for ease of handling. These are used for smaller mold sizes. The general ribbing designs are similar between gray iron and aluminum pallets. No-bake lines, where the molder needs to place the pallet

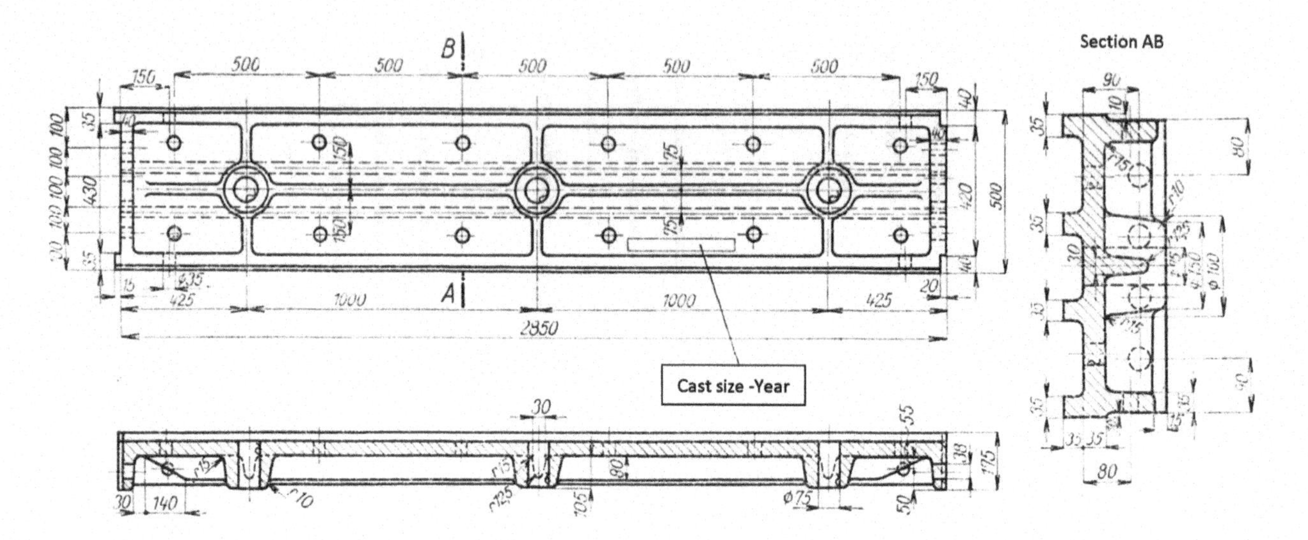

Fig. 4.3 Example of an assembled flask side wall

Table 4.8 Sizes and center distance of alignment holes

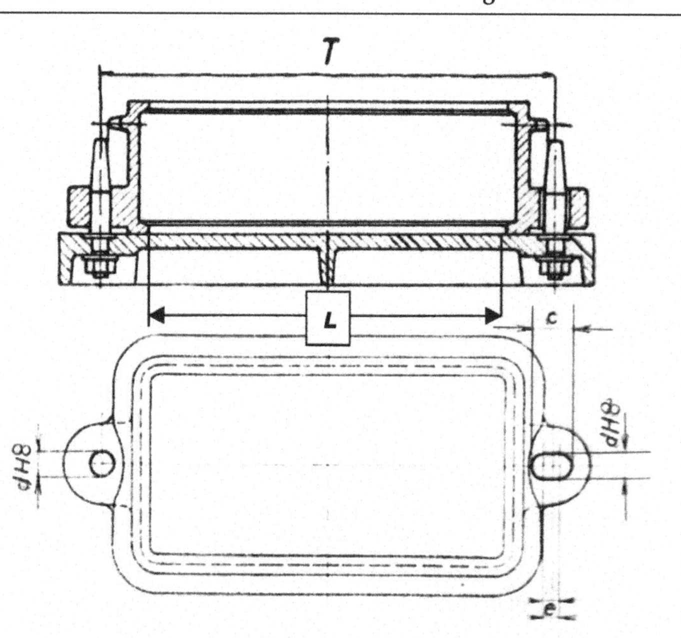

Length, L, mm	Center distance, T, mm	Hole dimensions		
		dH8	e	c
250	330	20	15	35
300	380			
350	440			
400	490			
450	560			
500	615			
600	750			
700	850	25	15	40
800	950			
900	1060			
1000	1180			
1100	1280	30	20	50
1250	1450			
1400	1600			
1600	1850			
1800	2050	35	20	55
2000	2250			
2250	2500			
2500	2750			

Table 4.9 Guide pin details

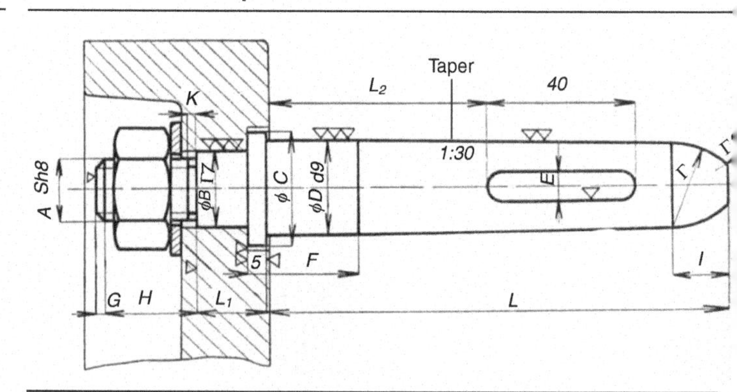

Diam., D	A	B	C	E	F	G	H	I	K	L(a)	L_1(b)	L_2(a)
20	M 16	18	24	5	25	2	20	10	4	75–225	13–44	25–60
25	M 16	20	30	8	25	2	25	15	4	75–225	13–44	25–60
30	M 20	25	35	8	25	2.5	30	15	5	75–225	13–44	25–60
35	M 24	30	40	8	30	2.5	35	15	6	75–225	13–44	25–60

Note: All dimensions in mm. (a) Dimensions depend on the height of the cope flask. (b) Dimension depends on the thickness of the pattern plate

over the back of the core-mold box before reversal and ejection, use wooden pallets for handling. The molds are transferred to cast iron pallets before pouring. Some casting facilities, especially those on no-bake molding lines, retain the wooden pallets but cover them with sand to protect against any damage due to metal spills.

4.3 Molding Machines

Molding machines have been developed over time, improving their productivity, mold size, mold quality, and the capability for automation. The workplace environment has significantly improved due to a reduction of noise and vibration. The sand filling, molding, and core setting are more automated in recent times. Real-time digital displays of the operations and troubleshooting directives are implemented in molding machines today.

Figure 4.6 provides an overview of different molding machines. There are many possible variations in machine features. Highly specialized machines based on electromagnetism and membrane molding are not included in this chapter.

Molding machines differ in one or more of these distinguishing characteristics:

- Sand compacting method—jolting, squeezing
- Pattern stripping method—pattern or mold movement up or down for separation of patterns and molds

Table 4.10 Steel 4140 or 4340 bushings hardened, tempered and nitrided

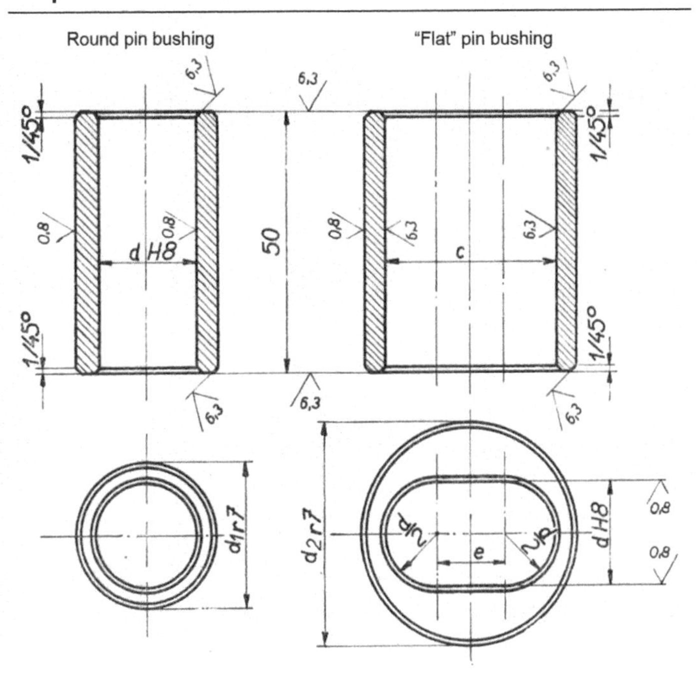

$\frac{L+B}{2}$	Pin bushing							
	Round			Flat				
	d_1r7	dHB	h	d_2r7	dHB	c	e	h
250–500	28	20	50	40	20	35	15	50
600–900	36	25	50	45	25	40	15	50
1000–1400	42	30	50	56	30	50	20	50
1600–2250	50	35	50	63	35	55	20	50

All dimensions in mm

- Flask filling method—gravity fill, blow fill, aeration fill
- Source of energy for working—pneumatic, hydraulic, a combination of pneumatic and hydraulic, electromagnetic, mechanical
- Number of stations for each cycle—1, 2, or 4
- Machine structure—1 column or 4 columns

Green sand molding machines are classified as:

- **Tight flask or flask** molding, where iron or steel flasks are used for molding
- **Flex-flask/flaskless** molding, where flasks are used for molding and the entire mold is closed and pushed out of the flasks before being placed on the pouring conveyor line
- **Flaskless molding machines**, where the machine sides and pattern plates function as retainers for sand compaction (instead of flasks) and the ejected molds advance to the pouring segment of the conveyor line

4.3.1 Comparison of Tight Flask Molding and Flaskless Molding

The use of molding flasks allows the mold cavities to be close to the wall of the flasks without the potential for metal leaks at the parting plane. Flasks also allow for clamping the two halves or providing for heavier weights over the mold, which prevents the lifting of the cope or the cores against the buoyancy of the liquid metal. Molding in flasks allows the molds to be moved to an adjoining bay for cooling thicker castings that need longer cooling times.

However, flasks are capital-intensive and require budgeting for periodic maintenance. Refurbishing by replacing worn-out bushings or resurfacing the damaged parting surfaces creates additional costs. Complex castings with intricate cores must be molded in flasks, and the sand sticking to the flasks needs to undergo regular automated cleaning. Generally, molds larger than 1500 to 1800 mm (59 to 71 in.) require molding in flasks.

Flasks must be punched out or vibrated to separate the castings from the molding and core sand, and then returned to the molding station for reuse. The automation for flask return requires additional capital and conveyor space.

Flaskless molding benefits from the ability of molds to enter a rotary shakeout unit without the need for mechanization to

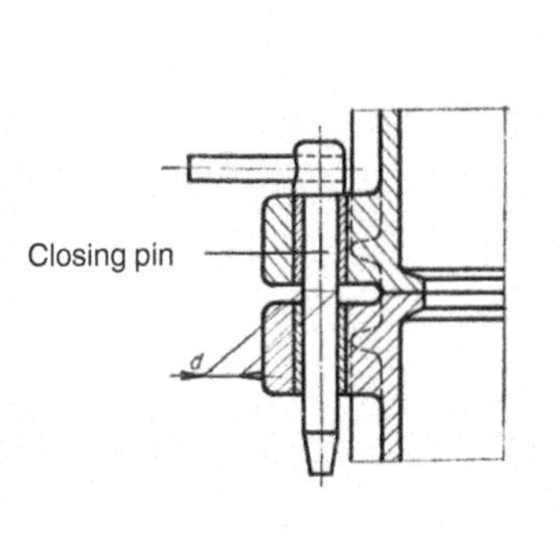

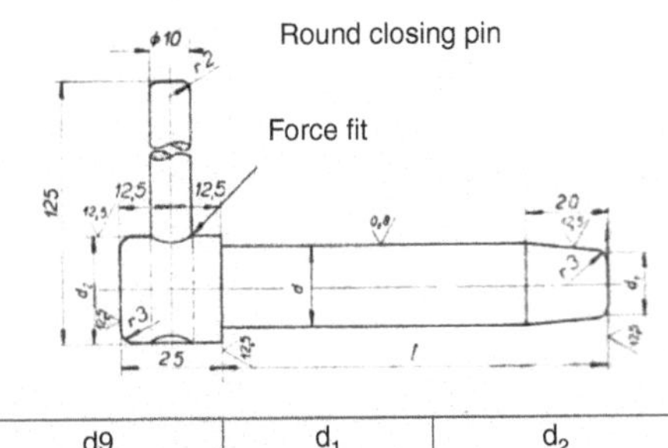

d9	d_1	d_2
20	25	16
25	30	20

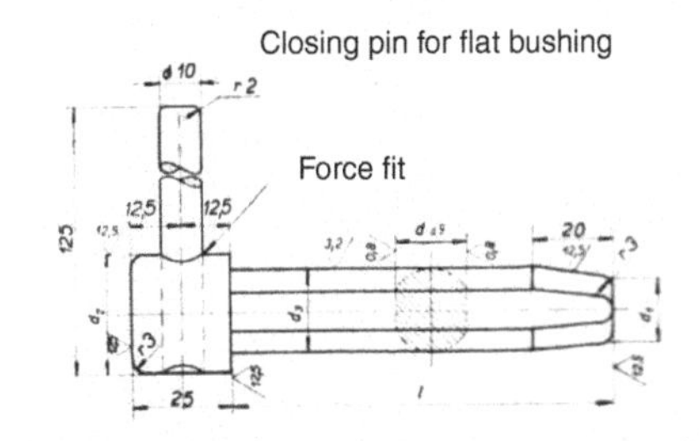

d d9	d_1	d_2	d_3
20	21.5	25	16
25	26.3	30	20

Fig. 4.4 Closing pin details

separate the molds from the flask. However, the pallet that supports the mold needs to be returned to the molding line. This necessitates space for a return conveyor.

Capital costs are reduced on no-bake lines due to the elimination of flasks and accessories and avoiding the cost of maintaining the flasks.

The operation of loading the molds with flat weights to withstand the buoyancy of the liquid metal needs to be automated. Some percentage of flaskless molds may leak at the parting plane if the cavity is too close to the edge or if the molds crack due to the loading weights. The potential for mold breakage due to the pressure of the molding weights or the possibility for mold leakage at the parting line is reduced by inserting a slip jacket to cover an area across the parting plane. The cycle times of the molding machine are impacted by the time needed for core setting, and this can be a limitation on the output, in both flask and flaskless installations.

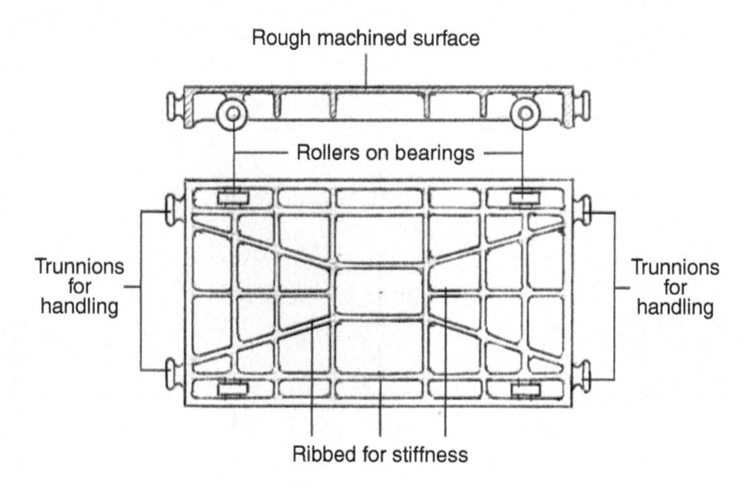

Fig. 4.5 Gray cast iron pallet

4.3.2 Pneumatically and Hydraulically Compacted Molding Machines

Molds produced in tight flasks are parted horizontally. These can be further classified based on the method of sand compaction into:

- Pneumatically compacted
- Hydraulically compacted

The pneumatically compacted machines can be classified into two groups based on sand compaction:

- Jolt-pneumatic squeeze
- Jolt-hydraulic squeeze

Jolt-pneumatic machines use air pressure for compaction from the bottom of the flasks and air pressure for squeezing from the top of the mold. Jolt-hydraulic machines use air for compaction from the bottom of the flask and hydraulic pressure to squeeze from the top of the mold. Jolt-hydraulic squeeze machines enable higher squeeze pressures than jolt-pneumatic or pneumatic jolt-squeeze machines.

The density of sand compaction decreases as the height of the flask increases. Squeezing from the top of the flasks makes up for the decrease in packing density over the height of the flask.

Hydraulic squeeze machines are usually constructed with four columns to withstand the higher pressures of the hydraulics.

Figure 4.7(a) is a schematic of a jolt-pneumatic molding machine, and Fig. 4.7(b) shows a jolt-hydraulic squeeze machine. Figure 4.7(c) shows some details of the jolt-squeeze machines.

The type of machine shown in Fig. 4.7(c) consists of two coaxial pistons housed inside a cast cylindrical casing, connected to the table platen. The jolt piston rises up as the air is let in. As the piston rises, it opens the air exhaust port, which causes the table to drop. This rising and falling motion occurs a preset number of times. A spring at the bottom of the piston cushions the

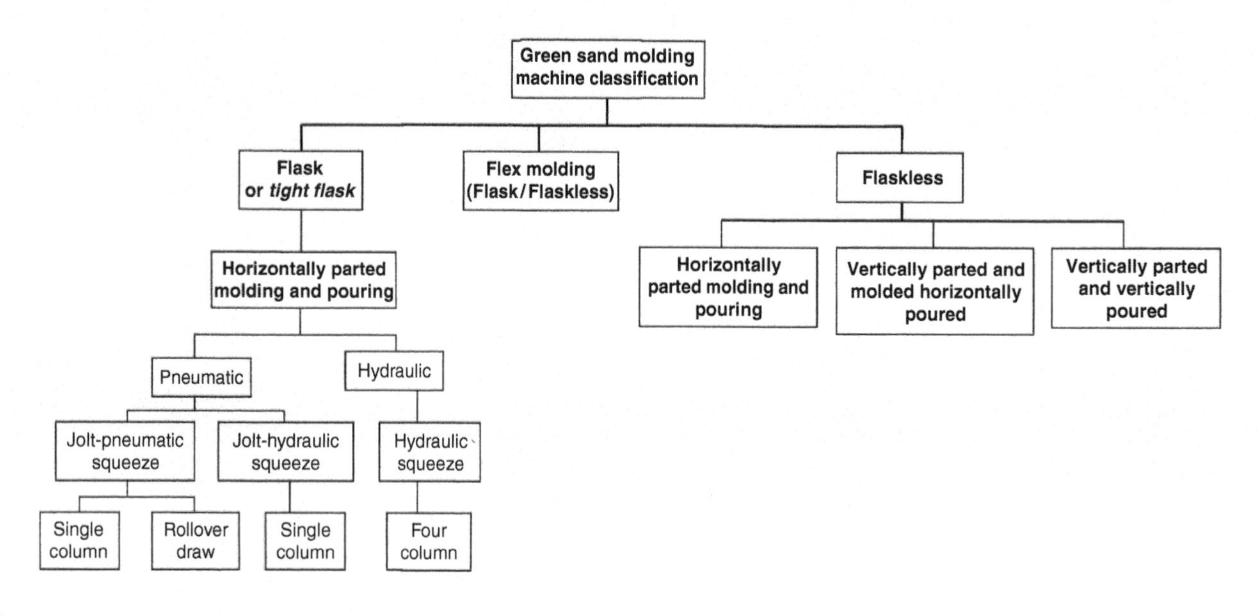

Fig. 4.6 Classification of green sand molding machines

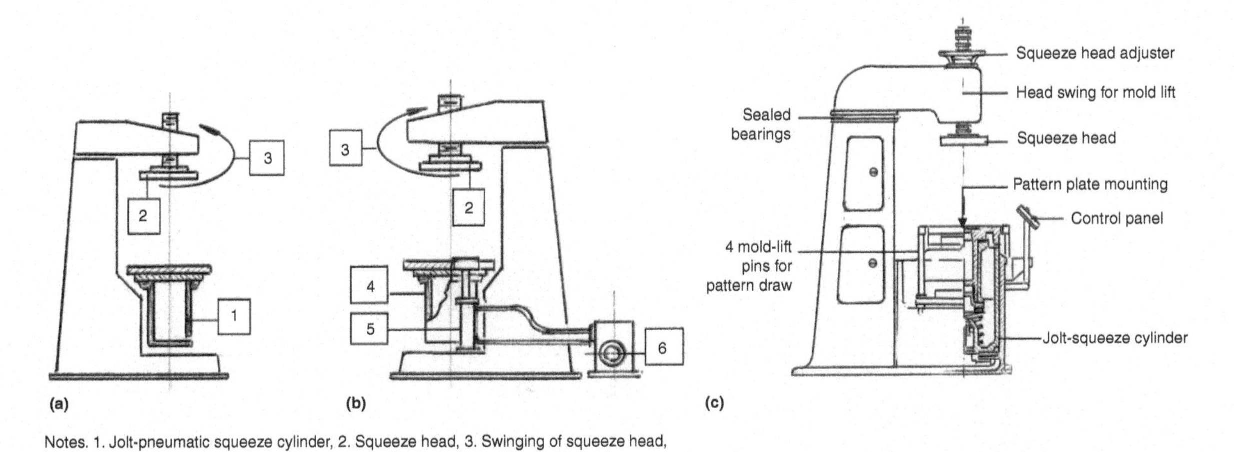

Fig. 4.7 Pneumatic machines; (a) jolt squeeze, (b) jolt-hydraulic squeeze, (c) jolt squeeze details

fall, which decreases the noise. The second piston, known as the squeeze piston, pushes the table up toward the squeeze head to squeeze the sand from the top of the flask. The squeeze head is engineered for manual adjustment based on the flask height. Some machines are available with power-operated squeeze heads for adjustments.

Over time, modifications and improvements to this process have been developed. A few patented modifications have decreased the noise. Machines are programmed for a pre-jolt even as the sand is filling the flask, resulting in more even distribution of the sand. Some machines are also designed for a simultaneous jolt and squeeze, producing molds of even compaction and hardness (Ref 4). Four ejector or stripping pins, one at each corner of the flask, raise the mold off the pattern. Vibrators actuated simultaneously assist in stripping the pattern without any mold damage. Some machines are offered with self-aligning bars to lift the mold. Many of the machines are provided with an adjustable two-speed draw, with an initial slow speed followed by faster speeds to reduce the cycle time.

Pneumatic jolt-hydraulic squeeze machines are designed with two hydraulic cylinders, one on each side of the machine, as shown in Fig. 4.7(b). These actuate to lift the table against the squeeze head with a force higher than that of the pneumatic machines. The squeeze head is swung out to the side to be able to lift the mold off the pattern mounted on the machine platen. The squeeze head structure is mounted on a ball bearing ring for swinging. Some machines are automated.

Jolt-pneumatic machines are classified based on their structural configuration and the method of drawing the pattern from the mold into:

- Single column
- Rollover-draw

The rollover-draw machines, used for medium-sized molds, use top compaction by a secondary jolt to compact the sand closer to the pattern.

Single column machines are offered as:

- Swing head
- Swing arm machines

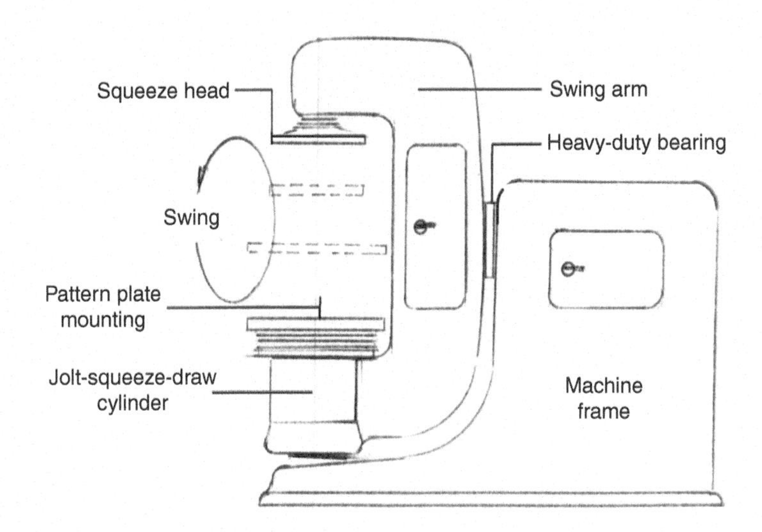

Fig. 4.8 Single-column swing-arm machine

The column structure is a modified I-beam for lightweight machines and a box-like structure for heavier machines that require more rigidity.

Figures 4.7(a) and (b) are schematics of swing head machines. Figure 4.8 displays a swing arm machine. The jolt-squeeze cylinder performs an additional function of drawing the pattern away from the mold, following the swinging operation. The stripped mold then moves to the conveyor.

Figure 4.9 is a schematic of a rollover-draw molding machine.

The jolt rollover-draw machine consists of an arm or a combination of arms that produce a crank action movement to move the compacted mold to the pattern draw and stripping station. The flask is clamped onto the machine platen using pneumatic clamps. The jolt table compacts the sand which descends from the overhead hopper. The top of the mold is compacted by laying weights on the top of the mold and jolting it. Alternatively, a squeeze head that moves in and out can also be engineered for squeezing from the top.

The compacted mold rolls over, as shown by dotted lines in Fig. 4.9. The mold rests on a conveyor, and the pneumatic clamps are released. A hydraulic cylinder or a scissors table lowers the mold off the pattern. The rollover arm swings the pattern back to the molding station.

Four-column machines have two of the four columns that act as hydraulic cylinders to apply squeeze pressure from the top. The other two posts guide the top platen. Because the columns restrict flask movement, this concept is used in designing multistation molding in flaskless molding.

4.3.3 Multistation Machines

The productivity of a molding machine is increased by reducing the cycle time. The sequence of operations in the molding process can be split by two, four, or multiple stations for cycle time reduction. Two- or four-station layouts can be operated by individual operators by using a shuttle or a turntable to move the same mold through individual operators to carry out simultaneous tasks on multiple molds. Figure 4.10 shows a schematic of the one-, two-, and four-station operations. The sharing of tasks can be modified depending on the complexity of the casting.

Multistation cell layouts are automated, reducing the manpower and increasing the consistency of the process elements. Engineering for multi-station operations is a productive way to reduce cycle time. The cycle time events are carried out simultaneously at adjacent locations, and the molds move through successive operations to complete a mold cycle. This layout design concept is valid for both tight flask lines and flaskless lines.

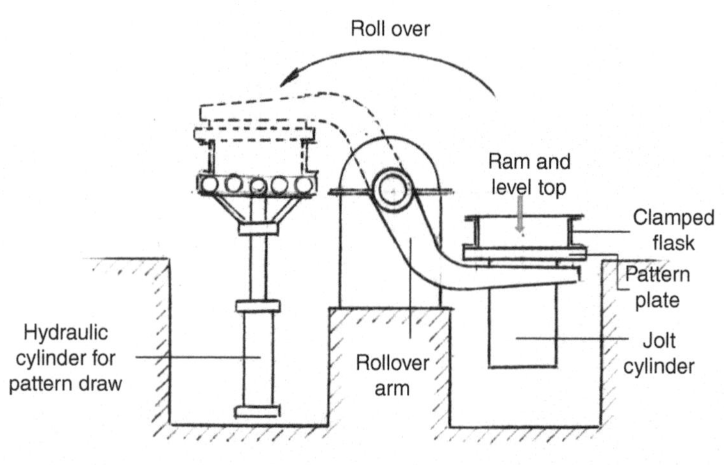

Fig. 4.9 Rollover-draw machine

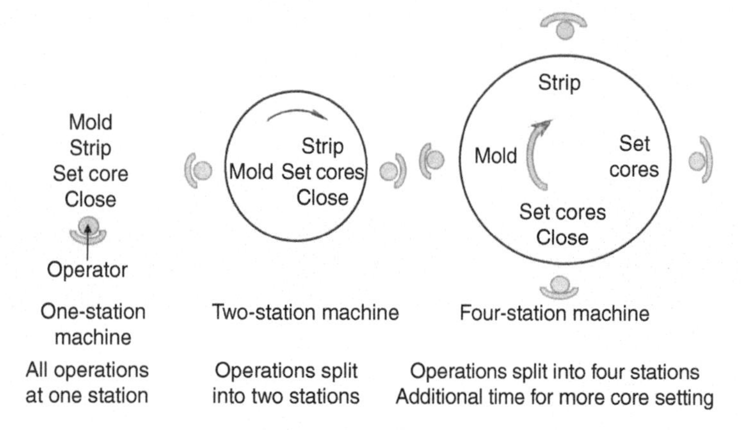

Fig. 4.10 Multistation cell concepts

Core setting time is a constraint for cycle time reduction. More time can be provided for core setting without increasing the cycle time by adding an additional station , as shown in the four-station layout in Fig. 4.10. Core-setting time can also be reduced by setting cores into fixtures or *masks* before the drag is compacted. The fixture or the mask is positioned over the mold, either manually or by using a manipulator, and the cores are released. The mold-closing pins or other pins provided in the core print enable accurate positioning.

High-production machines use a dedicated device such as a robot, or a combination of linkage mechanisms and pneumatic or hydraulic cylinders, to move the mask into position quickly and release the cores. Setting the cores into the masks can be done manually to reduce the capital cost.

Table 4.11 lists the operations that are executed sequentially on molding machines. The timing diagram shows the time saved by parallel processing of the drag and cope. Some operations are grouped for cycle time compression. Two separate production cells, one for the drag and another for the cope, can be engineered for maximum cycle time reduction. One single cell can be engineered to mold the drag and the cope alternately to reduce capital expenses, but this would not support total cycle time reduction to produce a full mold.

Figures 4.11 and 4.12 illustrate two concepts of multistation design layouts. Figure 4.11 shows an in-line layout, and Fig. 4.12 illustrates a rotary table layout.

The events at each station of the cell are listed at the top of each cell. A pick-and-place unit picks up the flask from the flask-return conveyor and places it on the pattern plate using the guide pins. An overhead dosing hopper is mounted on a trolley, to shuttle between the prepared sand hopper and the sand fill station. It is fitted with load cells to carry sand to the preset weight. The flask is filled to the established volume. The flask moves to the squeeze (or jolt-squeeze) station where the sand is compacted. The next station is equipped with a roll-and-draw machine. The mold is reversed, and the pattern is drawn out of the mold. The mold moves to the next station, where the cores are set manually or automatically. The drag mold moves to the next station, where the cope closes over the drag. The mold assembly is then clamped and moves to the pouring station.

These operations can also be planned using a rotary table, as shown in Fig. 4.12.

4.3.4 Molding Machine Features Enhancements

The development of molding machine innovations has improved mold quality, process consistency, and cycle times. The development of control systems has improved process monitoring and diagnostics capabilities. Some of these enhancements are:

- Hopper filling improvements—blow, aeration, and airflow
- Squeeze pressure balance control (to reduce pattern plate distortion)
- Mold height control—vertical parting

Table 4.11 Typical sequence of molding operations

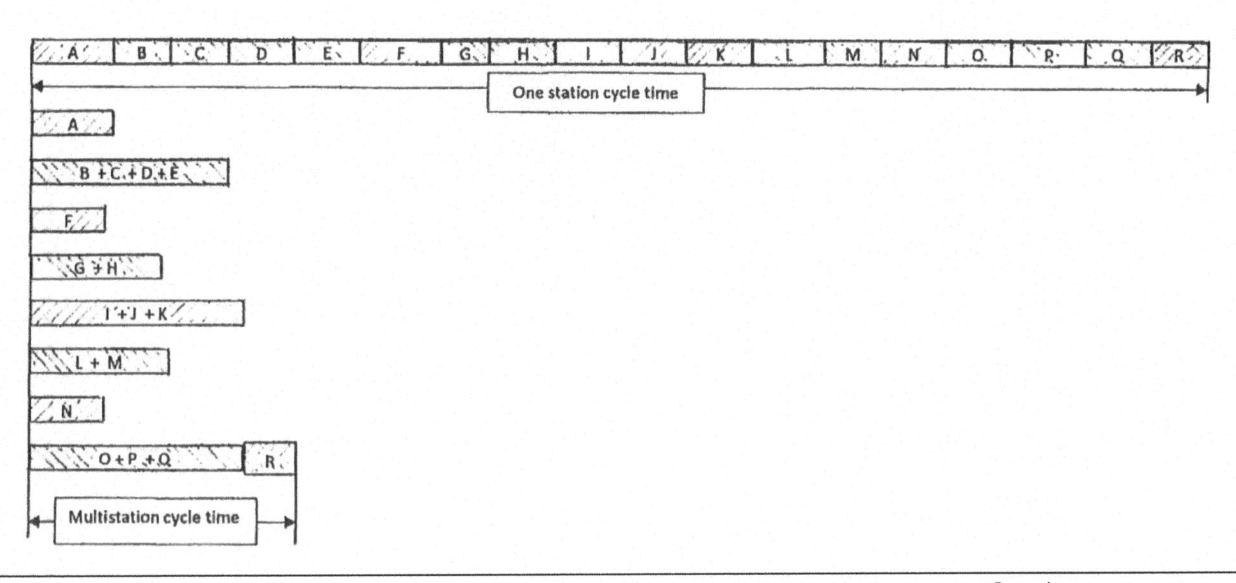

Designation	Operation
A	Fill sand into preset weight hopper
B	Set flask on pattern plate
C	Move drag flask with pattern plate below hopper
D	Apply mold release spray
E	Move flask over compaction table
F	Compact sand by jolt-squeeze or squeeze
G	Roll over and draw pattern out
H	Place drag on pallet
I	Move cope flask on pattern plate
J	Apply mold release spray
K	Move flask over compaction table
L	Compact sand by jolt-squeeze or squeeze
M	Open sprue and vents
N	Roll over and draw pattern out
O	Set cores in drag
P	Close cope over drag
Q	Clamp the flasks together with C-clamps or place weights over mold
R	Move the molds to pouring station

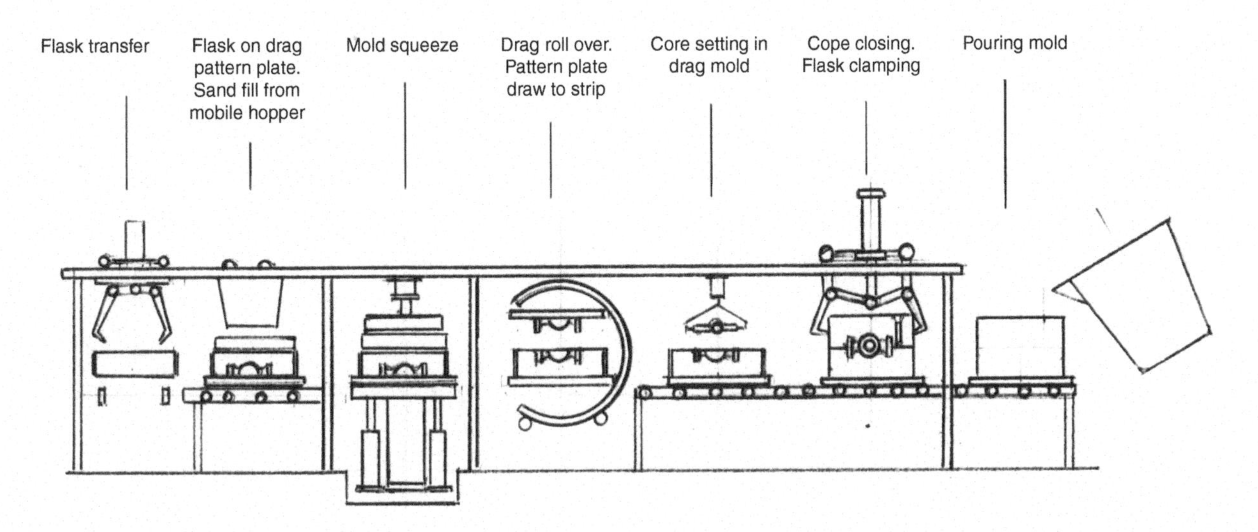

Fig. 4.11 In-line multistation layout

- No-spill sand—horizontal parting plane
- Pattern plate preheater
- Levelling frame pattern draw
- Multiram squeeze
- Pattern plate change safety feature

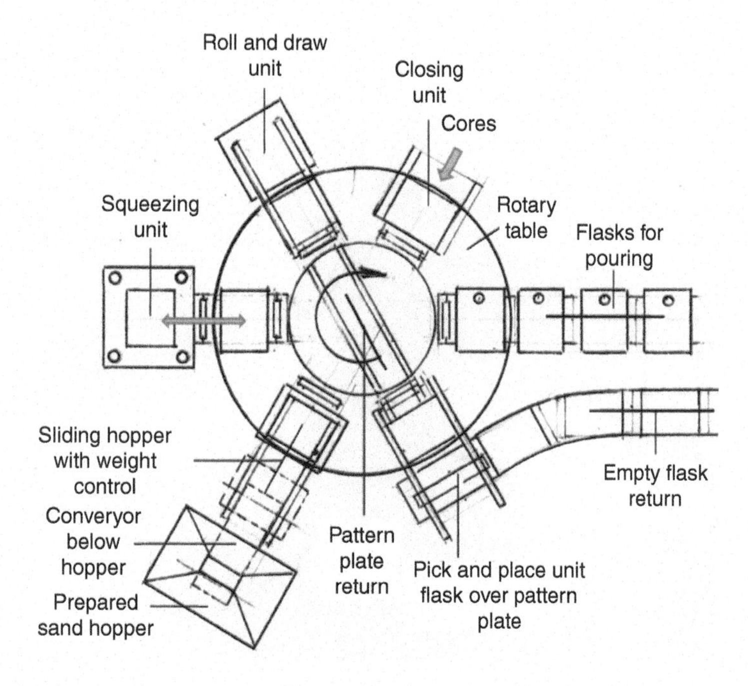

Fig. 4.12 Rotary table multistation layout

- Automated pattern plate change
- Mold machine control and diagnostics

4.3.4.1 Hopper Filling Improvements

The sand hoppers over the flasks on the machines used to be filled through bottom discharge clamshell doors. This method did not distribute the sand uniformly and required manual intervention. Automatic molding machines are now equipped with blow or aeration filling. Deep patterns benefit from the air-flow method for sand filling.

Figure 4.13 illustrates the sand-filling systems for molds.

4.3.4.1.1 Gravity Fill

The gravity fill method uses a bottom discharge hopper, usually with clamshell-style doors, which are operated pneumatically by a push button or manually by a pull chain. The prepared sand that is discharged tends to heap, based on the angle of repose of the mix. Operators start the jolt operation as the sand is being discharged, and this evens out the sand level. This least expensive system is adequate if the machine is manually operated, but it is slower than the other systems.

4.3.4.1.2 Blow Fill

The blow-fill system is suitable for automated molding machines. Air from a reservoir enters through a valve over the sand in the hopper. Sand is released through the openings into the mold cavity. This system is in operation on several automated machines with both vertically and horizontally parted molds. There is no need for manual intervention.

Gravity fill	Blow fill	Aeration fill	Air-flow
Sand hopper; Clam shell doors; Flask; Pattern plate; Jolt	Blow pressure 0.5 MPa; Air reservoir; Squeeze; Squeeze; Pattern plates	Aeration; Squeeze; Match plate; Air pressure 0.1 MPa	Squeeze; Squeeze; Vents; Pattern plates
Low capital Suitable for manually operated machines	Less capital Needs no intervention Suitable for vertically and horizontally parted molds	Increased sand flowability Dense molds, even density Lower noise level	Suitable for medium and large molds

Fig. 4.13 Alternative mold sand filling methods

4.3.4.1.3 Aeration Fill

Sand in the hopper is fluidized by blowing low-pressure air in high volume through ports provided in the lower area of the hopper. Fluidized sand enters the mold cavity on either side of the match plate. Vents are provided for the air to exit from the top of the mold cavity. This results in even distribution of the sand into pockets and produces uniform molds (Ref 5).

4.3.4.1.4 Airflow

Vents provided at the bottom of the mold cavity enable sand to fill in deep molds (Ref 5). Deep molds pose a challenge for uniform sand distribution problems, and air flow is an effective solution to address this issue.

4.3.4.2 Squeeze Pressure Balance Control

Vertically parted molds with a match plate are squeezed from both sides, as shown in Fig. 4.14 (Ref 5). The automatic balance of squeeze pressure minimizes match plate distortion. The advantages of this system are:

- Molds with minimum draft can be produced without damage.
- Widely differing pattern profiles on either side of the match plate can be accommodated without match plate distortion.
- Deep offsets of the pattern profiles do not pose any compaction issues.
- Thinner match plate thicknesses do not pose any distortion and mold breakage issues.

4.3.4.3 Mold Height Control — Vertically Parted Molds

The sand volume in the dosing hopper above the flasks will vary based on the targeted mold height. Process variations in sand volume, the squeeze board setting position, and degree of compaction by squeeze can influence the mold height. In an automatic mold height control system, the mold height is measured continuously, and the data is fed to a programmable logic controller (PLC). The squeeze board position is adjusted for the next mold based on the established relationship between mold height and squeeze board position. Figure 4.15 (Ref 5) is a schematic illustrating this concept.

4.3.4.4 Pattern Plate Heater

Pattern plates are made of aluminum or cast iron, both of which have good thermal conductivity. Prepared molding sand tends to stick to the match plate if the match plate is colder than the molding sand. This can be avoided by installing a radiant heater above the match plate to keep it warm. This approach is useful especially in the winter months.

4.3.4.5 No-spill Sand—Horizontal Parting Plane

The volume of sand is controlled by adjusting the position of the mold frame in the machine up or down. This helps to eliminate the potential for sand spills.

4.3.4.6 Leveling Frame Pattern Draw

The four-pin system has been used historically to lift the mold up from the pattern plate. A leveling frame is a significant improvement over the pin lift system to allow for even stripping without any mold damage. The leveling frame is adjusted by a leveling cylinder as illustrated in Fig. 4.16 (Ref 5).

4.3.4.7 Multiram Squeeze

A flat squeeze head tends to produce a non-uniform mold density, especially with a varied contour of the pattern. A squeeze head with a bank of individual heads to suit the pattern contour improves the ability to produce uniform mold density. Figure 4.17 illustrates this approach (Ref 5).

4.3.4.8 Light Curtain Safety Feature

The light curtain is a feature that protects machine operators during the changing of the pattern plates, ensuring that the machine does not automatically start during the changing procedure. The light curtain is a vertical light beam mounted on one side of the machine with a receiving sensor on the opposite side. This light beam is lit the moment a protective door is opened, and any interruption of the light beam by an operator stops the machine instantaneously, preventing injury.

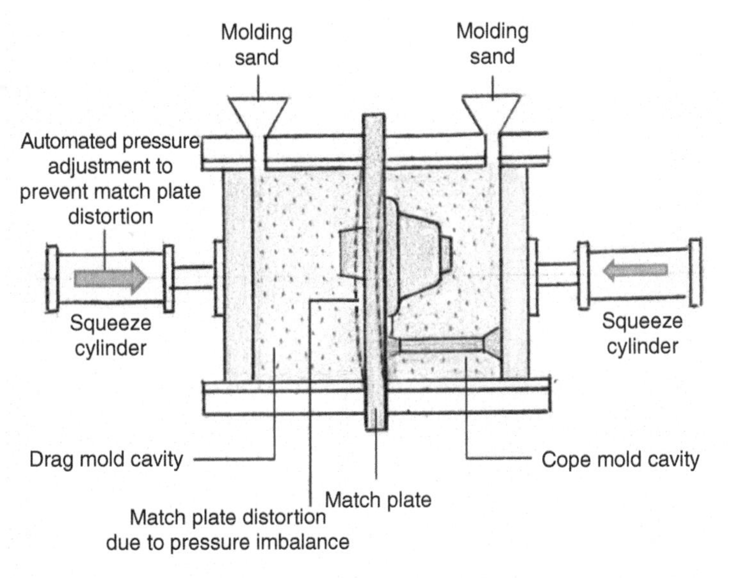

Fig. 4.14 Squeeze pressure balance control

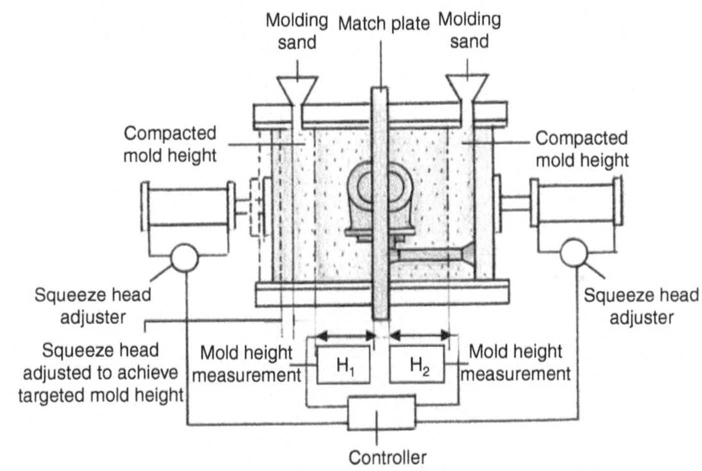

Fig. 4.15 Mold height control schematic

4.3.4.9 Automated Pattern Plate Change

The cope and the drag plates are mounted on a swivel mechanism so they can be indexed to the position alternately. The same concept is used when the pattern needs to be changed by mounting the next pattern on the empty station of the swing table. The next pattern indexes to the molding position automatically.

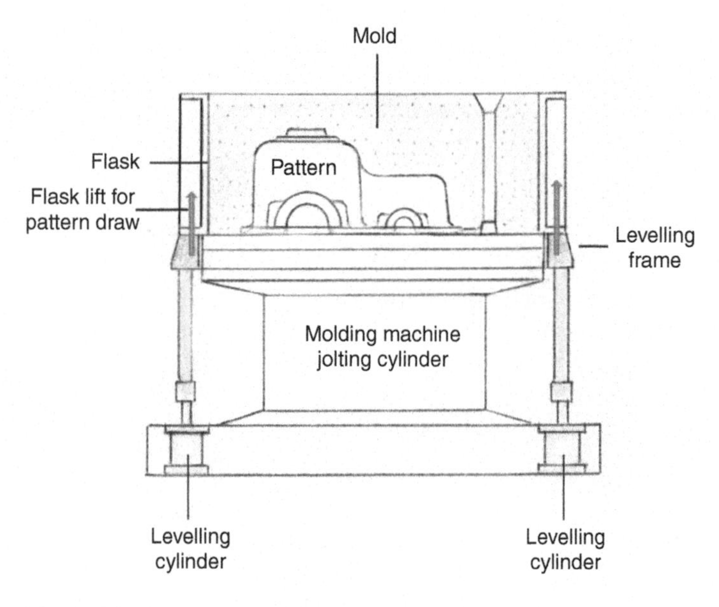

Fig. 4.16 Leveling frame for pattern draw

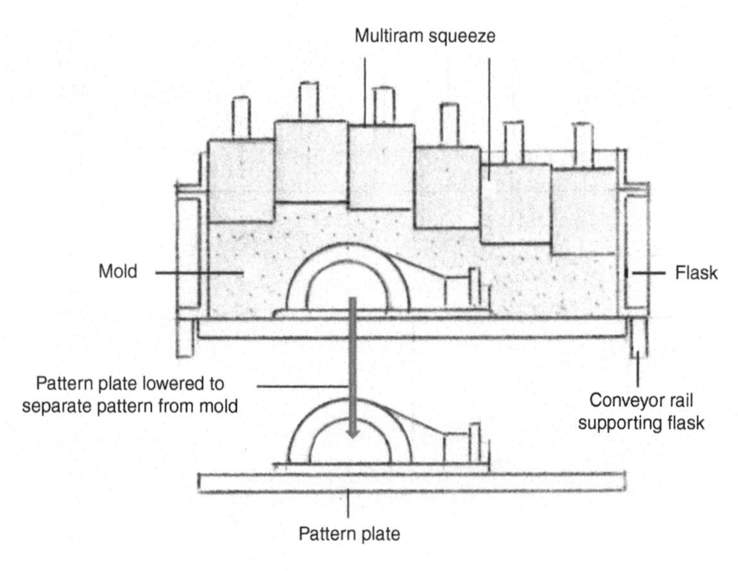

Fig. 4.17 Multiram squeeze concept

4.3.4.10 Molding Machine Controls and Diagnostics

Machines today have industrial grade, dust-proof, touch button controls and monitoring screens to indicate each operation. Troubleshooting software is also provided to locate machine malfunction and to enable quick diagnostics and rapid rectification.

Table 4.12 summarizes the features offered by one machine manufacturer (Ref 5). It provides an idea of the suitability for applications, flask sizes, and other technical features that contribute to mold quality.

4.4 Flaskless Molds

Flaskless molding with either horizontal or vertical parting is used in small and medium-sized molds for castings requiring simple or small-core assemblies. Highly automated molding machines that require minimum operator intervention for core setting are popular for high-volume captive foundries. These machines use hydraulic squeezing for mold production and are differentiated by the mode of sand filling.

Flaskless molds are classified into:

- Horizontally parted molds
- Vertically parted molds

4.4.1 Horizontally Parted Flaskless Molding

Figure 4.18 shows the sequence of operations in horizontal flaskless molding. When the stages of operations in the cycle are complete, the flasks are pulled up and out, leaving the flaskless mold assembly to be moved to the pouring line. The separation of the mold from the flasks is made easier by the activation of a pair of pneumatic vibrators that are mounted on the flasks.

In Fig. 4.18, station A shows a bottom flask placed on a machine table. Station B illustrates the positioning of a pattern plate over the flask. Station C shows the cope flask placed over the pattern plate. The assembly is rotated, and the drag is molded. At station D, a bottom board is placed over the drag mold. The assembly is rotated, and the cope is formed by squeezing (station E). The cope is lifted by a small hoist, and the pattern plate is lifted by the four pins on the machine table (station F). At station G, the cores are set, and the cope is closed. The assembly is clamped, and the cope and drag flasks are lifted out together, leaving the mold assembly on the bottom board (station H).

Table 4.12 Molding machine suitability and features

Suitability		Molds	Molding cope and drag		Sand fill type			Squeeze type		Draw type	
Speed and quality	Flask size, mm	/hr	Alternate	Simultaneous	Gravity	Aeration	Airflow	Level	Multiram	Standard	Levelling
Small and medium castings,uniform density with aeration, highly strong molds	700 by 650, min 1500 by 1200, max	150 max	X	...	...	X	...	X	...	X	X
Medium and large castings, high automation, highly flexible, suitable for accurate complex geometries	500 by 400, min 3000 by 2000, max	140 max	X	...	X	...	X	...	X	X	X
Medium sized castings, high automation, high speed, highly flexible, suitable for accurate and complex castings	500 by 400, min 1250 by 1000, max	250 max	...	X	X	...	X	X	X	X	X

Source: Ref 5

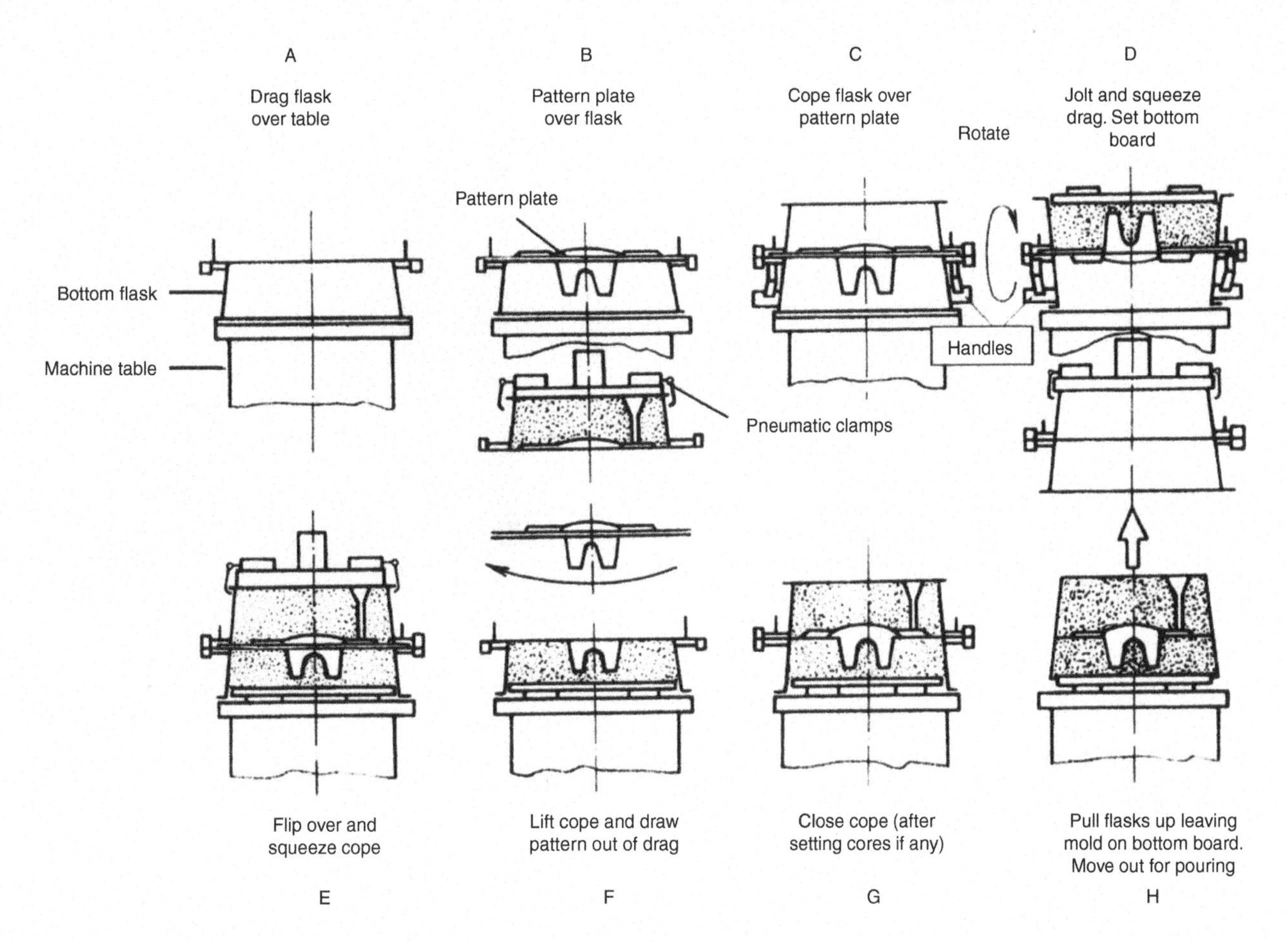

Fig. 4.18 Horizontal flaskless molding schematic

The release of the mold is assisted by using a vibrator on the flasks. The assembled mold moves to the pouring station, where a jacket is slipped in as a precaution against metal leakage at the parting plane and a mold weight is placed over the mold to prevent any mold lift due to the buoyancy of the liquid metal. The poured molds are now ready to cool and move to the shakeout station.

The taper on the walls of the flasks is about 9 to 10 degrees. A snap flask can be used where the side walls open slightly, making the ejection of the molds from the flasks easier. Figure 4.19 illustrates this concept.

Each of the drag and cope flasks is constructed with a hinge in one corner and a cam-actuated mechanism at the opposite end. A coil spring is inserted at the mating corner face to keep the two faces apart at the cam end. The cam lobes provided on each half are squeezed by a toggle lever to close the gap. More fully automated molding machines use a pneumatic cylinder to squeeze the two corners together in place of a manual toggle lever.

Highly automated high-production units have been in operation for many years using the horizontally parted flaskless molding process. Recent improvements and innovations include:

- Sturdier six-piece machine frame with accessible guide rods with linear bearings
- Open design concept for operator access from either side for easier maintenance
- Gravity fill hoppers on linear bearings to reduce maintenance costs
- Sealed linear bearings and magnetic cylinders without rods for reduced maintenance costs
- Unique cope lift and improved guidance for guide rods
- Innovative cope lift system to reduce cycle time
- Programmable pattern spray and pattern plate heating
- Blow or aeration mold fill, instead of gravity fill

Figure 4.20 (Ref 4) illustrates a high-performance molding machine using match plates.

Table 4.13 provides the sizes of the machines and the mold output per hour (without cores).

(Figure 4.21 (Ref 6) is a schematic of a high-production unit suitable for components with simple or small core assemblies.

In Fig. 4.21 the machine layout consists of three connected sections, each supported by pillars at the corners. Section 1 consists of a mounting for the cope flask at the top and a squeeze press at the bottom. Section 2 consists of the drag flask with pattern plate, both mounted on a rotating device to flip over the drag and the pattern plate from section 2 to section 1. Section 3 consists of a conveyor for feeding the bottom boards over the inverted bottom flask, as shown in A. An overhead hopper mounted on a trolley moves between sections 1 and 2 to distribute sand to the drag and cope by aeration filling. The closed hopper is pressurized for sand to flow through ports at the bottom of the hopper. The hopper is mounted on load cells programmed to

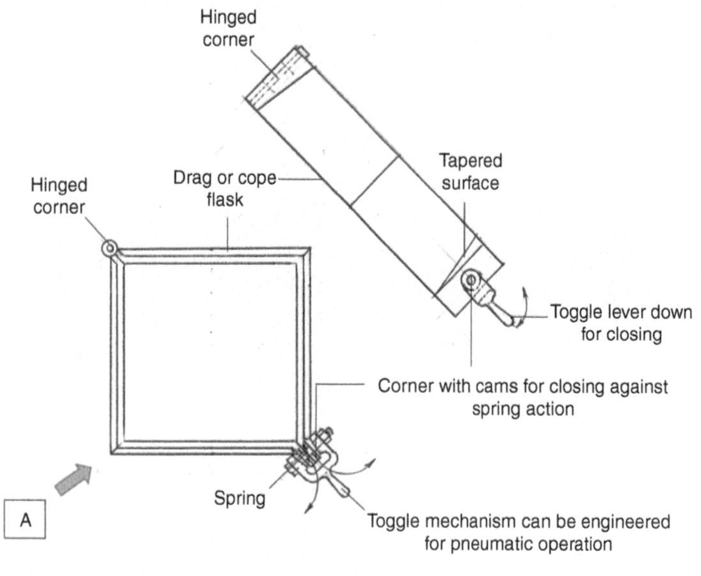

Fig. 4.19 Snap flask concept

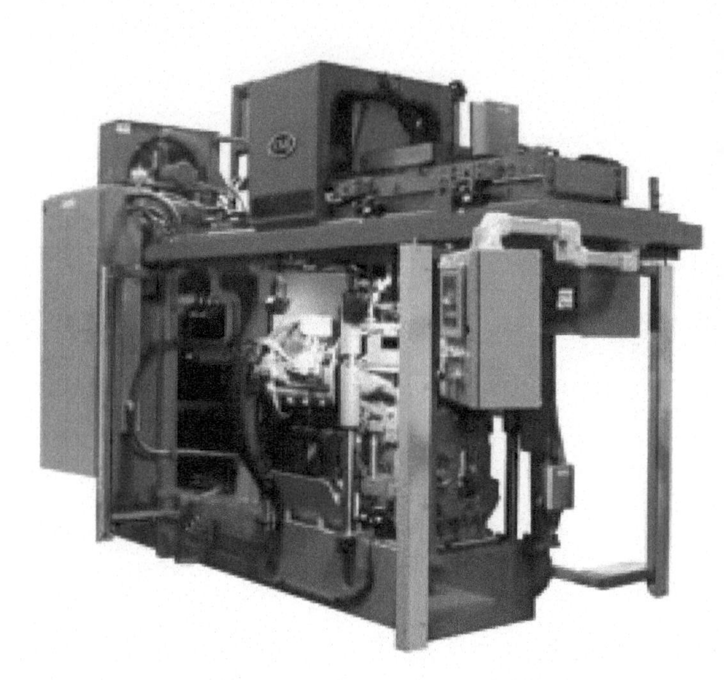

Fig. 4.20 Automatic match-plate molding machine. Source: Ref 4, EMI

Table 4.13 Automatic match-plate molding machine sizes

Model	1419	1620	2024	2026	2430	3032
Mold size, mm (in.)	14 by 19 (355 by 480)	16 by 20 (406 by 508)	20 by 24 (508 by 610)	20 by 26 (508 by 660)	24 by 30 (610 by 760)	30 by 32 (760 by 813)
Mold height, cope/drag, mm (in.)	5.5/4.5 (140/114)	6.5/5.5 (165/140)	6.5/5.5 (165/140), 8.5/7.5 (216/190)	12/11 (305/280)	10/9 (254/228), 12/11 (305/280)	12/11 (305/280)
Mold output rate, without cores	200	180	180	180	140	100

Source: Ref 8

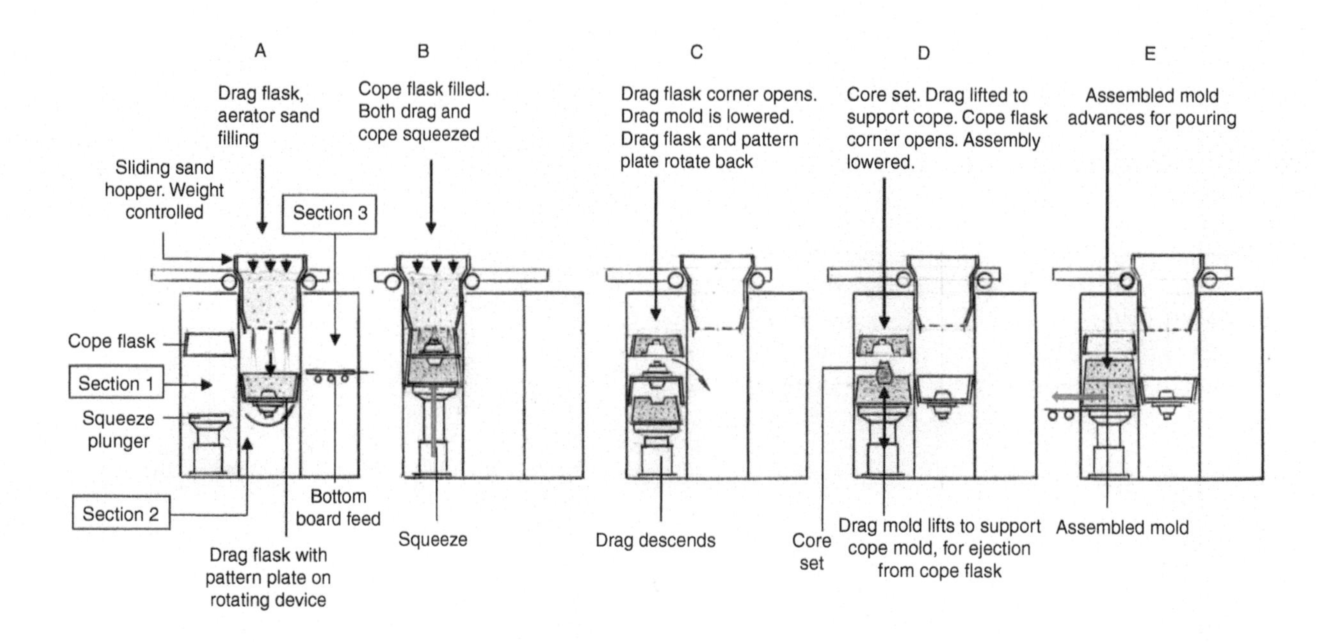

Fig. 4.21 Schematic of a high-production flaskless machine

unload only the targeted amount of sand. Both the flasks have pneumatic cylinders to unclamp the corner for releasing the molds.

The machine sequence is:

1. Events in A
   - The drag flask is inverted with the pattern plate mounted as shown in section 2.
   - The hopper fills the drag flask with prepared sand.
   - The bottom board is pushed to cover the drag flask filled with sand.
2. Events in B
   - The drag flask with sand and bottom board is inverted over to section 1.
   - The hopper moves over to section1 and engages with the cope flask in section 1.
   - The hopper is pressurized to fill the cope flask.
   - The hydraulic squeeze piston moves up to compact the sand in both the flasks.
3. Events in C
   - The drag flask-clamping cylinder is actuated to open the flask corner, releasing the mold.
   - The hydraulic plunger moves down, taking the drag mold with it.
4. Events in D
   - The drag flask rotates to section 2, making room for the drag mold.
   - The plunger moves up to the height suitable for core setting.
   - Cores are set.
   - The plunger moves farther, taking the drag mold to touch and support the cope mold.
5. Events in E
   - The pneumatic cylinder clamping the cope flask corner is actuated to open the flask, releasing the cope mold.
   - The plunger moves down, supporting the mold assembly, out of the cope flask.
   - The mold moves on the conveyor swung into position, taking it for preparation for pouring.
   - Jackets are fitted and weights are placed over the top of the mold before pouring.

One of the major machine developments is the blow fill mechanism, which is used to fill both the top and bottom flasks simultaneously (Ref 6). The time for drag transfer is saved, and simultaneous filling of both the top and bottom flasks reduces cycle time. This type of filling is suitable for deep mold cavities, as the potential for pattern plate distortion is reduced. Ref 6 provides the highlights of such machines with blow fill of both the drag and cope mold cavities simultaneously. Other innovations of these machines include:

- High-pressure hydraulic squeeze for more accurate molds
- Adjustable squeeze pressure and increased molding speed
- Mold height adjustment for varied parts
- Light guard or light curtain for operator safety

The ability to extend mold cooling time is important, depending upon the casting thickness and the casting weight. Figure 4.22(a) (Ref 6) shows an in-line installation of multi stacks for molds to increase the cooling time from 30 to 90 minutes, conserving space. Figure 4.22(b) (Ref 6) illustrates the development of a rotary table with multistacks (up to three stacks) for increasing mold cooling time after pouring. Jacket setting and mold loading are automated prior to the pouring operation.

Transfer mechanisms for stacking the molds after pouring and unstacking for moving to the shakeout station have been developed, saving space and for simplified fume extraction, resulting in an improved environment.

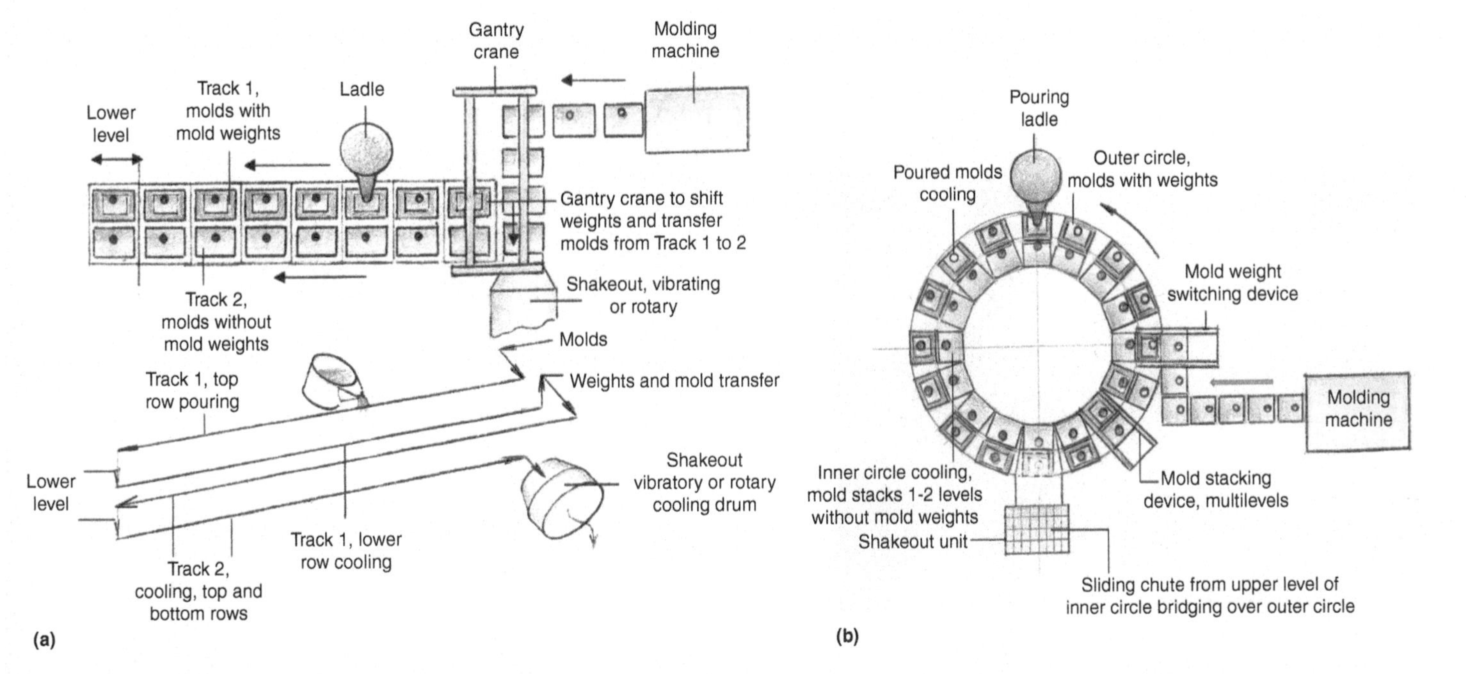

Fig. 4.22 Multistack rotaries; (a) conveyor for extended cooling time, (b) table for extended mold cooling

4.4.2 Flask and Flaskless Flexible Molding Machines

Section 4.3.1 addressed the benefits and drawbacks of tight flask and flaskless molding. One manufacturer (Ref 7) has engineered a flexible molding system that accommodates both options. The molds are punched out of the special flasks before pouring, which provides the advantage of loading a rotary unit for the mold shakeout operation.

When the system is used for molding in flasks, a punch-out unit and a vibratory shakeout system are engineered for separating the castings from the mold. The flasks return to the molding station. Figure 4.23 is a schematic illustrating the features of this system (Ref 7).

The flex line uses the same equipment for making the drag and the cope alternately, reducing the expense. The drag and cope pattern plates are mounted on a swing arm mechanism that permits the indexing of the drag and cope pattern plates alternately, as shown in A. The sequence of operations is:

A. The drag and cope pattern plates swivel into molding position alternately.
B. The dosing sand hopper is positioned over the drag flask, and sand is blown. The squeeze plunger moves up, compacting the sand.
C. The mold moves to the rollover and draw station, where the flask rolls over and the pattern plate is drawn out. The cope flasks get the sprue and vents drilled from the bottom.
D. Cores are set in the drag mold.
E. The cope mold is rolled over and closes over the drag, using the guide pins on the flask.
F. The molds are pushed out of the flasks for flaskless molding. For tight-flask molding, the molds stay in the flasks. Both the jacket and mold weight are placed automatically.
G. Molds are poured.
H. The jackets and weights are removed from the mold. Flaskless molds are tipped to enter the shakeout and cooling drum.
I. Castings are separated from the sand and move over a conveyor for shot blast cleaning. Molds produced in tight flasks are punched out, and castings are separated from the sand over a vibratory grid.

4.4.3 Vertical Molded and Horizontally Cast Molding Machine

Some casting configurations have a larger core or a heavier core assembly which needs adequate support. Figure 4.24 illustrates a machine concept addressing such a need (Ref 7). The two halves of the mold are simultaneously filled and squeezed, which reduces the cycle time and conserves capital.

The sequences of operation are summarized as:

A. The pattern plate is encased with the boxes, as illustrated.
B. The boxes are turned over by 90°. The sand hopper moves over the boxes to align the sand inlet ports, and it is pressurized to blow sand into the two boxes.
C. The sand inlet ports are closed, and the sand is squeezed in both directions to compact the mold.
D. The boxes are turned by 90° to the horizontal position. The cope mold is lifted off the pattern plate, which pattern plate is drawn off the drag mold.

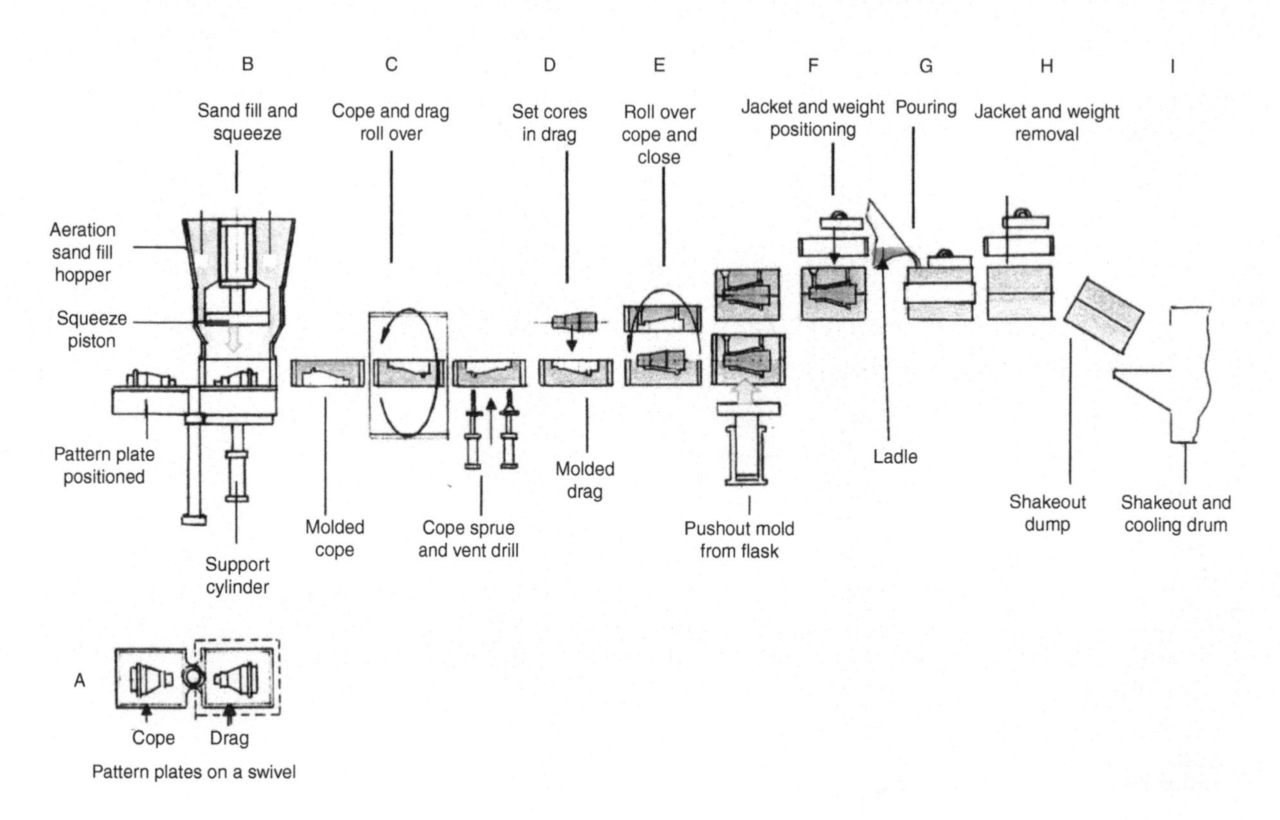

Fig. 4.23 Flex molding machine schematic

E. The cores are set into the drag mold, and the cope mold is closed.
F. The boxes are lifted off the mold. A jacket is slipped on, and the weights are placed over the mold. Then the mold is poured, the casting cools in the mold, and the mold moves to the rotary shakeout and cooling drum, where the castings and sand are separated.

4.4.4 Vertically Parted Molding Machines

Vertically parted molds offer high productivity with high automation; they are typically used for small to medium parts. Both un-cored and cored parts can be produced efficiently using this method. Many machines are in use for gray, malleable, and ductile iron castings. A smaller number of machines are used to produce brass and bronze castings. Aluminum castings can also be produced using vertically parted molding machines, and a few lines have been used for producing steel castings. These popular molding machines are often used for castings such as automotive, fittings, air conditioning, and cookware. Larger machines are typically used for truck markets, bigger municipal castings such as manhole covers, and parts for stoves. Vertically parted molding lines are used for long and short series operations because the pattern plates can be changed within 2 to 3 minutes, adapting to the just-in-time principle required for the shorter series.

An overview of a vertically parted molding line can be seen in Fig. 4.25(a) (Ref 7). The major elements of the molding line are indicated.

The details of the working principle of the machine are illustrated in Fig. 4.25(b) (Ref 7).

Two separate pattern plates are used for molding. One pattern plate is attached to the pressure plate (red semi-circle on the right side); this is called the pressure plate pattern. The other pattern plate is attached to the swing plate (red flat part on the left); this is called the swing plate pattern. In the vertical position, the swing plate is rigid to withstand the squeezing (compacting) operation and the squeezing forces from the pressure plate. The swing plate swings to the horizontal position after the mold has been shot and squeezed to enable the mold to advance.

The sequence of operations is:

1. **Sand shot:** Sand is blown into the molding chamber formed by the two vertical pattern plates and the molding chamber sides and top and bottom walls.
2. **Mold squeeze:** The mold is squeezed either by one of the pattern plates or both pattern plates moving toward the center of the molding chamber.

3A. **Moving forward in the chamber:** This optional operation is for moving the mold forward in the chamber after

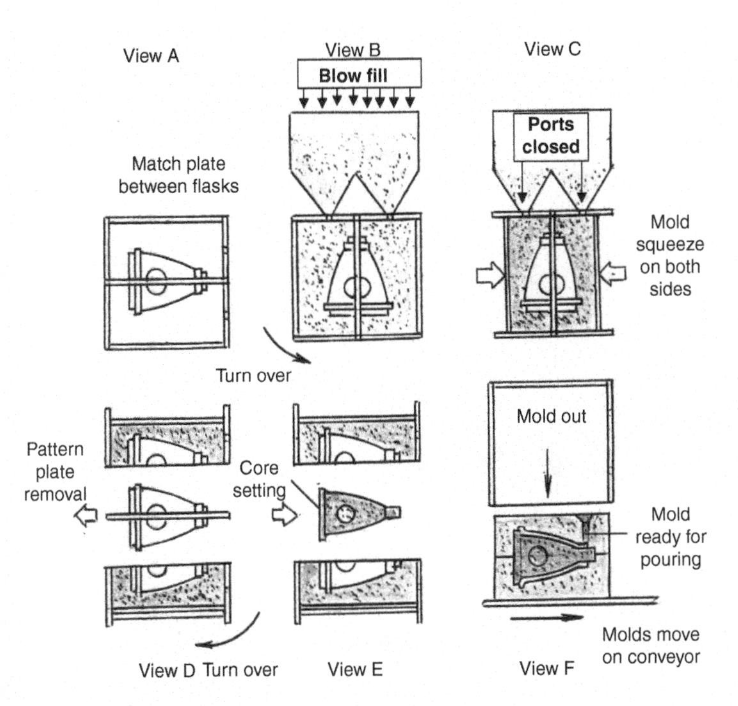

Fig. 4.24 Vertical molding and horizontal pouring concept

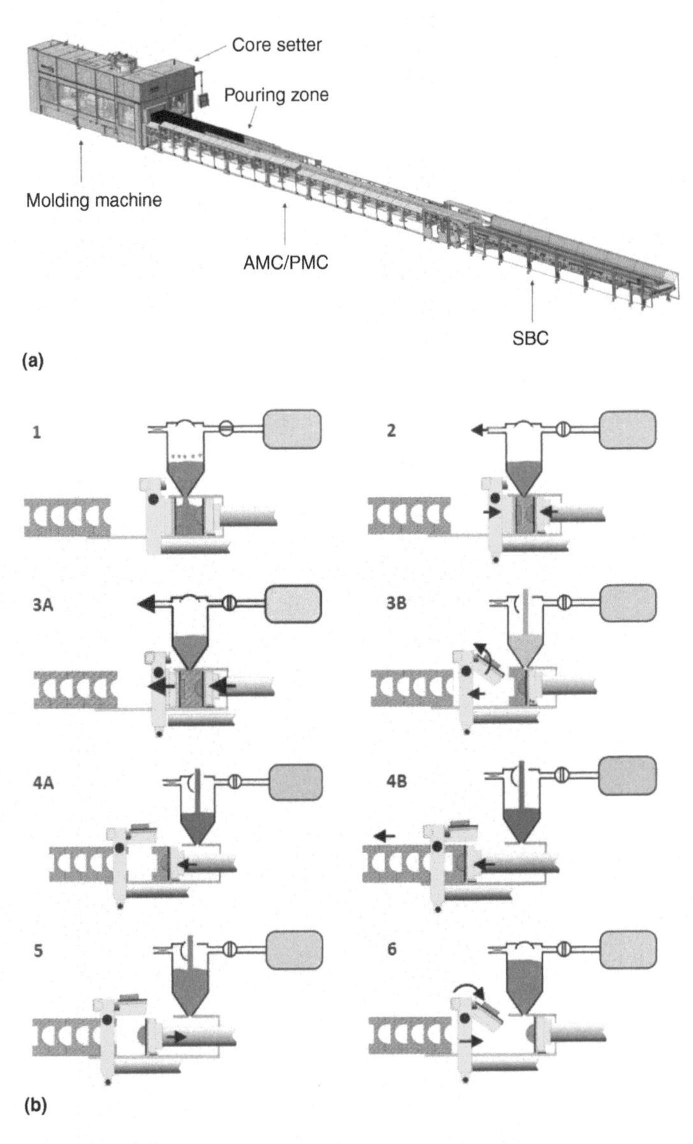

Fig. 4.25 Molding machine; (a) overview of vertical molding line, (b) working principle of vertically parted. Source: Ref 7, DISAMATIC

squeezing. This option ensures the reduction of possible loose sand in the swing plate side mold impression.

3B. **Stripping of the swing plate:** The hinged swing plate moves away to strip the mold and swings to the horizontal position to make room for the mold to advance.

4A. **Mold closing:** The pressure plate pushes the mold out of the chamber and moves it forward to mate with the previously produced mold.

4B. **Mold assembly transport:** The pressure plate advances farther, and the mold assembly is moved by one mold thickness to the left. The molds are assembled back-to-back; they create cavities for the liquid metal to be poured between each mold. Both sides of the mold are used for forming the mold cavity, which results in 101 mold halves to produce 100 molds. Closed molds are progressively poured at the same speed as they are produced.

5. **Stripping off the pressure plate:** The pressure plate withdraws after the pattern is separated from the mold.

6. **Closing the mold chamber:** The swing plate moves back and swings down to the vertical position, and the molding chamber is closed and ready for the next mold to be produced.

Figure 4.26 (Ref 7) provides some details of the machine construction and features. The main features of the machine are indicated.

The molding machine is built on a base frame onto which the molding chamber and other vital components are bolted. The swing and pressure plates are guided by heavy-duty tie rods or hydraulic cylinders similar to the pressure plate. Prepared sand is fed to the machine through the sand gate (5) seen on top right side in Fig. 4.26.

As mentioned previously, cored castings can be produced on vertically parted molding lines. In all cases, the core setting is performed by an automatic core setter (CSE). However, for very short series and prototypes or tests, the cores can be set manually.

The operation of the automatic core setter is presented in Fig. 4.27(a) (Ref 7).

The automatic core setter sets cores in the last produced mold while the molding machine is producing a new mold, without any loss of cycle time. Many automatic core setters use a *core mask*. Figure 4.27(b) illustrates a *core mask* with vacuum ports.

A core mask that holds the cores by vacuum is designed for each set of patterns. The positions of the cores must accurately match the positions of the cavities on the pattern plates. The core mask represents the negative impression of the swing plate pattern, or in other words, it represents the mold profile made by the swing plate.

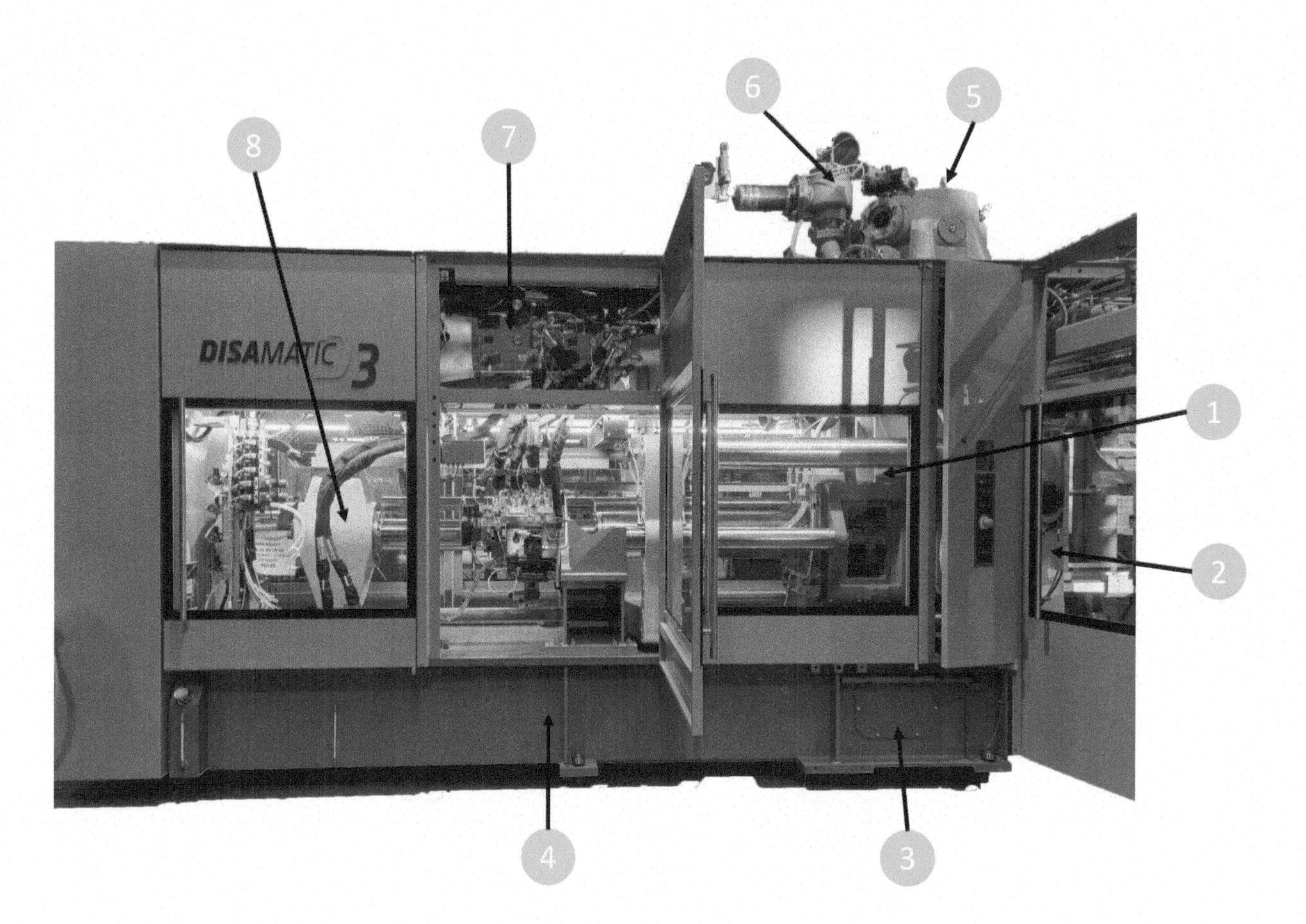

1. Molding chamber 2. Swing plate yoke 3. Sand spillage container 4. Bottom frame 5. Sand gate 6. Shot valve
7. Hydraulic pumps 8. Rear swing plate yoke

Fig. 4.26 Machine construction and features. Source: Ref 7, DISAMATIC

Hence, core masks are traditionally made by casting them in resin over the swing plate pattern, ensuring a perfect match with the positions of the cavities on the plate patterns. Different types of resins are used for casting core masks. The latest trend is to mill the core masks using CNC mills directly from the 3D CAD files with no or very little manual work, similar to patterns, instead of casting them in resin.

The core setter operates as the molding chamber is closed, with the last part of molding machine operation (6) overlapping with the first part of operation (3) (Figure 4.27(a).

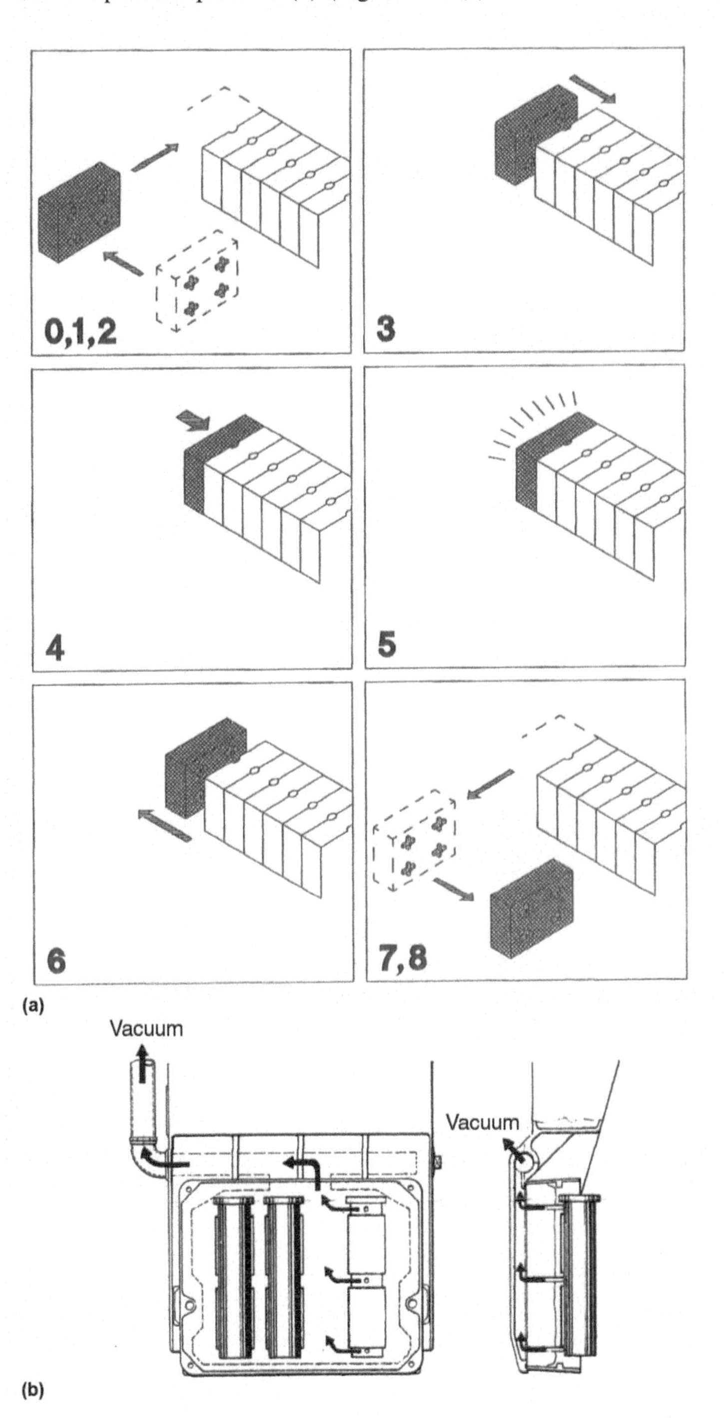

Fig. 4.27 (a) Working principle of the automatic core setter. (b) Core mask with vacuum ports. Source: Ref 7

The core setter operates on two sets of columns, making it possible for the sequence to move in the longitudinal direction of the mold assembly group and perpendicular to the mold string.

Referring to the stages in Fig. 4.27(a), which illustrate the working principle, the operation is:

Stage 0: The operator inserts the cores in the core mask and presses the start button of the core setter. The light curtain protecting the operator will be active from the moment the start button is pressed.

Stage 1: The core mask moves away from the light curtain.

Stage 2: When the molding chamber is closed, the core mask moves toward the molding line and stops behind the mold assembly row.

Stage 3: The core mask moves forward until it touches the last-produced mold.

Stage 4: The core setter sets the cores in the mold with a controlled setting force.

Stage 5: The vacuum is released for separation of the cores from the core mask.

Stage 6: The empty core mask moves away from the mold.

Stage 7: The core mask moves out toward its starting position. This makes it possible for the molding machine to close the mold where cores have just been set and continue the cycle of the molding machine.

Stage 8: The core mask moves forward toward the operator; the light curtain protecting the operator is switched off, and the operator can now insert cores in the mask for the next mold.

The movements of the molding machine and the core setter are synchronized. One unit will not start a new cycle until the other unit has completed one. This ensures that each mold is provided with only one set of cores.

4.4.4.1 Machine Sizes and Outputs

Table 4.14 (Ref 5) is an example of mold size options and corresponding machine parameters for the most widely used mold sizes for vertical molding. The machine offers five basic types: A, B, C, X, and Z. The available mold dimensions are shown in the table.

Vertical molding lines have chamber sizes from 500 mm (19.7 in.) wide by 400 mm (15.75 in.) high up to 1200 mm (47.24 in.) wide by 1050 mm (41.39 in.) high; they have a variety of speeds ranging from 150 up to 555 molds per hour, depending on the mold size and model.

4.5 No-Bake Molding Lines

The no-bake molding process is suitable for medium volume semicaptive markets for gray iron, ductile iron, compacted graphite iron, and steel. Cored castings with horizontally parted molds are good candidates for this process (Ref 4). There are, however, some exceptions. Some unique casting configurations lend themselves to molding horizontally but pouring vertically. The assemblies are clamped and turned by 90 degrees for pouring. Table 4.15 summarizes the major

Table 4.14 Typical vertical molding machine sizes and machine parameters

DISAMATIC D3		A	B	C	X	Z
Mold dimensions						
Height	mm	480	535	550	535	570
Width	mm	600	650	675	750	750
Thickness, rear sand slot	mm	150–395	150–395	150–405	150–395	150–405
Thickness, front sand slot	mm	120–395	120–395	120–405	120–395	120–405
Mismatch	mm	0.1	0.1	0.1	0.1	0.1
DISAMATIC D3-365						
Uncored	mold/hour(a)	365	365	365	365	365
Cored	mold/hour(a)	333	333	333	333	333
Cooling time, max	min(a)	77	77	77	77	77
Sand consumption, max	tons/h(b)	49	59	63	69	73
Power consumption	kW	55	55	55	55	55
Connected load	kVA	69	69	69	69	69
Compressed air use	Nm3/min	8	8	8	8	8
DISAMATIC D3-425						
Uncored	mold/hour(a)	425	425	425	425	425
Cored	mold/hour(a)	380	380	380	380	380
Cooling time, max	min(a)	66	66	66	66	66
Sand consumption, max	tons/h(b)	58	70	75	81	86
Power consumption	kW	55	55	55	55	55
Connected load	kVA	69	69	69	69	69
Compressed air use	Nm3/min	9	9	9	9	9
DISAMATIC D3-555						
Uncored	mold/hour(a)	555	555	555	555	555
Cored	mold/hour(a)	485	485	485	465	465
Cooling time, max	min(a)	49	49	49	49	49
Sand consumption, max	tons/h(b)	77	93	99	107	114
Power consumption	kW	60	60	60	60	60
Connected load	kVA	75	75	75	75	75
Compressed air use	Nm3/min	11	11	11	11	11
General parameters						
Conveyor length, max	m	86.5	86.5	86.5	86.5	86.5
Squeeze pressure	kp/cm2	1.5–16	1.5–16	1.5–16	1.5–15	1.5–15
Shot pressure	bar	0–4	0–4	0–4	0–4	0–4
Net weight	tons	21	21	21	21	21

(a) At 200 mm (7.9 in.) mold thickness, (b) at maximum mold thickness. Source: Ref 7, DISAMATIC

Table 4.15 Comparison of green sand and no-bake equipment and processes

Parameter	Green sand molding	No-bake molding
Annual volumes	Large and medium	Medium and small
Casting size	Small and medium	Medium and large
Complexity	Small and medium with less number of simple cores, preferred	Complex cores and core assemblies possible
Parting plane	Both horizontal and vertical	Horizontal mostly
Flask or flaskless	Both flask and flaskless	Flaskless mostly
Pattern equipment	Match plates and pattern plates made of cast iron and aluminum	Core-mold boxes made of wood and aluminum
Sand-binder mix	Sand + clay binder + water - mulled	Sand + binder + catalyst - mixed
Shelf life of sand mix	Hours depend upon rate of moisture evaporation	Minutes needed to compact sand before onset of setting
Mold stripping	Flask or mold moved away from pattern plate by machine	Clamped core-mold box moved away from mold by Roll-N-Draw machine
Cope reversal and cope closing	Roll over machine and Pick-N-place unit	Rollover and closing machine
Mold clamping and loading	Flasks clamped, flaskless molding-weights loading and unloading automated with jackets	Weights loading and unloading, automated
Shakeout	Flasks punched out plus sand broken on vibratory shakeout, easy breakdown of mold, flaskless molds tipped onto vibratory or rotary shakeout	Flaskless core molds tipped on to a vibratory or rotary shakeout. Lumps need more attrition for breakdown
Lump breakage	Easier	Harder, needs lump breakers
Reclamation	Dry or wet reclamation possible	Thermal reclamation, best choice

differences between the green sand and no-bake equipment and processes.

There are three distinct operations unique to no-bake lines, compared to green sand molding lines:

- Mold stripping—separation of the core mold from the core box by rollover and draw machines
- Mold handling—reversal of the cope and closing over the drag after core setting by mold manipulators or mold handlers
- Mold coating and drying—for steel castings

Mold stripping: The no-bake process uses a large core box to produce a core mold. A device is needed to turn over the core box and lift and separate it from the core box. After the core mold is stripped or separated, it is pushed onto a roller conveyor, and the machine brings the core box back to the initial position for sand packing in the next cycle. The machine typically used is known as a rollover draw machine.

Mold handling: The orientation of the core mold from the rollover draw machine is suitable for core setting. But the cope mold stripped by the rollover draw machines needs to be turned by 180° for closing over the drag. A mold handler or a mold manipulator is used for turning over and closing the mold.

Mold coating and drying: Core molds need to be surface coated with a refractory mold wash to prevent sand burn or metal penetration for thick-walled iron castings and many steel castings. (The high temperature of steel needs higher refractoriness to prevent the metal from penetrating in between the sand grains. Metal penetration makes surface cleaning very difficult).

4.5.1 Rollover and Draw Machines

Rollover and draw machines are configured in a few different ways. Figure 4.28 is a schematic of different types of rollover and draw machines. Each type has advantages and drawbacks:

- The single-column machine is suitable for relatively smaller mold sizes.
- The single-column swing type offers the advantage of moving from one conveyor to the other parallel conveyor. Small and medium mold sizes are configured using this type. This may suit some layouts; molds are lined up for pouring from a monorail or a gantry crane.
- The double column rollover and draw is suitable for medium-sized molds. But the machines of types in Fig. 4.28(d) and (e) are more popular.
- The full-barrel type is suitable for medium- to large-sized molds because the rigid frame is very stable for drawing larger molds.
- The open C-type is popular for medium-sized molds. The open side of the C-frame allows an option of molds exiting at 90° to the direction of entry. The C-frame also provides better access for servicing.

The mold or core box clamping system is variable and flexible to accommodate different mold or core box sizes. This is a significant advantage in jobbing and low-volume casting operations.

These can be integrated with existing or new roller conveyors.

Figure 4.29 provides the overall sizes of one unit of type Fig. 4.28(e) (Ref 8).

Figure 4.30 illustrates the working of the rollover and draw machine. Figure 4.30(a) shows the compacted core box with the bottom board inserted manually (or automatically for large-volume production). The clamping cylinders advance to clamp the core mold along with the bottom board. The machine rolls over, as shown in Fig. 4.30(b). The clamping cylinders are lowered to strip the core mold from the core box. The core mold moves onto the roller conveyor, and the machine is then rolled back in preparation for the next cycle.

Figure 4.31 is a large full-barrel-type rollover and draw machine (Ref 9). Table 4.16 lists the major dimensions of the available rollover machines (Ref 9).

4.5.2 Mold Manipulators

Mold manipulators are suspended on a gantry crane or a trolley for operator positioning. The handlers consist of two hydraulically operated grips for safe handling and an electric or hydraulic motor drive for rotation. They are usually operated by one operator for medium-sized molds and by two operators for large-sized molds. The supporting structure can be an A-frame or a beam. Figure 4.32(a) illustrates an A - frame type, and Fig. 4.32(b) is a typical beam-type handler. The beam-type handlers are also used for manipulating large flasks in sand molding by clamping on the trunnions and rotating.

Table 4.17 lists the A-frame and beam-type sizes of one manufacturer (Ref 10). Figure 4.33 lists the beam-type handler sizes from another manufacturer (Ref 9).

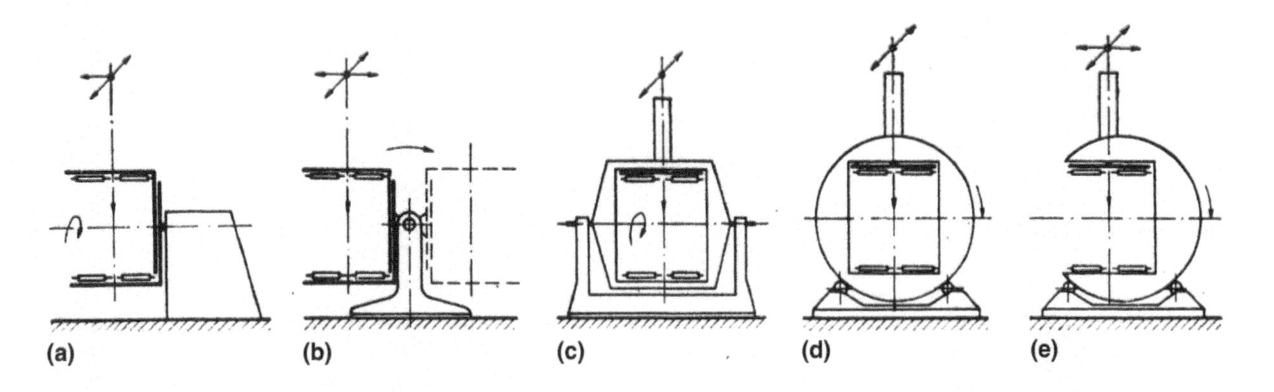

(a) Single column rollover and draw
(b) Single column swing-over and draw
(c) Double column rollover and draw
(d) Full barrel rollover and draw
(e) Open C- frame rollover and draw

Fig. 4.28 Types of rollover and draw machines

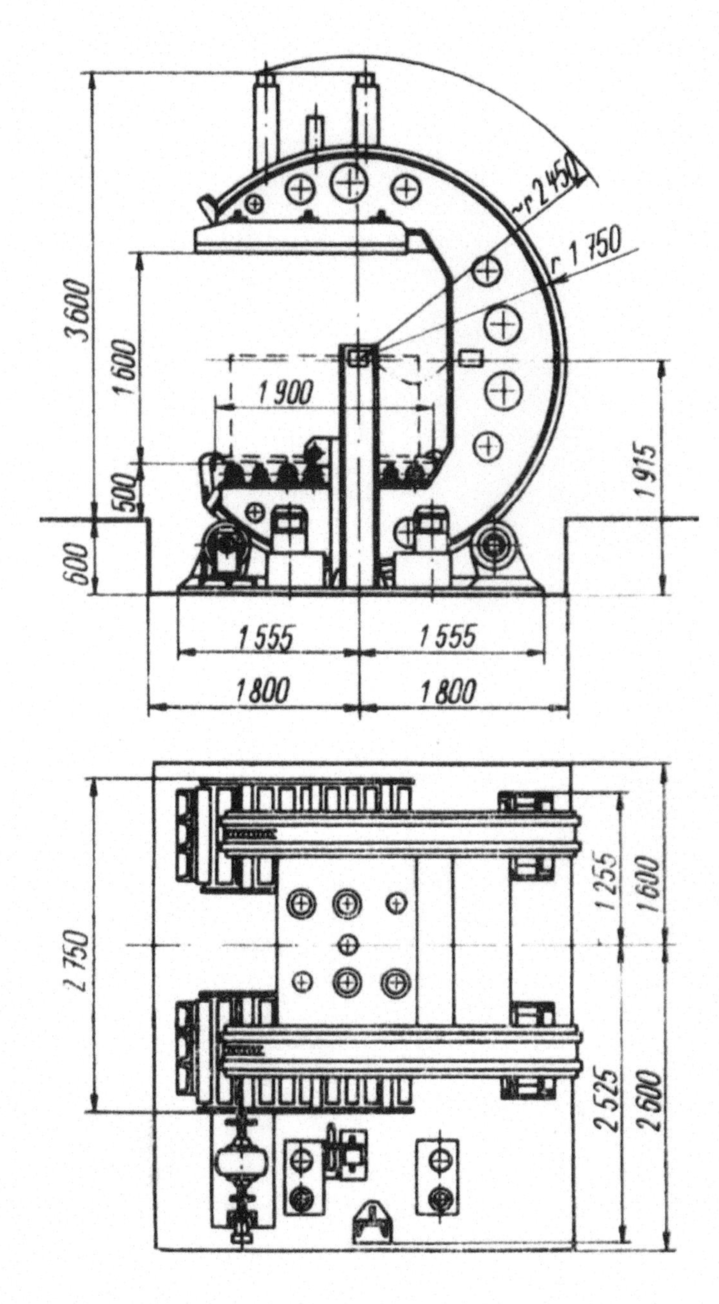

Fig. 4.29 C-type rollover and draw machine

4.5.3 Flow Coating and Drying

Molds can be spray coated or flow coated. Spraying requires considerably high skill for good coverage without excessive thickness. Flow coating requires less skill and demands less control of the viscosity. Flow coating has become more popular for coating applications in steel casting operations. Figure 4.34 is a schematic of tilt flow coating equipment.

Flow coating machines are designed to clamp the core molds, tilt, and position them to the core mold coating operator in an orientation that provides good access to varied cavity profiles ergonomically. These are designed as:

- Tilt flow coating
- Rotary flow coating

The molds are clamped by heavy-duty hydraulic cylinders for operator safety. Tilting is accomplished by means of two hydraulic cylinders, one on each side. The molds are tilted by nearly 90° for coating and about 110° to drain the coating. A drain trough catches excess mold coating and drains it back to the sump. A diaphragm pump delivers the coating to the operator's wand.

The diaphragm pump is equipped with an equalizer to ensure a steady mold coating flow without pulsation. A flow at a pressure of 0.1 to 0.2 bar flowing through a 12 mm (0.5 in.) diameter hose adjusted to reach 200 to 500 mm (7.87 to 19.7 in.) should be suitable for manual flow coating (Ref 11).

Mold coatings are water-based, and the core molds need to be heated in a drying oven to expel the moisture. Environmental regulations do not permit the use of the alcohol-based mold coatings used in previous years. Sprayed alcohol-based coatings can be dried by lighting them with a flame. Water-base coated core molds are dried in low-roof drying ovens with good temperature control and plenty of air for circulation. The oven is set to operate between 120 and 150 °C (248 and 302 °F). It is important to program the continuous ovens for slow initial heating to avoid the potential for trapping moisture below the skin of the coating.

Adequate coating draining time should be allowed by tilting the core mold to an angle of about 110° to drain the excess coating into the catchment trough and later into the sump. The core molds are then laid flat on the pallet to be moved into the drying oven.

Beam-type mold handlers can also be used for manipulating the core molds for coating. Dedicated equipment will ensure a steady movement of core molds to keep up with the production flow.

Rotary flow coating machines are similar to the rollover and draw machines but much simpler. Two sturdy guide rings connected by cross members are supported on rollers. The cross members support a vise for gripping the core mold. Urethane pads are used on the vise bars to grip the core molds.

The tilting of the core mold is accomplished by a heavy-duty hydraulic motor mounted with a reduction gear drive for smooth operation. Smaller molds can use pneumatic cylinders for tilting. Some manufacturers (Ref 11) provide dual tilting speeds.

The cores coated at an angle of 90° are usually tilted to an angle of 110° to 120° after the coating, allowing time for the excess coating to drain off. Coating manufacturers (Ref 12) recommend tilting the core during coating at an angle of 105° to 120° to achieve uniform coating thickness and to avoid tear drops.

Figure 4.35 shows a layout of a no-bake core molding line for a steel foundry producing medium-sized steel castings such as the valves used in the oil, natural gas, and chemical industries.

Core molds are produced using a continuous sand mixer. A few core boxes cycle in sequence around a square loop, as shown. (A large turntable with 6 or 8 stations replaces the square loop if a rotary loop is chosen). Ready-to-use sprue sleeves made of shell cores or compacted fiber and exothermic and insulating feeder cores are stored for ready use. The drags and copes are molded alternately. Packed core molds move in a square pattern using an automated transfer car.

The core molds harden in the square loop before entering a rollover and draw machine. The operator picks up a wooden

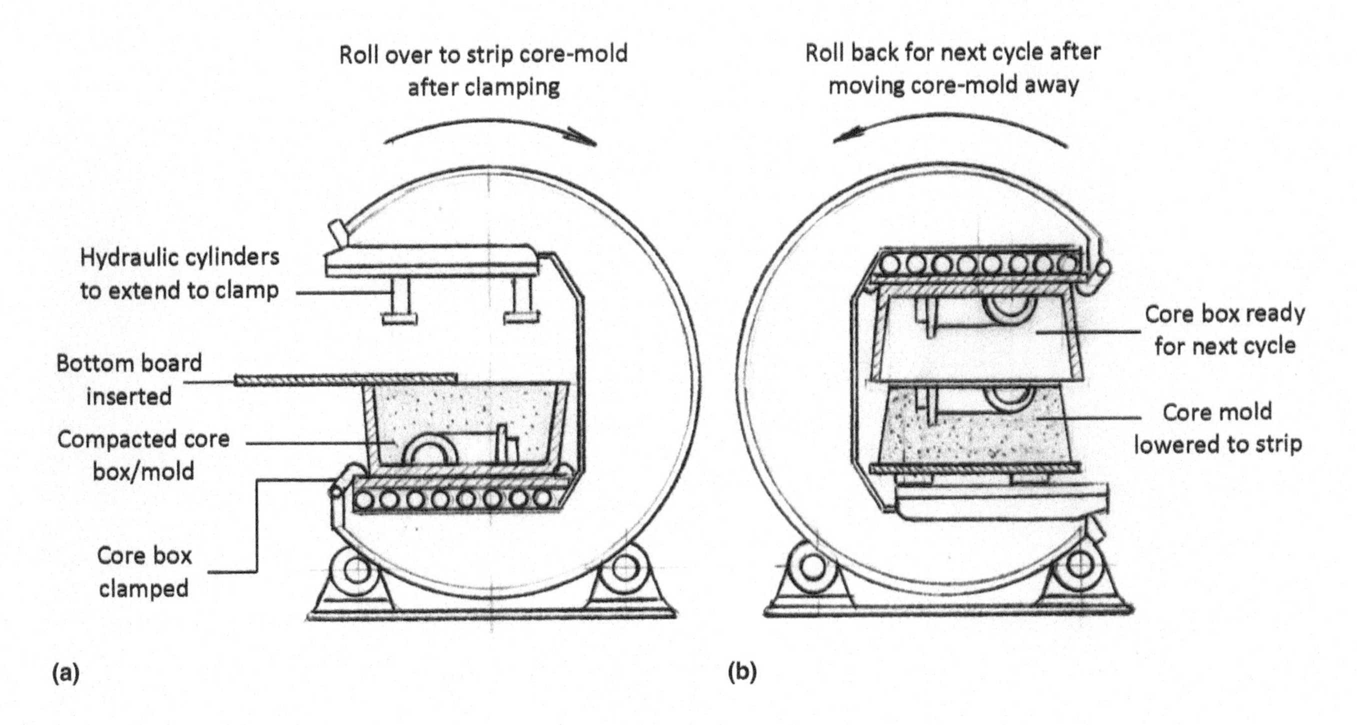

Fig. 4.30 Rollover and draw machine schematic

Fig. 4.31 Large full-barrel-type roller and draw machine. Source: Ref 9

pallet and covers the core mold before rolling and drawing. The core molds advance on the conveyor, which is programmed to reverse and deliver the used core box back to the loop to join the fleet. The core molds then move to the flow-coating machine, where an operator flushes them with mold coating, covering the areas that contact molten steel in the mold.

A mold manipulator acts as a transfer unit to switch the wooden pallets onto a cast iron pallet. The wooden pallets go back to the rollover and draw machine operator. Core molds on

Table 4.16 Roll over machine sizes

	Size							
Model	**JRD-1**	**JRD-2**	**2**	**3**	**4**	**5**	**6**	**7**
Length mm (in)	1000 (40)	1500 (59)	1000 (40)	1200 (48)	1600 (63)	2000 (79)	2500 (98.5)	3300 (130)
Width mm (in)	700 (28)	860 (34)	800 (31.5)	1000 (40)	1200 (47)	1400 (55)	1800 (71)	2000 (79)
Height mm (in)	250 (10)	350 (14)	350 !14)	425 (17)	475 (19)	550 (22)	675 (27)	750 (30)

Source: Ref 9

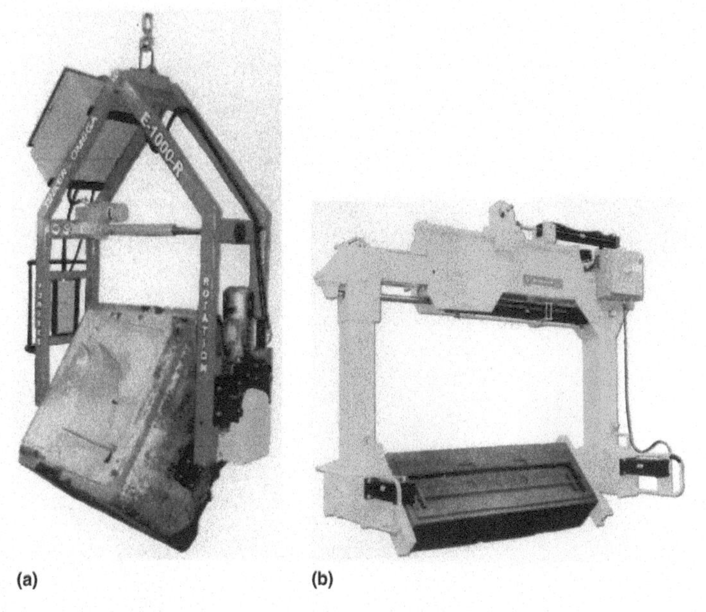

Fig. 4.32 Mold manipulators and handlers. Source: Ref 9, 10

the iron pallets enter the flow coating drying oven, where the moisture is expelled as the pallets move through the tunnel. Cores are set in the drag molds, and another mold handler closes the cope. A gantry crane automatically places weights over the mold. Molds are poured by an operator using a monorail to carry the ladle. The conveyor length is designed to allow for adequate cooling time.

The pallet is tipped to dump the poured core mold onto a vibratory shakeout. A jib crane assists in picking up the casting from the shakeout grid. Sand flows down to a return sand conveyor for further processing for reuse. A vibratory conveyor moves the castings through a cooling tunnel and into the cleaning room. Pallets pass through a cleaning station where they are cleaned by blowing and swiping. The pallets then return to the pallet switching station, as shown.

Table 4.17 Beam-type mold handler sizes

TN/TF model number	Capacity, kg (lb)	Typical operating envelope, dimensions, mm (in.)	Weight, kg (lb)
2000	907 (2000)	762 by 1270 by 762 (30 by 58 by 30)	590 (1300)
4000	1814 (4000)	1270 by 2134 by 1066 (50 by 84 by 42)	771 (1700)
6000	2721 (6000)	1422 by 2438 by 1270 (56 by 96 by 50)	1043 (2300)
8000	3627(8000)	1930 by 2794 by 1270 (76 by 110 by 50)	1270 (2800)
10,000	4536 (10,000)	2286 by 3353 by 1372 (90 by 132 by 54)	1492 (3300)
12,000	5443 (12,000)	2286 by 3353 by 1372 (90 by 132 by 54)	1497 (3300)

Source: Ref 10

4.5.4 No-Bake Core-mold Assembly

The top and bottom core molds are separately produced, and French core pins are used for suitable matching. Truncated conical stubs are mounted in each of the core mold boxes. Two separate cores (illustrated in Fig. 4.36) are set in the drag. The top core mold handled by the mold manipulator is aligned to match and close over the French cores set in the drag core mold, as illustrated.

4.5.5 Mold Clamping and Weight Loading

The top mold tends to lift due to the metal head and the buoyancy of the liquid metal displaced by the cores. The lifting force is countered by the weight of the cope mold and the weight of the cores. The net lifting force is countered by clamping the cope and the drag flasks together in tight flask molding. In the case of flaskless and no-bake molding, a loading weight is placed over the top of the mold.

Tight flask molds are secured in one of two ways:

- C-clamps holding the trunnions of the two flasks together
- Dovetail clamps holding the flask flanges together

Both methods require operators to clamp the molds before pouring and unclamp them before shakeout. There are a few casting producers who have automated this operation. Low-volume casting operations and jobbing foundries normally choose one of these methods, which allows maximum flexibility in the location of the sprue.

In the case of a no -bake core mold system, the weights on the top of the molds must be designed to leave room for the sprue and pouring cup. This requires a fixed location for the sprue, which may be a constraint in the pattern layout.

Pneumatic 'A' Frame Manipulator

Model	Max Load		Min Clamp		Max Clamp		Max Swing Radius		Height to Pivot		Box Pivot Available	Powered Rotation
P350	771 lb	350 kg	12"	300mm	30"	750mm	11"	275mm	37"	925mm	No	No
P1000	2205 lb	1000 kg	19"	475mm	55"	1390mm	25"	625mm	65"	1650mm	Yes	No

Electric 'A' Frame Manipulator

Model	Max Load		Min Clamp		Max Clamp		Max Swing Radius		Height to Pivot		Box Pivot Available	Powered Rotation
E1000R	2205 lb	1000 kg	16"	400mm	53"	1350mm	31"	800mm	67"	1700mm	Yes	Yes
E1250R	2756 lb	1250 kg	22"	550mm	71"	1800mm	32"	825mm	71"	1800mm	Yes	Yes

Hydraulic 'A' Frame Manipulator

Model	Max Load		Min Clamp		Max Clamp		Max Swing Radius		Height to Pivot		Box Pivot Available	Powered Rotation
H2000	4409 lb	2000 kg	22"	550mm	59"	1500mm	43"	1100mm	80"	2035mm	No	Yes
H3000	6614 lb	3000 kg	30"	750mm	75"	1900mm	47"	1200mm	90"	2275mm	No	Yes
H4000	8818 lb	4000 kg	39"	1000mm	93"	2350mm	59"	1500mm	112"	2835mm	No	Yes

Hydraulic Beam Manipulator

Model	Max Load		Min Clamp		Max Clamp		Max Swing Radius		Height to Pivot		Box Pivot Available	Powered Rotation
B1500	3307 lb	1500 kg	31"	775mm	57"	1450mm	31"	800mm	47"	1195mm	No	Yes
B3000	6614 lb	3000 kg	37"	950mm	78"	2000mm	30"	750mm	48"	1215mm	No	Yes
B5000	11023 lb	5000 kg	37"	950mm	78"	2000mm	43"	1100mm	77"	1962mm	No	Yes
B7500	16535 lb	7500 kg	49"	1250mm	132"	3350mm	55"	1400mm	94"	2375mm	No	Yes
B10000	22046 lb	10000 kg	53"	1350mm	136"	3450mm	55"	1400mm	94"	2370mm	No	Yes
B15000	33069 lb	15000 kg	73"	1850mm	148"	3750mm	58"	1465mm	102"	2590mm	No	Yes

Fig. 4.33 Frame and beam mold manipulator sizes and capacities. Source: Ref 9

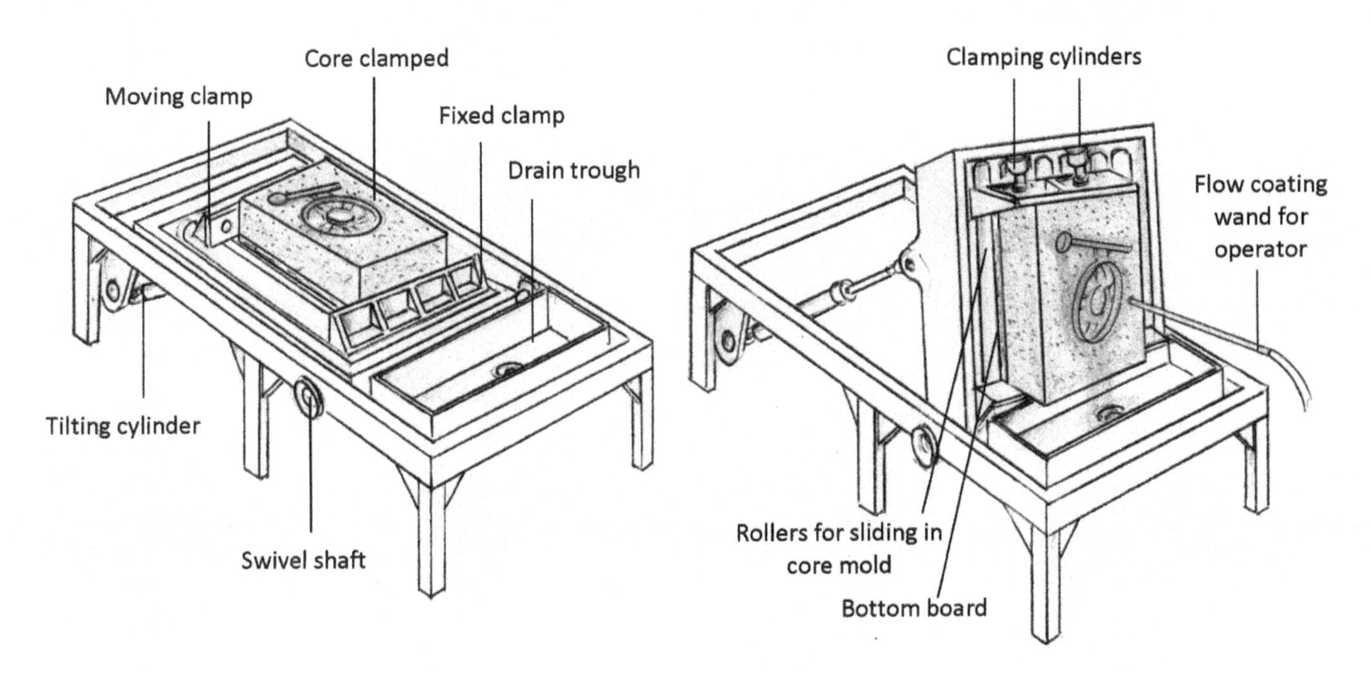

Fig. 4.34 Flow coating equipment

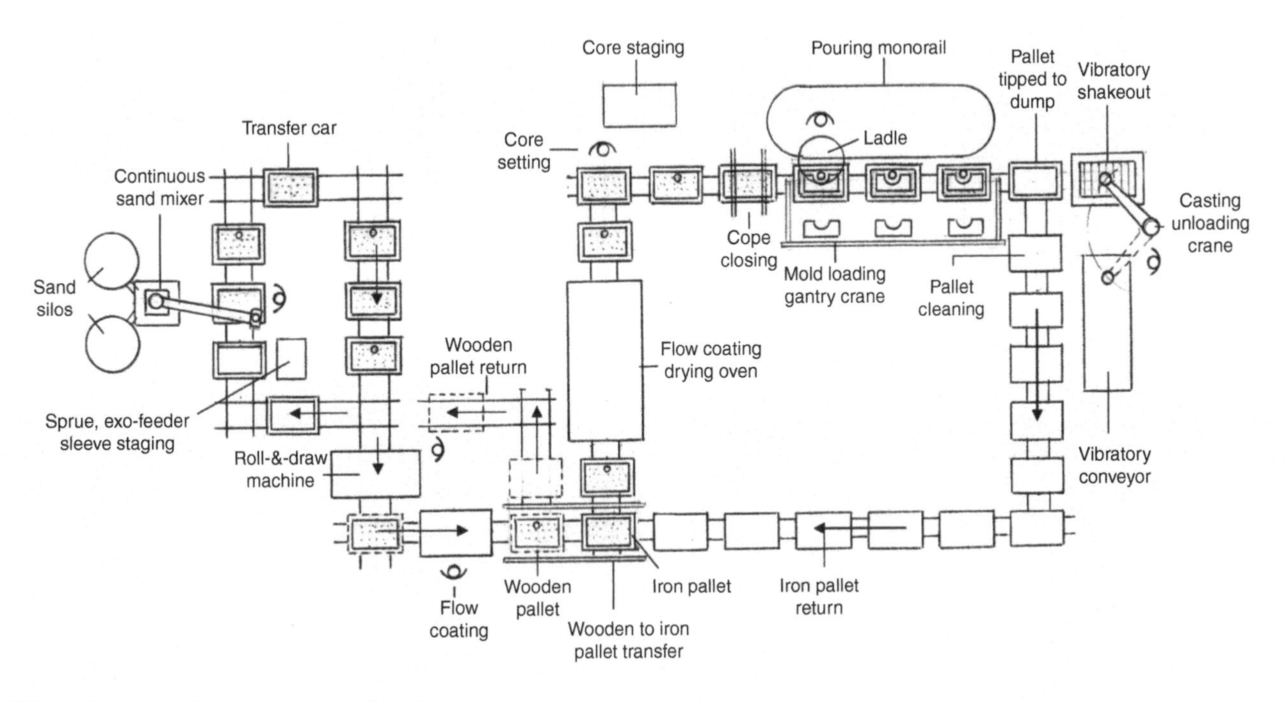

Fig. 4.35 No-bake line schematic for a steel foundry

Tight flasks as well as flaskless molds in green sand and no-bake molds can be loaded with a standard weight, either automatically or using a crane. The method of placing and removing weights can be automated. These methods lend themselves to casting facilities that produce standard parts or capital facilities that mass produce a few parts. In some cases, the dedicated pick-and-place unit places the jackets and the weights automatically.

The mold weight needed to resist cope-lifting is calculated in Eq 4.1–4.4, in the section 4.5.5.1 Mold Lifting Force Computation.

4.5.5.1 Mold Lifting Force Computation

The calculation of mold weights (Eq 4.1) is useful when the parts to be produced are known, as in captive foundries.

$$W_l = (1.3 \text{ to } 1.4) \times (L_m + L_c - W_c) \qquad \text{(Eq 4.1)}$$

where W_l is the weight on the molds to prevent the cope from rising and the metal from leaking at the parting plane. Normally a safety factor of 30 to 40% is used to account for approximations in calculations. Therefore, a multiplying factor of 1.3 to

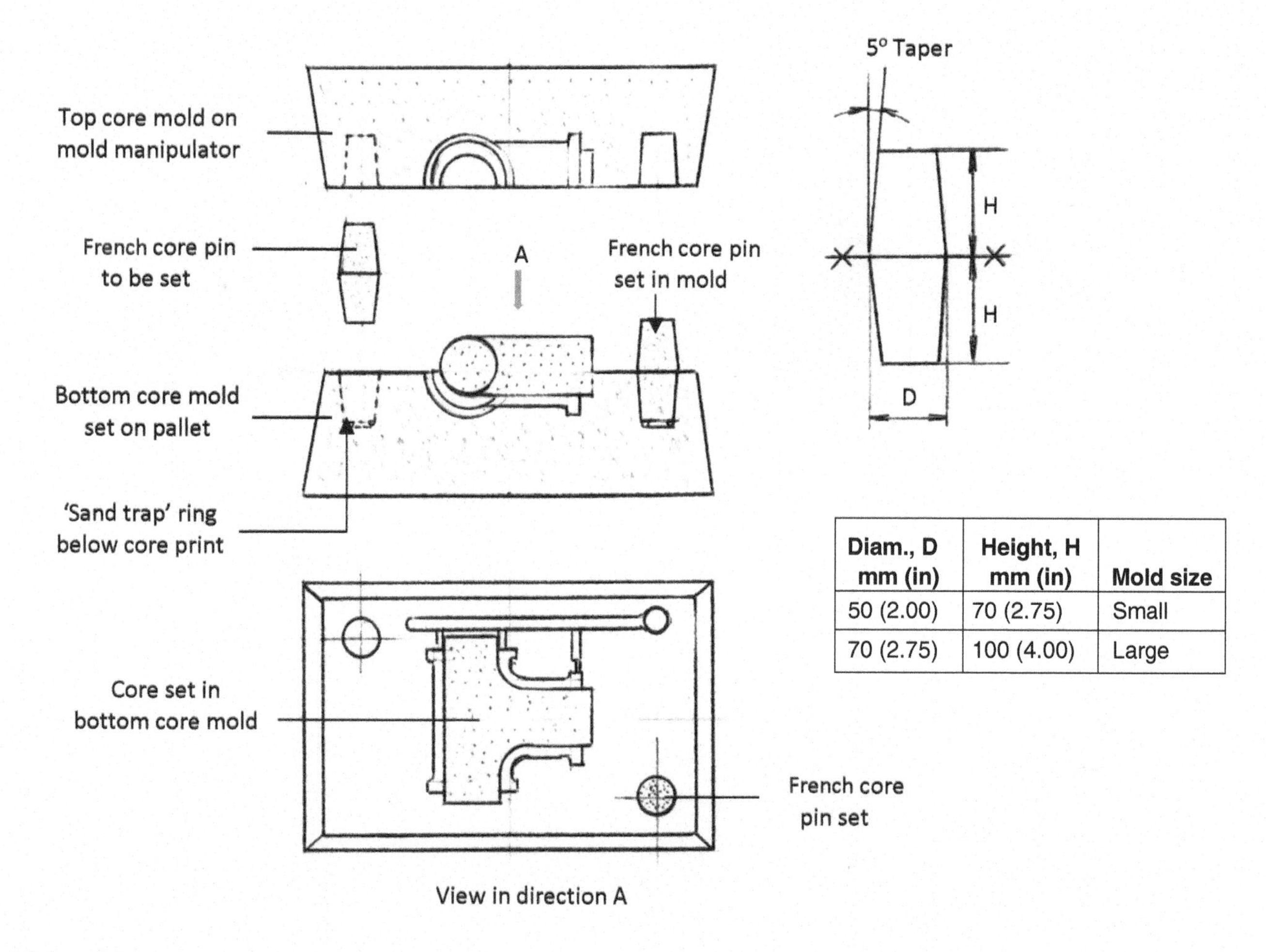

Diam., D mm (in)	Height, H mm (in)	Mold size
50 (2.00)	70 (2.75)	Small
70 (2.75)	100 (4.00)	Large

Fig. 4.36 French core pins for mold closing alignment

1.4 is used in the equation. L_m is the lifting force on the cope due to static head pressure, L_c is the lifting force on the core due to the buoyancy of the liquid metal, and W_c is the weight of the cope flask and the sand acting downward.

$$L_m = A \times H \times d_m \quad \text{(Eq 4.2)}$$

where A is the area of the casting at the parting plane in dm^3, H is the height of the cope or core mold above the parting plane in dm, and d_m is the average density of liquid metal in kg/dm^3.

For cast irons, the average liquid metal density is about 7.0 to 7.2 kg/dm^3.

$$L_c = V_m \times d_m - V_t \times d_c \quad \text{(Eq 4.3)}$$

where V_m is the volume of core surrounded by metal in dm^3 (excluding the core prints), d_m is the average density of liquid metal, V_t is the total volume of the core in dm^3 (including the core prints), and d_c is the density of the sand core in kg/dm^3 (about 1.4 to 1.6, depending on the sand).

$$W_c = W_s + W_f \quad \text{(Eq 4.4)}$$

where W_s is the weight of sand in cope, and W_f is the weight of cope flask (in no-bake molding there is no cope flask).

The weights must be retained over the molds until the castings are ready for shakeout after solidification. (See Fig. 3.22, 3.24, and 3.25 in Chapter 3 for shakeout time estimates.)

4.5.5.2 "C" Clamps for Mold Clamping

Tight flasks are clamped together using "C" clamps which hold the trunnions of the cope and drag flasks together. The "C" clamps are mostly forged or cast steel. Figure 4.37 (Ref 1) is an example of a group of "C" clamps for different flask heights. The dimension D in the figure is obtained by adding the heights of the cope and the drag.

4.5.5.3 Dovetail Clamps for the Flasks

Dovetail clamp systems are quite common for small and medium-sized flasks in medium-volume operations. The system consists of providing four-machined dovetails, tapered as shown in Fig. 4.38.

Cope and drag flasks are fastened together by sliding the tapered clamps and tapping them with a hammer. The clamps are removed by tapping out and recovered before punching the sand out of the flasks. This system, which is common in many casting operations, requires operators for clamping and

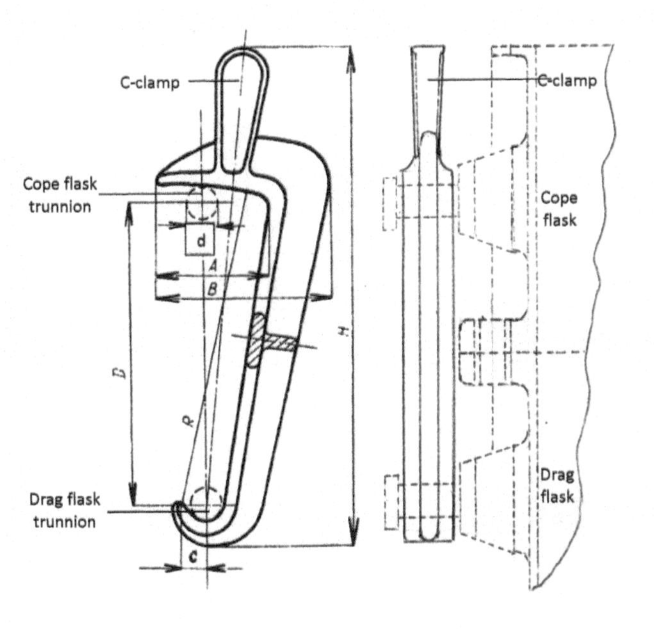

D mm (in)	100 (3.9)	160 (6.3)	250 (9.8)
d mm (in)	25 (1.0)	25 (1.0)	25 (1.0)
A mm (in)	100 (3.9)	100 (3.9)	100 (3.9)
B mm (in)	132 (5.2)	132 (5.2)	132 (5.2)
C mm (in)	17 (0.67)	17 (0.67)	20 (0.79)
R mm (in)	118 (4.6)	180 (7.1)	265 (10.4)
H mm (in)	170 (6.7)	280 (11)	375 (14.8)

Fig. 4.37 C-clamps for holding flasks together

unclamping. Very few casting facilities have automated this operation.

4.5.5.4 Automated Mold Weighting Systems

High volume green sand flaskless molding lines and medium-volume no-bake operations use automated systems for weighting down the molds or core molds. These systems need to be engineered in conjunction with the pouring monorails to avoid any interference. Two systems are illustrated:

- Automated mold loading conveyor
- Automated pick-and-place units

4.5.5.4.1 Automated Mold Loading Conveyor

The layout consists of a conveyor with loading weights suspended from an overhead chain. The conveyor height dips to place the weights onto the top of the molds or core molds. The conveyor height is designed to retain the mold weights on the top of the molds until solidification is complete and the molds have cooled to a temperature at which they can go to the shakeout unit. Figure 4.39 illustrates the layout of the conveyors for the molds or core molds, the mold loading conveyor, and the pouring monorail.

The weights lift off the molds before shakeout as the height of the conveyor is elevated. Molds are tipped onto a shakeout unit. The castings are then hooked out and sent to the cleaning room.

4.5.5.4.2 Automated Pick-and-Place Units

Pick-and-place units are becoming more popular due to their versatility. The cycle times are faster, and the capital

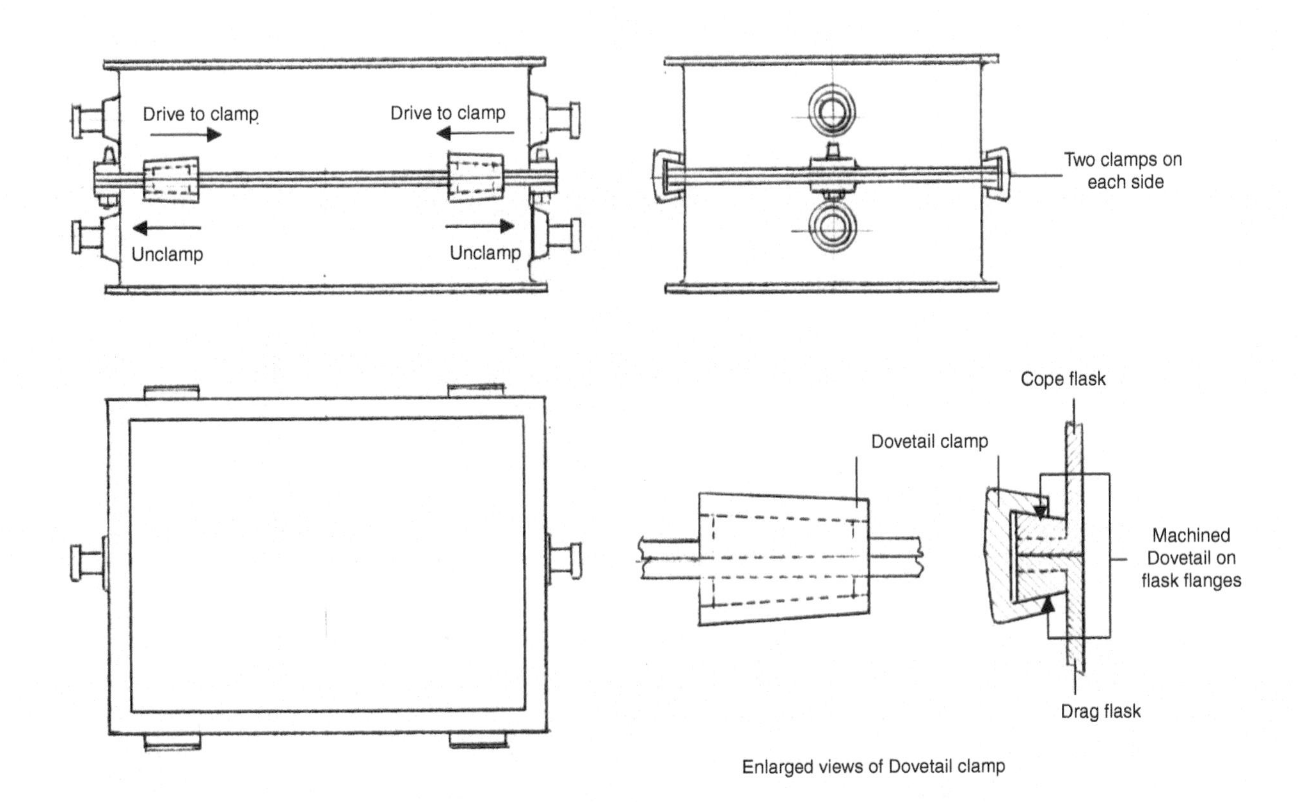

Fig. 4.38 Dovetail flask clamping system

costs are comparatively low. A trolley for transverse movement traverses back and forth, moving the weights on and off the mold assembly or core mold assembly. A pick-and-place unit (Figure 4.40) is mounted on a trolley with guide rails, and it is programmed to lower or lift the weight off the mold assembly. The lift unit may be hydraulic, and the transverse drive can be electric with a motor and a reduction gear drive.

4.5.6 Automated Core-Setting Systems

Manual core-setting is time consuming and can delay the cycle time of the molding machine, which affects cycle time

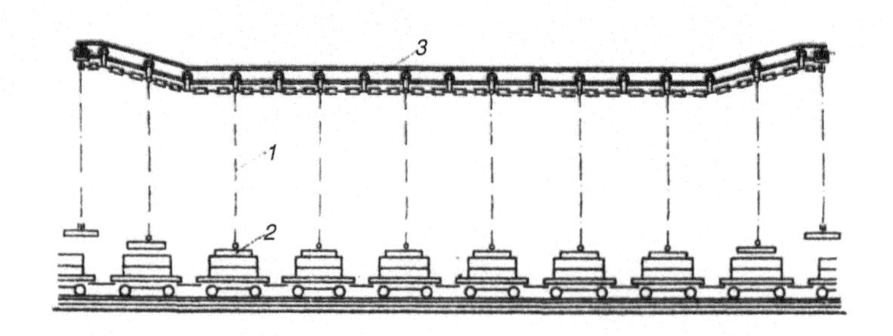

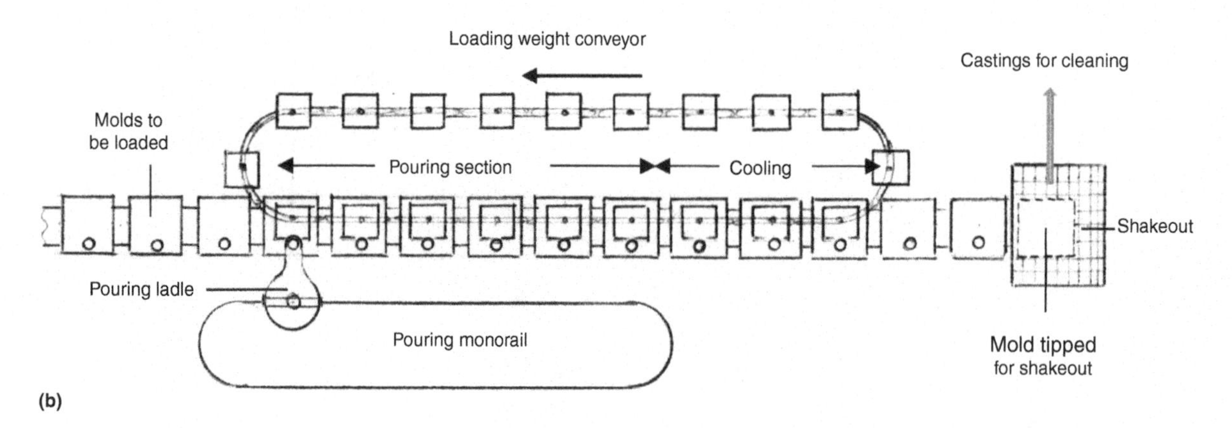

Fig. 4.39 (a) Automated mold weighting system. (b) Layout of conveyors for molds, mold weights and metal

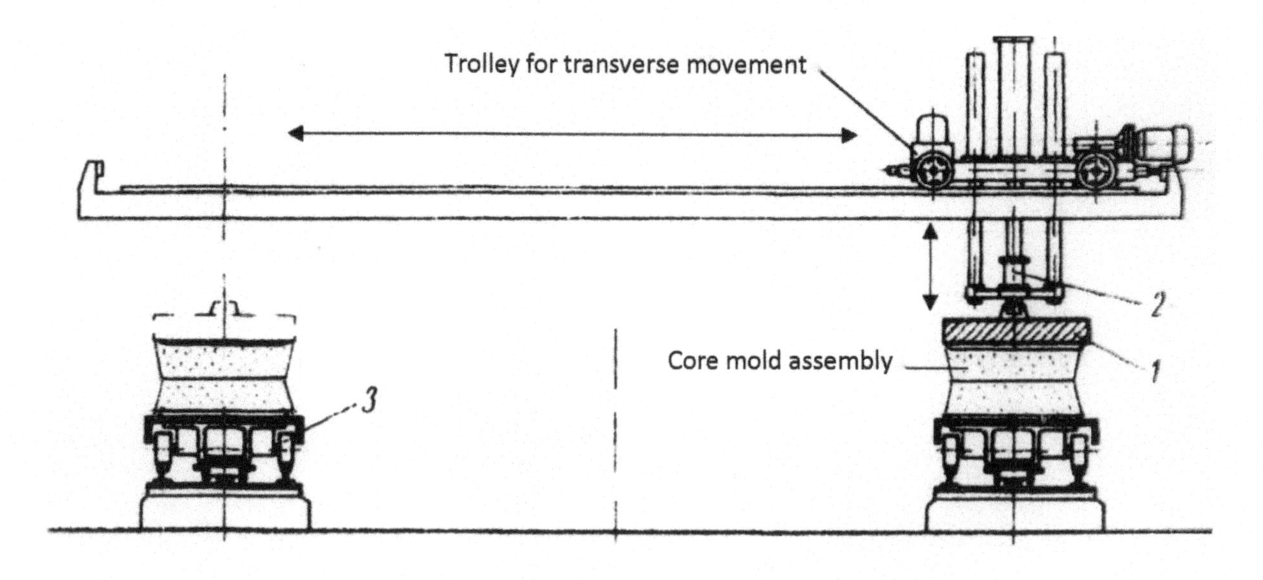

Fig. 4.40 Automated pick-and-place mold weighting system

consistency and productivity. Automation of core setting has an attractive payback with the reduction of manpower and improvement of machine uptime. In addition, it improves the dimensional consistency and reduces the potential for sand inclusions due to sand crush at the core prints which can occur in manual core-settings.

When a number of cores are assembled into a casting mold, there is a long core-setting time for the individual cores, which delays the molding cycle. The strategy used to reduce the core-setting time is to assemble the smaller cores into a base core or the main core, and use a core-setting fixture for a one-piece core-setting. A dedicated pick-and-place unit or a robot picks up the entire core assembly and sets the mold in a few seconds without delaying the molding cycle.

If there are several smaller mold cavities in a mold, for example in a molding line for pipe fittings or valves, the pattern layout is designed for combining the cores to simplify core-setting (Ref 4). Multiple fixtures can be designed for more than one operator to load a set of multiple fixtures without delaying the machine cycle. The multiple core-setters can be configured either as linear or rotary cells.

Pneumatically operated pick-up forks mounted on a sliding frame can be designed to pick up the core assembly and unload it precisely into the mold. When tight flasks are used, the guide pins of the drag flask can be used for accurate positioning. Other methods of picking up the cores include vacuum suction, the use of an inflatable bulb, or using an expanding collet to grab the cores. Flaskless no-bake molding uses either French cores (Fig. 4.36) or two alignment holes formed in the core molds and a guide rod for the assembly. Larger core assemblies are handled by robotic core-setters carrying a pick-and-place head.

Figure 4.41 (Ref 4) illustrates a rotary core-setting installation. The cores are set in fixtures mounted on a rotary table. The table indexes to position the core assembly over a core-setting machine. The drag moves in below the core-setting machine. The core assembly is lowered and the pick-up unit releases the cores. The drag mold moves out to a mold-closing station where the cope is closed. Such dedicated units are preferred in capital foundries where the part stays the same or nearly the same for a long time.

Figure 4.42 illustrates a robotic core-setter. The advantage of a robotic core-setter is that it lends itself to reprogramming or switching the core pick-up-and-place fixture for jobbing foundries where different parts are produced on different days. Linear layouts for core assembly feed conveyors are common.

Vertically parted molds use a core mask into which the operator loads the cores; a linkwork mechanism places the cores in the mold and applies mild pressure to lock the cores in position before the other half of the mold closes on the first half (see section 4.4.4 in this chapter).

4.6 Core-Making Equipment

One of the major advantages of castings is the ability to cast hollow shapes and to provide the material at locations in the casting where it is needed for functional performance. Sand cores that provide the shape during casting and collapse after casting are vital to this versatility in producing complex geometries.

Cores need to satisfy several requirements:

- Filling and formability during production
- Permeability for gas-hardening during manufacture
- Permeability for venting combustion gases during casting
- Collapsibility after casting for cleaning
- Surface integrity for needed finish and cleanliness
- Accuracy to maintain wall thicknesses
- Minimum cost
- Recyclability after use for environmental protection and affordability

Core-making machines focus on meeting these requirements. In addition, they are designed for:

Fig. 4.41 Automated core-setting unit. Source: Ref 4, EMI

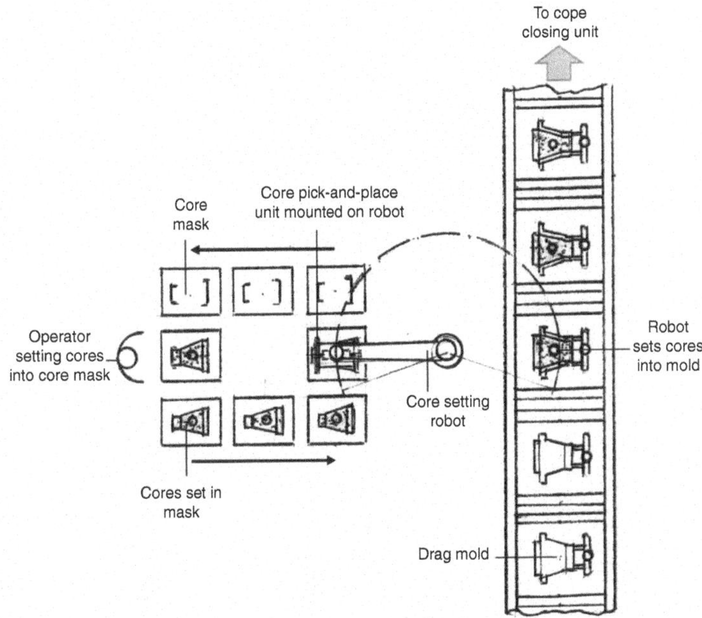

Fig. 4.42 Robotic core setter

- High productivity
- High automation to limit manpower costs and to achieve high consistency in quality
- Minimum operation intervention
- Multiple workstations for two or more types of cores to be produced simultaneously
- Quick tool change, especially in jobbing foundries
- Versatility to handle different core hardening systems

Cores are produced by bonding silica sand with organic or inorganic bonding agents. The tooling used is either at room temperature or at elevated temperatures, depending on the bonding system. The process using tooling at room temperature is called the cold box system, and the process using elevated temperatures is known as warm box or hot box, depending on the temperature range. Curing or hardening at room temperature is accomplished either by chemical reaction (air-set or no-bake) using activators or by using different gases for vapor curing.

Core box tooling for cold box systems uses either urethane or aluminum, depending on the life expectancy of the life of the part. Tooling for a hot box or warm box is usually made from heat-resistant steel or sometimes heat-resistant cast iron. Core sand is blown into core boxes using compressed air in core blowers or core shooters. Hot box shell core-making for medium-sized and larger cores uses mostly gravity to fill the core boxes.

Figure 4.43 is an overview of the different core binder systems.

Core-making machinery is designed to deliver high-quality cores at maximum productivity and minimum cost. The automation is geared toward assuring high uptime, consistent quality, and minimizing operator intervention. Machines are designed with flexibility for core box parting, both horizontal and vertical. Some machines are designed with multiple heads. Machines designed with dual heads require only one operator who works on both core boxes, alternating core curing (or gassing) time with blowing time. Some machines are designed with flexibility for either the cold box or warm box process. Machines designed for hot shell can also be used for the warm box process if necessary.

4.6.1 Cold Box Core Machines

Figure 4.44 illustrates a typical core-making machine (Ref 13) with flexibility for cold or warm hardening processes. The major machine features are:

- Flexibility for processes: cold or warm
- Core shooting volume range: 25 to 200 liters (26.42 to 211.39 quarts)
- Number of blow heads: single or multiple
- Number of process tools at a time: up to 6 parts
- Tool change: automatic
- Gas hood with integrated ejection plate
- Progressive central lubrication
- User-friendly control system with fault diagnostic capability

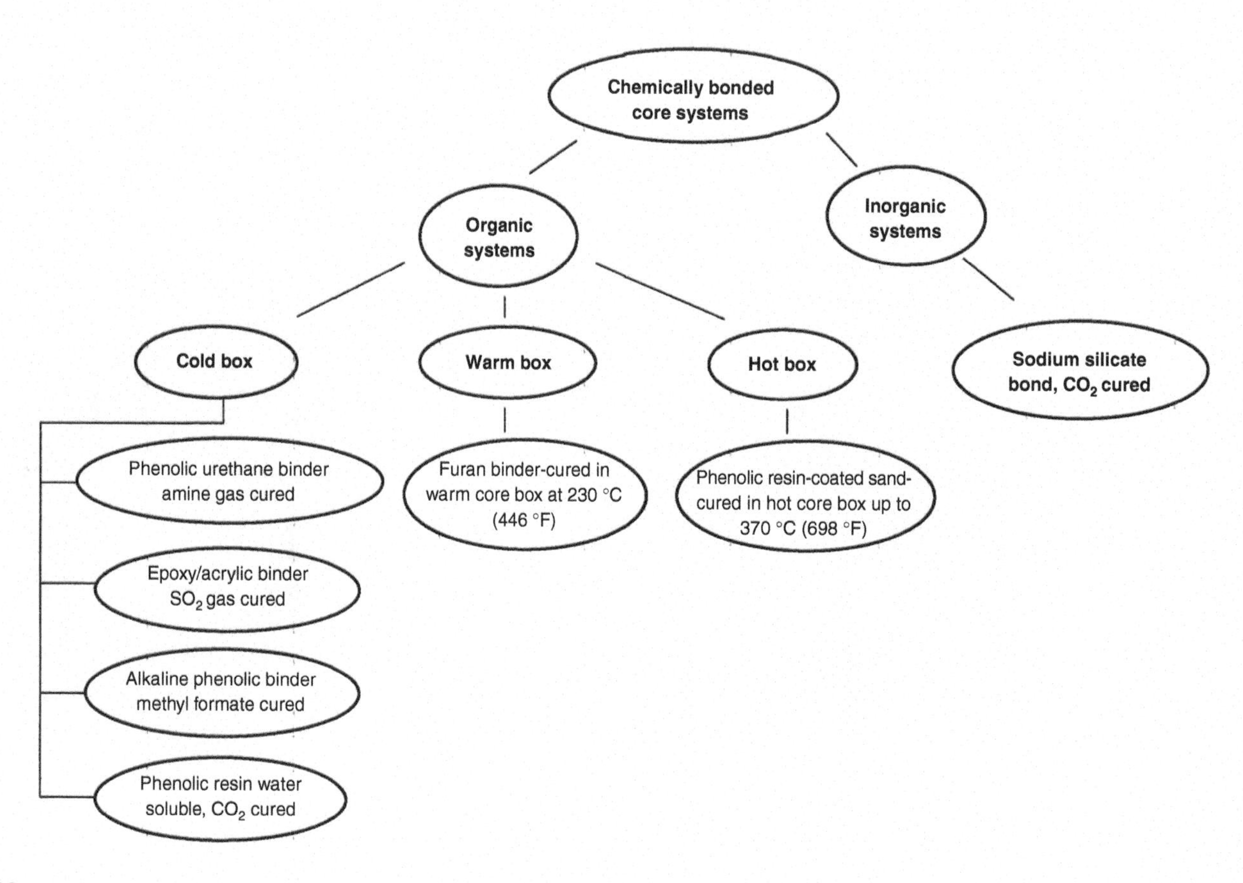

Fig. 4.43 Core sand systems overview

(a)

(b)

Fig. 4.44 Flexible cold box core-making machine, (a) with operator enclosure, (b) without enclosure to reveal details. Source: Ref 13, Laempe Reich

Table 4.18 (Ref 10) lists the major options for the machines available in this series.

Figure 4.44 is a schematic of a core-making cell, featuring the machine, the robot, the core wash, and the conveyor.

Figure 4.45 provides a concept of tooling with horizontally parted core boxes, as shown in the partial view in the direction of arrow A. The upper part of the core box engages with the blow head. The blow tubes are mounted on another slide (not shown in the figure). The bottom part of the core box has an ejector rack to eject the core once it is hardened by amine gas.

A robot picks up the core, reaching the chamber through a sliding door, and passes the core underneath a rotating brush whose mild abrasive action removes any parting plane flash. The core moves to an automatic dimensional inspection station where the critical dimensions are checked. Approved cores are dipped in the core wash to coat the part in selected locations of the core. The core is unloaded onto a stand in the core wash drying oven for one complete cycle of core-making. Cores whose wash has dried are picked up by the robot and placed on the conveyor for transport to the molding machine. Such cells can be designed with variations depending on the core design and if the need exists for selective core wash.

Manpower can be reduced if one operator manages two adjacent core-making stations at the same time. Figure 4.46 is an example of a dual-station core box machine.

Dual-station core box machines are designed to handle multiple core sand-hardening cold box systems, as shown in Fig. 4.46 (Ref 4). The salient features of these machines are:

- Produce high-quality cores
- Two different parts can be run at the same time

Table 4.18 Core making machine details

	LFB Series	
Core making machine, LFB model	25	50
Injection volume, liters	25 – 40	50 –200
Maximum injection range, mm	700 by 650	1000 by 1400
Standard injection range, mm	500 by 500	900 by 900
Cycle times (without gassing, injection, and venting), seconds	17	19
Tool height, mm	375–800	575–900
Vertical tool width, mm	200–600	300–1000
Horizontal tool width, mm	200–1150	300–1550
Tool depth, mm	300–1000	300–1600
Maximum tool weight, kg	1800	5000
Tool changer option, front or rear	Available	Available
Cleaning station	Available	Available

Source: Ref 13

- Suited for jobbing foundries where one part can have shorter duration runs while the other can have longer duration runs
- Quick tool change is an option
- Proprietary system for precise injection and minimum condensation for high-quality cores
- Capable of integration with mixers, gas generators, and automated core handling systems

Table 4.19 provides the major parameters of the available machines (Ref 4).

The cold box core systems that use gases such as amine or sulfur dioxide need gas generators for gas injection and scrubbers for purging the excess gas before opening (Ref 13).

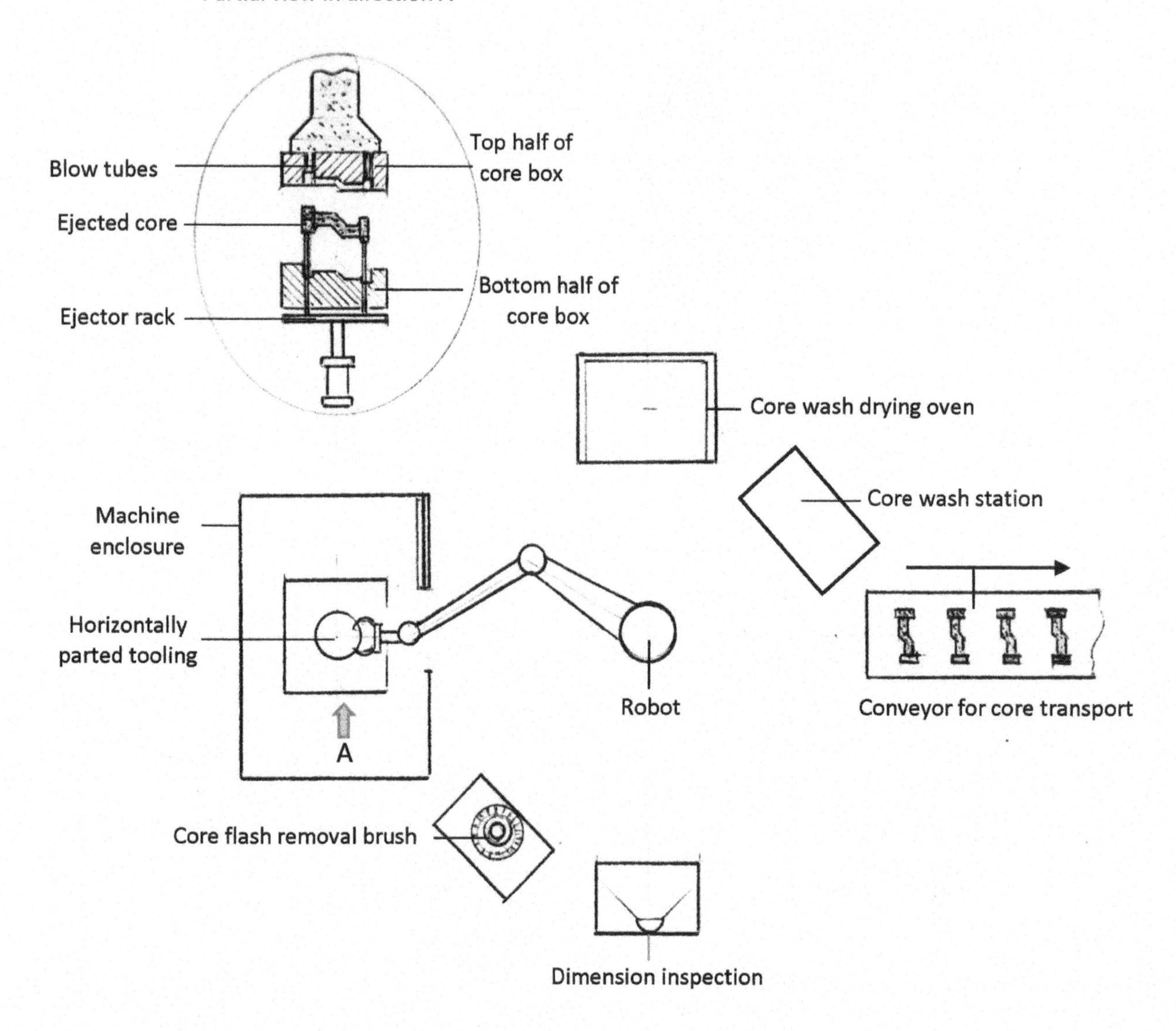

Fig. 4.45 Cold box core manufacturing cell concept

Fig. 4.46 Dual-station core-making machine. Source: Ref 12

Table 4.19 Dual cold box core making machine

Model	Blow capacity, Litres	Shooting area dimensions, cm (in.)	Tooling dimensions, cm (in.)	Dry cycle time, sec	Power
DC-1616	10	40.6 by 10 (16 by 4)	40.6 by 40.6 (16 by 16)	16	Pneumatic
DC-1818	15	45.7 by 10 (18 by 4)	45.7 by 45.7 (18 by 18)	18	Pneumatic
DC-2424	25	61 by 15 (24 by 6)	61 by 61 (24 by 24)	20	Pneumatic
DC-3030	30	76 by 15 (30 by 6)	76 by 76 (30 by 30)	23	Hydraulic

Source: Ref 3, EMI

4.6.2 Shell Core Machines

Shell core machines typically consist of a blow head chamber pressurized to blow the pre-coated sand into a heated two-part core box, mounted on two platens. Four tie rods connected across the two mounting platens enable the opening and closing of the core boxes. The tie rods and platen assembly are mounted on a rollover mechanism to rotate by 180°, which dumps any uncured sand into a container for reuse.

Another feature is a burner train or electrical heater cartridge to heat the core boxes from the back. An ejector rack enables pushing the hot shell out of the moveable half of the core box. The unloading of the core can be automated by either tilting the movable half or using a robot to grab the cores, because they are very hot to handle.

Figure 4.47 illustrates the main features of a shell core-making machine, for manual or robotic unloading.

Figure 4.48 (Ref 14) is a shell core-making machine suited for small and medium cores. The operations are sequenced and automated for running with little supervision. Larger machines for bigger cores are also available, and the main features are similar. View in Fig. 4.48(a) illustrates the main features. The sliding sand hopper moves over the top of the core box for blowing. Pressurized air is blown over the sand in the hopper, which forces the precoated sand into the core box cavity. Figure 4.48(b) shows the core box inverted by 180° to drain the cores of any unused sand. A catch funnel directs the drained sand into a container for reuse. This operation results in the formation of a shell.

Figure 4.48(c) illustrates the use of a pair of burners that swing into position to cure the inside of the shell. Figure 4.48 (d) shows the movable half of the core box tilted by 90°, actuating the connected cylinder. The ejector system on the movable core box half pushes the core onto a movable belt conveyor which is elevated to receive the cores. A blast of air cleans the core box halves, and the next cycle starts by reverting the tilted core box back to the vertical position and closing the core box.

4.6.3 3D Printed Cores

Complex cores with varied contours and undercuts benefit from a 3D printing approach. 3D printing of cores is done using CAD geometry without the need for a core box. The three-dimensional CAD model can be modified to generate the core profile. The tolerances are very close, and the drafts are minimal.

Figure 4.49 is a schematic of the process illustrating a core mold (Ref 4).

The furan binder jet printer is mounted on slides capable of traversing in X and Y directions, as illustrated. Each cross section of the core or core mold is defined by the X and Y coordinates. A thin layer of sand is spread, and the furan binder jet printer head traverses to harden the profile in the chosen cross section. Once the layer has hardened, the core former base is elevated by a small increment, and the second layer is built over it by the printing head. The entire core is built in successive layers. The uncured sand is removed by vacuum, and the sand is reused.

Figure 4.50 (Ref 15) is an example of a core with undercuts. The casting of a turbine runner produced using a 3D-printed core is pictured on the right.

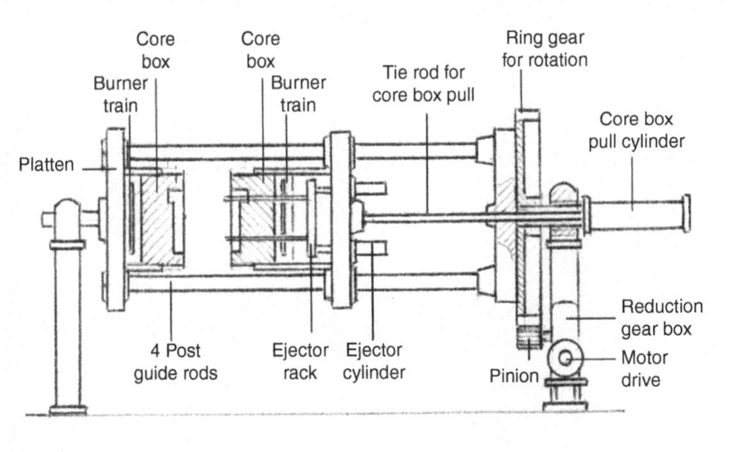

Fig. 4.47 Shell core machine concept

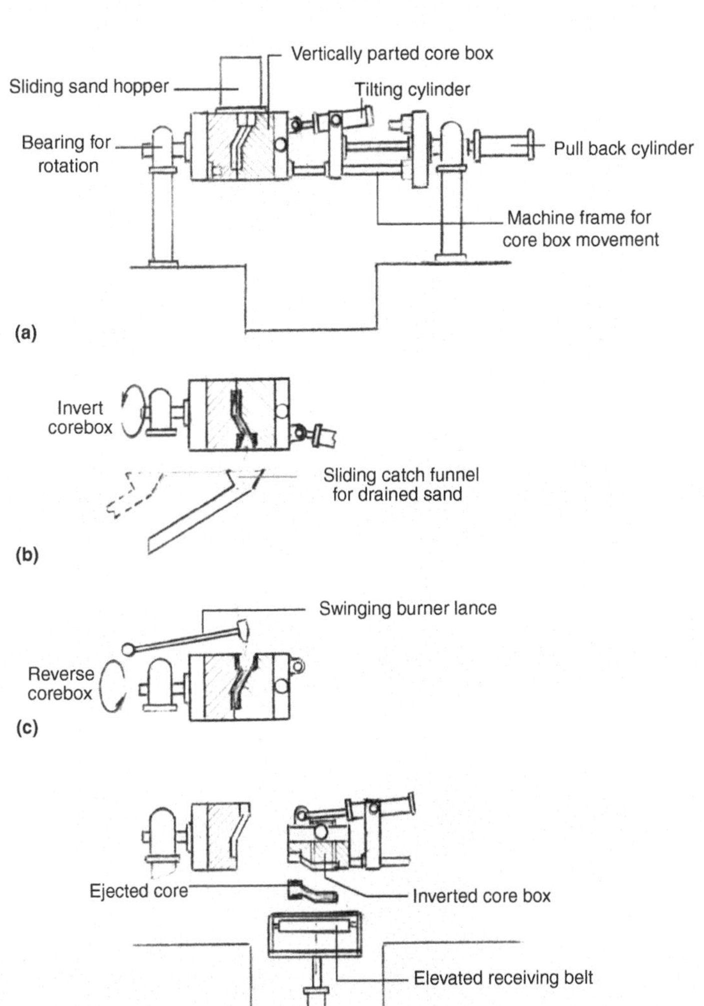

Fig. 4.48 Shell core machine concept for small and medium cores; (a) core blowing, (b) core draining, (c) inside flame curing, (d) core ejection

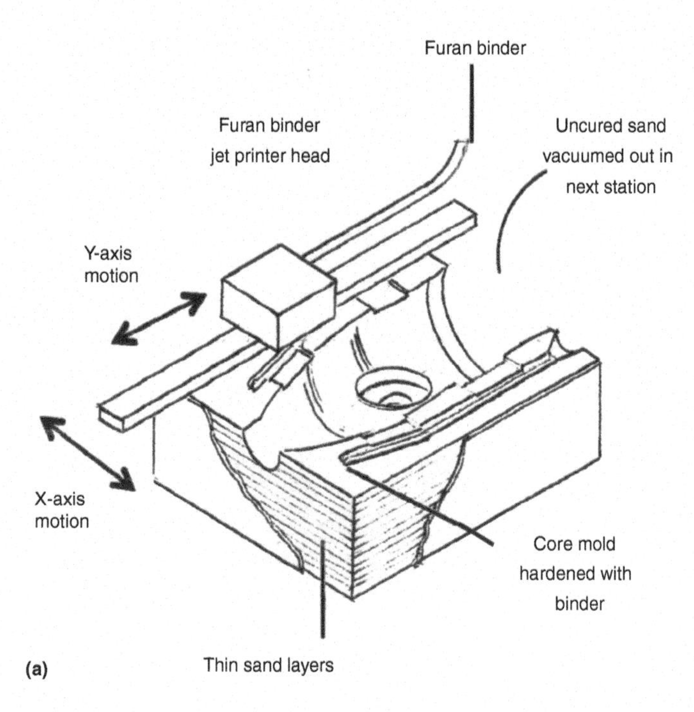

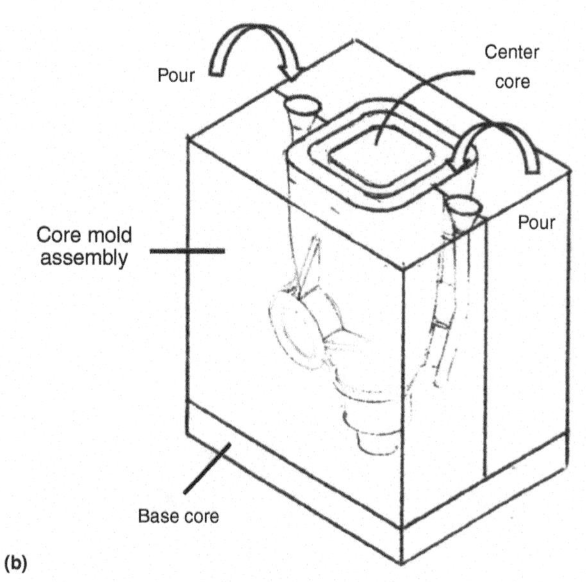

Fig. 4.49 Schematic of 3D core manufacturing; (a) core mold formation, (b) core mold assembly for pouring

Fig. 4.50 Example of a 3D-printed core and the casting. Source: Ref 15, Voxeljet

REFERENCES

1. R. Brabenec and S. Prokop, Molding Flasks (Formavaci Ramy) SNTL (Statni Nakladatelstvi Technicke Literatury) 1956
2. J. Nath, *Aluminum Castings Engineering Guide*, ASM International, 2018
3. Equipment Manufacturing International (EMI) Brochure, 2020 www.emiinternational.com
4. J. Nath, *Iron and Steel Castings Engineering Guide*, ASM International, 2022
5. Omega-Sinto Foundry Machinery Brochure www.omega sinto.com
6. Hunter Foundry Machinery Corporation Brochure www.hun terfoundry.com
7. DISAMATIC Inc. www.disagroup.com
8. J. Sirokich et al., *Foundry Equipment* (Zarizeni Slevaren) SNTL, 1968
9. Tinker-Omega Manufacturing Company Brochure, 2020 www.tinkeromega.com
10. Vulcan Engineering Co. Brochure www.vulcanengineering. com
11. ASK Chemicals Brochure 2020 www.infousa@ask-chemi cals.com
12. Palmer Manufacturing and Supply Inc. Brochure 2020 www. plamermanufacturing.com
13. Laempe Reich Brochure on Core Making Machines, 2020 www.laempereich.com
14. Kuo Kuen Industrial Co. www.kaokuen@ms41.hinet.net
15. Voxeljet Company Brochure 2019 www.voxeljet.com

Casting Equipment Engineering Guide
Jagan Nath
https://doi.org/10.31399/asm.tb.ceeg.t59370095

Copyright © 2023 ASM International®
All rights reserved
www.asminternational.org

CHAPTER 5

Casting Cleaning Operations

IRON AND STEEL CASTINGS require cleaning as they come out of the shakeout units to remove any burned sand and sand that may remain stuck from the mold. Blasting the castings with steel shot is the most common method for removing burned sand and sand adhering to the casting.

The cost of cleaning the castings represents approximately 20 percent of the total cost of manufacturing. The informed selection of the right equipment and its layout will maximize the efficiency and is vital to ensuring cost competitiveness and profitability.

Castings have gates, runners, and feeders which also must be removed. In gray and white iron castings, the gates and runners are normally detached from the casting during the shakeout process or on the vibrating conveyor following the shakeout. Flash occurs at the parting plane and the core prints in metals and alloys of high fluidity such as gray and white iron. The flash needs to be removed after shot blasting. Metals such as ductile iron and steel have less fluidity, so there is less flash in the parting plane and core prints compared to gray cast iron. The burrs that result from breaking the flash also need to be removed.

5.1 Casting Cleaning Operation Sequence

The casting cleaning process consists of several sequential operations and involves a range of equipment. The interconnected machinery must be laid out logically to maximize product flow.

Small and medium-sized castings are randomly oriented during the shakeout operation, which presents a challenge for automation. Castings from different molding lines may share a common shakeout unit, where the potential for mixing different castings is high. Devices to safeguard against mixing different castings need to be designed to prevent damage to the machinery, especially if the devices feeding the machines are automated. Captive foundries have the advantage that the shapes and sizes of castings are predetermined. Dedicated equipment can be cost justified, and the handling systems can be engineered to prevent mixing of different castiings.

Figure 5.1 shows the sequence of operations following the shakeout operation (Ref 1, 2).

Molds enter the shakeout units after allowing for a safe cooling time. The rotary media drums clean the castings as they traverse through the drum (Ref 3), and these units reduce the subsequent cleaning as much as 70 percent. As the castings exit the rotary drum onto an oscillating conveyor, operators separate the castings and the gating. Also, where multiple molding machines share a common rotary drum, the operators can separate the mixed castings. The disintegration of the cores (decoring) occurring in the rotary drum units helps to reduce the subsequent shot blasting time (Ref 3). Gates from iron castings are separated and cleaned. Effective decoring reduces the amount of new sand to be added to the return sand in green sand systems.

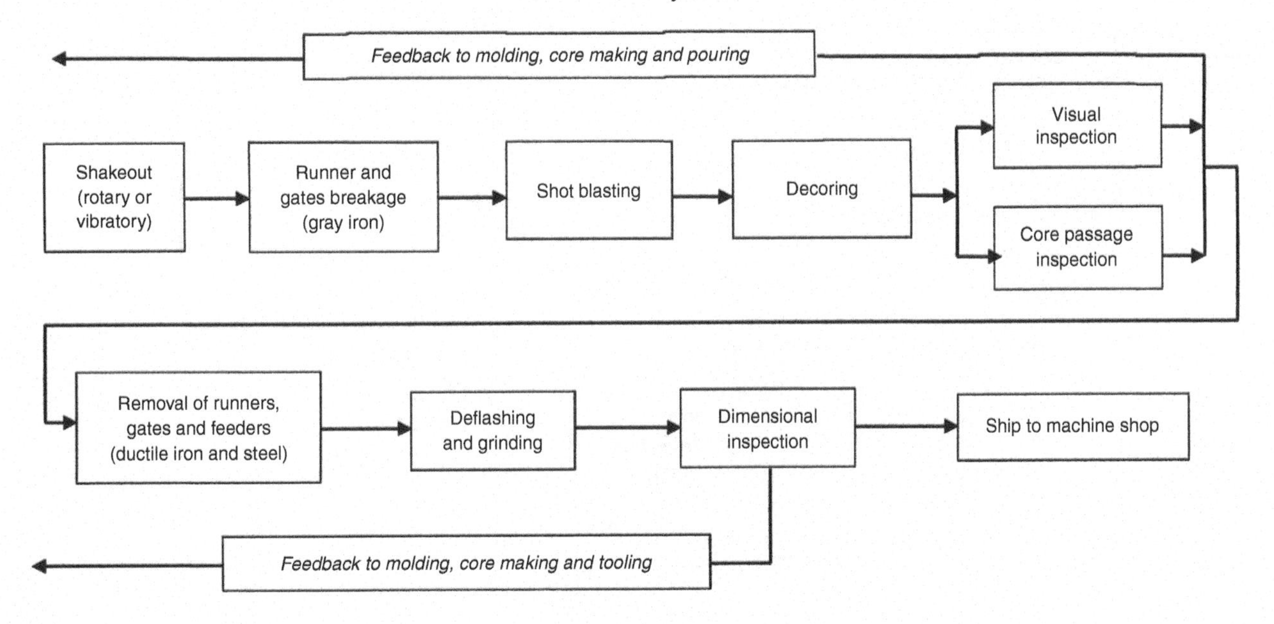

Fig. 5.1 Casting cleaning operations overview

Rotary shakeout systems are ideally suited for flaskless green sand and no-bake systems. Smaller volumes of larger sized castings are better suited for vibratory shakeouts, where crane handling and manual intervention are usually required.

Runners and gates from gray and ductile iron castings are separated in the rotary and vibratory shakeout units. The castings are discharged onto an oscillating conveyor. The feeders of ductile iron castings may not be automatically knocked out in the shakeout units. Operators are positioned on the side of the oscillating conveyor to knock out the feeders using a hammer or a suspended hydraulic wedge. The castings enter the shot blasting machines, and hardened steel shot propelled at high velocities by an impeller, impinges on the surface of the castings, removing any sand adhering to the casting surface.

The castings then enter a decoring station. Intricately cored castings such as cylinder heads and blocks require specially designed vibratory machines for thorough cleaning of the water jacket passages. Less intricately cored castings can be decored using vibratory decoring hammers.

Gray iron castings have thinner gates and normally do not have large feeders. The gates and vents are broken during the shakeout operation. Castings are visually inspected for any obvious anomalies, by the operators at the oscillating conveyor, as they separate the castings.

The castings then enter the degating station, where runners and feeders of ductile castings (such as steering knuckles and brake calipers) are removed using a hydraulic wedge to pry open the feeders attached to the casting. Smaller feeders for steel castings (up to about 150 mm or 5.9 in. diameter in carbon steel) are knocked off if they have a neck-down core (Ref 1). Standard and larger feeders of steel castings are removed using an oxy-acetylene torch or a circular saw.

Deflashing, or removal of the parting plane and core print flash, of medium castings is done by grinding on pedestal grinders. Large iron and steel castings are ground using a swing frame grinder which allows manipulation of the grinding wheel, while the large castings stay stationary. Smaller castings like iron fittings are also ground on pedestal grinders.

Bottom-discharge buckets containing smaller work pieces in bulk are ergonomically placed next to the operators for access through a hinged bottom door. Castings are then ready for a dimensional check. The dimensions are checked either by gaging or on a 3D coordinate measuring machine (CMM), and the results are conveyed to molding, core making, and tooling inspection for any remedial action. The castings are ready for shipment to the machine shop after metallurgical approval.

Details of these operations follow.

5.2 Casting Cleaning Options—Shot Blasting

Castings are partially cleaned in the shakeout units, depending on the type of the shakeout units being used. The rotary units used for small and medium-sized castings clean significantly better, reducing the shot blast time needed. The surface quality requirements of castings require them to be free from sand. They are shot blasted using steel shot impinging at high velocities to dislodge the sand sticking to the surface. Figure 5.2 outlines the overall categories of the shot blasting machines.

Each of these machines offers features that are appropriate for different size groups and types of castings. An analysis of the volume of production and size of the castings determines the type of machine that is most suitable.

The shot blasting machines are grouped into three main categories:

- Tumble blasters, where the castings tumble to expose the surface to the impinging shot stream.
- Table blasters, where the castings are placed on a rotating table to expose the castings to the impinging shot streams from multiple impeller heads. They are manually turned over for the next cycle of shot blast cleaning to expose the bottom surfaces.
- Cabinet blasters, where the castings are either suspended on a monorail that enters a cabinet with multiple impeller heads or they are placed on a rotating or stationary table.

5.2.1 Impeller Heads

The impeller head is the nexus of the shot blasting machines. Figure 5.3 (Ref 4) illustrates the major components of the assembly. The parts are produced from a special grade of high nickel chrome cast iron alloy called Ni-Hard (Ref 1), which makes them especially resistant to wear due to abrasion.

The impeller wheel is built with wear-resistant blades and is housed in a steel housing lined with wear-resistant plates. The components are designed to be highly resistant to abrasion, and also for quick replacement with minimum down time. Steel shot enters the assembly at the center through an elbow (Fig. 5.3b). The impeller and the control cage, along with the blade assembly, are driven by a motor Fig. 5.3(a). The control cage allows the amount of shot entering the radial blade assembly to be adjusted. Shot is thrown at high velocity to dislodge sand adhering to the castings.

The sand and shot are collected at the bottom of the machine and hoisted up by an elevator. The shot and sand mixture passes through a separator where sand is sucked into a cyclone or a bag house. The separated and cleaned shot is recycled as it enters the elbow through a funnel.

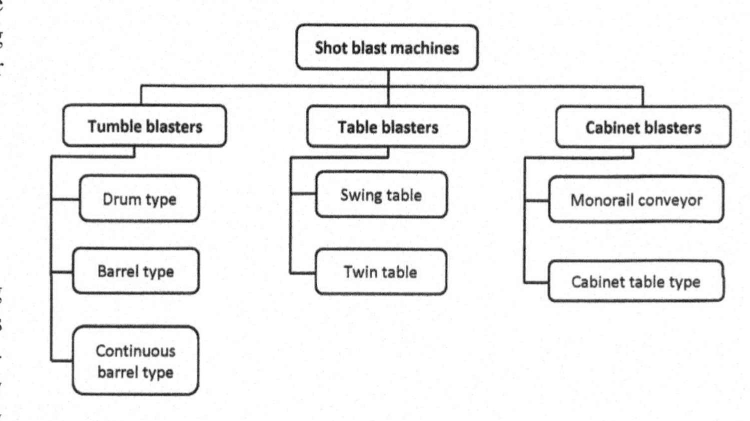

Fig. 5.2 Shot blasting machine types

(a)

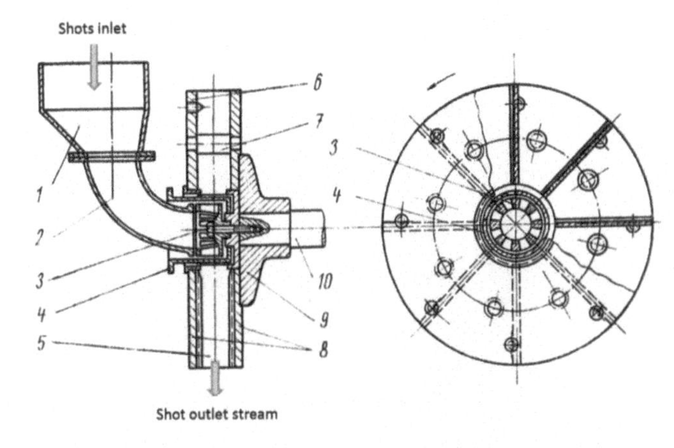

Notes: 1. Funnel 2. Feed spout 3. Impeller 4. Control cage 5. Blades 6. Blade anchor 7. Spacer stud 8. Side plates 9. Flange 10. Drive shaft

(b)

Fig. 5.3 Impeller (a) housing and components and exploded view, (b) wheel and infeed, cross section. Source: Ref 4

Separators for cleaning the steel shot from sand are available in several different designs. Some separators use air suction and other designs use centrifugal force.

5.2.2 Shot Blasting Machine Types and Suitability

All shot blasting machines have common features. They have one or more impellers that are mounted to provide for maximum exposure of the casting surface to the impinging shot (Ref 5, 6). The machines feature a bucket elevator to carry the steel shot up, and they use a separator to clean the shot. Finally, they feed the impellers through the center. The centrifugal force of the impeller throws the shot stream at high velocities to dislodge the adhering sand from the casting surface.

The major variances between machines are found in the loading and unloading of the castings, and the method of exposing the castings to the shot stream. The castings are exposed to the shot stream using barrel type tumblers, spinning hangers, or rotating tables.

Figure 5.2 presents an outline of the generic types of shot blasting machines. Further details are described in this section.

5.2.2.1 Tumble Blasters

Tumble blasters are divided into three groups:

- Drum-type tumble blasters
- Barrel-type tumble blasters
- Continuous barrel-type machines

5.2.2.1.1 Drum-Type Tumble Blasters

Figure 5.4 (Ref 7) is a schematic of a drum-type tumble blaster. The tumbling drum is made of thick (20 mm or 0.75 in.) highly abrasion-resistant manganese steel for high durability. The castings are subjected to motion in both longitudinal and radial directions as the drum rotates. The impeller is mounted on the cover, which allows loading and unloading of the castings using a front loader.

Figure 5.4(a) shows the loading position, Fig. 5.4(b) shows the cleaning position, and Fig. 5.4(c) shows the emptying position.

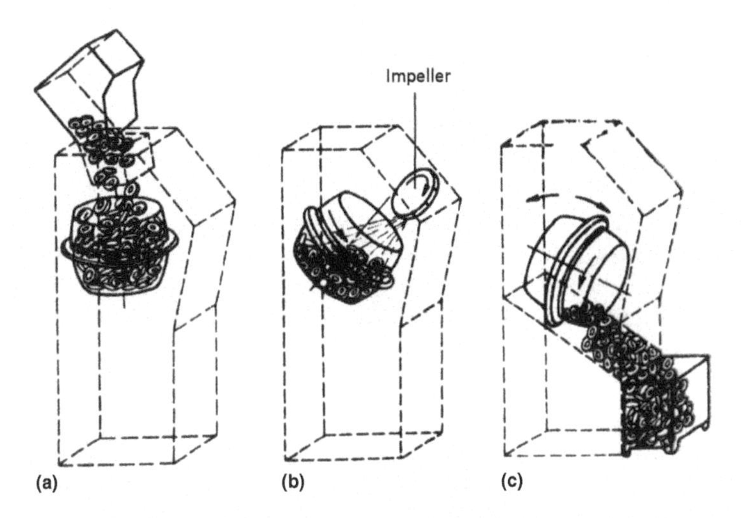

Fig. 5.4 Operation of drum-type tumble blasting machine, (a) loading, (b) cleaning, (c) unloading

The perforations at the bottom of the barrel allow the shot and the sand to exit the barrel. Sand and steel shot are separated, and the bucket elevator carries the steel shot up to the impeller for cleaning. The sand is then discarded.

Cleaned castings are dumped into a bottom discharge bucket that is transported and placed on a stand next to the grinding machine operator. The bottom opens, allowing the operator easy access to the castings. The operator performs a quick visual inspection and alerts the supervisor to any anomaly. Ground castings are transported to the machine shop in totes.

Drum-type tumble blasters are suitable for small iron and steel castings that have simple cores and those that do not get entangled due to their shape or get damaged due to tumbling. These machines are suitable for deburring, descaling, and surface preparation before painting.

Figure 5.5 (Ref 7) shows the complete installation including the drum, the cover with the impeller, and the bucket elevator.

Fig. 5.5 View of a drum-type tumble blaster. Source: Ref 7

5.2.2.1.2 Barrel-Type Tumble Blasters

Barrel-type tumble blasters consist of a heavy-duty metal conveyor made up of wear-resistant slats attached to form a barrel or trough, as shown in Fig. 5.6. The castings are loaded into the trough using a tilting tote in front, and they tumble as the conveyor rotates. The impeller wheel is mounted at the top of the machine. A sturdy door lined with wear-resistant plates covers the opening. The tilting tote is filled using a dump bucket mounted on the forklift or by using a front-loading bucket during the cleaning cycle of the previous batch, which eliminates machine down time.

Figure 5.7 illustrates the complete machine (Ref 8).

These machines are suitable for small to medium-sized iron and steel castings and are highly productive. Castings with projected arms or limbs that may get entangled are not suitable for cleaning using tumble blasting machines.

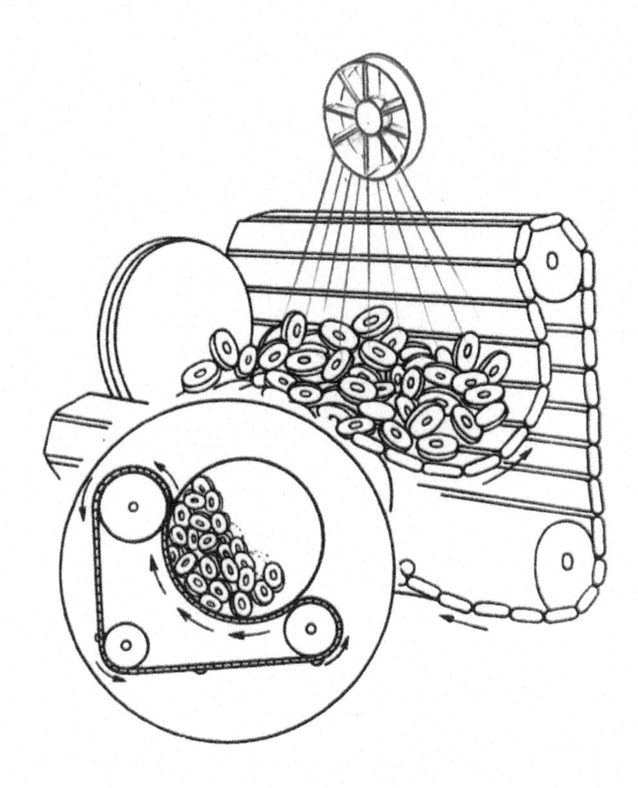

Fig. 5.6 Schematic of barrel-type tumble blaster

5.2.2.1.3 Continuous Barrel-Type Tumble Shot Blaster

The continuous barrel-type shot blasters are suited for large volume production with minimum operator intervention in loading and unloading. Figure 5.8 is a schematic showing the slat conveyor modification of the product flow. The castings are loaded on the left side through a volute casing to regulate the flow, and they exit on the right side. The impeller is mounted at the top. Two impellers can be mounted in line.

Figure 5.9 (Ref 2) is a sketch of the entire machine showing the volute intake, impeller, bucket elevator, and inspection doors. Sizes and styles slightly vary depending on the manufacturer.

5.2.2.2 Table Shot Blasters

5.2.2.2.1 Swing Table Blasters

Table-type shot blasters consist of one or two tables that swing out from the cylindrical cabinet to receive the castings, usually using an overhead jib or a gantry crane. Once the castings are laid flat, the table swings into the shot blast chamber , the doors close and the table starts to rotate.

Some designs provide a set of rotating tables mounted over the main table, as shown in Fig. 5.10(a). The individual sectional tables rotate in a direction opposite to that of the main table, improving the extent of exposure for the impinging shot stream. The impeller wheel is positioned above, for most efficient shot impingement. Upon completion of the cycle, the door of the chamber opens, and the table swings out. The castings are turned over to expose the bottom area to the blast cleaning.

Fig. 5.7 Barrel-type tumble blasting machine. Source: Ref 8

Figure 5.10(b) is a schematic showing the parts laid out on the rotary table for cleaning; it illustrates the loading and working positions of the table.

Swing table-type shot blasters are suitable for medium-sized castings where a crane is needed for handling.

5.2.2.2.2 Twin Table Blasters

The throughput of single swing table blasters is affected by the time taken for the table to swing out and the time to turn the

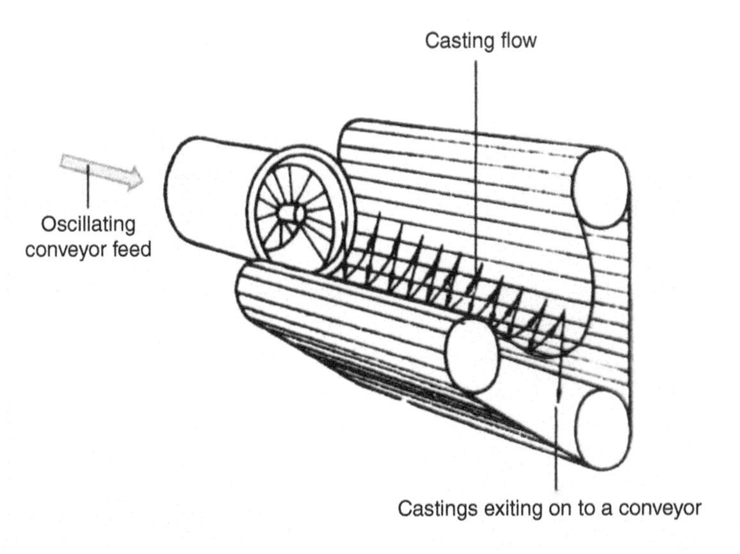

Fig. 5.8 Continuous barrel-type blaster concept

castings. The twin table machines enable the placement of the work on one table while the other table is inside the cabinet under the impeller wheel cleaning the castings, thus improving the productive time.

Figure 5.11 illustrates a twin table blaster which is suitable for medium-sized castings (Ref 8). The castings are laid out on one table while the castings on the other table are being cleaned by blasting. This configuration eliminates downtime because the machine permits the arrangement of the castings on one table while the second table is working inside the blast cabinet.

5.2.2.3 Cabinet-Type Blasters

Cabinet-type blasters are used for high throughput of medium- and large-sized castings. Medium-sized castings are suspended on hooks that rotate, giving good exposure to the impinging shot stream on the inside and outside of the casting. Large-sized castings are placed on tables inside the blast chamber, and the casting position is adjusted after each cycle.

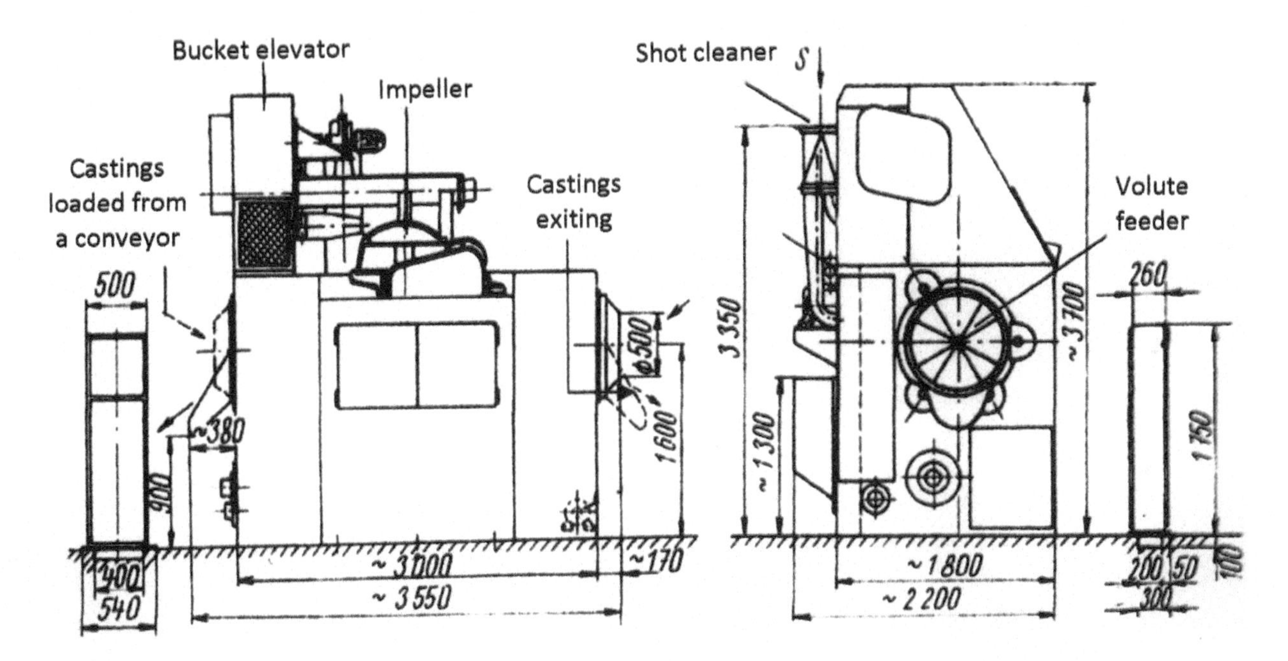

Fig. 5.9 Continuous drum blasting machine

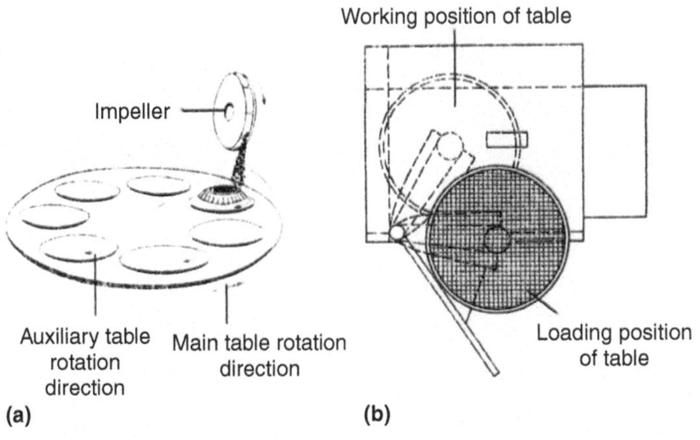

Fig. 5.10 Table-type shot blasting machine

Fig. 5.11 Twin table-type shot blaster. Source: Ref 8

5.2.2.3.1 Dual Chamber Spinner Blasters

The blasting unit in dual chamber spinner blasters consists of two sections positioned at 180° to each other. Some designs feature more than two sections. While one section is exposed to the shot stream for cleaning, the castings are hung on the opposite side hanger, thus saving loading time.

Figure 5.12 is a schematic of the setup, and Fig. 5.13 is an illustration of the machine (Ref 8).

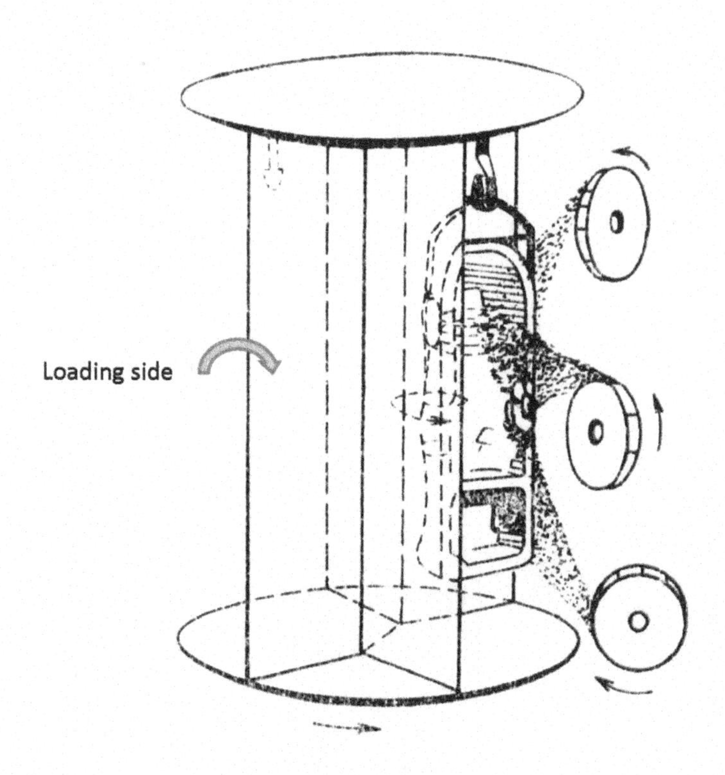

Fig. 5.12 Dual-chamber spinner shot blaster

Fig. 5.13 Multichamber spinner shot blasting machine. Source: Ref 8

5.2.2.3.2 Monorail Conveyor Cabinet Shot Blasters

Monorail conveyor cabinet shot blasters are versatile machines that can handle various mid-sized castings, which are suspended on hangers. The rotating hangers are suspended on a monorail that moves the castings into and out of the shot blasting cabinet. Multiple impellers at different heights and angles impinge the steel shot for high-quality cleaning.

Figure 5.14 is a schematic of a cabinet-type blaster with a monorail for transport into and out of the cabinet. The monorail extends out of the cabinet for continuous loading and unloading. These machines can be automated with robots. Figure 5.15 shows a complete machine (Ref 4).

Cabinet table-type machines are suitable for medium- and large-sized castings handled by an overhead crane. Multiple impeller heads mounted at different heights and angles are needed to shot blast the castings. Figure 5.16 shows the cabinet table-type machines.

5.2.3 Safety Features in Shot Blasting Machines

Shot blasting machines are designed with several operator safety features. The main safety features are:

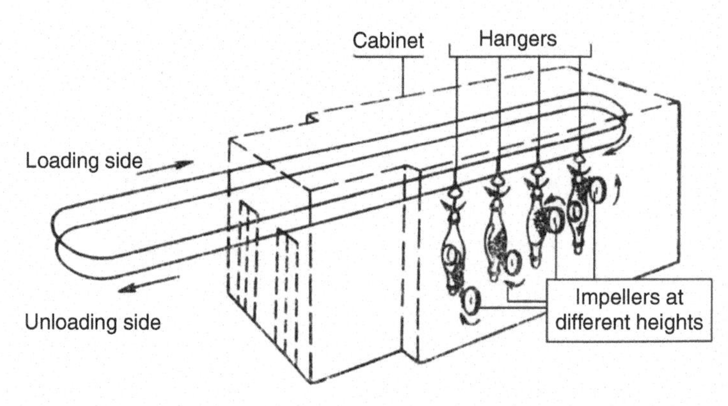

Fig. 5.14 Monorail hanger cabinet shot blaster schematic

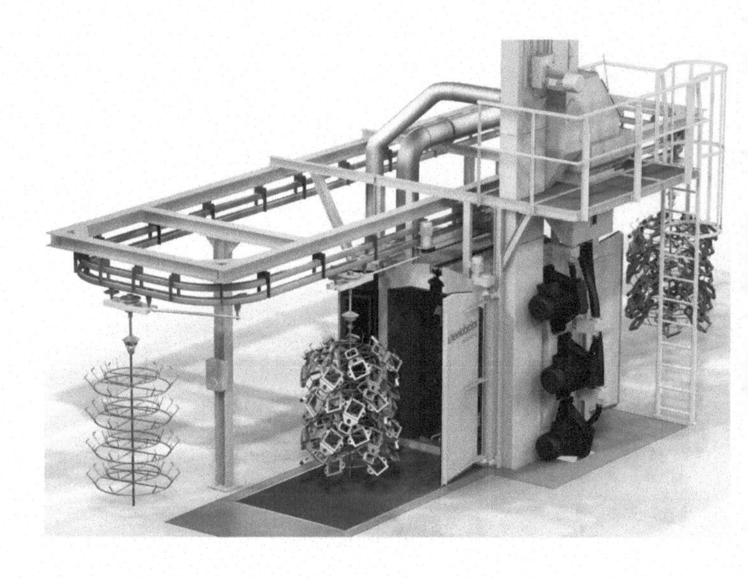

Fig. 5.15 Monorail hanger cabinet shot blasting machine. Source: Ref 4

- Safety light curtains on either side of the barrel shot blaster loading tote for forklift operator safety
- Protective grating on the side for mechanical protection against any castings falling out while unloading
- Control panel outside the safety curtains and protective grating
- Through belt for unloading into the baskets without operator intervention
- Dust collection system with filters and bags to suck up the sand and dust particles without injuring the operator
- Operators loading and unloading wear hearing protection
- Facilities where castings are handled by overhead cranes require the operators to wear hard hats.

5.3 Decoring, or Removal of Cores from Castings

The flow diagram in Fig. 5.1 shows that the decoring operation follows shot blasting. Heavily cored cast iron or compacted graphite iron castings such as cylinder heads and engine blocks produced in sand molds need an additional operation of decoring for sand removal from the intricately cored passages , following shot blasting.

Semipermanent aluminum castings with intricate sand cores ,such as cylinder heads, cylinder blocks, crank cases, and engine front covers produced by using steel permanent molds in gravity or low-pressure casting processes, are not shotblasted to prevent any surface damage on the soft external surfaces. The operations for decoring and cleaning the water passages are carried out soon after the castings are ejected out of the molds, skipping the shot blasting operation.

5.3.1 Decoring Machines

Cores produced with organic binders collapse easily at pouring temperatures, making their removal easy during the shakeout operation; the cores are removed by vibratory or rotary units. Cores that are used for water jackets of castings such as engine cylinder blocks, cylinder heads, and crank cases require manual assistance for thorough cleaning of the passages, after the shakeout. These passages require ultrasonic inspection to make sure that there are no core-butt remnants stuck in the passages. They require aggressive cleaning and washing, to ensure that the residue after the wash contains no suspended particles.

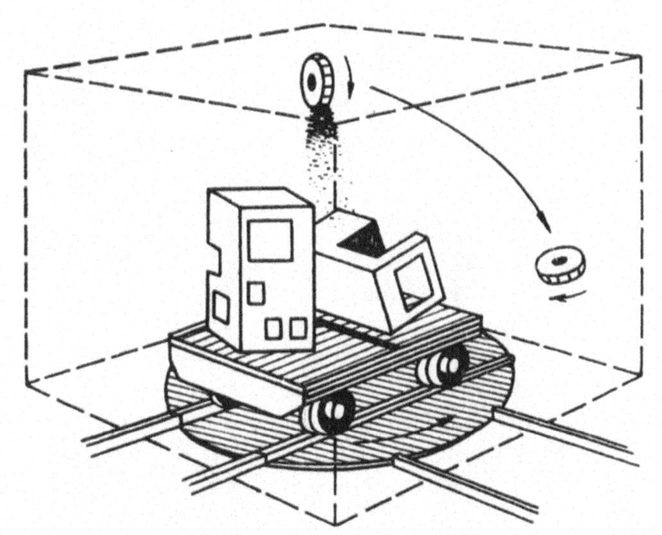

Fig 5.16 Cabinet table-type shot blasters for very large castings

Cores produced with inorganic binders such as sodium silicate require additives to the binder such as magnesium oxide, magnesium carbonate, or calcium fluoride to ensure that the cores lose their strength at elevated pouring temperatures. Ceramic cores used in investment casting require hydro-blasting for decoring.

Vibratory shakeout units (illustrated in Fig. 3.28 in Chapter 3 of this book) and rotary shakeout units (Fig. 3.29) are adequate to remove less intricate cores produced with organic binders. The core sand normally gets mixed with the green sand and is recycled. No-bake molds and cores disintegrate over the rotary shake-out units, and the castings are separated manually or robotically and hauled over a vibratory or oscillating conveyor. The castings are sorted, and the gating of gray and white iron is hammered out over the vibratory conveyor trough. The sand passing through the grating underneath is processed for recycling.

Castings produced in gray iron, compacted iron, and aluminum tend to be more intricately cored compared to castings produced in ductile iron and steel. This is because of the castability and thin wall capability of these materials. Intricately cored castings such as cylinder heads and engine blocks need special machines to remove the cores from the castings, and high acceleration is required to disintegrate the cores. It is also essential that these machines be engineered properly to enable cleaning without damaging or cracking the castings. Aluminum castings tend to have larger feeders over which the impact hammers impinge, transmitting the vibrations to the cored passages on the casting. Gray iron and compacted iron castings do not dent easily, and the impacting surfaces are chosen on the machined surfaces so that any indentations can be machined out. Sometimes an impact pad is added to the casting, which is machined off later.

Figure 5.17 is a schematic of a decoring machine for an aluminum cylinder head. Decoring machines have features such as:

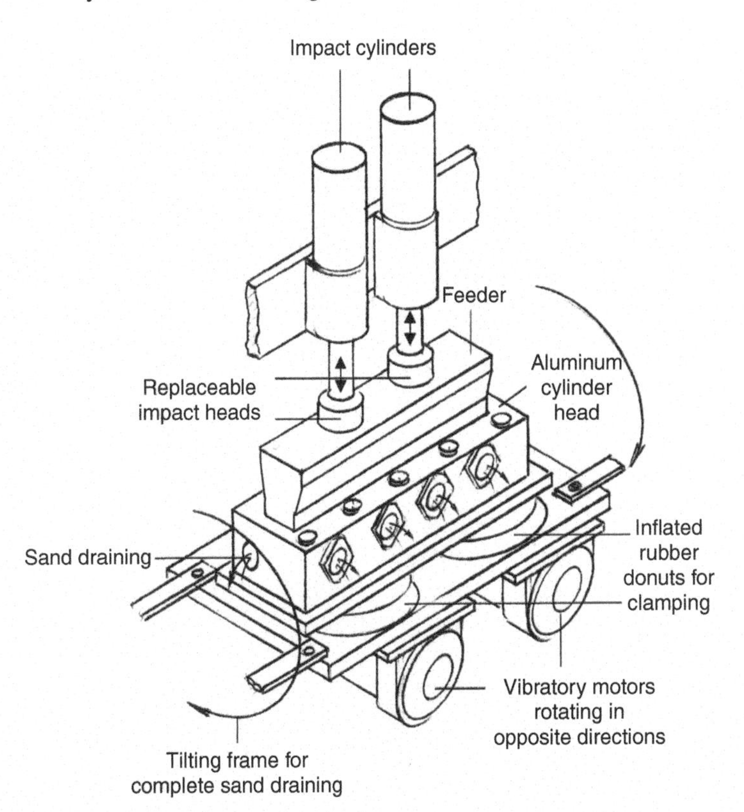

Fig 5.17 Decoring machine schematic

- A soundproof cabinet for isolating the noise. This is connected to an exhaust unit for dust control.
- A sturdy frame on which to mount the vibrating motors, which generate high acceleration to break the cores. The frame is designed for tilting to empty the core sand.
- A fixed frame that supports the high-frequency hammer heads (with replaceable heads) for separating the core from the casting. The hammers are capable of amplitudes up to 36 mm.
- A pair of air bags to hold the castings against the hammer heads
- Four isolators from which the frame is suspended
- A trough to collect the core sand and a conveyor to haul the sand out of the cabinet
- A robot for automatic loading and unloading for large volume castings in capital foundries

5.3.2 Inspection Instrumentation for Core Remnants Checking

It is critical that there should be no core sand remnants locked in the intricate cored cooling water passages after decoring. A borescope is a useful instrument for inspecting the inside of the cored passages. If the cores have cracked during pouring, there could be flash that would impede the flow of water during the operation of a casting.

5.4 Degating, or Removal of Runners, Gates, and Feeders

The secondary operations for the removal of runners, gates, and feeders should be engineered to incur minimum cost, no damage to the casting, and maximum safety to the operators.

The casting material influences the nature of the gates and feeders as well as the method for their removal. Iron castings typically have thinner and longer gates; they have small diameter vents and feeders, unless they are thick like ingot molds or rolls. Ductile iron castings have blind feeders, and the runners and gates are designed to feed into them. A relatively high-impact force is needed to remove the feeders from ductile castings. Special tools can be used to support their removal. Steel castings have larger sized feeders and shorter runners which cannot be removed by hammering, so they have to be cut out using oxy-acetylene torches or electric arc cutters.

5.4.1 Removal of Runners and Gates from Gray and White Iron Castings

Gatings in gray and white iron castings are brittle enough to be separated from the castings during shakeout. Any remaining pieces are manually hammered off as the castings pass through the vibratory or oscillating conveyor. Sometimes a small notch is provided at the gate-casting interface which provides a stress point and makes separation easier.

Rotary drum shakeouts separate the gates and runners effectively in gray iron castings. But it is advisable to break the runners into smaller pieces for ease of charging. Long runners can be crushed using a jaw crusher, illustrated in Fig. 5.18 (Ref 5). The crushing of gates and runners helps to increase the density of packing in the charging bucket, which helps reduce furnace charging time.

5.4.2 Removal of Runners and Feeders from Ductile Iron Castings

Runners on ductile iron castings are tougher, and more impact force is required to degate them. Feeders on ductile castings are separated from the casting using a hydraulic tapered wedge that pries the side feeder off the casting. Ductile castings such as steering knuckles, brake calipers, and differential carriers are produced in sizeable quantities that justify automation. In many of the ductile iron castings, the gates or runners feed into blind feeders. In such cases, the separation of the feeder from the castings is adequate.

Figure 5.19 (Ref 9) illustrates the use of a hydraulic tapered wedge to pry the blind feeder from a brake caliper casting. The gates and runners that are attached to the feeder are automatically separated from the casting.

Gates and feeders can be removed using circular saws or milling cutters in an automated operation. Circular castings are gripped in the center and rotated slowly. As the castings rotate, the feeders on the periphery pass through a roller which presses against the feeder, using hydraulic pressure. The feeder neck cracks under the pressure, and the feeder is separated from the casting without damaging the casting.

5.4.3 Removal of Runners and Feeders from Steel Castings

Manufacturers of medium-sized carbon and steel castings use a neck-down feeder. These feeders are approximately 120 to 150 mm (4.72 to 5.9 in.) in diameter and are knocked off with a hammer strike. Figure 5.20 (Ref 1) illustrates the use of standard feeders shown in view A and neck-down feeders in view B.

Standard feeders are removed using either an oxy-acetylene torch, an electric arc cutter, or a circular cutoff saw. Steel casting manufacturers use manual operations for feeder removal if the volumes and diversity of the jobs do not justify the capital for dedicated equipment and high automation.

Chapter 8, "Aluminum Die Casting Equipment," and Chapter 9, "Gravity Permanent and Semipermanent Mold Equipment," in this book discuss the degating of aluminum alloy castings.

5.5 Flash Removal and Automation

Flash forms in mold halves, at parting planes, and in clearances between the core print cavity and the core.

Flash is formed more easily in gray iron castings than in ductile iron or steel because of the high fluidity of gray iron. The thin flash that is formed cools very quickly, resulting in brittle white iron. Flash formed in gray iron castings is removed during the shakeout operation. The remaining flash is easily broken by light hammering. The extent of flash formed in ductile iron and steel castings is significantly less because of the lower fluidity of those

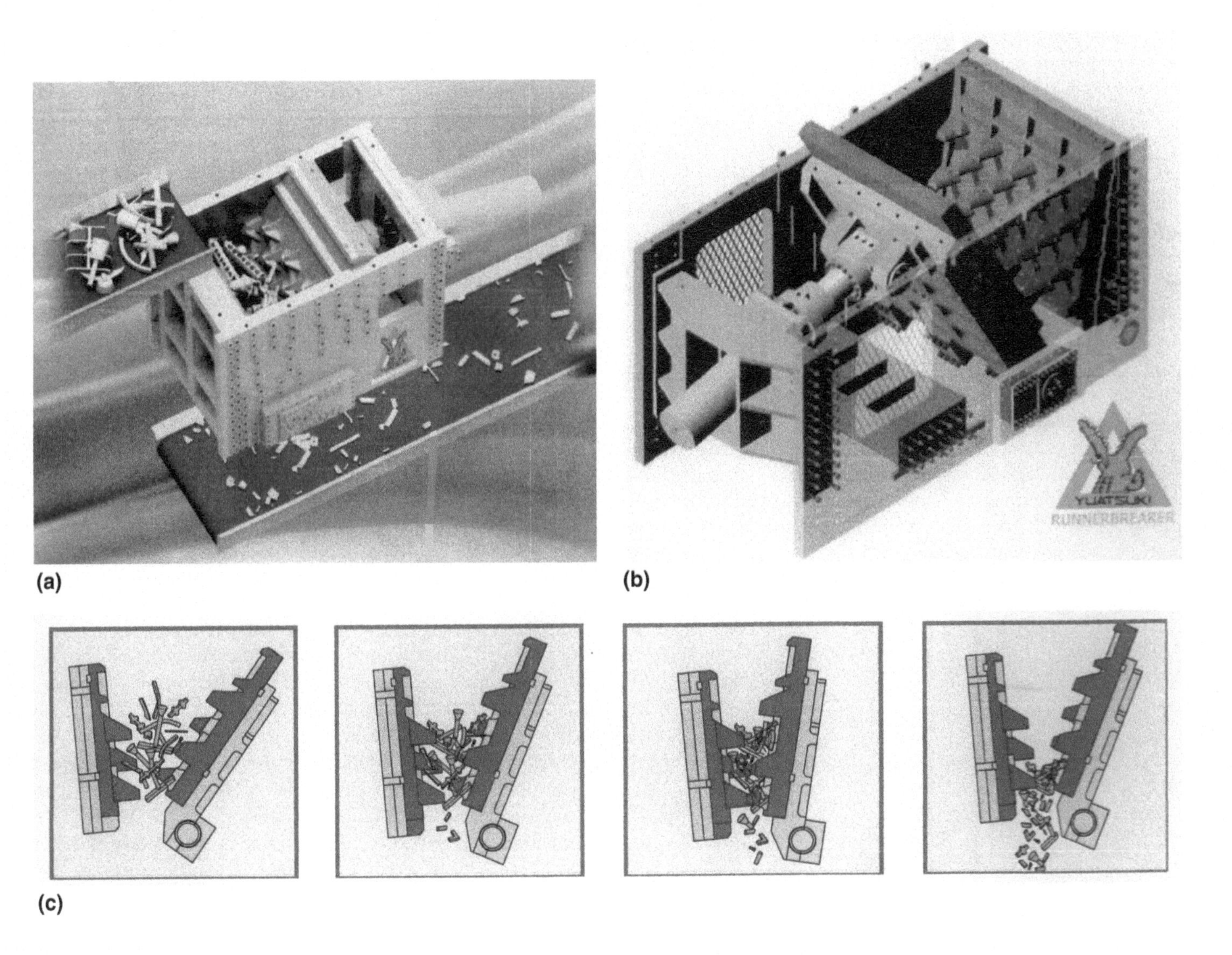

Fig. 5.18 Crusher for breaking runners and gates. Source: Ref 5

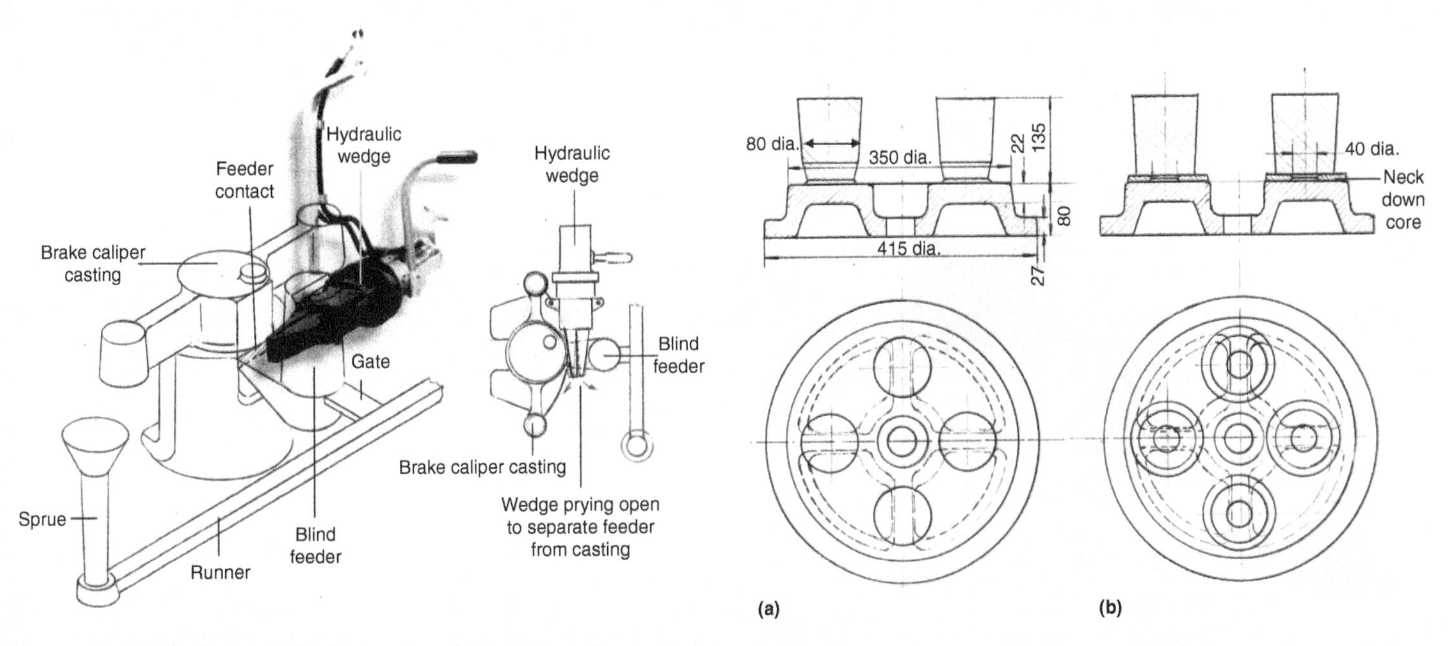

Fig. 5.19 Hydraulic wedge for feeder removal in a brake caliper casting. Source: Ref 9

Fig. 5.20 Examples of feeders, (a) standard, (b) neck-down

materials. Flash formed in ductile iron castings is tougher and needs to be hammered in small castings and chiseled in large castings. Flash formed in steel castings must be removed using an oxy-acetylene torch or by electric arc cutting.

5.5.1 Grinding Machine Types and Applications

Grinding machines use abrasive wheels which rotate at high speeds to remove the excess flash material and flash remnants. The wheels are driven by electric motors. Some hand-held grinders are pneumatically operated, which makes them lighter and easier to handle.

Grinders vary in style, depending on the applications and size of the castings. The extent of automation depends upon the annual volume of the type of castings. Figure 5.21 provides an overview of the different types.

Small castings, less than 10 kg (22 lb.), are manually ground on pedestal grinders, where the castings are manipulated and pressed against the rotating wheel for material removal. Heavier castings, about 50 to 100 kg (110 to 220 lb.), supported on stands, are ground by lightweight hand grinders operated by pneumatic air or by electric motors. Adjustable stands provide for ergonomics, which improves operator safety and productivity. The manipulation of castings is accomplished by a jib crane operated by the grinding operator.

Larger castings are ground using swing frame grinders, where the rotating grinding wheel is suspended on a frame that is controlled by the operator. The grinding wheels can be in-line or perpendicular. The turning and positioning of the castings is accomplished using a gantry or an overhead crane.

Automation is justified when small to medium-sized castings are produced in large volumes. Computer numerically controlled (CNC) machines are popular for such applications. Larger castings in medium volumes may benefit from robotic grinding where the movement of the grinding head and the orientation of the casting in a fixture are both computer-controlled.

Figure 5.22 (Ref 2) shows a large pedestal grinder for small to medium-sized castings in gray and ductile iron. The motors are housed in the middle casing. Suitable castings for this type of operation, are castings for agricultural implements, textile machinery, air compressors, and low-volume truck components. Smaller castings such as pipe fittings require smaller pedestal grinders. Such castings produced in large volumes are good candidates for automated tumbling for flash removal.

Figure 5.23 illustrates pneumatic and electric hand grinders. These are used for middle- to large-sized castings that can be handled by a jib or gantry crane to position the castings on an adjustable working table.

Adjustable height worktables or scissor tables help position the castings at a height suitable for the operator. Figure 5.24 shows one design for an adjustable worktable.

Swing frame grinders allow the grinding of large steel castings for a variety of heavy industries such as crushers for ores, large valves and fittings for chemical industries, and fluid handling. The grinders are suspended and balanced to enable the operator to

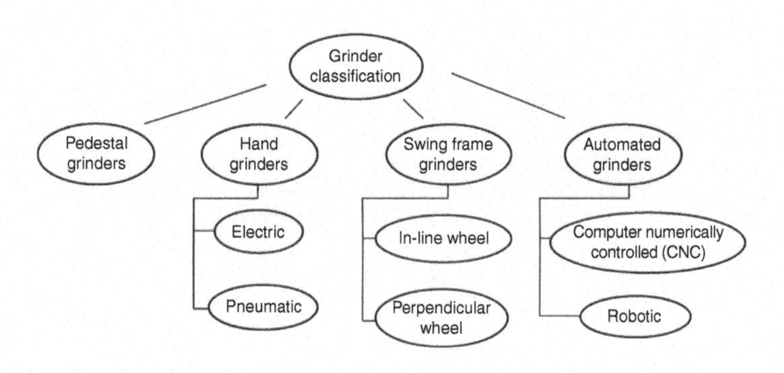

Fig. 5.21 Classification of grinders

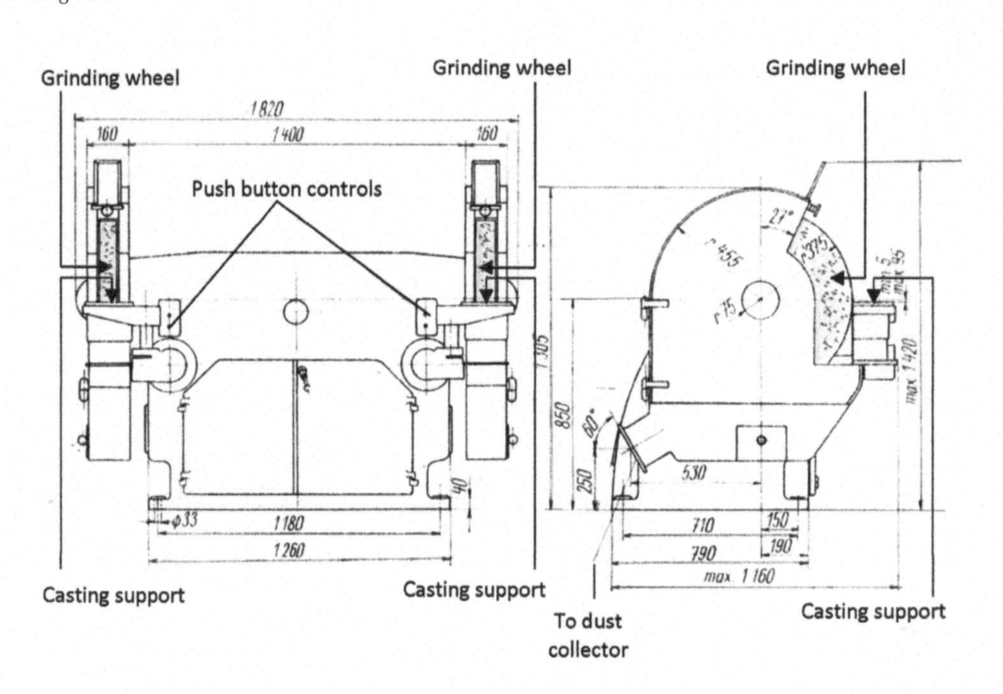

Fig. 5.22 An example of a pedestal grinder

manipulate the grinder easily. Figure 5.25 (Ref 2) shows a swing-frame grinder with the grinding wheel in line with the swing arm. Design with the grinding wheel perpendicular to the swing arm axis offers an improved view of the area needing to be ground.

5.5.2 Automated Grinders

Automation of grinding is justified when the annual volume of the castings is high. Examples of when automated grinders are practical include gray and ductile iron automotive castings, such as steering knuckles, control arms, and brake calipers. CNC machines provide the best solution for these high-volume medium-sized jobs.

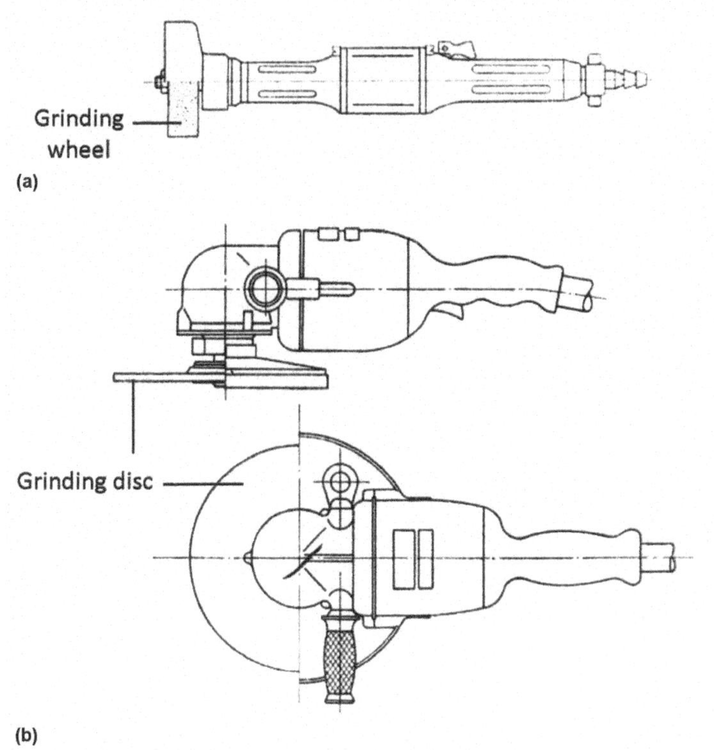

Fig. 5.23 Hand grinders, (a) pneumatic, (b) electric

Another area suitable for automation is when the medium- to large-sized casting is too heavy for manual positioning and the products are repetitive and produced in medium annual volumes.

5.5.2.1 Computer Numerically Controlled Machine Grinding

Computer numerically controlled (CNC) machines are used for castings that are produced repetitively in large annual volumes. They are versatile because they can be programmed for different parts. Examples of where the CNC machines are used include gray and ductile iron automotive castings, steel valve bodies, tractor components, and highway vehicle components.

Figure 5.26 (Ref 10) illustrates the casting method and subsequent CNC machining concept for burr removal in a ductile iron brake caliper.

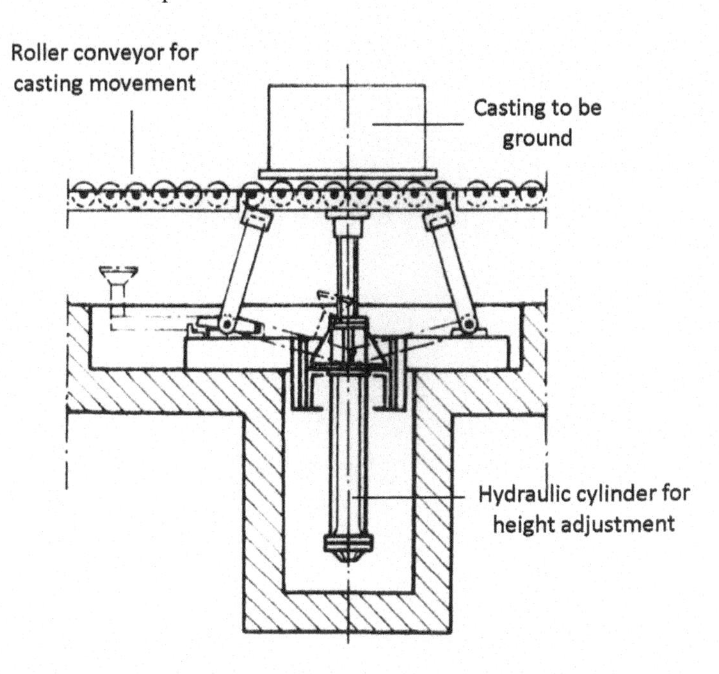

Fig. 5.24 Hand grinding adjustable table

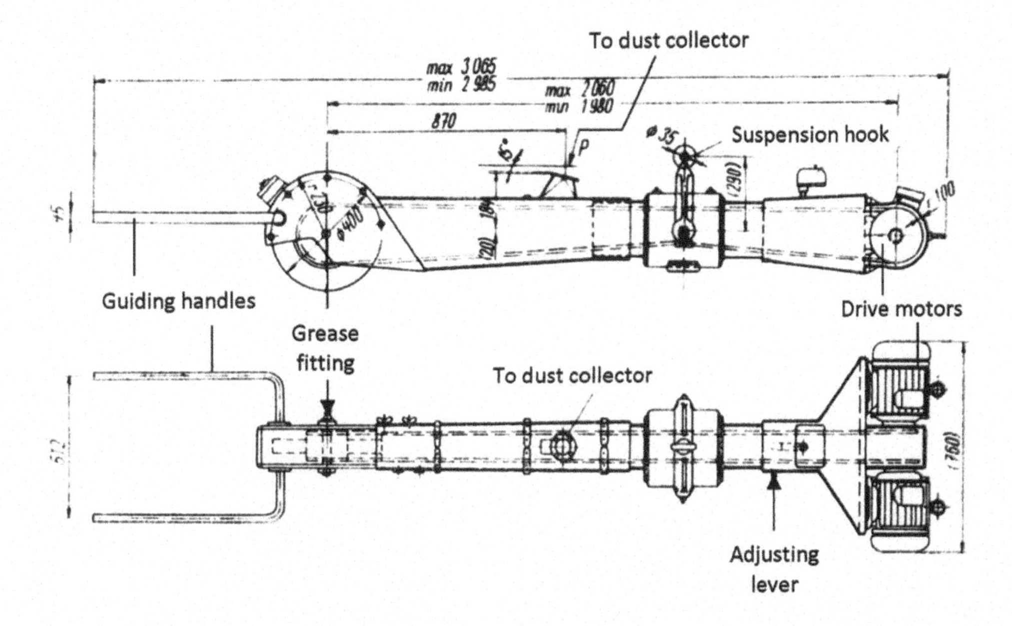

Fig. 5.25 Swing-frame grinder

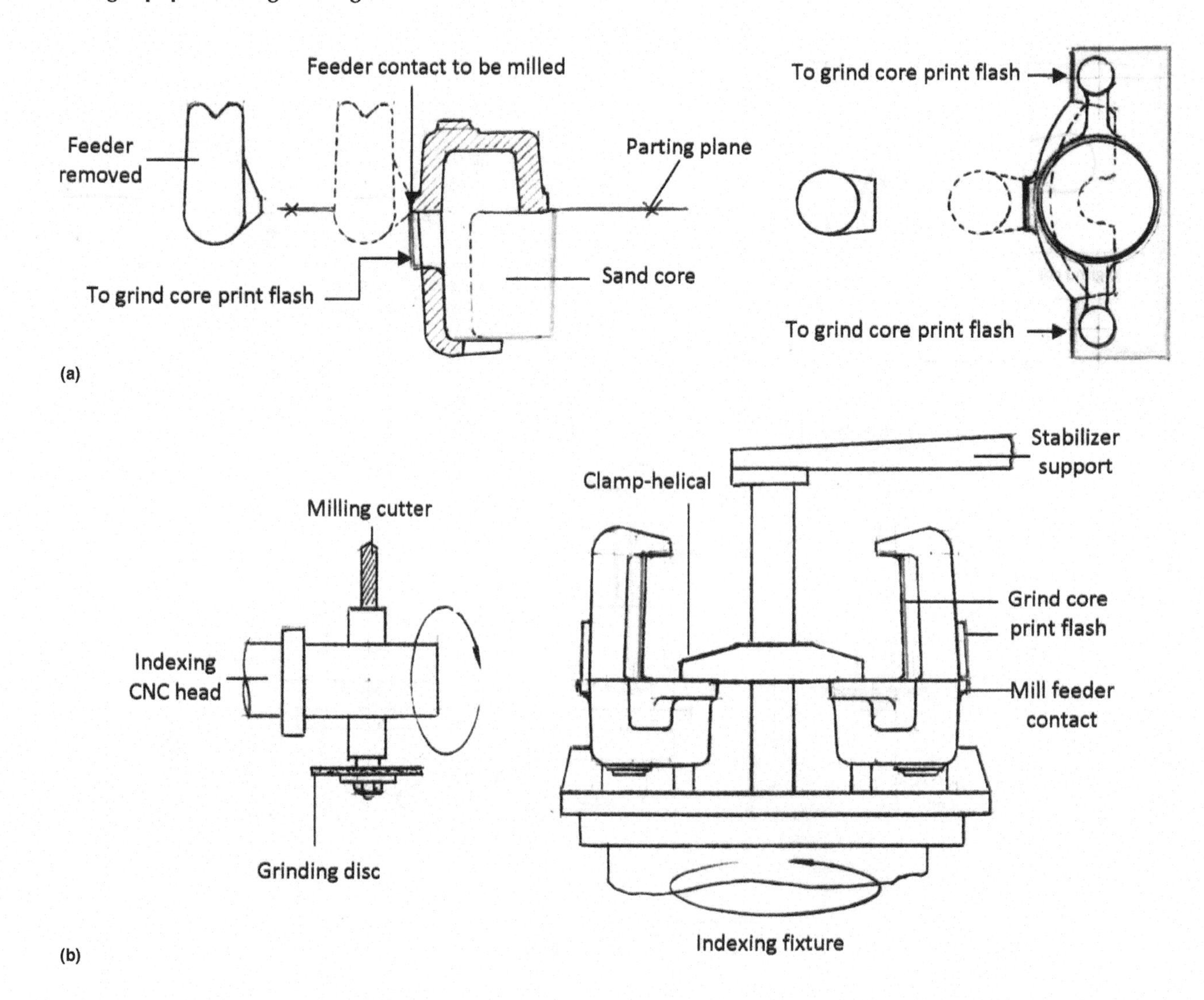

Fig. 5.26 CNC finishing concept for a brake caliper

Figure 5.26(a) shows the casting concept. The caliper is parted between the cup and the bridge. One core combines the cup and the cooling window opening. Flash occurs around the core prints, as shown in red, and the remnants of the flash on the casting need to be ground. The feeder contact-face may be jagged after removal of the feeder and if so, it requires milling.

The fixture for two calipers that face each other is mounted on an indexing table to grind or mill one caliper at a time. The CNC head carries a disc grinder on one side and a milling cutter on the opposite side. The disc grinder and the milling cutter are indexed to grind or mill alternately.

Some manufacturers offer turnkey solutions and machines with different sizes of envelopes and maximum payloads. Figure 5.27 (Ref 10) illustrates the turnkey concepts of these options. Table 5.1 (Ref 10) lists the available product portfolio.

5.5.2.2 Robotic Grinding

Medium-sized jobs (1500 to 2000 kg or 3307 to 4409 lb.) in medium annual volumes justify the use of robotic grinding and feeder removal. Steel castings for railroad applications, gear wheels, and large valve bodies are typical candidates. Figure 5.28 illustrates a concept of robotic grinding where the part is oriented by the fixture and the tool head is manipulated by the robot.

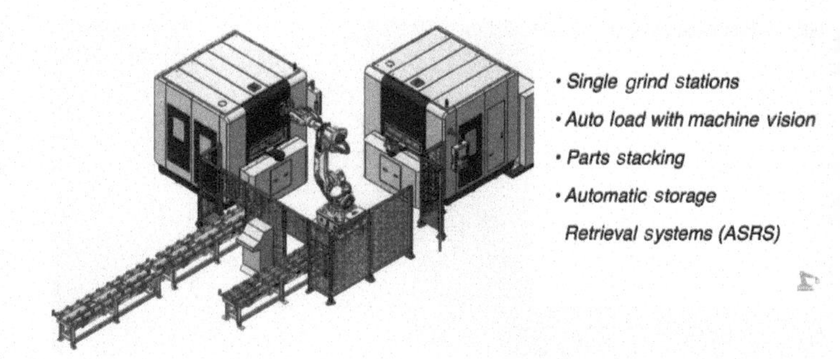

Fig. 5.27 CNC turnkey solution offerings. Source: Ref 10

5.5.3 Guidelines for Grinding Wheel Selection

A grinding wheel is a cutting tool with multiple cutting edges and bonded adhesives, dispersed with pores. The pores help to discard the metal chips generated during grinding. Grinding wheels contain two main components: abrasives and bonds.

Figure 5.29 summarizes the four types of abrasives and the three types of bonds. Different alloys and varied applications require specific combinations of the abrasive grain and the bond.

Table 5.2 (Ref 11) summarizes the factors that influence the selection of grinding wheels and the significance of these factors.

Table 5.3 (Ref 12) lists the general BHN hardness ranges of cast ferrous materials that are ground or cut during the cleaning operations.

Table 5.4 provides general information on the abrasives and bonding materials for different ferrous alloys.

Table 5.1 Maximum payloads and work envelope of machines offered

Property	Model					
	GS 40	GS70	GS80	GS120	GS160	GS200
Work envelope(a), diam × H, mm (in.)	550 by 300 (22 by 12)	750 by 400 (29.5 by 15.7)	1000 by 550 (40 by 22)	1300 by 750 (51 by 29)	1600 by 900 (62 by 35)	2000 by 1000 (78 by 39)
Payload(a), max, kg (lb)	50 (110)	100 (220)	250 (550)	400 (880)	1000 (2200)	1500 (3300)

(a) Part plus fixture. Source: Ref 10

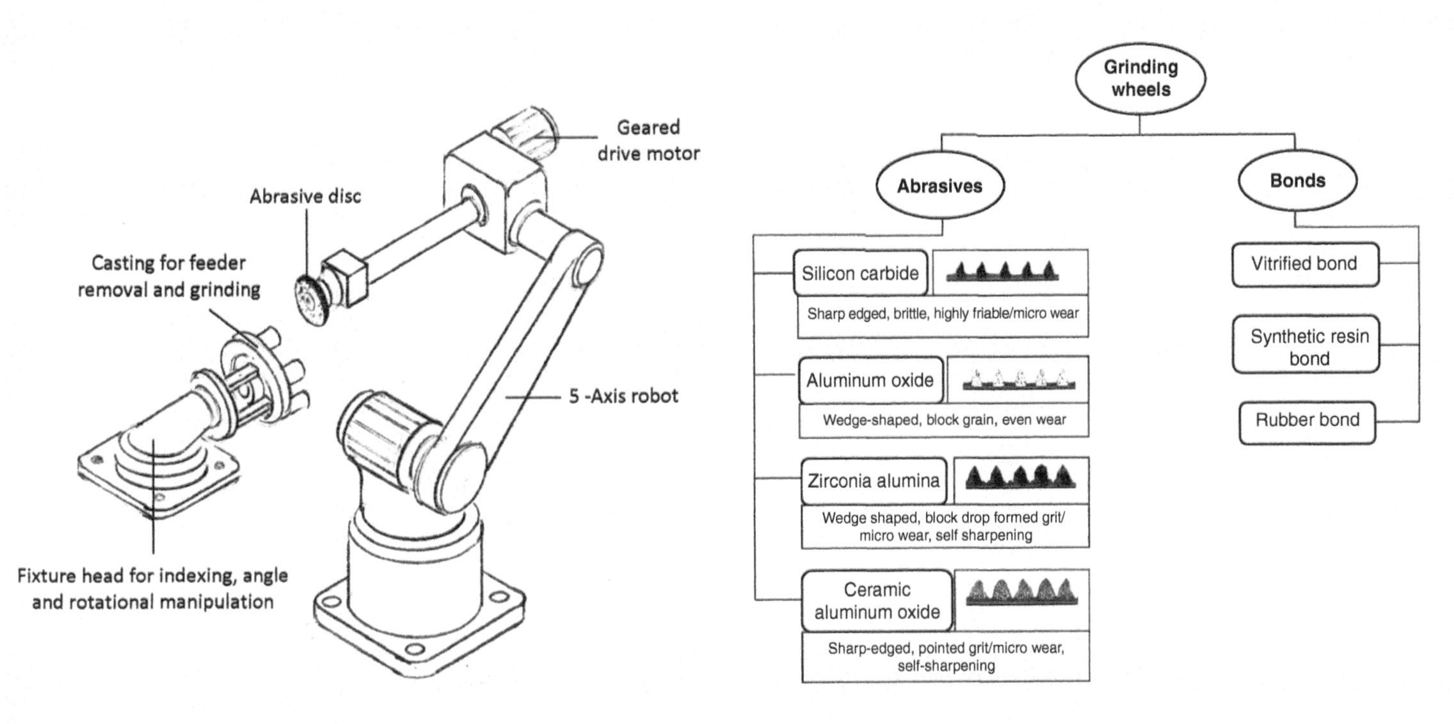

Fig. 5.28 Robotic feeder removal and grinding concept

Fig. 5.29 Grinding wheel constituents

Table 5.2 Wheel selection factors

Factor	Significance
Material hardness	The abrasive and bond depend upon the hardness of material. In general, the harder the material, the softer the wheel grade.
Stock to remove	The grit or size of grain depends upon the amount and rate of material removal. The coarser the grain, the larger the material removal rate. Soft grade wheels remove material faster.
Surface finish and shape	Synthetic resin bonds and higher wheel speeds produce higher finishes. Rubber bonded wheels produce very high finish.
Horsepower and machine type	High horsepower is needed for hard grade wheels for longer life. High grade wheels and higher horsepower needed for a narrow surface contact area. Vitrified wheels are used at < 6500 feet per minute (1981 meters per minute). Organic bonded wheels work from 6500 to 9500 feet per minute (1981 to 2896 meters per minute).
Wheel speeds and feeds	It is important to follow manufacturers' guidelines for safety and efficiency.
Size and hardness of contact area	Use for large contact area use – coarser grit and softer grade. Fine grit wheels work best for grinding hard materials. Medium and coarse grit wheels are best for softer materials.
Wet or dry grinding	Wet grinding is needed for higher surface finishes. Dry grinding requires dust extraction.
Severity of grinding	High severity of grinding may require coated special abrasive wheels.
Dressing method	Worn wheels need to be conditioned, shaped, and re-sharpened using a dressing tool. The dressing method influences the choice of the grinding wheel.

Source: Ref 11

5.6 Heat Treatment

Heat treatment following casting cleaning is an important process for property enhancement and stress reduction. Heat treatment objectives are to:

- Relieve internal stresses and improve dimensional stability
- Reduce hardness, eliminate carbides, and improve machinability
- Improve wear resistance
- Increase strength and hardness
- Increase ductility and toughness

Each of these objectives involves heating the casting below or above the critical temperature and holding the temperature for a set duration for homogenization, followed by controlled cooling to modify the microstructure. The temperature range depends upon the type of heat treatment and the alloy content. The holding times and targeted microstructures depend upon the section thicknesses.

5.6.1 Stress-Relieving of Cast and Ductile Iron Castings

Internal stresses develop in castings due to uneven cooling of different section thicknesses and the resistance of mold and cores to casting contractions. The internal stresses affect dimensional stability because they are naturally relieved over the course of several days after casting. Typical examples of this are long machine tool beds and engine cylinder blocks. The stress-relieving operation relaxes the internal stresses and imparts dimensional stability. Precision in machining and dimensional stability are both achieved by stress-relieving. Figure 5.30 (Ref 13) provides an overview of the stress-relieving cycles as shown in A.

Table 5.5 (Ref 13) provides guidelines for times and temperatures for different wall thicknesses and material grades of gray and ductile iron castings

Table 5.3 General Brinell hardness numbers (BHN) of castings

Material	Condition	BHN range
Gray iron grades	As cast	170 – 269
Alloyed cast iron	As cast	260 – 350
Chilled cast iron	As cast	400 – 500
Austenitic heat resistant cast iron	As cast	120 – 248
Ni-hard	As cast	550 – 650
Malleable iron	Air quenched	163 – 217
Malleable iron	Liquid quenched	187 – 302
Ductile iron	As cast, < 60% pearlite	187 – 255
Ductile iron	As cast, > 80% pearlite	217 – 269
Ductile iron	Quenched and tempered	240 – 300
Austempered ductile iron (ADI)	Grade 1	269 – 341
ADI	Grade 2	302 – 375
ADI	Grade 3	341 – 444
ADI	Grade 4	388 – 477
ADI	Grade 5	402 – 512
Carbon and low alloy steels	Normalized and tempered	179 – 241
Carbon and low alloy steels	Quenched and tempered	241 – 311
High strength steel -tooling	Quenched and tempered	415 – 477
Austenitic manganese steel	Quenched and tempered	350 – 450
High strength structural steel	Normalized and tempered	207 – 255

Source: Ref 12

5.6.2 Annealing of Gray Iron Castings

Castings are annealed to improve machinability by modifying their microstructure from pearlite to ferrite and graphite. Carbides in castings (seen in the microstructure of chilled edges) due to unintended chemical composition variances or high cooling rates can result in tool breakage, which can shut down automated machining lines. The different annealing cycles used are illustrated in Fig. 5.30 B, C, and D (Ref 13). The annealing cycles are designated as:

- Ferritizing annealing or subcritical ferritizing
- Medium or full annealing
- Graphitizing annealing
- Normalizing annealing

Table 5.6 lists the different types, objectives, and temperature ranges for heating and cooling rates of the annealing cycles.

5.6.3 Annealing and Normalizing of Ductile Iron Castings

The practice of annealing ductile iron castings is similar to that of gray iron. Table 5.7 summarizes the types of cycles and their objectives.

Normalizing results in a fine pearlitic matrix microstructure with good wear resistance, good machinability, and excellent flame hardening or induction hardening capability. Heavy casting section thicknesses may require alloying during casting or need for faster cooling rates in heat treatment, for the targeted microstructure. Fast cooling with fans may induce internal stresses which require a stress-relieving operation. Large complex castings with varied section thicknesses may be unsuitable for normalizing because of higher induced stresses and varying hardness in casting sections.

Table 5.4 Abrasives and bonding of abrasive wheels for cast ferrous materials

		Abrasives				Bonds		
Casting material	Rough grinding high stock	Silicon carbide	Aluminum oxide	Zirconia alumina	Ceramic aluminum oxide	Vitrified	Synthetic resin	Rubber
Gray iron and chilled iron		X				X		
Gray iron, high material removal	X			X			X	
Carbon steels, alloy steels, malleable iron			X				X	
Steels, cut off operations				X			X	X (for high finish)
Alloy steels, precision grinding					X		X	

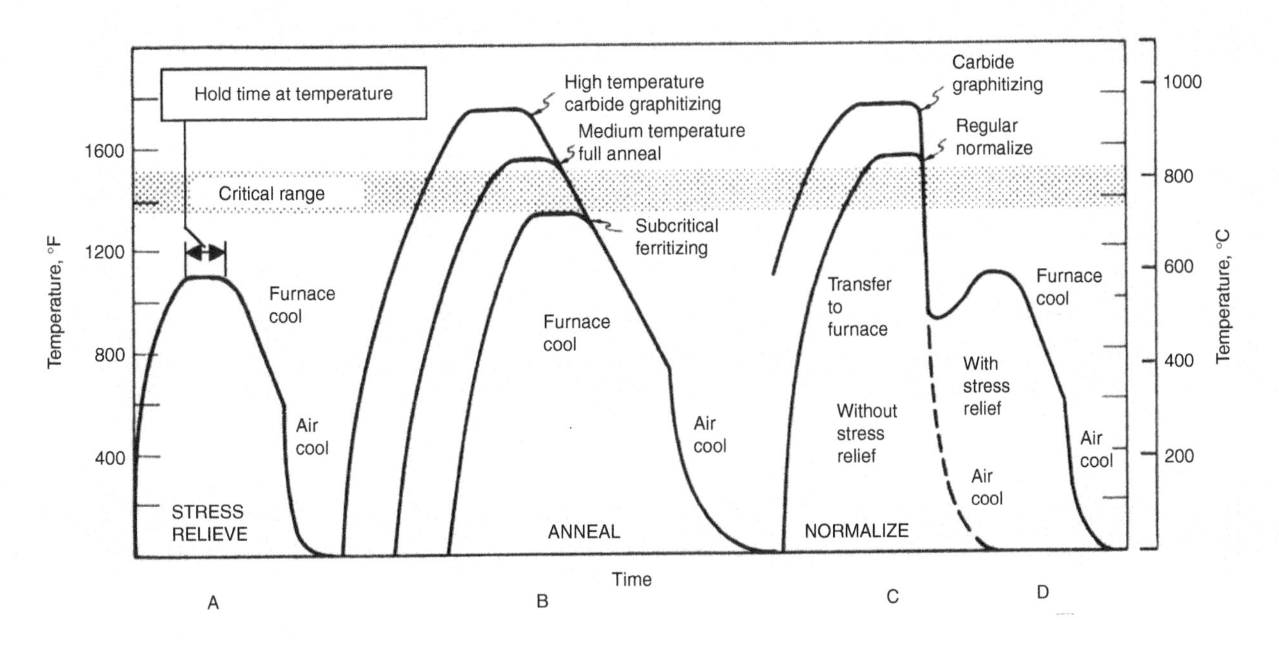

Fig. 5.30 Stress-relieving cycles (A), annealing cycles (B), and normalizing cycles (C, D). Source: Ref 13

Table 5.5 Stress relief cycle guidelines

Type and class of casting	Section thickness range, mm	Stress relieving cycle(a), °C (°F)
Unalloyed gray iron Class 20 and 30	Up to 50	2 h at 565 to 579 (1050 to 1075)
	50 to 100	1.5 h per 25 mm thickness at 565 to 579
	Over 100	6 h at 565 to 579 (1050 to 1075)
High strength and low alloy irons Class 40 to 60 and ductile iron	Up to 50	2 h at 565 to 579 (1050 to 1075)
	50 to 100	1.5 h per 25 mm thickness at 565 to 579
	Over 100	6 h at 565 to 579 (1050 to 1075)
High alloy irons	Up to 50	2 h at 595 to 649 (1100 to 1200)
	50 to 100	1.5 h per 25 mm thickness at 595 to 649
	Over 100	6 h at 595 to 649 (1100 to 1200)

Stress relief cycles A. (a)All temperatures listed are those of actual castings. Source: Ref 13

Table 5.6 Types of annealing cycles and their objectives

Type of annealing	Purpose	Temperature range, °C (°F)	Duration	Cooling rate
Ferritizing or subcritical ferritizing	Improve machinability by converting pearlite to ferrite and graphite	700 to 760 (1300 to 1400)	45 minutes per 25 mm casting section thickness	Furnace cool to 315 °C (600 °F) at 55 °C/h, cool in air from 315 °C to room temperature
Medium or full	Improve machinability by converting pearlite to ferrite and graphite - of alloy cast iron	815 to 900 (1500 to 1650)	60 minutes per 25 mm casting section thickness	Furnace cool to 315 °C (600 °F) at 55 °C/h, cool in air from 315 °C to room temperature
Graphitizing	Achieve maximum machinability by eliminating massive carbides and converting pearlite to ferrite and graphite	900 to 950 (1650 to 1750)	1 to 3 h plus 1 h per 25 cm of casting section thickness	Furnace cool to 315 °C (600 °F) at 55 °C/h, cool in air from 315 °C to room temperature
Normalizing	Eliminate massive carbides. Retain pearlite for maximum strength and hardness	870 to 950 (1600 to 1750)	1 to 3 h plus 1 h per 25 cm of casting section thickness	Air cool to below 480 °C (900 °F) as shown in C, subsequent stress relieving may be needed as shown in D

Annealing cycles B, C, and D. Note: Temperature ranges shown are actual casting temperatures. Source: Ref 13

5.6.4 Heat Treatment of Steel Castings

Steel casting heat-treating objectives are:

- Achieve targeted mechanical properties
- Stress-relieving
- Obtain corrosion resistance
- Avoid difficulties in finishing processes

Figure 5.31 (Ref 12) is a schematic showing the different thermal treatments and the modes of heating and cooling.

Table 5.7 Annealing cycles of ductile iron

Type of annealing	Purpose	Temperature range(a), °C (°F)	Duration	Cooling rate(b)
Ferrtizing or low temperature	Obtain grades 60-45-12 and 60-4-18 in irons with no carbides	720 to 730 (1325 to 1350)	1 h per 25 mm casting section thickness	Furnace cool to 345 °C (650 °F) at 55 °C/h, cool in air to room temperature
Full, low silicon irons	Obtain grades 60-45-12 and 60-4-18 in irons with no carbides with maximum low temperature impact strength	870 to 900 (1600 to 1650)	Until equalizing at control temperature	Furnace cool to 345 °C (650 °F) at 55 °C/h, cool in air to room temperature
Graphitizing, high temperature	Obtain grades 60-45-12 and 60-4-18 in irons with no carbides	900 to 925 (1650 to 1700)	Minimum of 2 h	Furnace cool to 700 °C (1300 °F) at 95 °C/h, hold at 700 °C for 2 h, furnace cool to 345 °C at 55 °C/h, cool in air
Two-stage annealing, graphitizing plus ferritizing	Obtain grades 60-45-12 and 60-4-18 in irons with no carbides, in castings where rapid cooling is practical	870 to 900 (1600 to 1650)	1 h per 25 mm casting section thickness	Fast cool to 675 to 700 °C(1250 to 1300 °F) at 2 h per 25 mm of casting section, reheat to 730 °C, air cool.
Normalizing and stress relief	Obtain grades 100-70-03 and 80-55-06 in irons with no carbides	900 to 925 (1650 to 1700)	Minimum of 2 h	Fast cool with fans to 640 to 650 °C (1000 to 1200 °F), furnace cool to 650 °C, furnace cool at 55 °C/h to 345 °C, cool in air

Note: (a) Temperature ranges shown are for actual castings; (b) Slow cooling from 540 to 315 °C (1000 to 650 °F) reduces residual stresses. Source: Ref 13

There are three basic methods of cooling castings that have been austenitized:

- Slow cooling in furnace — for full annealing
- Air cooling in atmosphere — for normalizing
- Rapid cooling by quenching in a liquid medium — for quench hardening

The most frequent heat treatment processes are:

- Normalizing followed by tempering
- Quench hardening followed by tempering
- Annealing or full annealing

5.6.4.1 Austenitizing

Castings are heated above the upper critical transformation temperature to transform pearlite and ferrite to austenite. The transformation temperatures depend upon the carbon and alloy contents of carbon and low alloy steels. High alloy steels need higher temperatures for the complete dissolution of carbides.

The time required for the castings to reach the targeted austenitizing temperature depends upon several factors, such as heat-treating equipment, furnace loading practice, and maintenance. Furnace loading practice depends upon the casting size and loading density. It is a good practice to attach thermocouples to the castings to register the actual casting temperatures instead of just using the thermocouples measuring the temperature of the furnace. The temperatures vary between the inside periphery and center of the furnace. The holding time shown in Fig. 5.31 refers to maintaining the temperature at a constant level for the center of the castings to reach the targeted austenitizing temperature, as measured by the thermocouples.

5.6.4.2 Normalizing

Normalizing consists of:

- Heating the castings to a temperature above the upper critical temperature
- Holding for a duration to complete transformation to austenite
- Removing the castings from the furnace
- Cooling in still air with air access around each casting (without forced-fan cooling)
- Tempering as needed

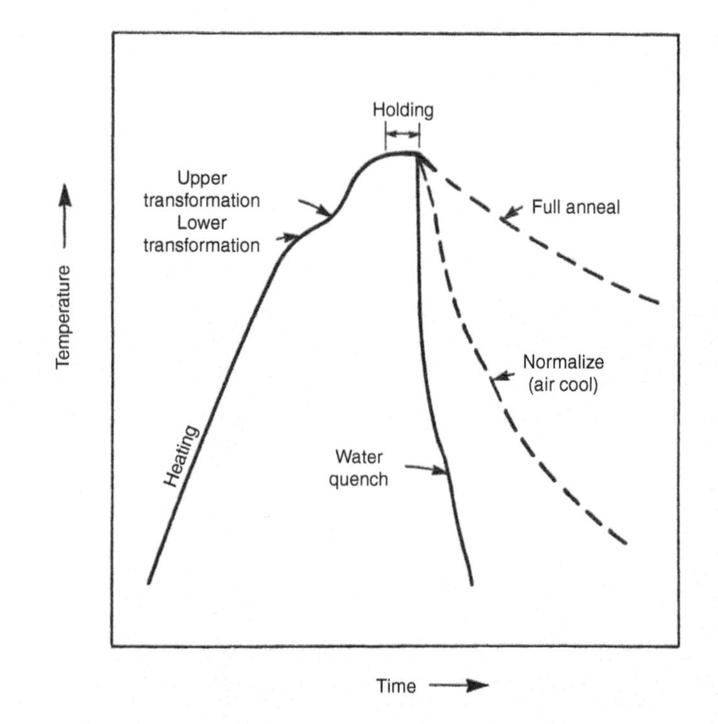

Fig. 5.31 Heat treatment modes. Source: Ref 12

Figure 5.32 (Ref 12) illustrates the temperature bandwidth required for austenitizing.

The resultant microstructure after normalizing is a mixture of pearlite and ferrite with good tensile strength, low residual stresses, no distortion, and good machinability. The normalizing process is economical and attractive for enhancing the mechanical properties of the castings.

The extent of cooling is important. Insufficient circulation of cooling air may result in annealing. Excessive forced-fan cooling may quench the castings. Normalizing is followed by tempering.

5.6.4.3 Quench Hardening

Quench hardening consists of:

- Heating the castings to a temperature above upper critical temperature, like normalizing

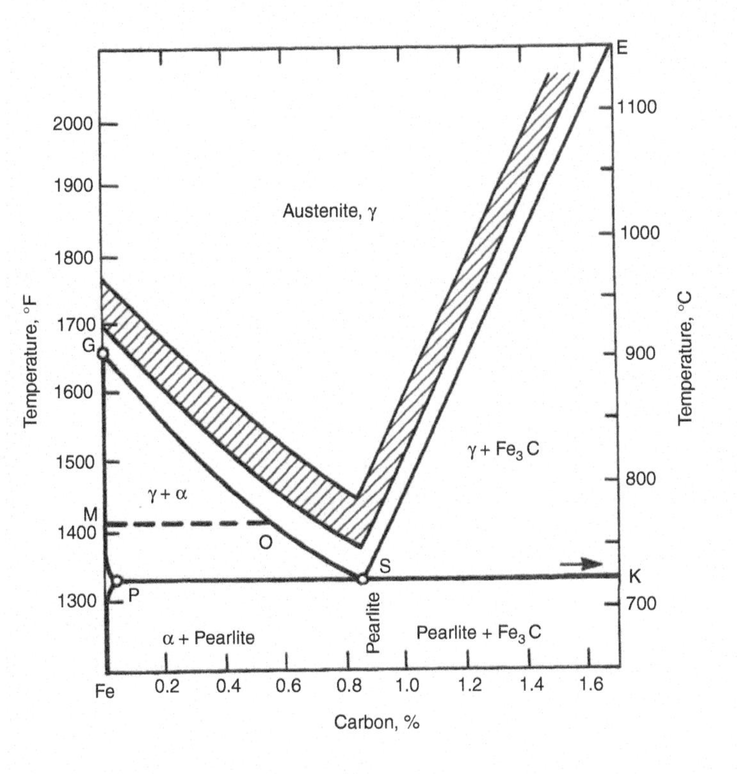

Fig. 5.32 Normalizing temperatures for carbon steels. Source: Ref 12

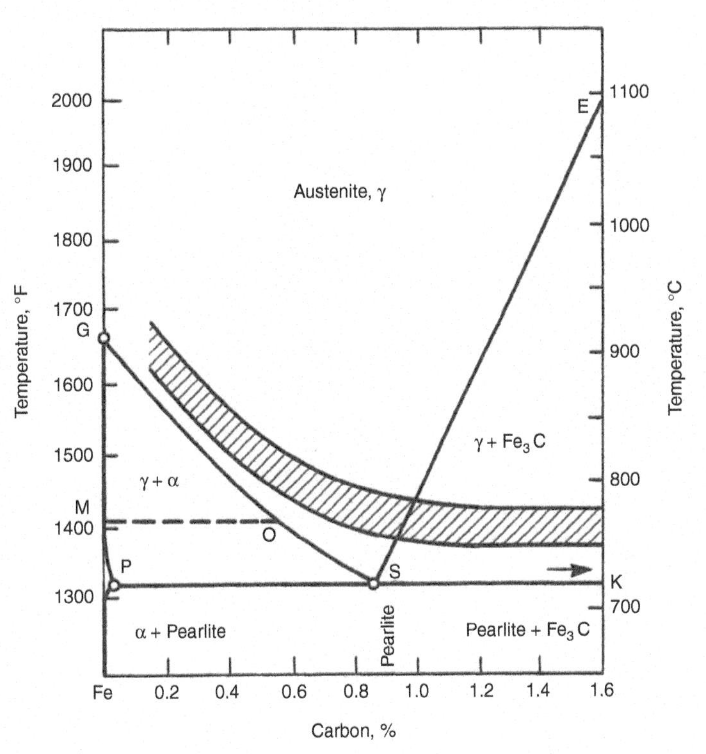

Fig. 5.33 Hardening temperatures for carbon steels. Source: Ref 12

- Holding for a duration to complete the transformation to austenite
- Removing the castings from the furnace quickly
- Rapidly quenching in a liquid medium fast enough to form martensite without the formation of ferrite and pearlite. (Note: the cooling curve should avoid the nose of the time-temperature transformation diagram, Ref 2)
- Tempering the castings

Figure 5.33 (Ref 12) illustrates the temperature bandwidth for hardening.

Quenching also prevents the re-precipitation of undesirable carbide phases that lower the corrosion performance. The hardenability of the alloy depends upon the chemical composition.

Alloys that are less *hardenable* require higher rates of quenching to obtain martensite. As the section size increases, the alloy content needs to be higher to ensure adequate quench rates in the center of the casting cross section. Quench tanks are provided with effective agitation systems to prevent the formation of vapor enveloping the casting. Maintaining the quenchant at the appropriate level for quench severity requires control of water bath temperature and a good agitation system.

Quench severity is influenced by several factors, such as quenchant temperature, turbulence of the agitation system, viscosity of the quenchant, and the mass of the load of castings in relation to the quench tank volume.

The quenching medium is mostly water. High carbon and deep-hardening steels need oil to reduce the quench rates and prevent cracking, but oil quenching creates a fire hazard. The addition of polymers to quench water can reduce the cooling rates to levels similar to using oil. The disadvantage is that the polymer coats the castings and an additional rinsing is required to clean them. Also, the composition of the quenchant changes, requiring replenishment. Certain low alloy steels in very thin sections can be air hardened.

The quenched castings are tempered to achieve high strength combined with toughness.

5.6.4.4 Tempering

The main purpose of tempering after normalizing and quench hardening is to reduce any residual stresses that may have developed during transformation and cooling.

The process consists of:

- Heating the castings to below the transformation range where austenite begins to form
- Holding for a specified duration (from 30 minutes to several hours)
- Cooling in air

Hardness decreases as the tempering temperature increases. Figure 5.34 (Ref 12) provides the relationship between hardness and tempering temperature. Figure 5.35 (Ref 12) shows the relationship between hardness, tempering temperature, and tempering time.

Carbon and low alloy steels are tempered at temperatures ranging from 177 to 704 °C (350 to 1300 °F). The temperature and time are interrelated for targeted hardness. Tempering at temperatures below 593 °C (1100 °F) results in temper embrittlement in certain steels. Tempering temperatures below this level should therefore be avoided.

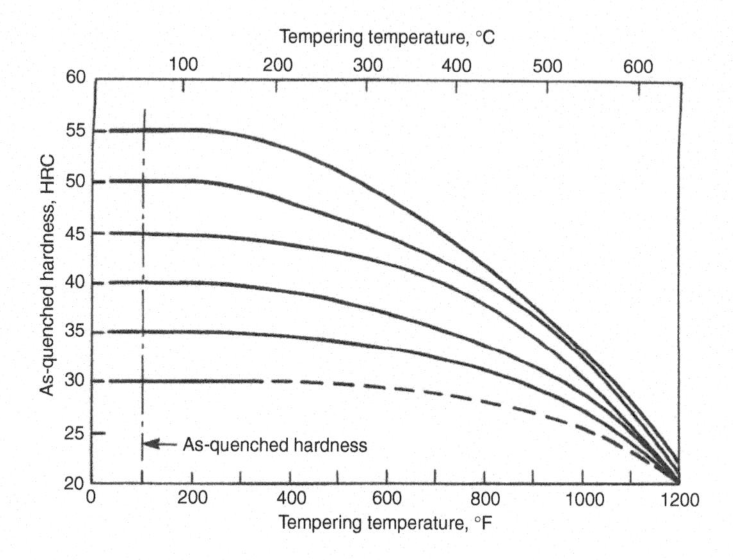

Fig. 5.34 Relationship between tempering temperature and hardness. Source: Ref 12

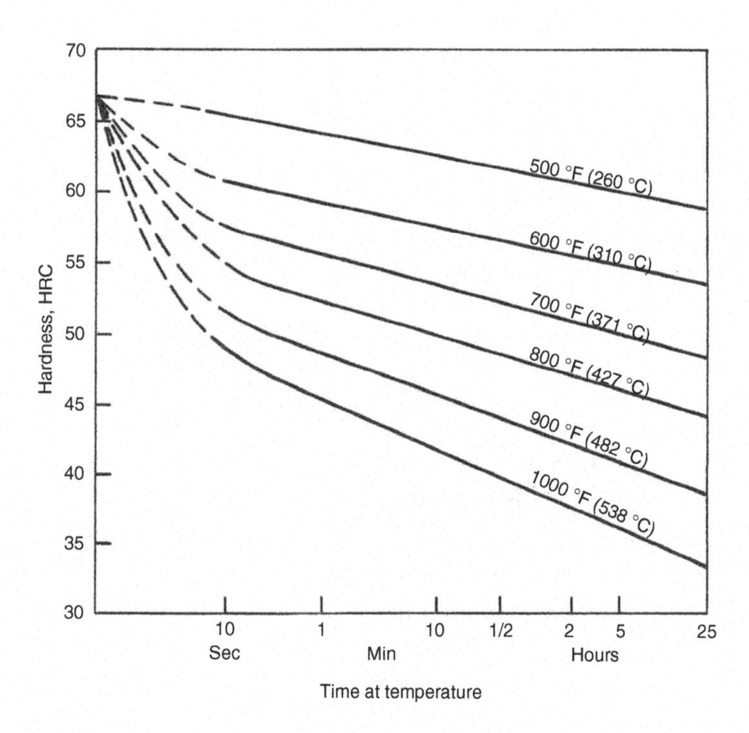

Fig. 5.35 Relationship among hardness, tempering temperature, and time. Source: Ref 12

Tempering relieves internal stresses, and usually there is no need for subsequent stress-relieving. If operations such as induction, flame hardening, or excessive grinding or welding are carried out, then a stress-relief operation may be needed.

5.6.4.5 Annealing or Full Annealing

Low carbon steels are annealed to obtain a soft and machinable structure with high ductility and lower strengths. Low alloy steels are annealed for good machinability and are heat treated after machining to obtain the desired mechanical properties. The full annealing process involves:

- Heating the castings above the upper critical temperature
- Holding the temperature adequately high to achieve complete transformation to austenite
- Cooling the castings in the furnace at a controlled rate by reducing the heating or turning the burners off
- When the castings reach 427 °C (800 °F) below the lower critical temperature, castings are air cooled, freeing up the furnace for the next cycle.
- A pearlite-ferrite structure with high ductility and high machinability is obtained.

5.7 Product Quality Control

Product specifications are established to ensure fitness for function. Quality checks are performed to confirm that the product meets the specifications and to verify that the design is validated. Quality verification checks are done before machining and assembly.

Quality checks are grouped into three categories:

- Material checks
- Product checks
- Functional checks

Figure 5.36 provides an overview of the generic quality checks. The quality checks for gray iron, ductile iron, and steel differ. Sections 5.7.1, 5.7.2, and 5.7.3 detail the relevant quality checks for these three materials.

5.7.1 Gray Iron Castings

Quality checks applicable to gray iron are:

A. Material checks

- Chemistry: C, Si, Mn, P, and S
- Microstructure: Graphite shape and size (per ASTM A247), pearlite, and ferrite percentages on separately cast test bars of 30 mm (1.18 in.) diameter poured from the same metal along with the molds
- Mechanical properties: BHN hardness on designated areas of a small number of castings each shift, tensile strength machined out from separately cast test bar of 30 mm (1.18 in.) diameter, as per ASTM standard E8M or E8 and measured on 3 bars

B. Product checks

- Critical dimensions designated by the designer (samples and pilot production runs)
- Surface Integrity: to check 100% of castings for visual anomalies per acceptance standards established
- 100% product check for core breakage in air or water passage blockage, if sand cores are used

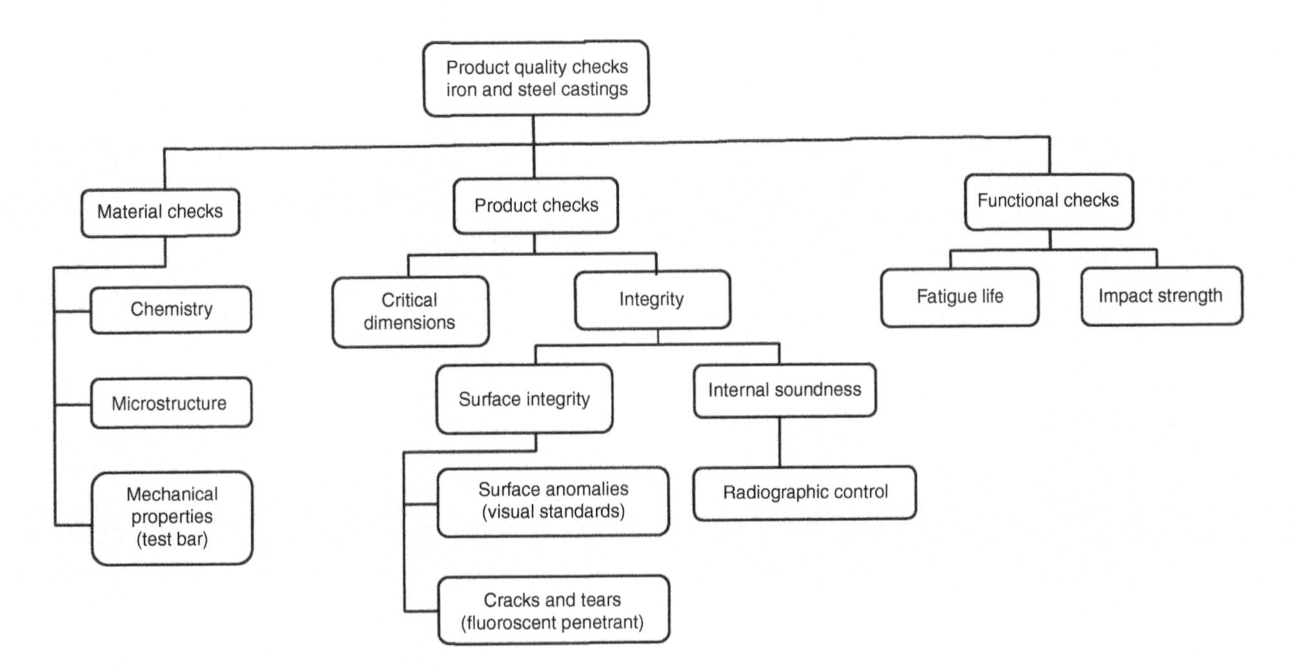

Fig. 5.36 Overview of quality checks for iron and steel castings

5.7.2 Ductile Iron Castings

The quality checks that apply to ductile Iron are:

A. Material checks

- Chemistry: C, Si, Mn, P, S and Mg, and Mo, Ni and Cu, if alloyed
- Microstructure: Nodularity (measured on the test bar cast over the last mold from each treated ladle). Nodule count, pearlite, ferrite, and carbide percentages on a test bar cast poured from the last mold cast from each treated ladle
- Mechanical properties: Tensile and yield strengths, elongation percent and BHN on separately cast test bars cast from test bar molds placed over the last pour of every treated ladle (average of three test bar values)

B. Product checks

- Gaging of critical dimensions identified for statistical capability
- 100% nodularity check on every critical safety structural casting such as a knuckle, brake caliper, or control arm (preferably automated) by ultrasonic testing and 100% Brinell check on every critical safety structural casting
- 100% nodularity check on all critical components such as valves for steam and chemical industries by ultrasonic testing
- Surface integrity: Check 100% of all castings for visual surface anomalies
- 100% product check for core breakage in air or water passage blockage, if sand cores are used

C. Functional checks

- Conduct fatigue and impact tests on a sample basis, on critical structural parts that are subjected to impact loads or fatigue loads in service. Approve batch based on the success of the sample test.
- Verify weight variance on a regular schedule basis for conformance.

5.7.3 Steel Castings

The quality checks that apply to steel castings are:

A. Material checks

- Chemistry: C, Si, Mn, S, P, and all alloying elements
- Microstructure: Pearlite and ferrite in test coupons
- Mechanical properties: Tensile and yield strengths, percent elongation and BHN on test bar machined from a keel block cast from the same ladle (ASTM A370)

B. Product checks

- Critical dimensions if any, by gaging on sample basis
- 100% visual checks for surface anomalies and surface finish
- Check 100% by magnetic particle inspection or dye penetrant inspection (ASTM E165, ASTM E443) for cracks or tears on structural products subjected to fatigue or impact loading (ASTM E109, ASTM E138. Use reference photographs as per ASTM E125).
- Check on the sampling basis or 100%, depending on the product function, for internal defects such as shrink porosity or gas porosity using radiographic inspection (ASTM E94, ASTM E142, ASTM E446).

C. Functional checks

- Conduct fatigue and impact tests on a sample basis, on critical structural parts that are subjected to impact loads or fatigue loads in service. Approve batch based on the success of the sample test.

- Verify weight variance on a regular schedule basis for conformance.

REFERENCES

1. J. Nath, *Iron and Steel Castings Engineering Guide*, ASM International, 2021
2. J. Sirokich et al., *Foundry Equipment* (Zarizeny slevaren), SNTL, 1968
3. Didion International Inc.brochure, www.didion.com
4. Wheelabrator, a Norican Company brochure, www.wheel abrator.com
5. Viking Technologies, Inc., www.vikingtechnlogies.com
6. Blast Cleaning Technologies brochure, www.bct-us.com
7. USA Pangborn brochure, www.pangborn.com
8. Viking Blast and Wash Systems, www.vikingcorporation.com
9. LGT Manufacturing Co. Info@lgtmanufacturing.com
10. Lianco Technologies and Del Sol Industrial Services. www.liancotechnologies.com
11. Norton -Saint-Gobain brochure, www.nortonsaint.com
12. P.F. Wieser, *Steel Castings Handbook* (5th Ed), Steel Founders' Society of America, 1980
13. C.F. Walton and T.J. Opar, *Iron Castings Handbook*, Iron Castings Society, 1981

Casting Equipment Engineering Guide
Jagan Nath
https://doi.org/10.31399/asm.tb.ceeg.t59370115

Copyright © 2023 ASM International®
All rights reserved
www.asminternational.org

CHAPTER 6

Iron and Steel Melting Furnaces

THE MELTING FURNACE is a critically important piece of equipment in a foundry; it influences the balance between the metal required by molding machines and the melting capacity, as well as the foundry's capability for high-strength and thin-walled castings. The competitiveness of a foundry is impacted by the selection of the right kind of furnace and its melting capacity to meet the needs of the parts being molded. Environmental regulations impact the choice of melting equipment, as the cost of emission controls must be factored into the total cost.

6.1 Furnace Types and Applications

There are many types of furnaces, and they offer a variety of features. The selection of a furnace depends upon:

- Casting metal or alloy
- Metal grade
- Types of charge material
- Minimum wall thickness of casting
- Metal treatment needed
- Metal composition limits
- Limits of trace elements

Table 6.1 summarizes the major factors influencing the selection of furnace type.

Figure 6.1 shows the different types of furnaces used for iron, steel, and aluminum foundries.

Different furnaces offer individual features and technical and cost advantages that make them suitable for melting selected alloys. Furnaces are broadly classified into two groups:

- Furnaces suitable for iron and steels
- Furnaces suitable for aluminum alloys

The furnaces that are suitable for cast iron and steels are:

- Cupolas
- Electric induction furnaces
- Arc furnaces

Furnaces suitable for aluminum alloys are:

- Stack melting furnaces
- Reverberatory furnaces
- Resistance crucible furnaces

The furnaces suitable for aluminum alloys are addressed in Chapter 7, "Aluminum Melting, Holding, and Dosing Furnaces," in this book.

Furnaces suitable for copper-base alloys are:

- Crucible furnaces (oil or gas fired)
- Electric induction furnaces
- Electric channel-type furnaces

Crucible furnaces are suitable for bronzes and brasses. Gas-fired crucible furnaces are the most popular type. Electric induction and channel types are suitable for bronzes, brasses, and copper-nickel alloys. Some processes use a combination of both induction (for melting) and channel (for holding) furnaces to reduce temperature fluctuations that result from frequent charging.

Cupolas are used for gray cast irons and white irons, where a continuous supply of one or two grades of metal is needed. Typical examples of this type of operation are pipe manufacturing by centrifugal casting, ingot mold manufacturing, and pipe fitting production. Cupolas, lined with basic refractories in the hearth

Table 6.1 Selection factors for furnace type

Factors	Explanation
Casting metal or alloy	Gray iron casting foundries can use either a cupola or an induction furnace for melting. Alloy cast iron, malleable iron, and ductile iron foundries use induction furnaces. Steel foundries use either induction furnaces or arc furnaces.
Metal grade	Gray iron foundries producing the higher grades need a higher steel percentage in the charge. It is possible to use a hot-blast cupola, but induction furnaces may be a better choice.
Types of charge materials	If a high percentage (> 20%) steel scrap is needed to target low carbon in cast irons, the induction furnace is a good choice. If the sulfur level is to be kept low to produce ductile iron, or if alloying is needed in cast irons, the induction furnace is the best option. The basic lined cupola is also feasible.
Minimum wall thickness of castings	Thinner walls need higher superheat temperatures. For gray irons, the choice is between a hot-blast cupola and an induction furnace.
Metal treatment needed	An induction furnace is the best choice for melts needing the introduction of magnesium to produce ductile iron.
Metal composition limits	If low levels of sulfur and phosphorus are required and if lower levels of carbon and silicon are needed, an induction furnace is the only choice.
Limits of tramp elements	Keeping elements that damage the iron's microstructure at low levels requires close control of the charge materials. An induction furnace is the best choice for all irons.

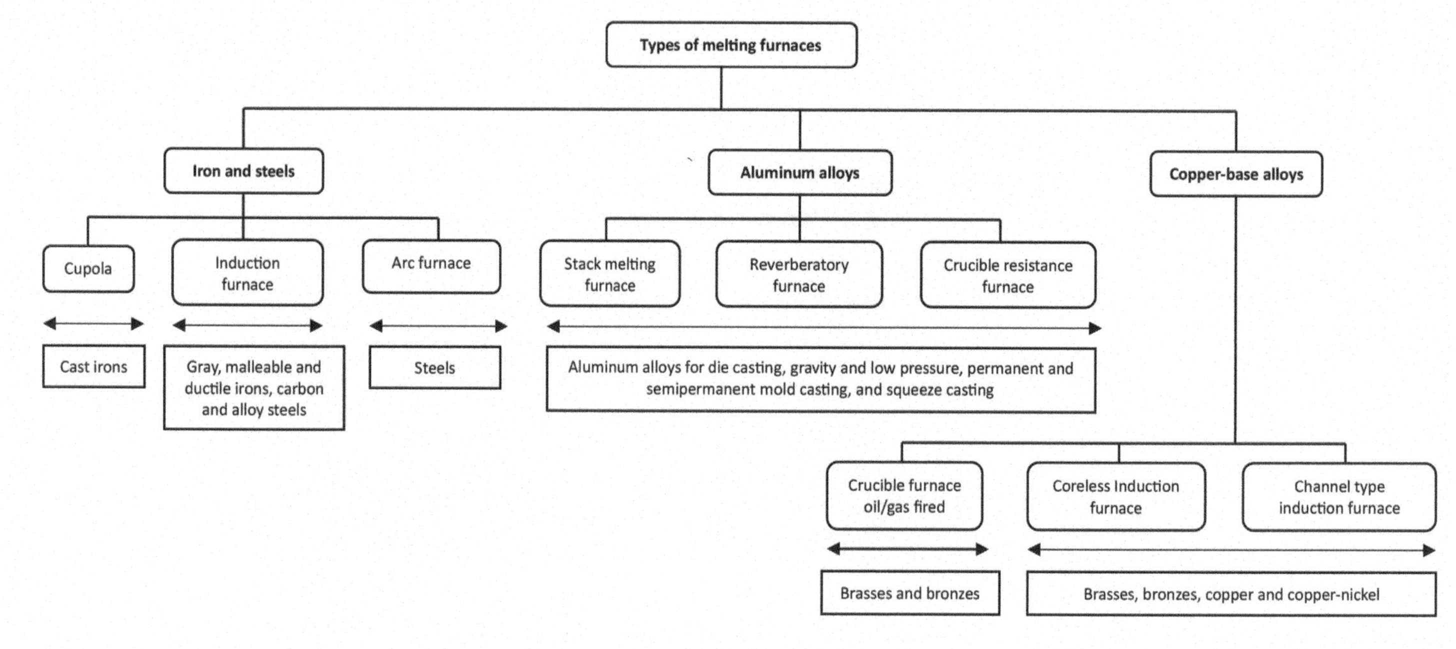

Fig. 6.1 Types of melting furnaces

region, produce low-sulfur iron suitable for ductile iron production. The various types of cupolas offer several cost advantages.

Coreless induction furnaces have a tubular copper coil embedded in a refractory lining through which a current is passed. The magnetic field induced in the charge inside the cylindrical lining generates heat to melt the metallic charge.

These induction furnaces are able to provide precise composition control for low phosphorus and low sulfur levels and the ability to alloy with nickel and chrome, both of which have higher melting points. Additionally, these furnaces allow for the close control of the superheat temperature. The flexibility of using varied kinds of scrap makes the induction furnace a popular choice for automotive engine components and alloy cast iron grades for several applications. However, the capital investment is higher for an induction furnace than for a cupola.

The second type of induction furnace, called the channel-type furnace, has one or two coils wound around metal channels, which are formed in the refractory lining at the bottom of the furnace. These form a *heel* for the furnace which is maintained for furnace operation. The metal channels are connected to the main cylindrical portion of the furnace. The molten metal, due to the induction currents in the heal, sets up convection currents to heat the entire body of the metal. Such furnaces are capable of very close temperature control and are often used as holders.

Coreless induction furnaces are versatile for melting gray cast irons, white irons, alloy cast irons, and the melts suitable for treatment to produce ductile iron, carbon, and alloy steels.

Arc furnaces are used exclusively for melting steel scrap of large sizes and different grades. The large furnaces deliver huge quantities of metal for producing large steel castings. Three carbon electrodes contact the charge placed at the bottom of the furnace and strike an arc when supplied with electrical power, which generates the heat.

6.2 Cupolas—Types, Advantages, and Applications

The cupola is the oldest iron melting device. The first cupola was installed between 403 and 221 B.C. in China, followed by integration of the blast furnace between 202 B.C. and 220 A.D. A modern version of the cupola was introduced in France in 1720.

The cupola offers several advantages, including:

- Simple construction
- Low operating expense
- Continuous delivery of large quantity of metal
- Simple operation
- Uninterrupted service
- Simple charge make-up systems and charging equipment

There are some disadvantages to using a cupola furnace, including:

- Inability to change the grade of metal at short notice
- Limited capability for superheating
- Inability to hold the chemistry to close limits
- High alloys of chrome and nickel are not possible
- Limited raw material diversity
- Need for high grade coke
- Requirement of a basic lining for low sulfur iron

Some of the disadvantages have been addressed in recent years. For example, duplexing with an electric furnace allows alloying, and preheating the blast or recuperative blast provides higher superheating temperatures. Automation of charging and charge makeup has reduced variations in chemistry and improved the ability to change the grade of the metal.

Figure 6.2 is an overview of the main types of cupolas. Cupolas are grouped based on:

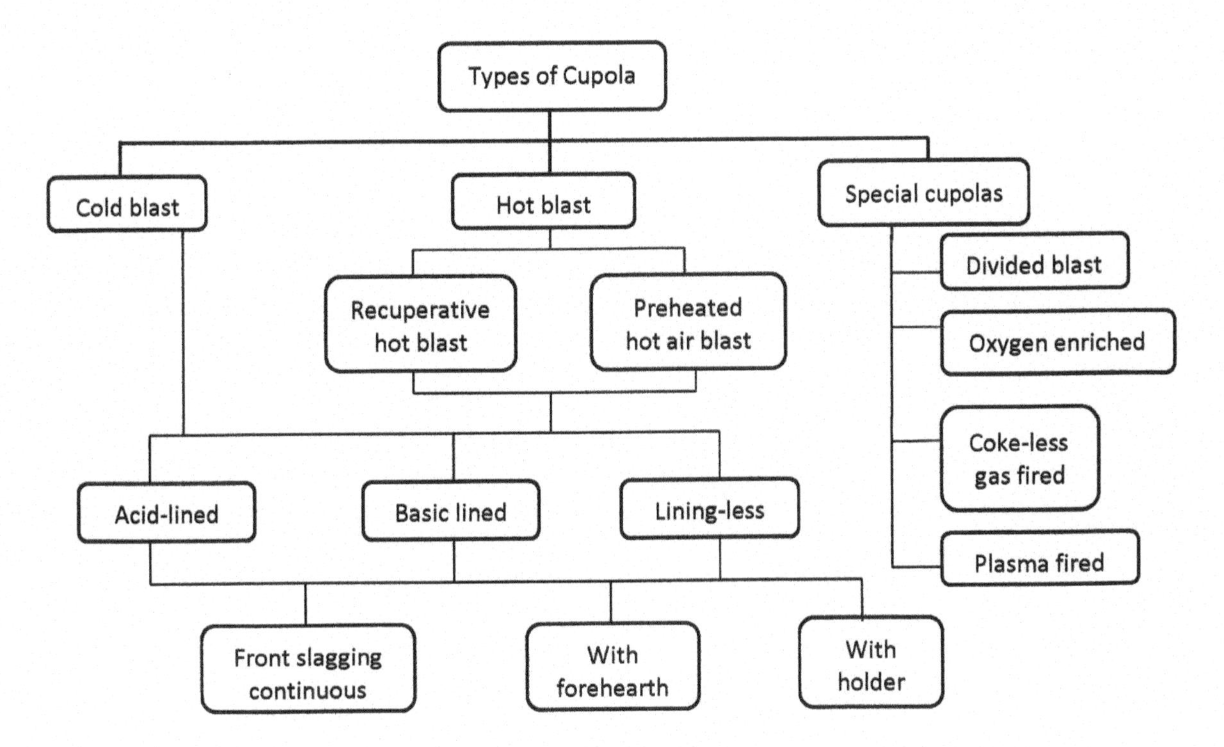

Fig. 6.2 Overview of cupola types

- Type of air blast —cold or hot blast
- Hot blast system — recuperative hot or preheated air blast
- Kind of lining — acid, basic, or liningless

In each type, the cupola can be batch or continuous in operation. The cupola can be designed with a forehearth for homogenizing the chemistry. It can also be designed for a duplex operation with an attached holder for alloying and superheating. Special cupolas are those that use natural gas (or cokeless) or are plasma-fired to generate the heat.

The air blown for combustion of coke can be at room temperature or heated to increase the melting temperature. The higher melting temperature permits the use of a higher percentage of steel scrap in the charge. The air blast can be heated externally in a preheater using natural gas, which heats a refractory-lined structure through which the air is passed before it enters the wind box of the cupola. Another alternative is to use the heat of the exhaust gases from the cupola with a heat exchanger to exchange the heat of the exhaust with the air blast that enters the wind box. Both hot blast versions have been proven reliable over the years and are popular because they need little maintenance, and the investment costs are easily justified.

6.2.1 Cold-Blast Cupolas

Figure 6.3 (Ref 1) shows the overall features of a typical cold-blast cupola installation. Table 6.2 (Ref 1) provides the major dimensions for reference.

The major highlights of the cold-blast cupola presented in Fig. 6.3 include:

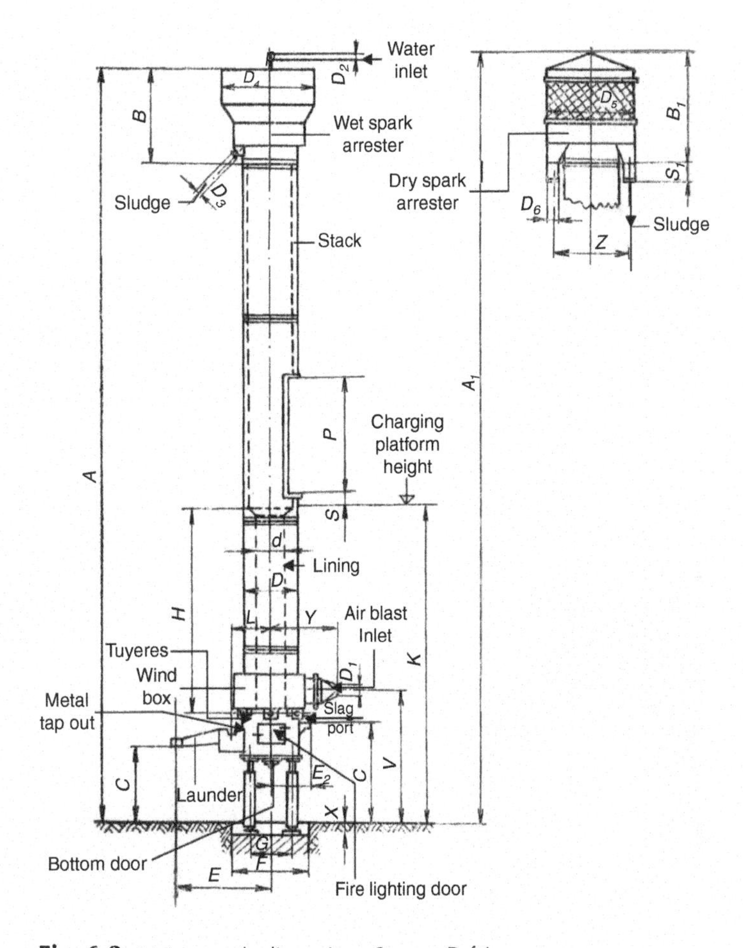

Fig. 6.3 Major cupola dimensions. Source: Ref 1

- High cylindrical structure with the refractory lining in the melting area
- Annular wind box which surrounds the melting zone and distributes the air through individual tuyeres to provide air for combustion
- Slag port to drain the slag which floats over the metal bath
- Tap hole that is opened by a spike or closed shut using a conical clay bot
- A launder to direct the tapped metal into a ladle
- A bottom door used for access to the melting zone to patch the worn refractory
- A door to light up the coke bed and close for preheating before melting
- Two alternative methods to arrest the sparks: the wet method which uses water, and the dry method which uses a heavy-duty mesh and a canopy

Table 6.2 provides the major dimensions required for planning the support structure. Table 6.3 (Ref 2) provides the cupola dimensions for various outputs, based on the ratio of the metal: coke from 6:1 to 12:1.

Table 6.3 provides the volume of air needed for various sizes of cupolas.

Table 6.2 Major dimensions of a cupola, mm

Output t/hr metal:coke 6:1	2.2	3	4	5	6.5	7.5	8.5	10.5	12.5
d	600	700	800	900	1000	1100	1200	1400	1600
A min	900	1000	1100	1100	1250	1250	1300	1350	1500
B min	2150	2300	2400	2400	2570	2570	2600	2850	2900
D	1220	1320	1420	1520	1620	1720	1820	2020	2220
E	1350	1350	1450	1450	1680	1680	1900	2150	2300
F	350	350	450	450	550	550	600	700	700
G max	800	825	900	900	950	950	980	1100	1120
H min	4400	5025	5700	6300	6950	7550	8180	9500	10720
L min	1475	1590	1800	1800	2070	2070	2170	2450	2500

Source: Ref 1

6.2.2 Hot-Blast Cupolas

Hot-blast cupolas utilize a hot air blast by heating the combustion air using the heat of the exhaust flue gases or by using an external air heating furnace. The flue gases consist of a high percentage of carbon monoxide (CO), and further combustion in a heat exchanger results in a hot blast temperature of 400 to 600 °C (752 to 1112 °F).

The advantages of this type of system are:

- Ability to use a higher percentage of steel scrap in the charge
- Higher superheat temperatures
- More consistent chemical compositions
- Environmentally preferred
- Higher thermal efficiency

The two basic types in are:

- Recuperative hot blast — recuperating the hot exhaust gases to preheat the air blast used for combustion
- Preheated hot air blast — using a separate heating chamber where a natural gas burner heats a maze of refractory brickwork. Air for combustion is heated as it passes through the heated brickwork before it enters the wind box.

6.2.2.1 Recuperative Hot Blast Units

The recuperative hot blast cupolas have a heat exchanger that can be mounted in one of two ways:

Table 6.3 Melting rates in tons per hour for lined cupolas

Cupola Size	Shell Diam.	Lower Lining Thick-ness**	Diam. Inside Lining	Area Inside Lining Sq. In.	Theoretical Air Flow thru Tuyeres SCFM	Recommended Blower Cap'y.		METAL TO COKE RATIO (Carbon Content of Coke 90%)						
								6:1	7:1	8:1	9:1	10:1	11:1	12:1
						Volume ICFM	Pressure oz./in.2	333#/ton 99ft3/#c	286#/ton 107ft3/#c	250#/ton 112ft3/#c	222#/ton 118ft3/#c	200#/ton 123ft3/#c	182#/ton 128ft3/#c	167#/ton 132ft3/#c
0	27"	4½"	18"	254	573	640	8	1.1	1.3	1.4	-----	-----	-----	-----
1	32"	4½"	23"	415	937	1040	16	1.9	2	2.2	2.4	-----	-----	-----
2	36"	4½"	27"	573	1293	1430	20	2.6	2.8	3.1	3.3	-----	-----	-----
2½	41"	7"	27"	573	1293	1430	20	2.6	2.8	3.1	3.3	-----	-----	-----
3	46"	7"	32"	804	1815	2000	24	3.7	4	4.3	4.6	4.9	-----	-----
3½	51"	7"	37"	1075	2426	2700	24	4.9	5.3	5.8	6.2	6.6	-----	-----
4	56"	7"	42"	1385	3126	3450	24	6.3	6.8	7.4	7.9	8.5	9	-----
5	63"	9"	45"	1590	3589	4000	28	7.3	7.8	8.5	9.1	9.7	10.3	-----
6	66"	9"	48"	1810	4085	4500	32	8.3	8.9	9.7	10.4	11.1	11.7	12.4
7	72"	9"	54"	2290	5170	5750	32	10.4	11.3	12.3	13.1	14	14.8	15.7
8	78"	9"	60"	2827	6380	7100	32	12.9	13.9	15.2	16.2	17.3	18.3	19.3
9	84"	9"	66"	3421	7720	8600	36	15.6	16.8	18.4	19.6	20.9	22.1	23.4
9½	90"	9"	72"	4072	9190	10200	36	18.6	20	21.9	23.4	24.9	26.3	27.8
10	96"	9"	78"	4778	10790	11900	36	21.8	23.5	25.7	27.4	29.2	30.9	32.7
11	102"	12"	78"	4778	10790	11900	36	21.8	23.5	25.7	27.4	29.2	30.9	32.7
12	108"	12"	84"	5542	12510	13900	36	25.3	27.3	29.8	31.8	33.9	35.8	37.9

Source: Ref 2

- Directly over the stack
- Between two cupolas sharing one recuperator

Recuperators directly mounted over the stack consist of an annular chamber built out of heat resistant steel. Flue gases passing through the stack impart heat to the shell. Spiral louvers provided in the annular space increase the heat exchange.

Figure 6.4 is a schematic of a recuperator shared by two cupolas. Figure 6.4(a) shows the schematic layout, and Fig. 6.4(b) shows the detail of another design of the heat exchanger elements made of heat-resistant steel.

6.2.2.2. Preheated Air Blast Units

Preheated air blast units utilize a separate heating chamber, which usually contains a natural gas burner and a maze of refractory bricks which are heated. The air blast passes through the heated brickwork before entering the wind box. Figure 6.5 (Ref 1) is a schematic of this equipment.

6.2.3 Cupola Classification According to the Lining

Cupolas are classified as:

- Acid refractory-lined
- Basic refractory-lined
- Liningless

A majority of cupolas are acid refractory-lined, or acid-lined. This type of lining is less expensive than the basic lining. But the basic lining produces the low-sulfur iron which is needed to produce ductile iron. The life of basic refractory lining is improved by the addition of water cooling around the lining.

The acid lining consists of bricks made of silica or fire brick. A typical composition of the acid lining is: silica 52 to 62%, alumina 31 to 43%, and titanium oxide 1.5 to 2.5%. Basic linings are compounds of calcium and magnesium oxides, such as dolomite or magnesite. The combination of magnesite and chromium oxide is common in the linings. Liningless cupolas do not have any refractory lining in the shaft or the stack area. However, they have acid lining below the level of the tuyeres. The advantages of liningless cupolas are longer duration of working, lower heat loss, and reduced coke consumption.

Figure 6.6 is a schematic showing the cooling of the shell to extract heat from the basic lining. Although water jackets have been

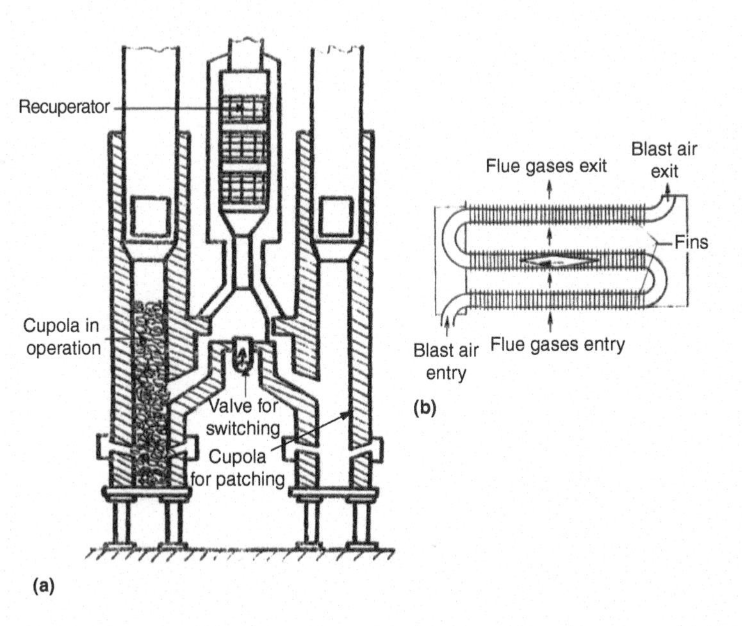

Fig. 6.4 Hot-blast cupola schematic, (a) shared recuperator, (b) heat exchanger details

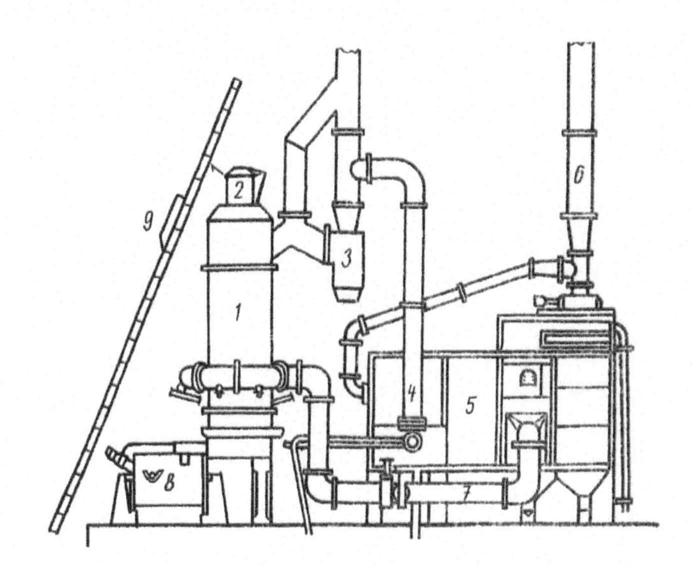

Notes: 1. Cupola 2. Charge hopper 3. Dust collecting cyclone 4. Blast entry 5. Fired recuperator 6. Exhaust for flue gases 7. Heated blast 8. Fore hearth 9. Skip charger

Fig. 6.5 External preheated blast unit

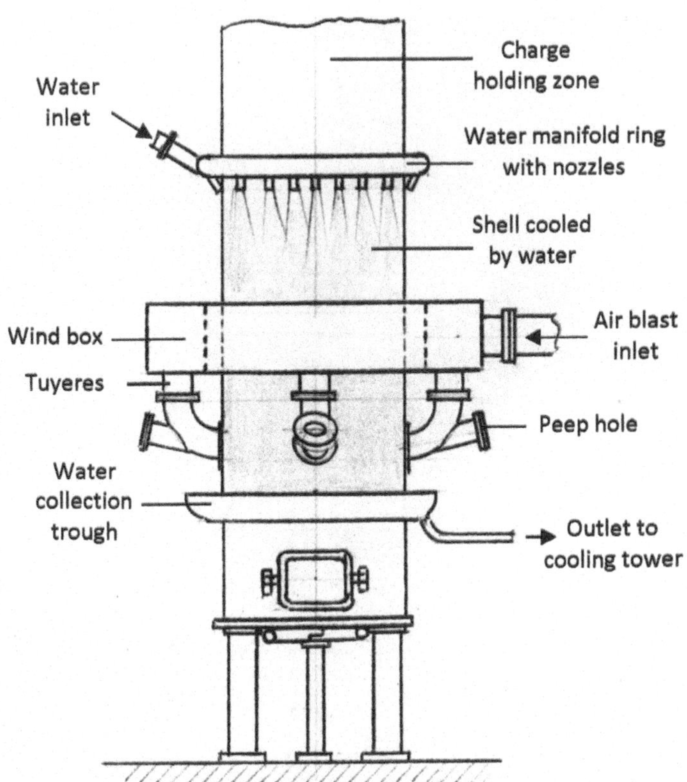

Fig. 6.6 Cupola shell cooling schematic

tried, a series of nozzles that produce a shower around the shell are simple and safe to maintain. Table 6.4 lists the melting rates and major dimensions of liningless hot-blast cupolas (Ref 2).

6.2.4 Continuous Operation of the Front-Slagging Cupola

Casting facilities that work around the clock and those that use large volumes of liquid metal benefit from designs that do not need to open and close the metal taphole and the slag port. Casting operations for pipe casting, ingot molds, and other steel plant equipment are typical examples of such facilities. The front-slagging cupola provides for continuous metal and slag flow.

Figure 6.7 is a schematic of the front-slagging cupola. The slag is skimmed off by a slag dam. It flows out through the notched opening, and the liquid metal flows through the launder into the ladle. This system requires that the charging system be engineered for automation for an uninterrupted charge supply.

The slag can be directed by means of a sloping launder at an inclination of 1:10 while water is percolated to granulate the slag. The volume of slag that is generated can be around 0.6 to 0.8 cubic meters per ton of metal melted.

6.2.5 Cupolas with a Forehearth

The amount of metal required fluctuates depending on the individual parts being molded at any given time. The forehearth provides a buffer to handle the temporary fluctuations of the metal demand. The metal from the cupola flows into another cylindrical chamber with a refractory lining. The metal is tapped from the forehearth instead of from the cupola. It is essential to preheat the forehearth adequately before the start of melting to avoid excessive temperature loss. Some cupolas provide an additional gas burner in the forehearth to maintain the targeted temperature of the tapped metal. These burners can heat the forehearth to between 800 and 1000 °C (1472 to 1832 °F). Figure 6.8 (Ref 1) shows the forehearth of a cupola.

6.2.6 Cupola with a Holding Furnace or a Holder

A holding furnace, also known as a holder, adds flexibility to cupola operations for alloying, superheating, and holding to accommodate fluctuations in metal demand. In addition, holding furnaces can be used to change the grade of the material by adding steel scrap of 6 to 12 mm thickness. Holding furnaces can be tilted to pour into the ladles. Usually, two cupolas share one holding furnace.

Holding furnaces were formerly gas-fired. But line frequency (50 or 60 Hz) electric induction furnaces are now popular and are generally preferred. The magnetic field created by the coils forms eddy currents inside the liquid metal, resulting in good mixing, which homogenizes the metal. This also eliminates the temperature gradient from the top to the bottom of the metal bath. The power consumed is low, about 25 to 30 kWh/t. Smaller holding furnaces of between 2 and 15 tons have one coil, as shown in Fig. 6.9(a), and larger furnaces (over 15 tons) have two coils, as shown in Fig. 6.9(b). Holding furnaces installed in conjunction with cupolas are called duplex furnaces.

6.2.7 Special Cupolas

Several innovations have been introduced that make cupolas more thermally efficient and able to increase the melting rate or the tapping temperatures. Examples include:

- Divided-blast cupolas
- Oxygen-enriched blasts
- Cokeless gas-fired cupolas
- Plasma-fired cupolas

Table 6.4. Sizes and melting rates of liningless hot blast cupolas

Cupola I.D. of Melt Zone	Cupola Area Sq. In.	Theoretical Air Flow thru Tuyeres SCFM	Recommended Blower Cap'y.		METAL TO COKE RATIO (Carbon Content of Coke 90%)					
			Volume ICFM	Pressure oz./in2	5:1 400#Coke/T on 86ft3/#c	6:1 333#Coke/T on 99ft3/#c	7:1 286#Coke/T on 107ft3/#c	8:1 250#Coke/T on 112ft3/#c	9:1 222#Coke/T on 118ft3/#c	10:1 200#Coke/T on 123ft3/#c
54"	2290	5170	6000	48	10	10.4	11.3	12.3	...	...
60"	2827	6380	7100	48	12.4	12.9	13.9	15.2	...	...
66"	3421	7720	8600	48	15	15.6	16.8	18.4	19.6	...
72"	4072	9190	10500	56	17.8	18.6	20	21.9	23.4	...
78"	4778	10790	12000	56	20.9	21.8	23.5	25.7	27.4	29.2
84"	5542	12510	13900	56	24.2	25.3	27.3	29.8	31.8	33.9
90"	6362	14360	16000	64	27.8	29	31.3	34.2	36.5	38.9
96"	7238	16340	18000	64	31.7	29	35.6	38.9	41.5	44.3
102"	8171	18440	20300	72	35.7	37.3	40.2	43.9	46.9	50
108"	9161	20680	22800	72	40.1	41.8	45.1	49.2	52.6	56
114"	10207	23040	25500	80	44.7	46.5	50.2	54.9	58.6	62.4

Source: Ref 2

6.2.7.1 Divided-Blast Cupolas

Divided-blast cupolas divide the blast between two sets of tuyeres:

- One set is the conventional tuyeres below the wind box.
- The second set is 800 to 1050 mm above the conventional tuyeres.

The blast is divided between them as lower:upper equals 70:30 or 50:50.

The advantages experienced are an increased melting rate and reduced coke consumption.

6.2.7.2 Oxygen-Enriched Cupolas

Oxygen-enriched cupolas inject oxygen through (a) the air blast or (b) into the well. A 2% enrichment of the air blast with oxygen increases the metal temperature by 10 °C with a 0.4% carbon pickup.

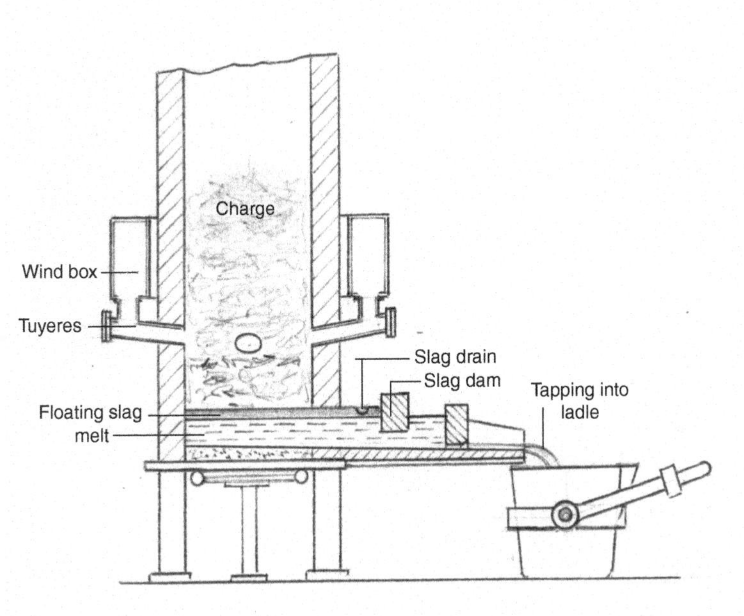

Fig. 6.7 Front-slagging cupola schematic

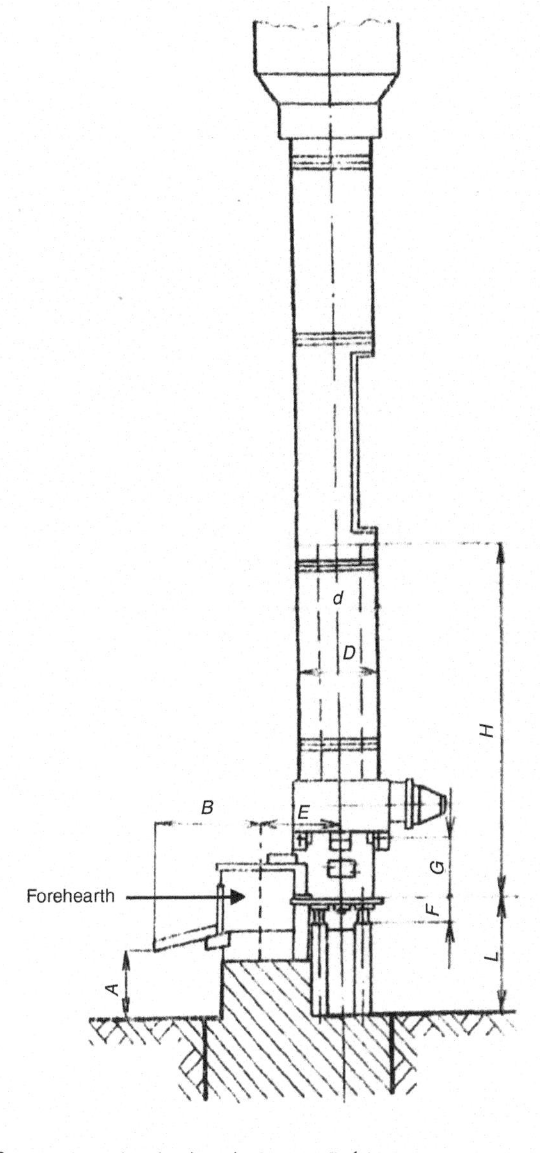

Fig. 6.8 Cupola with a forehearth. Source: Ref 1

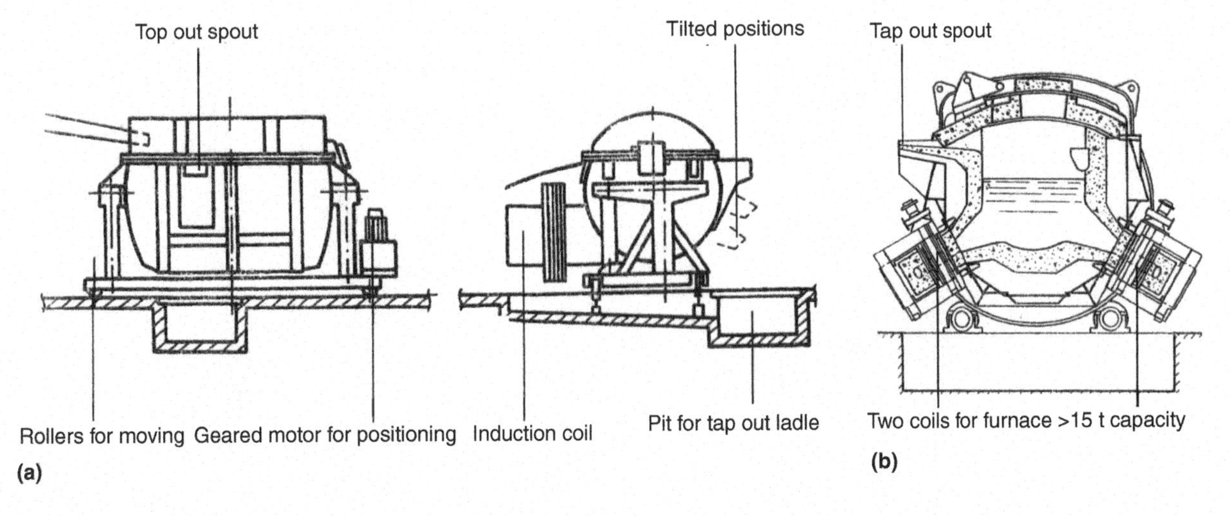

Fig. 6.9 Metal holding furnaces with (a) one, and (b) two induction coils

Instead, a 2% oxygen injection into the cupola well increases the metal temperature by 70 °C with a 0.2% carbon pickup.

6.2.7.3 Cokeless Gas-fired Cupolas

The cokeless cupola was invented in 1972 in the UK (Ref 3). It uses gas-fired burners for melting instead of conventional coke. The cokeless cupola offers several advantages over conventional coke-fired cupolas, such as:

- Eco-friendliness — effluent emissions contain a low amount of flue gases (SO_x and NO_x) and suspended particle matter, resulting in the ability to use less expensive dust removal systems (Ref 4).
- No sulfur pickup because no coke is used. The metal is suitable for ductile iron production.
- Higher thermal efficiency because the volume of slag generated is less compared to conventional coke-fired cupolas
- Closer chemistry control compared to conventional cupolas

Cokeless cupolas are combined with a line-frequency holding furnace for melt temperature control and homogenization. Carbon needs to be added to the melt because there is no carbon pickup in gas-fired cupolas. It is best to carburize the melt by adding synthetic graphite or natural graphite to the holder. For installations without a holder, carburization is accomplished by additions to the bath in the cupola. Carburization in the holders, however, is more consistent. Figure 6.10 illustrates a duplex installation of a cokeless cupola combined with a line-frequency holding furnace. Figure 6.11 is a 3D view of a similar arrangement (Ref 5).

6.2.7.4 Plasma-Fired Cupolas

Plasma is a more efficient way of achieving melting temperatures compared to the conventional method using coke. The plasma torch positioned at the bottom of the cupola converts electrical energy to thermal energy by boosting the temperature through gas ionization. The high temperatures of nearly 1371 °C (2500 °F) reduce the amount of coke needed as well as the combustion air blown into the cupola (Ref 6).

The plasma-fired cupola was a joint development in 1983 by Westinghouse Electric Corporation, Modern Equipment Company, and General Motors Central Foundry, to enable the use of low-cost scrap such as borings and turnings and reduce coke consumption (Ref 7). The trials confirmed the feasibility of melting borings and turnings up to 70% of the charge using a hot-blast and a shell-cooled cupola, with a 2 MW Mark II plasma torch at the bottom.

The plasma-fired cupola offers several features, including:

- The plasma generated provides high energy for melting, reducing coke consumption and combustion air.
- The low gas velocities through the cupola shaft permit the charging of fine aggregates of cast iron turnings and borings. Trials with 70% of charge comprising cast iron borings were successful. Only a small amount of air, typically about 100 scfm, flows through the plasma torch.
- Oxidation loss of silicon, chrome, and manganese are minimized due to the reducing conditions in the lower sections of the cupola. Silicon loss is reduced by 30%. Silica sand which has been added is reduced to silicon at these temperatures.
- The amount of iron oxide in slag is reduced substantially.
- The reducing atmosphere permits the use of oxidized scrap such as mill scale.
- The cupola can operate at various levels of electrical energy input to the plasma torch, independent of the coke combustion. Increases in the melting rate of 50% have been reported in trials.

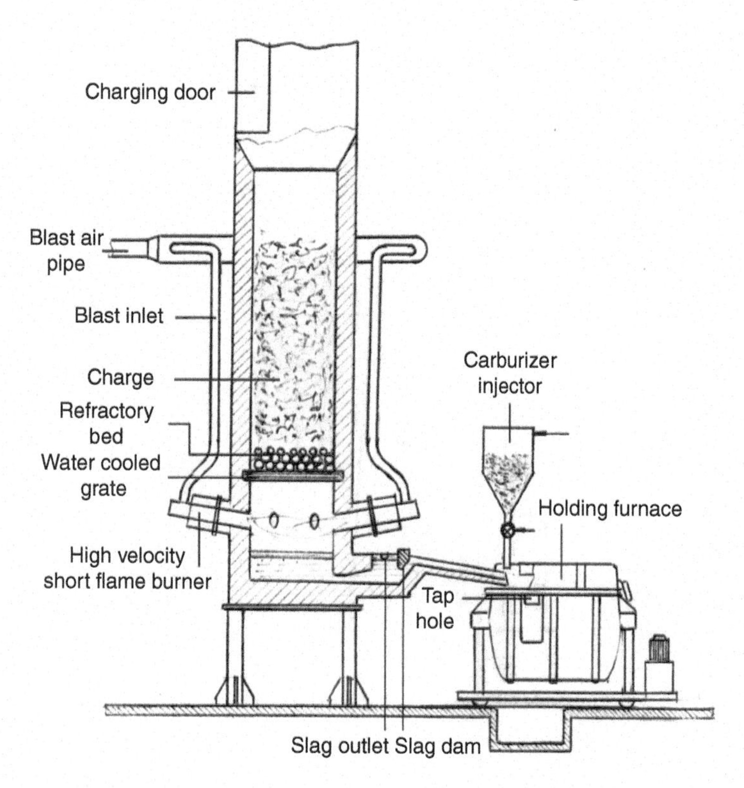

Fig. 6.10 Cokeless cupola schematic

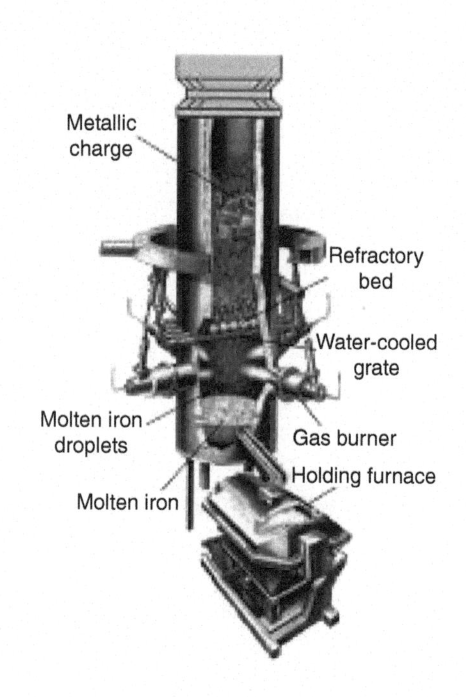

Fig. 6.11 Cokeless cupola 3D view. Source: Ref 5

- Lower coke consumption and reduced air velocities result in reduced need for emission control.
- Lower quality coke can be used.
- Lower operating costs are reported.

Figure 6.12 is a schematic of a plasma-fired cupola. The main features are:

- A front-slagging cupola (rated at 2.5 tons per hour) with a plasma torch of 2000 kW at the bottom of a cupola
- A screw feeder to feed cast iron turnings and borings into the cupola, close to the plasma torch
- An air line capacity of 100 scfm feeds the plasma torch
- An airline blast is adjacent to the torch
- A hot-blast system capable of handling a temperature of 1370 °C (2500 °F) with a cyclone separator
- A wet scrubber system for emission control
- A charging door permits feeding nearly 30 to 50% of the charges like a conventional cupola

6.2.8 Charging Systems Mechanization

Continuous operation of a cupola requires a dependable and consistent charging system for charge makeup and transport of the material to the charging door.

The charge storage and handling systems are addressed in Chapter 2, "Sand and Metal Charge Storage and Handling," in this book. Figure 2.1 illustrates the bunkers or hoppers for storage of charge materials. Handling of the ferrous metallic charges has been simplified by using electromagnets with load cells that fill the charge buckets. The dosing of the coke and limestone is accomplished by calibrating the feed belts underneath the storage hoppers. A system should be established to record the amounts of the charge materials automatically. Additionally, a system of communication among the charge-makeup operator, furnace operators, and metallurgist should be designed up front.

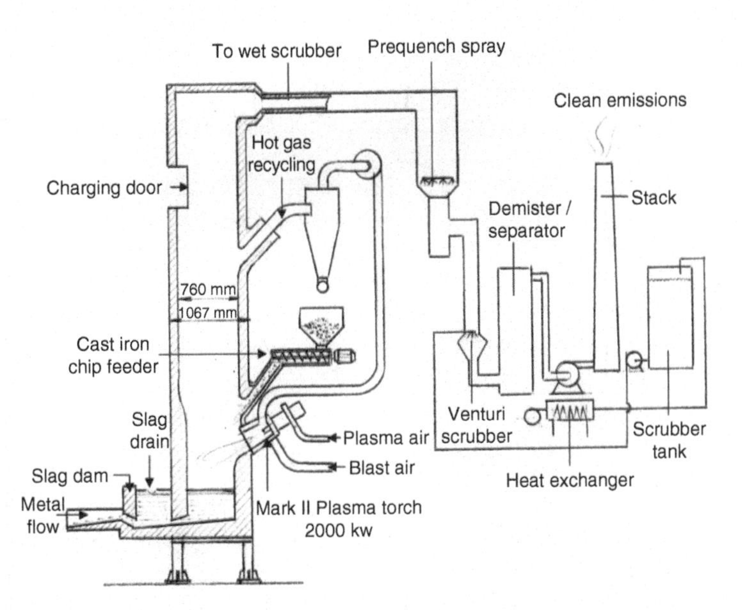

Fig. 6.12 Plasma-fired cupola schematic

The skip hoist which was used for many years suffered from premature wear of the lining due to the impact of the charge on the inner refractory lining. Bottom discharge buckets became more popular because they unload the charge in the center of the cupola. There are many variations in designs, and two of the more common types are presented here.

Figure 6.13 (Ref 1) illustrates a modification of the skip hoist with a bottom discharge bucket instead of a skip bucket.

Conventional cupolas are operated every other day, allowing for cleanup and refractory patching on nonworking days. Two cupolas share a common charging system which is indexed to the cupola in operation for the individual day.

Figure 6.13 shows the same bucket going up the incline and returning. The charge makeup is done when the bucket returns down after emptying the contents. The concept is similar to that of the skip bucket, except the bottom discharge bucket is used instead.

Figure 6.14 (Ref 1) is a schematic of the installation where more buckets are accommodated on the conveyor for continuous charging. The cupola does not wait for the charge, even when there is a change in the charge makeup. The initial investment for this system is higher compared to the system illustrated in Fig. 6.13, but the cupola uptime is also higher.

As in Fig. 6.13, common charging equipment serves the two cupolas in Fig. 6.14 alternately. This system uses bottom discharge buckets. There are two types of these buckets: Fig. 6.15(a) clamshell opening, and Fig. 6.15(b) bottom cone opening.

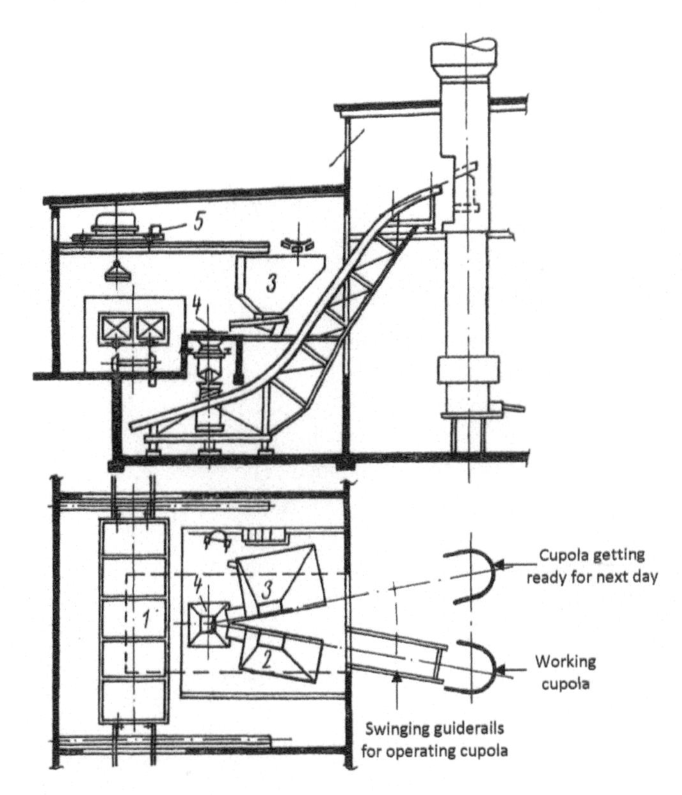

Fig. 6.13 Cupola charging system (1). Source: Ref 1

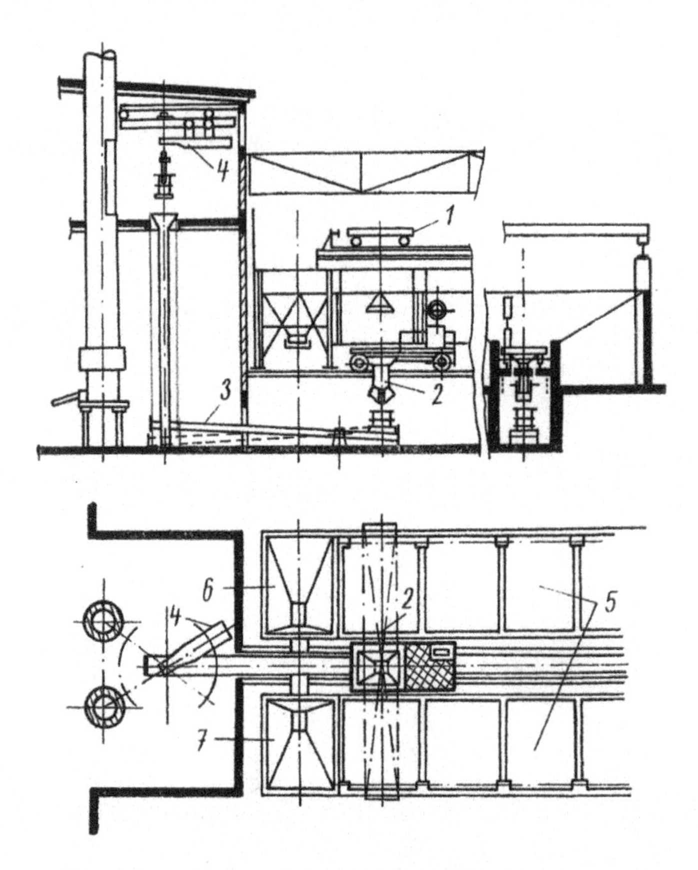

Notes: 1. Crane with electromagnet 2. Mobile weighing scale 3. Conveyor for buckets 4. Elevator with indexing mechanism 5. Storage hopper for metallic charge materials. 6. Hopper for coke 7. Hopper for limestone

Fig. 6.14 Cupola charging system (2). Source: Ref 1

Figure 6.15 is a schematic highlighting these features. The clamshell doors Fig. 6.15(a) open as the bucket sits on a support ring in the stack at the charging door. In Fig. 6.15(b), the bottom cone opens as the bucket sits on the ring.

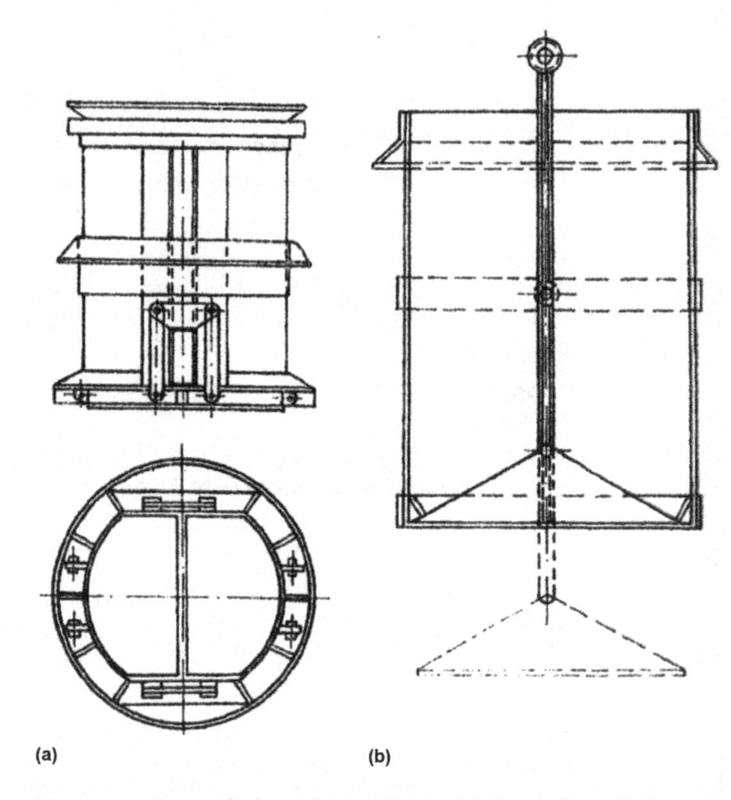

Fig. 6.15 Bottom-discharge bucket designs, (a) clamshell, and (b) bottom cone openings

6.2.9 Charge Height Monitoring Systems

It is difficult to determine the height of the charge in the cupola, because of its high stack and to confirm if charging is required. The amount of metal tapped from the cupola depends upon the number of molding machines in operation and the weight of the parts being molded. Therefore, variances in the metal as it is tapped occur, and the mechanization of the charging system should be designed to adjust for metal needs *just in time*. Sensors placed on the cupola stack are engineered to guide the charge preparation crane operator.

There are two types of charge height indicator systems:

- Contact type — where the contact between the charge and the sensor is used as a guide
- Noncontact type — where a gamma ray tube is used to detect the charge height and serve as a guide

6.2.9.1 Contact Height Monitoring

The contact method relies on a metallic charge that bridges across the diameter of the cupola to establish contact between two sets of electrical contacts embedded in the lining. Four contacts are embedded in each half of the diameter as seen in Fig. 6.16 (Ref 1).

These contacts are embedded below the charging door. When the charges connect, a relay signal is sent. A green light indicates that charging can continue, and a red light indicates that charging should stop. The orange color indicates that the charge bucket is in transit.

6.2.9.2 Noncontact Height Monitoring

The noncontact method of height monitoring is based on the charge interrupting the rays of a radio isotope source such as Cobalt 60 from reaching a detector located on the opposite side of the cupola diameter. Figure 6.17 shows the radio isotope and a detector both mounted over the shell diametrically opposite to each other, at a height a little below the charging door sill.

Figure 6.17(a) indicates that the charge has interrupted the beam from the radio isotope to the detector, sending a red-light signal to the crane operator to stop charging. Figure 6.17(b) indicates a condition where a charge can be added; in this instance, the light indication is green. Figure 6.17(c) shows another configuration using one source and two detectors, suggesting to the operator to watch after one or two more charge loads.

6.3 Induction Furnace—Types and Applications

The basic principle of induction heating was discovered in 1831 by Michael Faraday. Coreless induction furnaces for

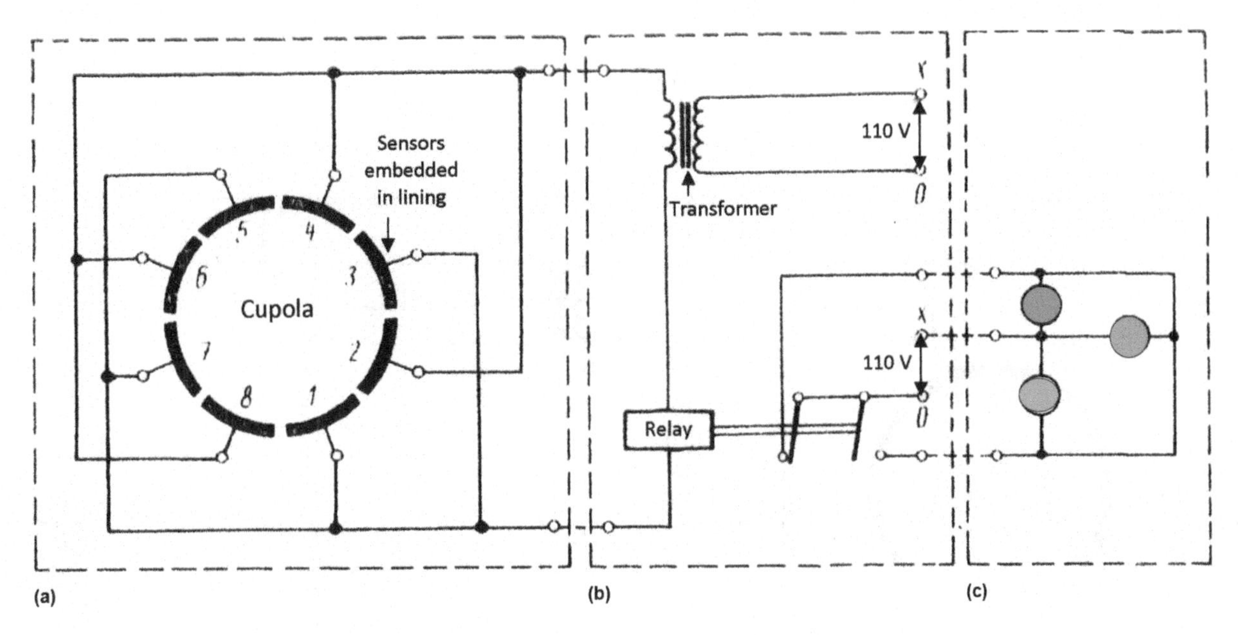

Fig. 6.16 Cupola charge level control, (a) measuring circuit, (b) relay circuit, (c) indicator lights. Source: Ref 1

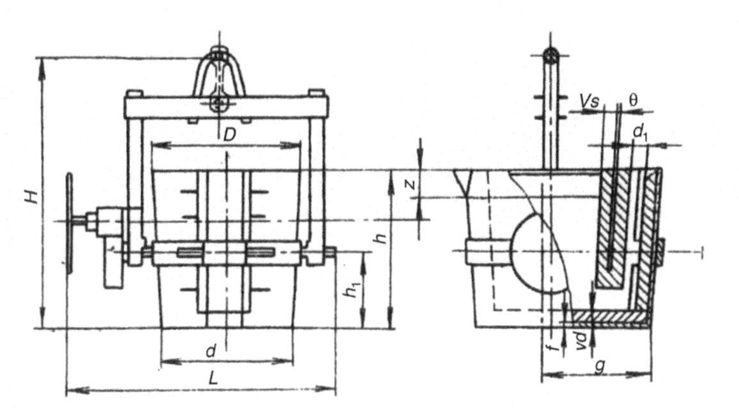

Fig. 6.17 Cupola charge height monitoring using radio isotopes, (a) stop charging, (b) continue charging, (c) watch after one more charge

melting became popular in 1930. An induction furnace consists of a water-cooled copper coil wound around a rammed refractory crucible that holds the metallic charge and liquid metal. A high-wattage current through the copper coil creates a reversing magnetic field inducing an eddy current which flows through the charge. The electrical resistance of the charge creates sufficient heat to melt the charge. The molten metal is stirred by the eddy currents.

The furnace frequencies depend upon the type of metal, the rate of melting, and the volume capacity of the melt. The frequencies begin at 50 to 60 Hz and range up to 400 Hz and higher. The frequency also influences the depth of penetration, which is a function of the volume or the diameter of the lining in the furnace. Smaller diameters need higher frequencies, and lower frequencies stir the metal more vigorously.

Typically, smaller furnaces of one ton capacity can melt the cold charge in one hour. The power required depends upon the furnace capacity. For melt sizes from 20 kg to 65 tons, the power required ranges from 10 to 200 kW.

The advantages offered by induction furnace are:

- Versatility —suitable for gray iron, white iron, alloyed cast iron, ductile iron, alloyed ductile iron, carbon, and alloy steels
- Energy efficiency
- Close chemistry control capability
- Low emissions, resulting in low emission-control costs
- Reduced silicon loss of 10 to 15%, compared to silicon loss in cupolas of 25 to 30%

The disadvantages of induction furnace are:

- Lack of refining capability
- Charge materials must be free from oxides
- Loss of alloying elements during melting needs to be compensated

Induction furnaces are classified in two general categories:

- Coreless induction furnaces
- Channel-type furnaces

Coreless induction furnaces are used extensively for melting a variety of metals, particularly all varieties of cast iron, carbon, and alloy steels. The primary water-cooled copper coil induces a current in the metallic charge which generates enormous heat. These furnaces are very versatile. The charge melts quickly, and the furnace can be used to hold the metal for extended durations. The efficiency in power utilization for coreless induction furnaces is high, about 75%, while the efficiency of channel-type furnaces is even higher, about 95 to 98%.

The channel-type induction furnace generates heat using the same principle as a transformer. The current passes through a primary coil surrounding an iron core. The secondary circuit is the molten metal channel surrounding the primary coil. The current passing through the primary coil induces a larger current in the secondary molten metal loop. The resistance of the metal creates the heat for melting.

Channel-type furnaces are typically used for holding the metal for long durations with minimum expenditure of energy. Duplex installations with both coreless induction furnaces for melting and channel-type furnaces for holding are also economical. They offer the flexibility of induction melting in the off-peak hours and holding and distribution to the molding lines during the busier day shifts.

6.3.1 Coreless Induction Furnaces

The furnace body of a coreless induction furnace has a heavy-duty rolled steel framework, which is rigid enough to withstand the weight of the coil, lining, and metal, in vertical and tilted positions. Figure 6.18 (Ref 8) illustrates a coreless induction furnace body. A hydraulic cylinder tilts the furnace coil and the rammed refractory crucible that holds the metal. A hydraulic pump (with a back-up pump connected in parallel) actuates the hydraulic cylinder. Safety is insured by a strong steel shell body surrounding the coils and the refractory lining.

The coils are made of heavy-duty extruded copper. Each coil is supported by welded studs attached to vertical posts to ensure rigidity against vibration and provide leak-proof connections for the cooling water (Ref 9). The refractory life is extended, and the temperature gradients are evenly spread by the top and bottom cooling coils. Normally, the refractory lining is provided 20 to 25% above the height of the coils to provide the freeboard room to stir. Figure 6.19 (Ref 10) shows a furnace cross section and relevant details. The copper and cooling coils are identified. Item 8 identifies the earth rod which monitors metal leakage that could occur through minor cracks in the lining.

Fig. 6.18 Furnace body of an induction furnace. Reprinted with permission from Ref 8

The deionized coil cooling water is in a closed circuit that passes through a heat exchanger. The furnace is charged through vibratory conveyors.

Larger induction furnaces for melting steel have stainless steel coils at the top and bottom; these coils have proven to extend the life of the lining (Ref 11). A dome-shaped swinging lid with refractory lining is common for all furnaces to cover the metal after charging, during melting, and during holding processes; this lid serves to conserve heat and reduce temperature loss. Table 6.5 (Ref 1) lists the overall sizes of small and medium-sized furnaces. Table 6.6 (Ref 1) lists the power, melting times, and cooling water needs for different furnace capacities. The table provides information for planning,

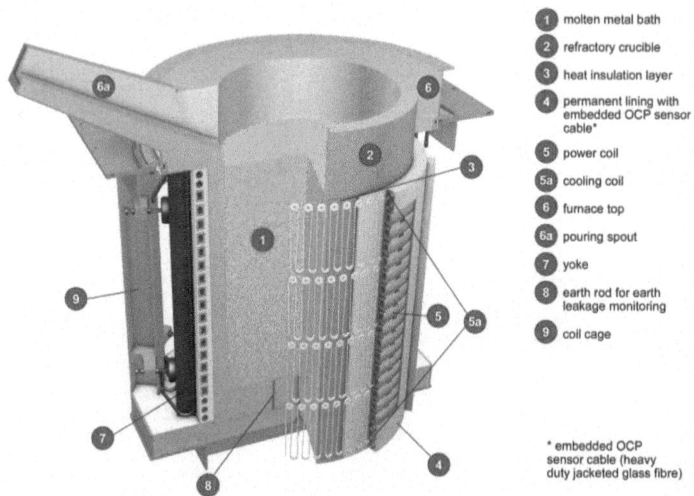

Fig. 6.19 Furnace cross section showing details of coils. Reprinted with permission from Ref 10

Table 6.5 Overall sizes of induction furnaces

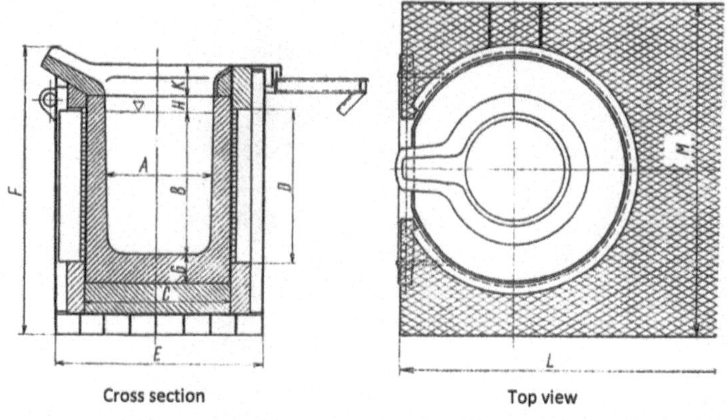

Furnace capacity kg	Small to medium sizes, mm										
	A	*B*	*C*	*D*	*E*	*F*	*G*	*H*	*K*	*L*	*M*
100	230	345	330	400	680	820	105	65	84	850	1050
250	300	450	440	510	706	1050	145	80	103	1350	1250
500	400	540	530	620	856	1188	170	85	120	1420	1380
1000	520	680	730	750	1210	1770	190	115	150	2105	2250
2000	650	850	910	900	1384	1925	210	150	190	2330	2360
3000	800	950	1060	1010	1630	2310	220	170	210	2900	3290
6000	970	1200	1250	1140	1860	2820	250	220	340	3300	3610

Table 6.6 Power requirements of Induction Furnaces

	Power requirements							
Furnace size, kg	Rating, kW	Frequency, Hz	Voltage, V	Instantaneous rating, kW	Melting time, min	Average consumption, kWh/t	Cooling water, m^3/h	Furnace weight, kg
100	100	2000	800	125	60	900	0.85	535
250	300	1000	1 500	380	40	800	3.0	1450
500	500	1000	1500	600	75	780	3.1	2000
1000	800	600	3000	1000	60	750	6.7	5100
2000	800	600	3000	1000	120	740	6.7	8870
3000	1600	600	3000	2000	70	670	11.7	16,300
3000	560	50	380	630	70°/12 min.	Holding	5.0	16,300
6000	1500	50	1200	1600	2 t/h	550	10.8	21,800
6000	200	50	380	240	100°/12 min	Holding	2.5	21,800

although there may be some variances based on differences between furnace manufacturers.

6.3.1.1 Impact of Power Supply Frequency

The optimum frequency for operation is influenced by the furnace size, power input, and extent of stirring needed. The frequency and the furnace size have an inverse relationship: a small furnace operates best at high frequencies, and a large furnace operates best at lower frequencies. The stirring increases if the furnace is operated at lower than the ideal frequency. Excessive stirring results in increased slag inclusions and gas pickup, in addition to excessive lining wear.

Stirring is also a function of the power input. For the same furnace size and frequency, increased power results in more stirring. Therefore, there is a three-way relationship between furnace size, frequency, and power input to produce optimum stirring, as shown in Fig. 6.20 (Ref 12).

6.3.1.2 Frequency and Furnace Size

The furnace frequency influences the depth of field, and this bears a relationship to the size of the furnace for an adequate rate of heating. There are optimum frequencies for different furnace sizes, considering the depth of field and extent of stirring (Ref 12).

Table 6.7 (Ref 12) shows the relationship among the furnaces sizes, the ideal frequencies, and the acceptable operating range for good stirring.

Frequencies that are too high or too low can have harmful effects. Operating the furnace at a frequency lower than the optimum may produce excessive stirring, resulting in slag, refractory gas inclusions, high melt loss, and excessive lining erosion. Furnaces operated at frequencies higher than the optimum may experience insufficient stirring, uneven temperatures in the melt, non-homogeneity of alloy distribution, and excessive temperatures at the periphery of the bath.

6.3.1.3 Frequency and Melting Time

The melting time is influenced by furnace frequency and the melting material. Figure 6.21 (Ref 1) represents the relationship between furnace frequency and the minimum melting time for iron, steel, aluminum, and copper.

6.3.1.4 Influence of Furnace Size

The size of the furnace influences:

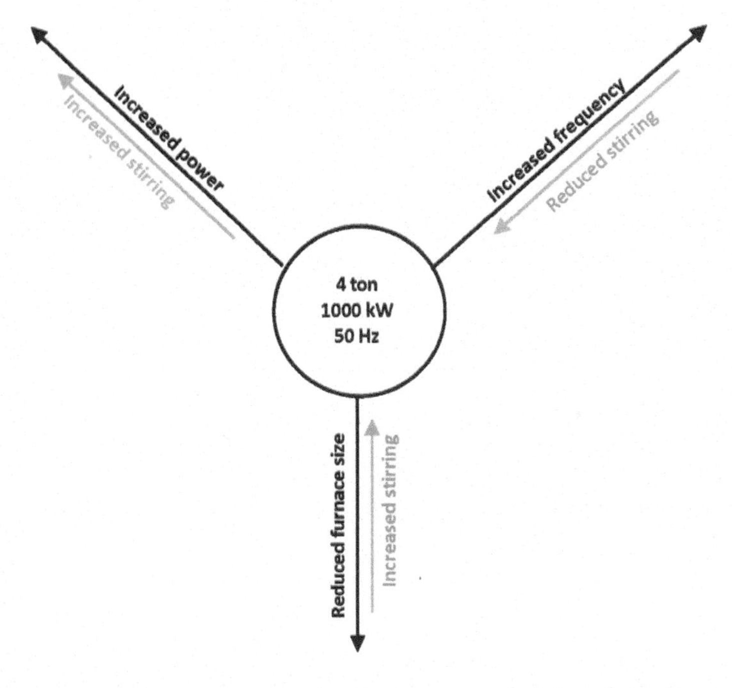

Fig. 6.20 Furnace capacity, frequency, power, and stirring relationships. Source: Ref 12

Table 6.7 Relationship between furnace size and frequency induction furnaces

Furnace size, kg	Ideal frequency, Hz	Acceptable range, Hz
50	5000	2500 to 10,000
250	2000	1000 to 4000
500	1,025	500 to 2500
2500	400	300 to 1000
5000	300	150 to 600
25,000	120	50 to 300
50,000	80	50 to 150

Source: Ref 12

- Rate of heating for different metals being melted
- Power (kW) needed for different levels of stirring of the bath
- Power needed to hold the metal at a targeted temperature

The choice of the furnace size depends upon the melting and superheating rate and the peak demand of the molding machines. The rate of melting is influenced by the level of stirring as the liquid metal starts to form.

6.3.1.5 Rate of Bath Temperature Rise

Figure 6.22 (Ref 1) shows the rate of bath temperature rise for different frequencies of 50, 500, 1000, 2000, and 10,000 Hz and different metals, depending on the size of the furnaces.

The figure also provides data on the rate of temperature rise for channel type holding furnaces operating at 50 Hz.

6.3.1.6 Power and Furnace Size

The power needed for the furnaces is proportional to furnace capacity, but the frequency influences the stirring of the metal bath. While strong stirring reduces the time required to melt, lower frequencies are preferred for holding the bath after melting and superheating. Figure 6.23 (Ref 1) illustrates the relationship among these variables.

The average power consumption for melting different materials include:

Material	Energy, kWh/ton
Cast iron	550 to 600
Ductile iron	550 to 650
Carbon and alloy steels	600 to 680
Aluminum alloys	650 to 700
Brasses	360 to 410

The average values are for general orientation and computation of furnace capacities and other factors. The superheat temperature of the alloys influences these average values.

6.3.1.7 Holding Power

The power needed to hold the superheated melt while waiting for the molds to be prepared is important for both coreless induction and channel-type holding furnaces. The furnace size influences the power required to hold the metal. Figure 6.24 (Ref 1) demonstrates this relationship. As the furnace capacity increases, the power required to hold the metal also increases.

6.3.2 Channel-Type Holding Furnaces—Cast Irons

This section addresses the holding furnaces primarily for gray and ductile irons. Holding furnaces are less common for steels because of the high temperatures and superheat needed to pour steel castings. The development of a pressure-pour unit for steel is addressed later in this chapter. Superheated metal can be held

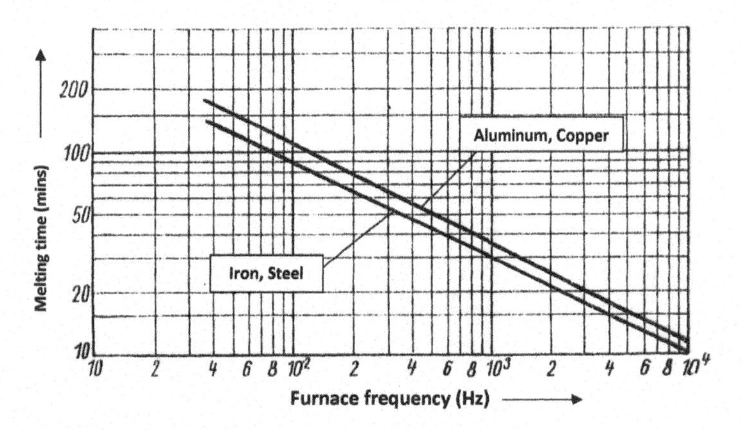

Fig. 6.21 Relationship between furnace frequency and melting rate

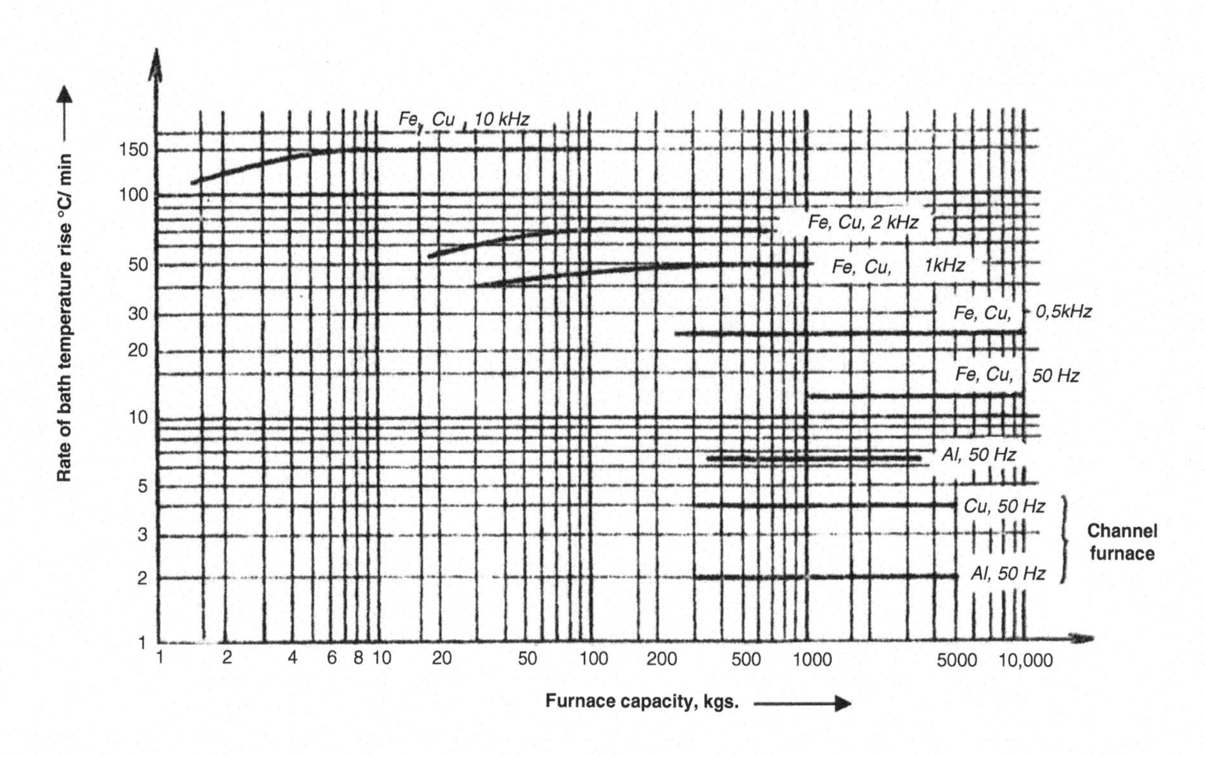

Fig. 6.22 Rate of bath temperature based on furnace size and power frequency. Source: Ref 1

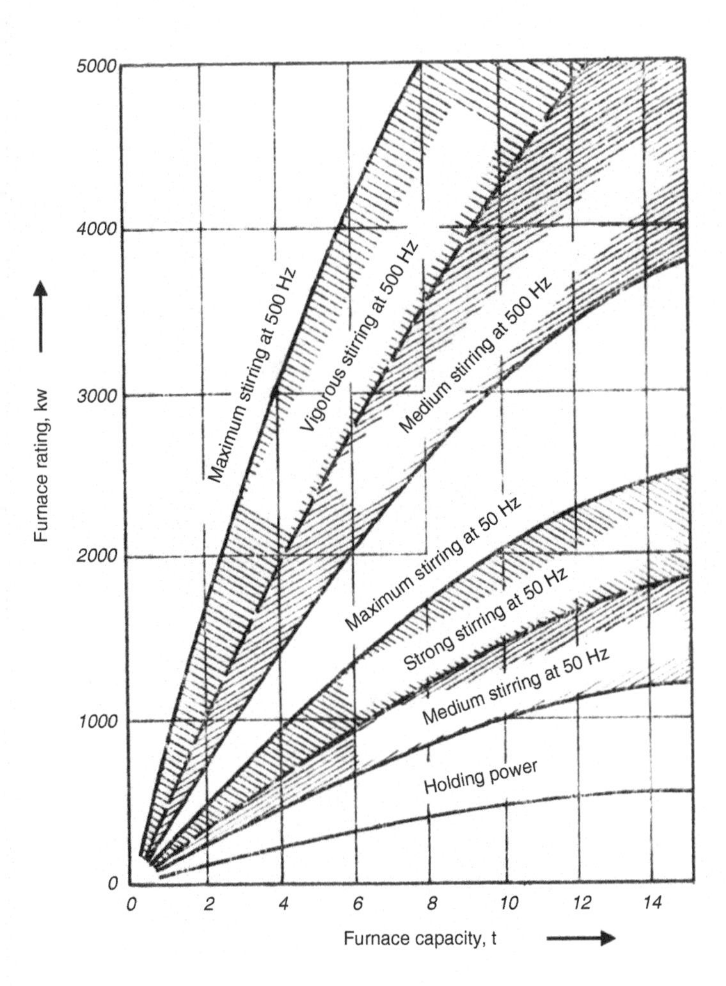

Fig. 6.23 Relationship among furnace capacity, power, and stirring. Source: Ref 1

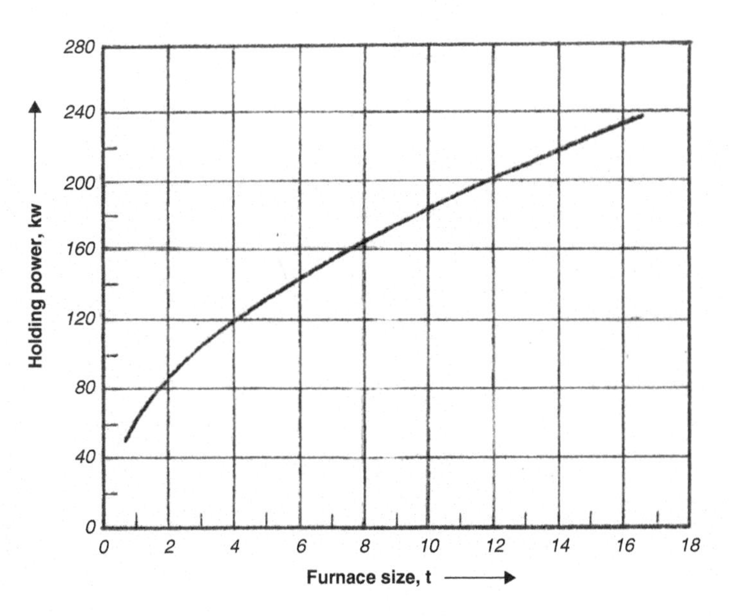

Fig. 6.24 Relationship between furnace size and holding power. Source: Ref 1

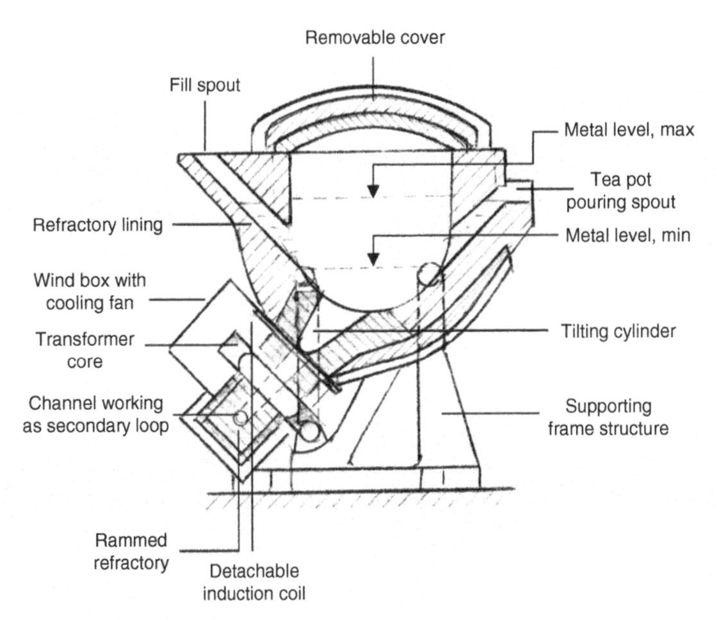

Fig. 6.25 Channel-type holding furnace with one induction coil

at the targeted temperature after melting in a coreless induction furnace. Channel-type holding furnaces are more suitable for holding metal for longer durations (weeks to months) economically with low power consumption.

Large and continuously operating installations such as pipe-making and steel plant equipment use duplexing, combining a coreless induction furnace for melting with a channel-type furnace for holding. Regions offering incentives of lower power rates during the evenings schedule the melting during off-peak hours using a coreless induction furnace, duplexing the metal to a holding furnace to pour the molds produced in the morning shifts.

6.3.2.1 Channel Furnace Operation

Channel furnaces operate on a principle similar to that of a transformer; the primary circuit is the inductor coil, and the secondary circuit is the metal in the channel encased within the coil, along with the protective refractory lining. An alternating current through the primary inductor induces heavy circulation currents in the molten metal (secondary) in the channel. The heat generated in the melt sets up convective currents to heat the entire liquid metal in the furnace. Figure 6.25 is a representative cross section of a channel furnace showing the major highlights of the process.

A hemispherical shell supported on a structure with a hinge and a tilting cylinder is the major hardware feature of a channel furnace. The lining in the main body is built of refractory bricks. The teapot spout and the fill spout are lined with rammed refractories. The inductor is bolted to the bottom of the furnace, forming the channel which acts as the secondary of the inductor. Some furnace designs feature a side fill instead of a teapot fill. The refractory-lined dome-shaped cover can be moved for patching and repairs only. A slag-skimming door is designed above the maximum level of the liquid metal. Recent designs provide for back-slagging, which involves tilting the furnace back, opposite to the pouring spout, and removing the slag through the skimming door. Some designs permit tilting the furnace to replace the induction coil.

Table 6.8 provides the major dimensions of a single inductor channel holding furnace. Furnaces rating 300 kW for iron and

Table 6.8 Channel holding furnace parameters

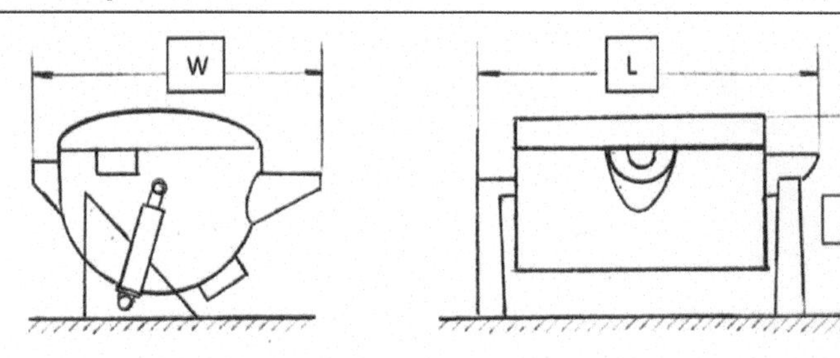

Available capacity, tons	Total capacity, tons	Length, L, cm	Width, W, cm	Height, H, cm	Tilted height, cm	Typical rating, kW	Holding power at 1350 °C	Holding power at 1500 °C	Power for superheating, kW	Superheating rate, 100 °C / h	Energy consumption, kW/100 °C
4	6	4200	3300	2200	3200	200	91	105	94	3.75	55
6	10	4300	3400	2900	3400	400	120	140	235	9.52	44
10	14	4500	4500	4300	4100	470	147	162	255	10.0	41
15	20	4600	4600	4400	4600	600	156	187	311	12.2	42.2

steel have one induction coil. Furnaces rating 600 kW have two induction coils. Figure 6.26 (Ref 1) shows a large holding furnace with two induction coils.

6.3.3 Additional Furnace Features

Several features have been added to furnaces for emergency handling, relining time reduction, improved charging, easier deslagging, charge preheating, and emission control.

6.3.3.1 Engineering Induction Furnaces for Emergency

There are three critical provisions to deal with emergencies:

- Power shut-down
- Furnace tilting pump failure
- Coil water overheating due to heat exchanger deterioration

Figure 6.27 illustrates several provisions to address these emergencies. The figure shows a concrete pit in front of the furnace lined with firebricks. If the power shuts down and will not resume for several hours, the furnace needs to be emptied. A hydraulic pump for tilting is provided with a diesel generator to supply the needed power. The same diesel generator supplies the power needed for the coil water circulating pump. A forklift truck with a drum ladle carries the metal from the furnace to cast into the ready-made pig mold. An ingot mold with a diameter less than the inner diameter of the furnace is set into the pit for rapid transfer.

Another mode of failure is the potential for the hydraulic pump (plunger or vane pump) to fail, even when there is power. Another pump is connected in parallel with two-way valves interconnecting the two units. The valve positions are designed to switch, providing time for refurbishing or replacement of the defective pump.

Salt build-up in the heat exchanger tubes reduces the capacity for heat transfer, increasing the temperature of the coil-cooling water. The tubes need to be cleaned using mild acid or by deep hole reaming. Another heat exchanger connected in parallel with two-way valves allows the two units to be alternated for maintenance. Some systems provide spare circulation pumps in parallel, although the potential for their failure is not as acute.

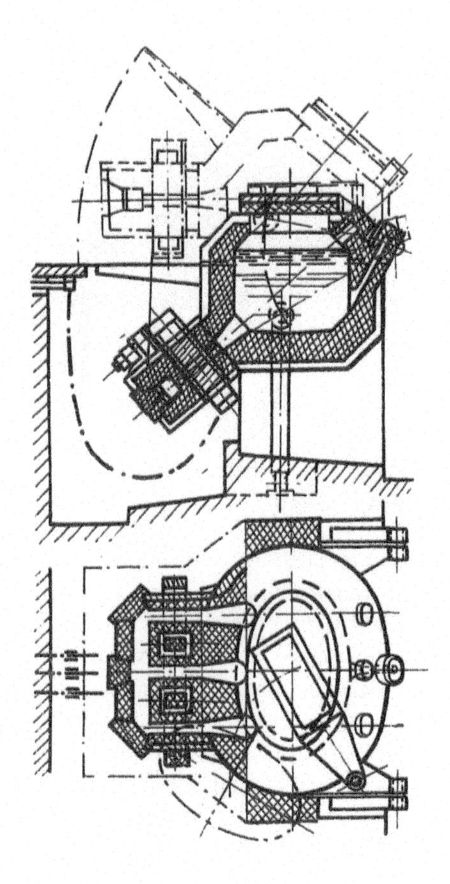

Fig. 6.26 Channel-type holding furnace with two induction coils

6.3.3.2 Relining Time Reduction

One of the more time-consuming and labor-intensive operations is the removal of a worn out lining. A hydraulic ram with a pusher block at the bottom of the lining reduces labor and improves furnace uptime (Ref 11). Such a device also improves the working environment, minimizing operator exposure to the dust that is generated while the lining is being chipped out.

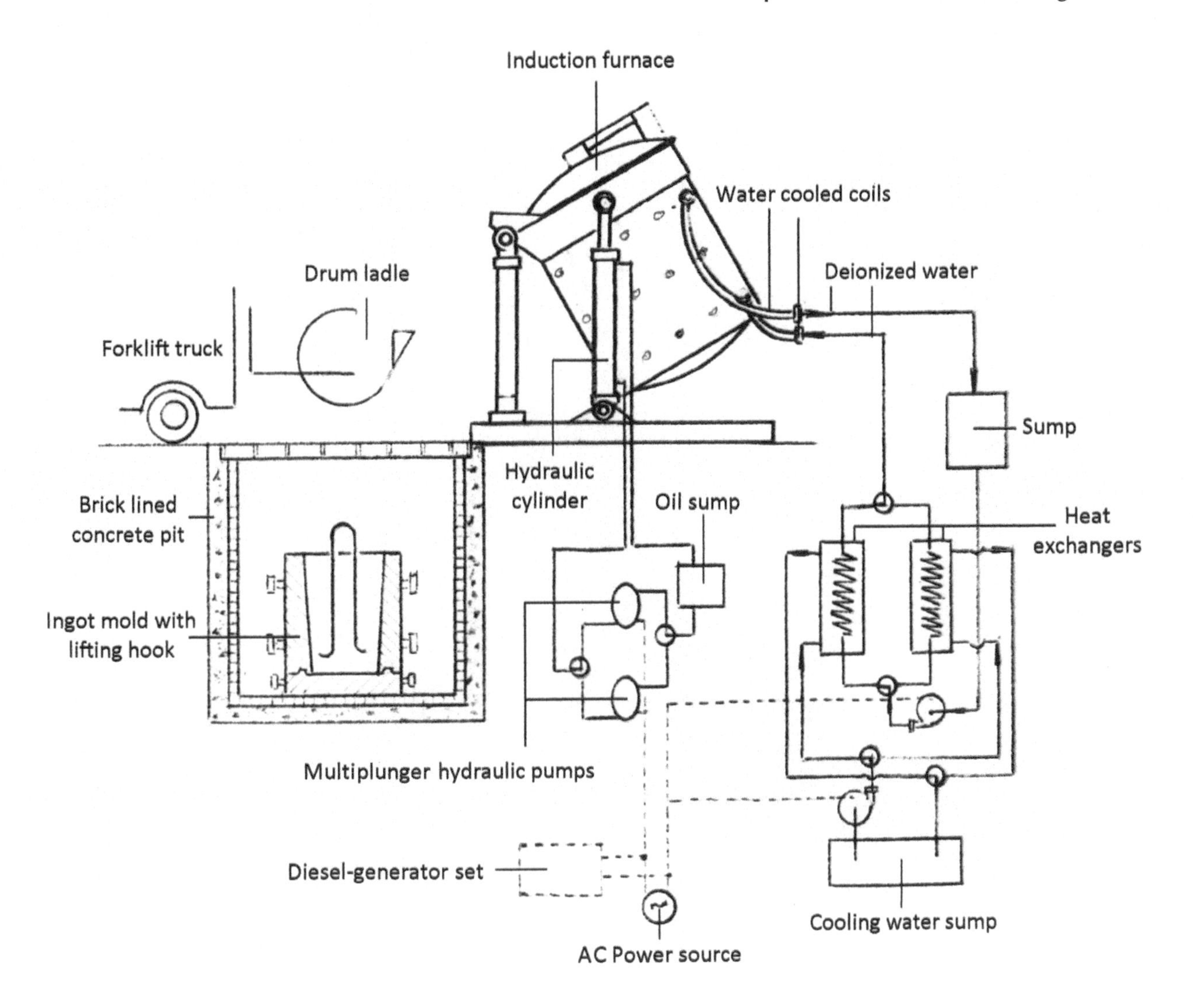

Fig. 6.27 Induction furnace emergency backup system

6.3.3.3 Mechanized Charging Systems

One of the advantages of ferrous charging materials is the ability to use an electromagnet for lifting and dosing for charge makeup. The major equipment consists of vibratory conveyors and bottom discharge buckets. While the mechanization for charging is being designed, care must be taken that there should be no interference with the fume extractor hood, and the furnace operator's access to skim the slag should not be impaired. Figure 6.28 illustrates one of the systems for mechanized charging. It illustrates the crane magnet depositing the charge in the hopper. The vibratory feeder below transports the material to a bottom-discharge bucket. The filled bucket is hauled by another crane into the melting bay and positioned centrally over the induction furnace. The charge is then unloaded as the bucket is slowly lifted.

Two types of bottom-discharge buckets, as well as the hopper and vibratory conveyor are illustrated in Fig. 6.28 to 6.30:

- Figure 6.28 shows a bucket with the bottom made of pie-shaped segments with hinges at the circumference (view A). These segments are tied with a rope, while the bucket is laid on a stand. The bucket filled with the charge is lifted and moved centrally over the furnace. The rope that is holding the pie shaped segments together burns, and the charge is unloaded gradually as the bucket is raised.
- Figure 6.29 (Ref 13) illustrates the hopper and the vibratory conveyor.
- Figure 6.30 (Ref 10) represents the bottom discharge bucket with clamshell doors. The bucket is moved centrally over the furnace. The bucket is slowly raised, during which the clamshell doors open and the charge is gradually unloaded.

6.3.3.4 Furnace Design for Easier Deslagging

Newer induction furnaces (Ref 11) are designed for back-slagging. Furnaces are designed to tilt backward by an angle of 33° toward the rear for easier drawing out of the slag through the rear of the furnace into a slag hopper. The tilt frame is provided with an extended hydraulic cylinder to enable the back-slagging process.

6.3.3.5 Charge Preheating

Preheating the charge reduces the energy costs by around 10% (Ref 11) and decreases the melting time. Additionally, it helps to improve operator safety by evaporating any moisture

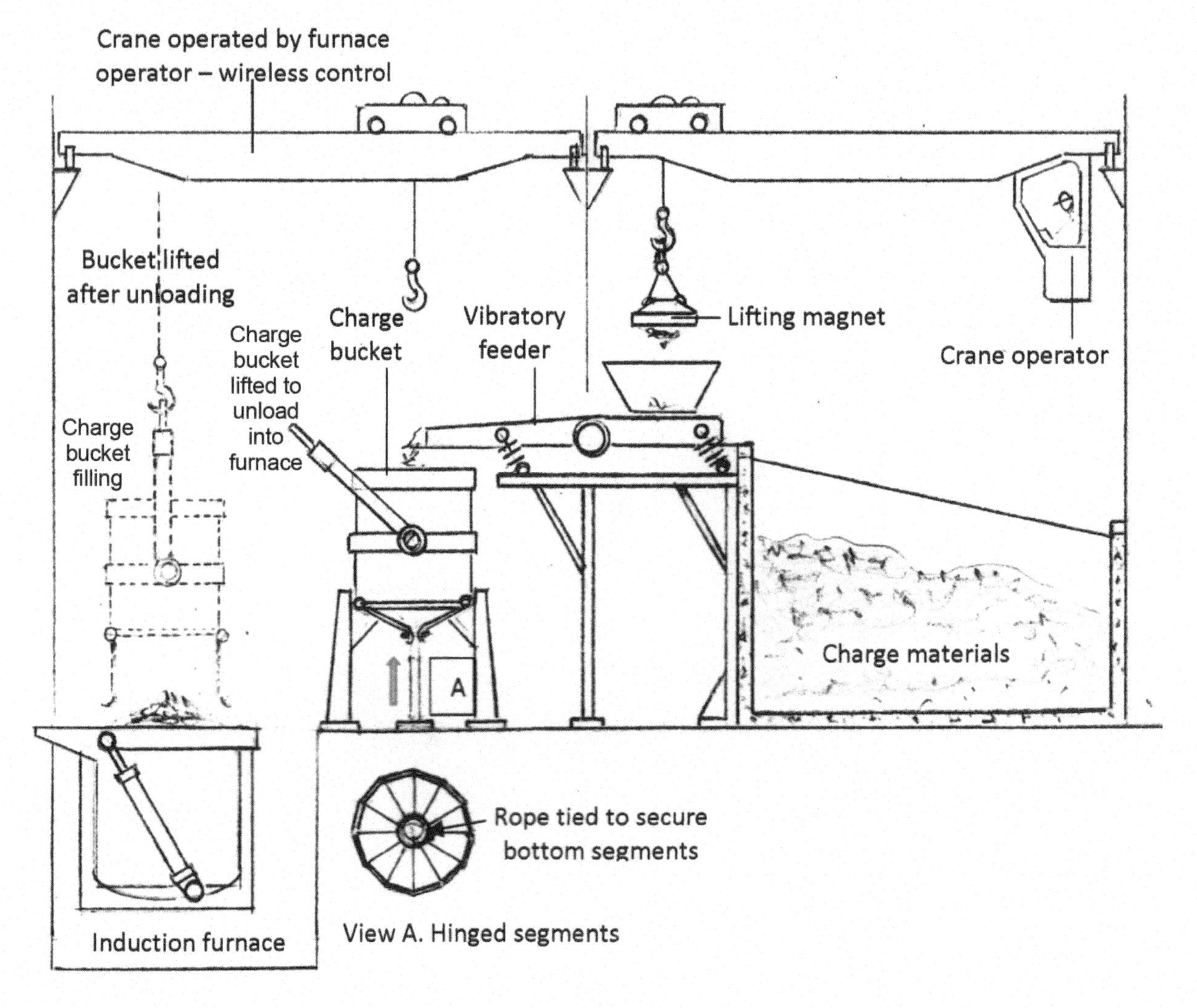

Fig. 6.28 Induction furnace charging system

Fig. 6.29 Vibratory charge feeder. Reprinted with permission from Ref 13

Fig. 6.30 Bottom-discharge charging bucket with clamshell bottom doors. Reprinted with permission from Ref 10

contamination of the charge materials prior to charging in the furnace. The residual moisture could form steam, resulting in a violent reaction or an explosion and potentially endangering the operator and the surrounding environment.

The equipment consists of installing a high-velocity reducing flame-train over the charge materials. Precise air-to-fuel ratios are controlled by microprocessors to ensure a reducing flame that avoids oxidation of the charge materials.

The heated charge is discharged from the vibratory conveyor into the charge bucket.

Another system consists of a burner train below the charge bucket positioned on the stand. The radial bottom segments shown in Fig. 6.28 are modified to provide for a central hole of 125 to 150 mm for the preheating burner flame. The individual segments are tied together using steel wire instead of a rope.

6.3.3.6 Emission Control

Induction furnaces release a lower volume of emissions than cupolas and arc furnaces. Recent EPA limits have prompted the retrofitting of many induction furnace installations with dust suction hoods and bag houses. The fume and dust suction hoods are designed to swing out for charging and furnace servicing. These hoods also help in dust collection when the lining is being chipped during maintenance procedures.

6.4 Arc Furnace Features and Applications

The arc furnace can melt a variety of steel scrap (especially low quality, cheap scrap) and refine it to meet composition limits. Carbon or graphite electrodes contacting the charge create an arc to generate heat, and the radiant energy melts the steel. As the metal melts, the electrodes generate heat more rapidly. This type of furnace is more prevalent in steel industries that recycle steel and in miniature rolling mills.

The advantages of the arc furnace are:

- 100% of the steel can be recycled. Many re-rollers for merchant mills install arc furnaces.
- The furnaces can be started and stopped based on operating needs, which affords greater flexibility.

The disadvantages of the arc furnace are:

- They have huge power demands, which potentially strains the power grid; melting needs to be done during night shifts.
- The cost of emission control is very high compared to emission controls for the induction furnace. Some casting facilities are replacing their arc furnaces with induction furnaces for this reason.
- The creation of the arcing for melting is very noisy, making hearing protection essential in the melting shop.
- The amount of cooling water needed for the body and the roof is high.
- The bath temperature gradient is unfavorable because the heating source is at the top of the melt.

6.4.1 Construction Features

The steel furnace shell can be cylindrical or clam-shaped with a domed cover. The furnace shell has openings for the metal spout and the deslagging door. The furnace body and cover are water-cooled, using embedded pipes. The lining is built of shaped refractory bricks. Three graphite electrodes dip into the furnace through the cover, and they are held in position by water-cooled copper clamps. The clamps are designed to apply uniform pressure on the electrodes. Hydraulic cylinders move the electrodes up or down.

Figure 6.31 shows the overall view and the cut section of the furnace. The furnace shell rides on curved shoes for tilting. Alternatively, some manufacturers hinge the front of the furnace and have one or two tilting cylinders to tilt the furnace by about 40 to 42° for pouring and 15° backward for deslagging. Tilting is accomplished by one or two hydraulic cylinders, depending on the capacity of the furnace. Some manufacturers use electric motors and gear drives for tilting instead of hydraulic cylinders.

Exhaust ducts for the furnaces are provided in the cover. When the cover swings out, the ducting is disconnected. The furnace is charged by buckets with clamshell doors at the bottom, and sometimes electromagnets can be configured to load the furnace. Care must be taken to ensure that finer material is located at the

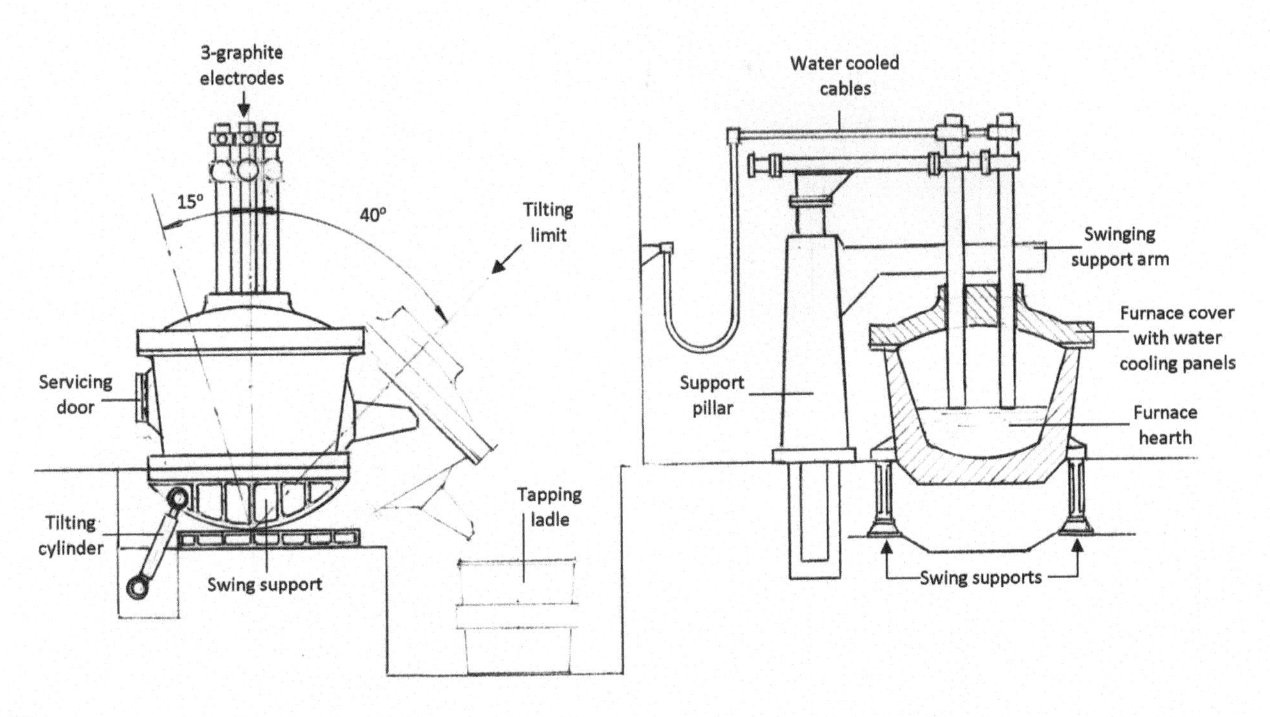

Fig. 6.31 Arc furnace schematic

top of the charge, because arcing will melt the finer material more easily.

Table 6.9 (Ref 1) provides information on the kVA rating, electrode consumption, and other parameters for small arc furnaces (from 3 to 10 tons). Table 6.10 offers data on melting time and cooling water consumption for arc furnaces with a capacity of 6 to 30 tons (Ref 1).

6.4.2 Bath Agitation Improvement

One of the drawbacks of the arc furnace is the non-homogeneity of the bath temperature from top to bottom. This is more significant in larger furnaces (about 30 tons capacity). Alloys added to the furnace bath are liable to segregate due to inadequate mixing or stirring of the bath. Furnace designers have devised an inductor at the bottom of the furnace to create agitation, which can result in improved temperature and alloy distribution.

Figure 6.32 illustrates the inductor attached to the bottom of the furnace. This is essential to providing a non-magnetic base for the shell. The inductor is designed for a low frequency of 0.5 to 1.2 Hz.

The arrows in Fig. 6.32 indicate the direction of metal circulation. Furnaces with a capacity of 30 tons and above that melt carbon steel, and furnaces with a capacity of 10 tons and above that melt high alloy steel have realized a quick payback for the investment in this feature.

6.5 Cost and Other Considerations in Furnace Selection

The selection of the right kind of furnace depends upon a number of factors. Table 6.1 provides a basic overview of these considerations. Cupolas had been very economical and popular for manufacturing gray, white, and ductile irons for many decades. Improvements have been made over several years to improve their efficiency and capability. Similarly, large steel foundries have been using arc furnaces for many years. The increased emphasis on environmental concerns has positioned the induction furnace as the preferred choice for gray, white, and ductile irons, as well as steels.

Several factors influence the selection of electrical furnaces:

- Cost
- Metallurgical factors
- Kind of available steel scrap
- Environmental factors
- Grid power factors
- Furnace size
- The skill level of the furnace operators

Figure 6.33 summarizes these different factors.

Table 6.9 Arc furnace capacities and parameters

Parameter	Unit	Size 1	Size 2	Size 3
Nominal capacity	t	3	5	10
Permitted overload capacity	t	4.5	7.5	15
Transformer capacity	kVA	2250	3250	5000
Secondary voltage	V	95–220	95–220	95–240
Number of steps	Number	4	6	12
Secondary amperage	A	5950	5950	12050
Dimensions of melting zone, diameter	mm	2000	2350	3000
Dimensions of maximum metal level	mm	1760	2000	2850
Depth of bath	mm	350	400	500
Outer dimension of cover	mm	3450	3750	4900
Outer dimension of shell	mm	3200	3500	4600
Dimension of electrode envelope	mm	800	900	1100
Diameter of electrode	mm	250	250	350
Electrode consumption per 1 ton of metal	kg/t	7–8	7–8	7–8
Cooling water requirement	m^3/h	16	18	22

Table 6.10 Furnace and kVA ratings of arc furnaces from another manufacturer

Parameter	Unit	Size 1	Size 2	Size 3
Nominal capacity	t	6	10	30
Permitted overload capacity	t	8	13	36
Transformer rating	kVA	3000	5000	12000
Secondary voltage	V	75–240	95–260	95–320
Dimension of metal bath, diameter, max	mm	3300	3900	5000
Electrode diameter	mm	250	300	450
Tilting for pouring, max	degrees	40°	40°	40°
Tilting for deslagging, max	degrees	15°	15°	15°
Electrode consumption per 1 ton of steel	kg/t	5–6	6	6 –7
Cooling water requirement	m^3 / h	10	12	20
Melting time (heavy charge)	min	100–110	100–120	130–140

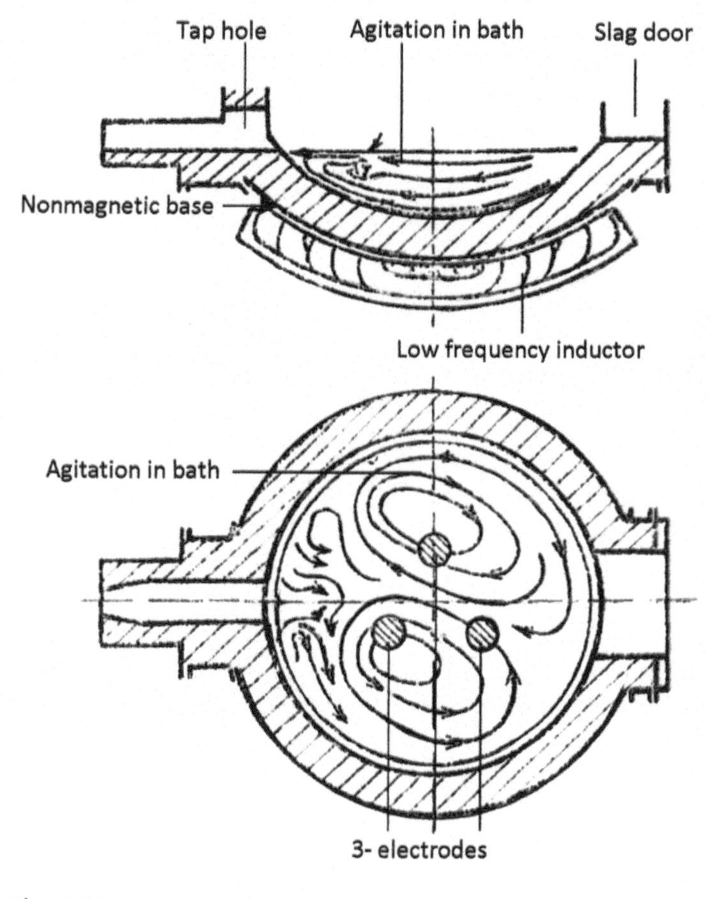

Fig. 6.32 Arc furnace melt agitation improvement

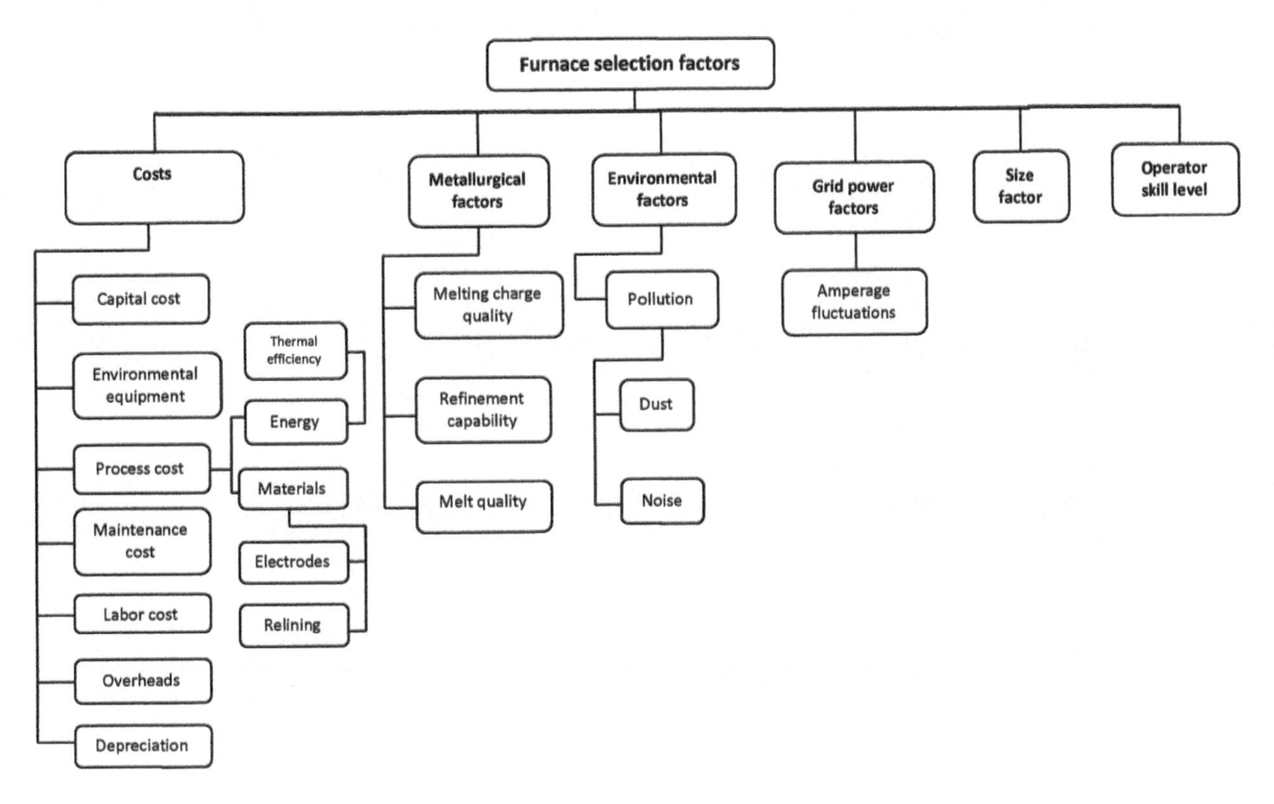

Fig. 6.33 Overview of factors determining furnace selection

6.5.1 Costs

Cost is one of the most critical factors that impact competitiveness and profitability of a furnace operation. Costs are subdivided into:

- Capital costs
- Environmental equipment costs
- Process costs
- Maintenance costs
- Labor costs
- Overhead
- Depreciation

The process costs are subdivided into:

- Energy costs impacted by thermal efficiency
- Material costs, such as arc furnace electrodes and induction furnace relining

Capital costs are determined by the selection of the options, the degree of mechanization, and the sophistication of the controls. Sometimes the base costs may be very attractive, but the cost of the additional equipment to comply with environmental regulations may be very high, as in the case of arc furnaces compared to induction furnaces.

The degree of automation and mechanization depends upon the annual volumes and the volumes per tool setup. Automation that improves the quality of castings and replaces manual labor is worth the investment. Investment in updated and sophisticated controls is less expensive and is recommended especially if the updates improve process control, problem solving, and maintenance alerts.

Process costs consist of:

- Energy costs, which are influenced by thermal efficiency
- Material costs, such as arc furnace electrodes and the cost of re-lining an induction furnace

Furnaces have different energy needs (kWH/t) depending on the materials, which affects the cost. Maintenance costs can be estimated by examining the records of similar in-house equipment. The cost of maintaining spare parts in-house should also be accounted for. Labor costs are calculated based on the skill level and competitive wage rates in the region. The number of operators needed should be calculated with an additional provision for absenteeism and vacations (approximately 8 to 15%). The allocation of overhead costs depends upon the controller in the company, and there are variances among different companies. The depreciation of the equipment is normally calculated over an assumed lifespan of 7 to 10 years for molding and melting equipment.

6.5.2 Metallurgical Factors

Although cost is a major factor, metallurgical factors can outweigh cost considerations in many cases. Metallurgical capability is a critical factor in the selection of the right kind of furnace that can accept the available charge materials and deliver melt to the targeted specifications.

Examples of metallurgical factors include:

- The conventional cupola (which is acid refractory lined) is not suitable for ductile iron production because of its tendency to pick up sulfur from the coke; the sulfur adversely affects the graphite modification by magnesium additions to the melt. Sulfur has a pronounced affinity for magnesium, forming magnesium sulfide and limiting the availability of free magnesium for the melt, for graphite shape modification.

- A basic lined cupola (lined with magnesite or dolomite) permits low-sulfur melts because of fluxing oxides and sulfides in the melt with a basic slag.
- Liningless cupolas can produce low-sulfur melts with low-sulfur pig irons (such as Sorel) with low sulfur contents between 0.006 and 0.012%.
- Malleable iron foundries require low-carbon melts (with 2.16 to 2.90% carbon and 0.9 to 1.9 % silicon) to solidify as white iron. White iron is subsequently heat treated to obtain cluster graphite nodules with a targeted matrix. The low-carbon and low-silicon melts necessitate a high percentage of steel scrap, requiring the installation of hot-blast cupolas.
- Ductile iron melts requiring low sulfur prefer coreless induction furnaces, as they provide the ability to melt a high percentage of steel scrap to achieve low-sulfur melts.
- Casting operations for alloy cast irons, alloyed ductile irons, and austempered ductile irons require coreless induction furnaces for the melting of alloys with a high melting point.
- Large steel operations producing low-carbon and low-alloyed steels employ electric arc furnaces because of the ability to use low-quality steel scrap. They refine the melts by means of desulfurization, dephosphorization, and decarburization.
- Small and medium-sized steel foundries with varied parts (jobbing) prefer the short-time start and stop capability of coreless induction furnaces. The precise control of carbon and capability for low nitrogen are additional advantages, especially for alloyed steels.

6.5.3 Environmental Factors

- Environmental factors have a significant impact on the selection of furnaces. Casting operations face many challenges including:
    - The operating conditions in casting facilities are hot, dusty, and noisy.
    - Casting operations compete with local industries to attract and retain qualified workers.
    - Pricing pressures from tier 2 and tier 3 suppliers prevent casting operations from offering higher wages than other local industries.
    - a. The availability of skilled work force is limited. There are not many training programs for the required skills.
- Different melting units have varied emissions that need to be controlled to comply with local and federal regulations.
- The capital and operating expenses of emission-control equipment need to be quantified and taken into consideration when selecting the casting equipment.
- Different melting units have varying levels of noise and dust. The arc furnaces for steel melting are very noisy because of the arcing. The melt refinement by lancing and fluxing produces a considerable amount of fumes and dust.

6.5.4 Power Requirements and Grid Power Fluctuations

The power requirements for melting and holding are important considerations while selecting the melting equipment. Casting facilities must be equipped with transformers and switch gear to handle the power required. Power companies also charge casting facilities energy rates based on the peak power and power surges. Arc furnaces have power surges that strain the power company grids. Induction furnaces have more consistent power requirements and can operate around the clock, unlike arc furnaces which must melt at night to take advantage of reduced nighttime energy rates.

6.5.5 Size Factor

The size of the melting operation affects the melting cost per ton, as costs are averaged over the melt tonnage. Capital operations of tier one suppliers tend to be larger, with steady and predictable power demands. Such operations will select large furnaces that operate around the clock. Jobbing operations with varied products and metal specifications tend to choose smaller units in multiple numbers for flexibility, to enable them to meet the variable demand.

6.5.6 Operator Skill Level

Some operations such as arc furnace melting need higher skills for operation during multiple shifts. The availability of skilled labor is diminishing, especially in the casting industry. Some operations, such as melt refining or fluxing and degassing, cannot be easily automated.

6.5.7 Alternative Furnace Options

Project designers for new foundries and for foundries planning on expanding capacity need to list the alternatives and make a structured analysis of the suitability, costs, advantages, and drawbacks before arriving at a consensus as to the right choice. Table 6.11 lists the available alternatives of melting units for different materials. A comparison of two alternative units, an induction furnace and an arc furnace, is tabulated in Table 6.12.

6.6 Automatic Pouring Units

Casting facilities that use high-production molding machines find it useful to have automated pouring units to keep up with the productivity of the molding machines.

Automatic pouring units offer several technical, quality, and productivity advantages, including:

- High uptime of molding lines due to constant metal supply
- Consistent pouring rate and controlled pouring temperature
- Slag-free metal and constant metal head by the stopper-nozzle or slide gate mechanism
- Consistent pouring volume regulated by camera control
- Reduced scrap due to improved process consistency
- Reduced labor requirements and improved work environment

Table 6.11 Alternative furnace options

Material	Furnace Options	Comments
Gray iron	Cupola with a recuperator or coreless induction furnace	Cupola has lower capital option. But if cast iron borings and chips are >15% of charge, Induction furnace is a better choice.
Alloy cast iron	Coreless or channel induction furnace	Coreless induction furnace is suitable for batch operation with alloying. Channel furnace is good for one alloy at a time and continuous operation.
Ductile iron	Basic lined cupola or coreless induction furnace	Larger operations may find a basic lined cupola economical, but coreless induction furnace offers flexibility in charge material and closer temperature control.
Steel	Arc furnace or coreless induction furnace	Arc furnace requires lower capital investment, but the environmental equipment is expensive. Surges in amperes drawn, noise, and dust are disadvantages. But arc furnaces allow less expensive charge materials and permit refining.

Table 6.12 Comparison of induction and arc furnaces for steel castings

Factor	Element	Coreless induction furnace	Electric arc furnace
Cost	Capital	Higher (to obtain from quotes)	Lower (to obtain from quotes) High power system costs
	Environmental equipment	Lower (nearly 85% of arc furnace environmental equipment)	Higher. Pollution control equipment cost about 15% of capital costs
	Process cost	Relining costs are high. Thermal efficiency 60 to 75% Energy 650 to 775 kwh/t	Steep increase in graphite electrode prices Lower energy 470 to 520 kwh/t
	Maintenance cost	Higher costs	Lower costs
	Labor cost	Lower	Higher because of greater skill needed for operation .
Metallurgical factors	Charge material quality	Needs cleaner scrap, as no refining is possible	Can accept dirtier scrap, as refining is possible: desulfurization, decarburization, lower nitrogen
Environmental factors	Pollution	Less pollution	Higher noise level, slag disposal, higher dust levels
Electrical grid factors	Transformers and switch gear	No power surges	Power surges with arcing variation
Size factor	Economy vs. size	Smaller units for more flexibility of material grades. Size < 10 t/h	More economical in large units
Operator skill level	Availability and cost	Lower	Higher

6.6.1 Pouring Units for Cast Irons

Units fitted with an induction coil work as a channel holding furnace for improved homogenization of the alloy with improved superheat control. Alignment of the pouring nozzle over the sprue is achieved by using hydraulic drives for *X-Y* axis adjustment. These units are suitable for gray iron and ductile iron, and they can be also used for steel.

Figure 6.34 (Ref 10) illustrates an induction coil fitted underneath the pouring unit to maintain a constant temperature. Figure 6.35 (Ref 10) shows the unit with a stopper rod.

6.6.2 Low-Pressure Pouring Units for Steel

Many automotive castings are being converted to aluminum from iron and steel because aluminum (with its density of nearly one-third of iron and steel) offers an opportunity for vehicle weight reduction. But the strength and stiffness of aluminum alloys are significantly less than the strength and stiffness of steel. If steel castings could be manufactured with thin walls (from 2.0 to 2.5 mm up to a length of about 1200 cm), steel could compete with aluminum and gain market share. Thin-walled structural castings such as knuckles, control arms. and differential carriers may be typical candidates for casting in steel. The concept of using low-pressure pouring has made this approach possible, as shown in Fig. 6.36.

The furnace consists of a refractory lined channel-type furnace with an inlet fill port that is sealed pressure-tight after each furnace

Fig. 6.34 Induction coil below the pouring unit. Reprinted with permission from Ref 10

Fig. 6.35 Automatic pouring unit with stopper-rod control. Reprinted with permission from Ref 10

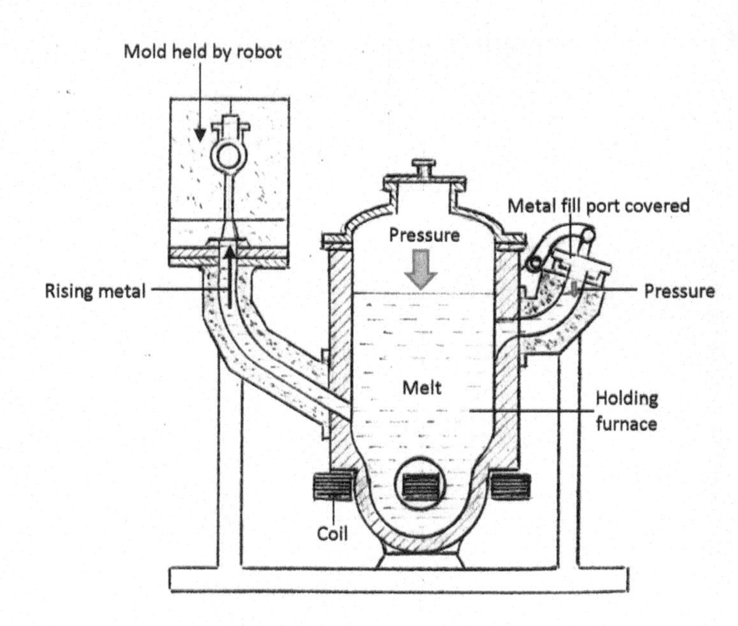

Fig. 6.36 Low-pressure pouring of steel

fill. When the top of the furnace and the inlet port top are pressurized simultaneously, molten steel fills non-turbulently up the pouring spout into the no-bake mold, which is held in place by a robot. The cell is integrated with the no-bake core mold manufacturing. Figure 6.37 (Ref 8) is an illustration of the pouring unit.

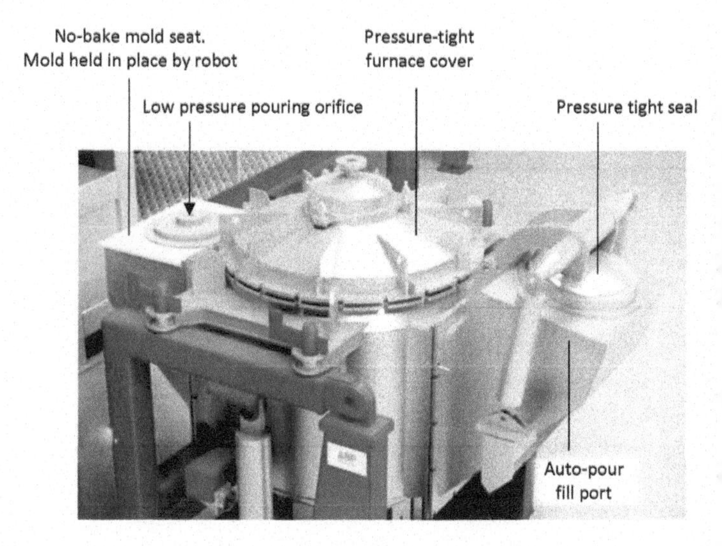

Fig. 6.37 Low-pressure pouring of steel castings. Reprinted with permission from Ref 8

6.7 Engineering for Efficient Metal Distribution and Accurate Pouring

Effective engineering for timely metal distribution and precise pouring influence both productivity and quality. The distribution system needs to keep pace with the demands of the molding machines in real time without any downtime. The pouring systems should ensure that the molds are poured with the correct metal volume at the required superheat temperature without any turbulence or slag entrainment.

6.7.1 Metal Distribution Systems

Metal from the melting units needs to be tapped, delivered, and poured into the molds in the shortest time practical and with reliable consistency. The way the system is designed and the extent of automation chosen depend upon the nature of the production (mass production or jobbing), and the average and the maximum size of the casting being produced.

Large foundries producing medium- to large-sized castings using ladles handled by overhead cranes may design the layout of the cranes to overlap, as shown in Fig. 6.38. The metal tapped from the furnace into a large ladle is handled by an overhead crane and placed on platforms that are built with refractory bricks in each molding bay. The overhead crane of the individual molding bay picks up the ladle to pour the castings.

Casting operations producing medium-sized castings can be grouped as medium- or large-volume. Medium-volume operations may choose a lift truck to transfer the ladle from the melting furnace to the molding bay. Table 6.13 illustrates a drum

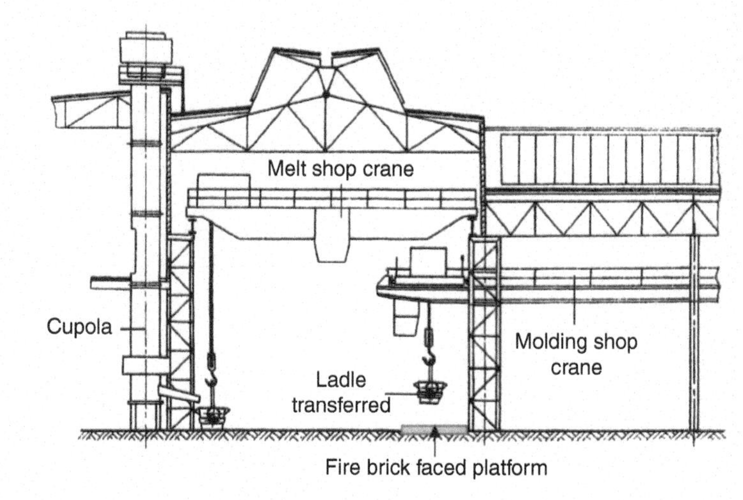

Fig. 6.38 Cross-bay transfer of metal from melting to molding

Table 6.13 Drum ladle sizes

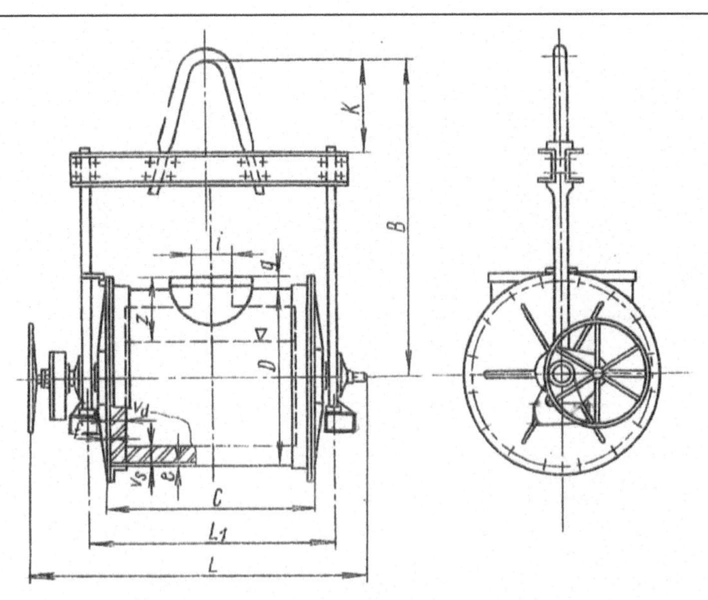

Ladle volume, dm³	Main dimensions, mm													Ladle weight, kg	
	D	C	L	L_1	B	K	Z	i	g	v_d	v_s	e	f	Steel shell	With lining
63	560	650	1170	800	1090	305	220	160	35	70	60	12	30	260	430
100	630	740	1260	890	1170	325	240	150	40	75	65	13	32	300	500
160	760	870	1400	1050	1300	365	265	170	42	85	70	14	35	480	860
250	830	1000	1620	1220	1505	420	305	195	48	95	80	16	40	658	1158
400	960	1130	1750	1350	1625	455	330	210	52	105	88	18	44	780	1440

ladle suspended by a lift truck used for transferring metal from the melting bay to the molding bay. Some lift truck manufacturers offer a design where the drum ladle is supported for direct pouring into a mold instead of transferring to a crane in the molding bay. The figure also provides overall dimensions for general guidance.

Manufacturing operations for larger volumes of medium-sized castings use semi-automatic pouring systems. Figure 6.39 (Ref 14) illustrates this type of installation. Pouring molds are situated on a conveyor line with a large ladle that is carried on monorails; the operator rides on the carriage and pours the molds.

Captive foundries can realize a good payback from using totally automated systems of pouring. These units offer many technical advantages for large-volume production. The advantages include:

- Highly accurate pouring based on *teach and playback* control technology
- Consistent pouring volume, eliminating overpours, underpours, and assuring targeted yield and quality
- Close control of metal trajectory by simultaneous control of the stream and height by *X*, *Y*, and *Z* axis manipulation

Fig. 6.39 Semiautomatic pouring of different medium-sized castings. Reprinted with permission from Ref 14

Table 6.14 (Ref 15) lists the main features of an automated system for small and medium-sized parts. A similar product offering is available for medium to large parts. The additional features offered are:

- Capability of inoculation in stream
- Unit to measure metal stream temperature
- Detection of molding line speed to synchronize pouring
- Automatic pigging of leftover metal in ladle
- Automatic changing of ladles
- Detection of the start of the pour (important when mold heights vary)

Figure 6.40 (Ref 15) is a view of an automated pouring unit.

Table 6.14 Features of automated pouring units

Specification	Model					
	FVN-I	FVN-II	FVN-III	FVN-IV	FVNX-II	FVNX-III
Ladle capacity, kg (lb)	250–400 (551 –882)	450 –650 (992 –1433)	700 –900 (1543 – 1984)	1000 –1300 (2205 –2866)	450 –650 (992 –1433)	700 –900 (1543 –1984)
Molding speed, sec/mold	18	25	30	30	25	30
Weight control	By load cells					
Pouring time, sec	< 8	< 13	< 18	< 18	< 13	< 18
Pouring speed, kg (lb)/sec	2 –5 (4 –11)	2 –10 (4 –22)	2 –12 (4 –26)	2 –12 (4 –26)	2 –10 (4 –22)	2 –12 (4 –26)
Driving units	AC servo drive and inverter drive					
Functions	Teaching mode, poured mold detection, auto weight measuring, pouring speed setting, automatic selection of pouring position, synchronous travel with molding, pouring start detection					

Source: Ref 15

Fig. 6.40 Automated pouring unit in operation. Reprinted with permission from Ref 15

6.7.2 Ladles for Metal Distribution and Pouring

6.7.2.1 Standard Ladles

Ladles play an important part in contributing to casting quality. Their design should permit fulfillment of these functions:

- Enabling uninterrupted pouring at the targeted rate
- Requiring minimum effort from the pouring operator
- Permitting ladle rotation to maintain the metal stream trajectory
- Minimizing temperature loss of superheat

Many ladles have lids to prevent rapid radiation loss from the top of the metal surface. The refractory of the lining withstands the heating and cooling associated with pouring. The ladles are preheated by gas burners and are washed with a refractory paint during periodic reconditioning. Some ladles are specially insulated with refractory blankets between the shell and the refractory lining to reduce the rate of temperature fall as the ladles are emptied.

Standard ladles are classified as:

- Lip-pouring ladles with a spout
- Teapot ladles for arresting slag
- Bottom-pouring ladles with stopper rods (common for pouring steel castings)
- Bottom-pouring ladles with slide stoppers (common for pouring steel castings)

Special ladles include:

- Pouring ladles tipping around the spout
- Ductile iron treatment ladles

6.7.2.2 Lip Pouring Ladles with a Spout

Table 6.15 lists the main dimensions of standard lip-pouring ladles. Table 6.16 provides the sizes of larger lip-pouring ladles.

6.7.2.3 Teapot Ladles

Teapot ladles are designed to skim the slag that floats on the metal surface, preventing it from getting into the castings. The maximum sizes are smaller than those of the lip-pouring ladles, as their pouring rates are limited. Table 6.17 lists the main dimensions of teapot ladles.

6.7.2.4 Bottom-Pouring Ladles

Bottom-pouring ladles are used exclusively for pouring steel castings. These are not well suited for cast irons, as the high fluidity of the melted iron causes liquid metal leakages. Bottom-pouring ladles have either stopper rods or sliding gates to start and stop the pouring process. Table 6.18 shows the major dimensions of medium-sized ladles, and Table 6.19 provides the major dimensions of larger ladles with stopper rods.

6.7.3 Special Ladles

6.7.3.1 Ladles Tipping around Spout

Ladles that tip around the spout offer high precision in targeting the sprue and maintaining a metal stream of low free height. Figure 6.41 is a schematic illustrating this design concept.

Table 6.15 Standard lip pouring ladle sizes

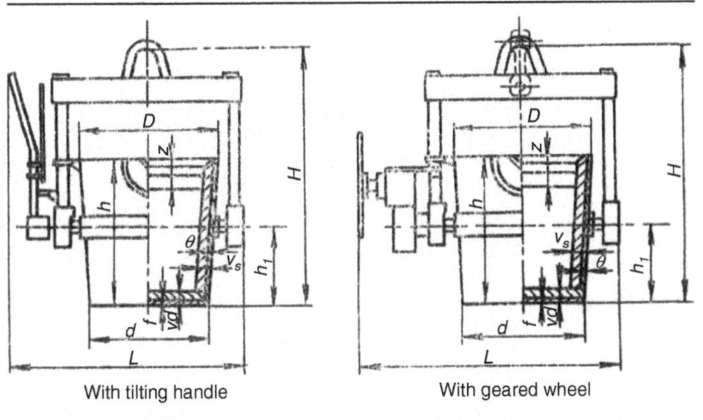

Ladle volume, dm^3	Main dimensions, mm: Ladle D	d	h	With hardware H	L	Ladle weight, kg: Without refractory	With refractory
16	345	300	360	778	845	65	105
25	400	342	415	850	868	70	110
40	460	400	480	920	930	107	157
63	530	460	553	1106	1080	172	252
100	620	540	647	1267	1160	228	328
160	720	650	760	1740	1360	500	670
250	830	750	872	1842	1470	600	830
400	944	850	1040	2060	1615	840	1220
500	1050	950	1100	2155	1720	926	1406
630	1120	1000	1170	2360	1890	1200	1750
800	1218	1100	1300	2663	2055	1550	2250
1000	1324	1200	1365	2672	2160	1700	2600
1250	1400	1300	1460	2800	2230	2045	3305

Table 6.17 Teapot ladle sizes

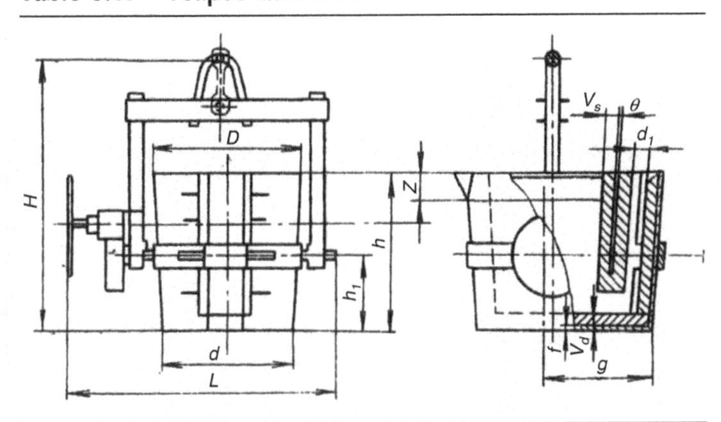

Ladle volume, dm^3	Main dimensions, mm: Ladle D	d	d_1	h	Handling hardware H	L	Ladle weight, kg: Without lining	With lining
40	530	460	65	553	1058	1080	171	311
63	620	540	70	647	1187	1146	216	416
100	720	650	80	760	1750	1360	540	860
160	830	750	85	872	1852	1470	618	1088
250	944	850	90	1040	2060	1615	900	1480
400	1050	950	100	1100	2182	1720	1000	1920
500	1120	1000	105	1170	2490	1890	1260	2240
630	1218	1100	110	1300	2723	2055	1630	2810
800	1324	1200	115	1365	2913	2160	1820	3700
1000	1400	1300	120	1460	3148	2240	2310	4550
1260	1496	1350	130	1574	3212	2340	2662	5170

Table 6.16 Large lip-pouring ladles

Ladle capacity, tons	Permitted overload, tons	Main dimensions, mm: d	D	h	H	L	Weight, kg: Without lining	With lining
10	1.9	1400	1600	1600	3355	1800	2650	2100
16	2.3	1600	1800	1835	3730	2060	4500	2700
20	3.0	1700	1940	1950	4030	2200	5500	3100
25	3.7	1800	2100	2070	4250	2360	7000	3500
32	4.3	2000	2265	2265	4515	2600	8300	4250
40	4.6	2100	2450	2450	4600	2750	10,500	4700
50	5.8	2200	2600	2600	4850	2960	12,850	5500

Table 6.18 Medium-sized ladles with a stopper rod

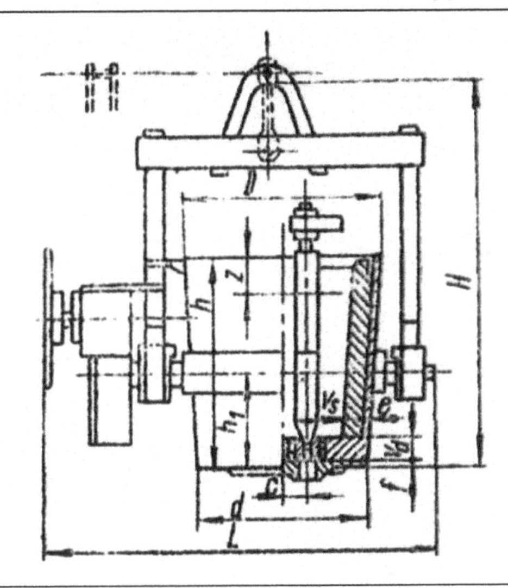

Ladle volume, dm^3	Main dimensions, mm: Ladle D	d	h	Handling hardware H	L	Ladle weight, kg: Without lining	With lining
250	944	850	1040	2060	1615	960	1420
400	1050	950	1100	2170	1720	1050	1780
500	1120	1000	1170	2490	1890	1325	2225
630	1218	1100	1300	2723	2055	1753	2733
800	1324	1200	1365	2736	2160	1880	3480
1000	1400	1300	1460	3087	2240	2295	3980
1250	1496	1350	1574	3202	2340	2642	4842

6.7.3.4 Tundish Ladles for Ductile Iron

Special ladles for ductile iron processing feature a pocket at the bottom for depositing magnesium and ferrosilicon for ductile iron inoculation to modify the graphite shape. The top cover of the ladle provides an orifice through which the metal is poured (Ref 16). When an adequate level of metal builds up, it interacts with the treatment alloy, producing a reaction due to magnesium vaporization. The treated alloy is poured through the teapot into the mold.

Figure 6.42 provides a schematic along with overall available sizes of tundish ladles. Figure 6.43 (Ref 17) lists the thickness of the linings and the free top space.

Table 6.19 Large-sized ladles with a stopper rod

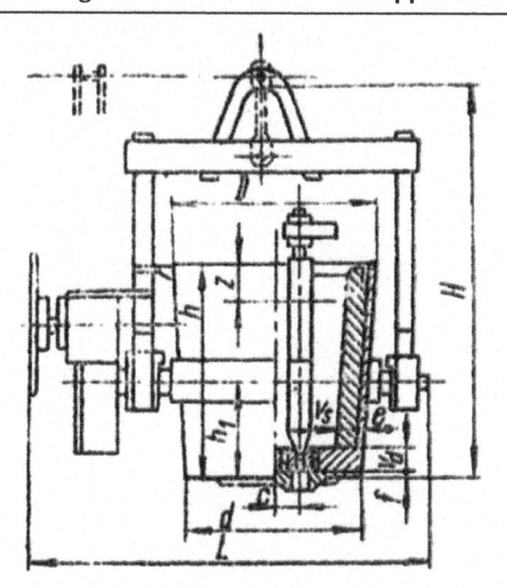

Ladle size, t	Permitted overload, t	Major dimensions, mm				Ladle weight, kg	
		d	D	h	L	Without lining	With lining
10	1.3	1350	1600	1600	1850	1700	1800
16	2.1	1550	1830	1830	2100	2900	3500
25	3.5	1800	2136	2115	2400	4000	5050
40	4.8	2100	2490	2435	2900	7000	8006
50	5.0	2200	2644	2600	3000	8000	9600
63	5.6	2400	2850	2850	3300	10,000	11,500
80	7.5	2600	3080	3150	3600	14,000	15,500
100	10.5	2800	3300	3400	3900	18,000	19,600
130	10.8	3050	3600	3600	4200	23,000	24,500
160	12.2	3200	3800	3950	4500	28,500	30,000

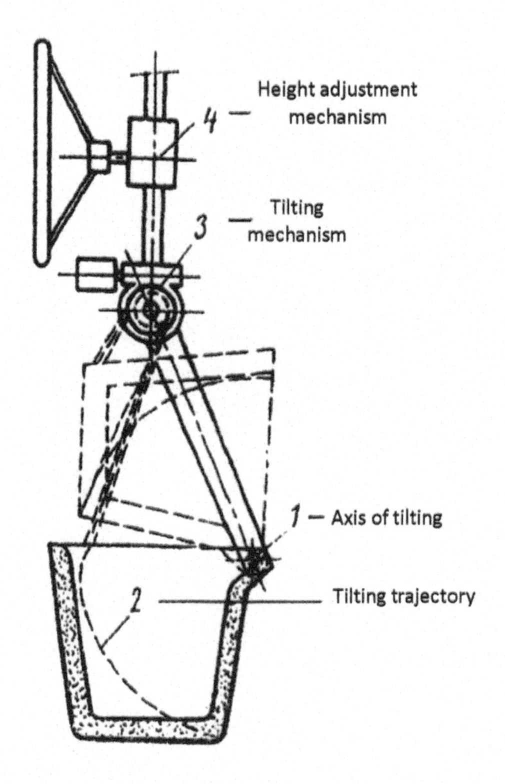

Fig. 6.41 Ladle tilting around the spout

SERIES 1189

Gear	CAP	Side Lining	Bottom Lining	Free Top Space
A	500	2 1/2	4	14
C	1000	2 1/2	4	14
C	1500	2 1/2	4	14
D	2000	2 1/2	4	14
E	2500	3	5	14
E	3000	4	5	18
E	3500	4	5	18
E	4000	4	5	18
E	4500	4	5	18
F	5000	4	6	18
F	5500	4	6	18
F	6000	4	6	18
F	6500	4	6	18
F	7000	4	6	18
F	7500	4	6	18
G	8000	4 1/2	7	20

Fig. 6.42 Tundish ladle refractory and freeboard details. Reprinted with permission from Ref 17

6.8 Pouring Volume Control

The volume of liquid metal poured into the mold is important for quality control, process consistency, and housekeeping. The height of the feeders is designed to ensure adequate feeding to eliminate shrink porosity in the steel castings. Overpours will result in *mushrooming* of feeders, which affects feeding and housekeeping.

The level of the metal in the mold is measured by triangulation using a laser sensor. Figure 6.44 (Ref 17) is an illustration of the laser sensor controlling the level of metal in the mold. It can be set to control the metal in the launder or pouring cup or the height of the open feeders, depending on the metal being poured.

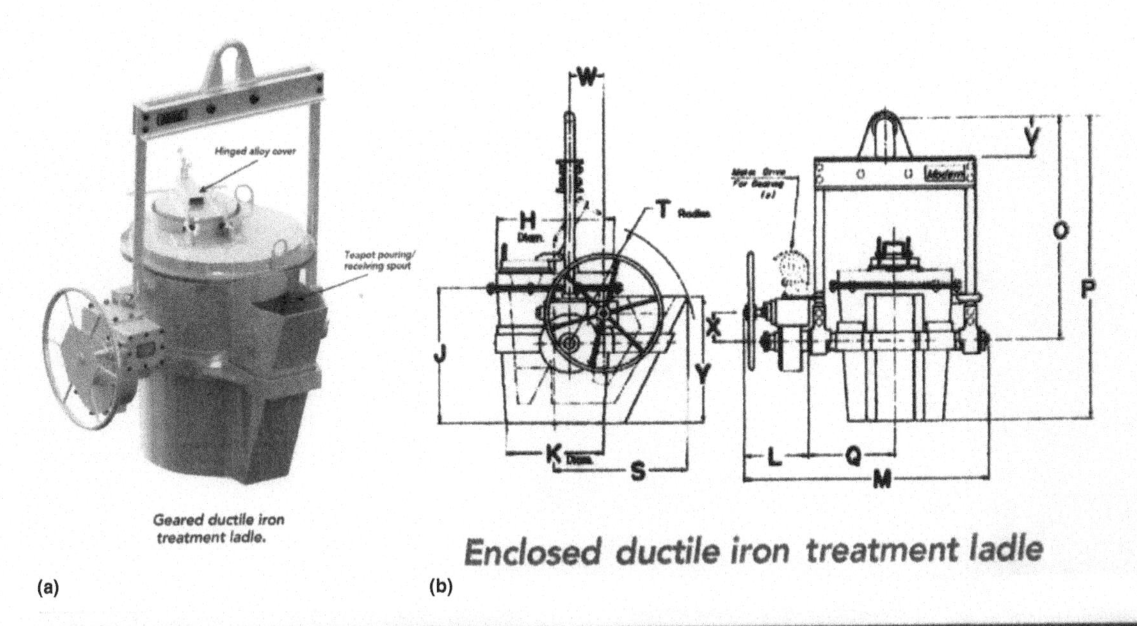

SERIES 1189

CAP	H	J	K	L	M	O	P	Q	S	T	V	W	X	Y
500	24	31	20 1/2	10 7/8	47 1/2	50	68 1/2	17 1/2	27	28 1/2	5 1/2	5 3/4	5	27 1/2
1000	28 1/2	33	25	12 5/8	55 1/4	56	76	20 5/8	29	28 1/2	8	7 1/2	6 1/2	30 5/8
1500	29	39	24	18	60 3/4	60	82 1/2	20 3/4	30 1/2	33	8	9 1/4	8	36
2000	32 1/2	40	28	18	64 3/4	62	84 1/4	22 3/4	32	33 1/2	8	9 1/4	8	38 1/4
2500	34	43 1/2	29	19 1/2	70 5/8	66	90 1/4	24 5/8	33 1/2	37	8	10 1/4	10	40 1/4
3000	36 1/2	48 1/4	31	19 1/2	74	73	97 1/2	26 1/4	35	39	11	10 1/4	10	45
3500	38	50	33	19 1/2	75 1/2	75	100	27	39	44	11	10 1/4	10	47
4000	39 1/2	52	34	19 1/2	76	76	102	27 1/2	41	45 1/2	11	10 1/4	10	49
4500	39 1/2	55	34	19 1/2	76	78	106	27 1/2	42 1/2	46 3/4	11	10 1/4	10	52
5000	42	55	36	21	81	80	108	29	45	49	15	11 3/4	12 1/4	52
5500	43	55 1/2	37	21	83	81	109 1/2	30	46	50	15	11 3/4	12 1/4	52 1/2
6000	44 1/2	56 1/2	38	21	84	82	113	30 1/2	46	51	15	11 3/4	12 1/4	53 1/2
6500	45 1/2	57	39 1/2	21	85 1/2	83	113 1/2	31 1/8	48	53	15	11 3/4	12 1/4	54
7000	46 1/2	58	40	21	86 1/2	83	114	31 3/4	49	54	15	11 3/4	12 1/4	55
7500	47 1/2	58 1/2	40 1/2	21	87 1/2	83 1/2	114 1/2	32 1/4	50	55 1/2	15	11 3/4	12 1/4	55 1/2
8000	49	63	42	19 1/2	89 1/2	93	125	34	52	58 1/2	20	16 1/4	20 1/4	60

Note: All dimensions in inches.

Fig. 6.43 Ductile iron treatment ladles, (a) geared and (b) enclosed. Reprinted with permission from Ref 17

Fig. 6.44 Automated level control for pouring metal. Reprinted with permission from Ref 17

REFERENCES

1. J. Sirokich et al., *Foundry Equipment*, SNTL, 1968
2. Whiting Corp.
3. R.T. Taft, Cokeless Cupolas Ltd. 1972
4. S. D. Singh et al., "Eco-friendly Cupola," *Indian Foundry Trade Journal*, April 1998
5. William Powell et al., *Cupola Furnaces*, ASM International, 2008
6. The EPRI Center for Material Production, Industrial, Agricultural and Technical Services, Carnegie Mellon Research Institute, Pittsburgh, Pennsylvania
7. Jim Gary (GM Power Train), Mark Darr, and S.V. Dighe (Westinghouse Electric Co.), "Plasma Cupola Operation at General Motors AFS Casting Congress," May 1998
8. ABP Induction systems GMBH brochure
9. Ajax TOCCO Magnethermic Corp. brochure
10. Otto Junker GmbH brochure
11. Inductotherm Inc. brochure
12. Michael Fanz-Huster, "Selecting the Right Unit for Efficient Induction Melting," *Foundry Management & Technology*, January 2021
13. Foundry Projects Unit, Staffordshire, UK
14. Acetarc.co.uk brochure
15. Sinto America/Roberts Sinto brochure, "Econopour" automatic pouring systems
16. Modern Equipment Co. brochure
17. Precimeter Inc. brochure

Casting Equipment Engineering Guide
Jagan Nath
https://doi.org/10.31399/asm.tb.ceeg.t59370145

Copyright © 2023 ASM International®
All rights reserved
www.asminternational.org

CHAPTER 7

Aluminum Alloy Melting, Holding, and Dosing Furnaces

MELTING, HOLDING, AND DOSING equipment for aluminum alloy melting equipment are very similar for aluminum sand casting, die casting, gravity permanent and semipermanent mold casting, low-pressure permanent mold casting, counter pressure casting, and squeeze casting. Nearly 50% of the energy consumed in a casting operation is in the melting, holding, and dosing processes. The selection of the right equipment for these operations is critical for energy efficiency and minimizing metal loss due to oxidation. The cost of aluminum alloys for these processes varies, ranging from 45 to 65% of the total manufacturing cost. Therefore, it is important to select the right melting equipment for minimum melting loss.

Achieving high mechanical properties requires that the metal charge should contain a large percentage of primary metal in the form of T-bars or ingots, to limit the deleterious effects of iron and other tramp elements. Secondary metals such as sows or ingots influence charging methods that are dependent on the furnace design. Metal quality is also dependent on the types of products and the processes used. This has a bearing on the types of holders, metal treatment, and dozers.

The yield, which is a ratio of the weight of the casting divided by the amount of metal poured, is determined by the processes that are used. The difference in yield among different processes is primarily due to the percentage of weights of gates and feeders. The sprues, runners, gates, and feeders are recycled in all processes, and these influence the amounts of T-bars, sows, and ingots that are used. The mechanical properties required and type of procured charge materials such as T-bars, sows, and ingots are important considerations in the choice of suitable melting equipment. The decision regarding whether to install a centralized melting unit or to use individual melting units for the different production cells depends on the volume of the output from each cell and the diversity of the alloys cast in the facilities and frequency of casting multiple alloys.

The emission control systems required for reducing global warming and protecting the operator's wellbeing impact capital outlay and operating expenses. The growing market for electric vehicles has caused market adjustments and downsizing of powertrains. The increased demand for lighter vehicles has compelled material shifts toward the use of aluminum, replacing the former preponderance of iron and steel.

This chapter highlights the different melting, holding, and dosing furnaces that are available for the alternative casting processes used for the production of aluminum castings. Generic equipment features, unique capabilities, and suitability for different types and sizes of operations are detailed. Factors influencing the choice of the melting, holding, and dozing equipment are presented. Equipment for recycling machined chips is also addressed.

7.1 Melting Furnace Overview

Aluminum melting furnaces are currently either gas-fired or electric. Both kinds are suitable for batch or continuous operations. Crucible furnaces are suitable for smaller batch operations, where multiple grades or varied chemical compositions are needed in a day's production schedule. In general, stack melting furnaces (stack melters) and coreless induction furnaces are suitable for medium-sized continuous operations, while reverberatory furnaces (or reverbs) are used for large continuous operations. Figure 7.1 presents an overview of the different common furnaces.

Melting furnaces are grouped into four broad categories:

- Crucible furnaces
- Reverberatory furnaces (or reverbs)
- Stack melting furnaces (also called stack melters, tower melting furnaces, or jet melting furnaces)
- Electric coreless induction furnaces

7.2 Crucible Furnaces

Crucible furnaces are suitable for smaller volumes of liquid metal. They are useful for jobbing work and situations where multiple grades or chemistries are needed. The investment and operation costs are relatively low, but labor costs are high.

Crucible furnaces can be classified as gas-fired or electric. Figure 7.2 illustrates crucible furnaces; Fig. 7.2(a) shows a gas-fired crucible furnace and Fig. 7.2(b) illustrates an electric crucible furnace. Both consist of a crucible made of a highly heat-conducting material such as silicon carbide or graphite placed over a refractory tile and surrounded by an insulated cylindrical shell. A gas burner-train lights the flame at the bottom, and a fan supplies air for combustion. The flame swirls around the crucible, melting the charge inside. The exhaust gases exit at the top, as shown. A heat exchanger can be added to utilize the heat of the exhaust gases, improving the thermal efficiency.

Figure 7.2(b) illustrates the resistance coils that are mounted along the inside circumference of the insulated shell of the electric furnace. As power is supplied, the hot resistance coils radiate heat to the crucible, melting the charge inside. The rate of heating is slower than in a gas-fired furnace, but there are fewer emissions.

Only ingots, casting returns (sprues, runners, gates, and feeders), and scrap can be charged into the crucibles. T-bars and sows are too large for crucible furnaces.

7.2.1 Comparison of Gas-Fired and Electric Resistance Crucible Furnaces

Table 7.1 summarizes the attributes of gas-fired and electric crucible furnaces (Ref 1, 2, 3). The shorter meltdown times in gas-fired furnaces are accompanied by lower fuel costs but higher melt loss and a hotter operator environment. The melting time, melt loss, and energy consumption are critical factors which influence the selection of the right kind of furnace.

The crucible furnaces can be configured in various ways. Figure 7.3(a) shows a lift-out crucible type of furnace. After the metal is melted, the crucible is lifted using a pair of tongs. It is then set into a ladle handling ring and secured using metallic wedges (made of steel tubing). The crucible is used to pour the casting directly without any transfer to another ladle. The lift-out crucible is used in less mechanized jobbing foundries for smaller jobs. Figure 7.3(b) illustrates a ladler crucible furnace where the crucible is static and liquid metal is ladled out

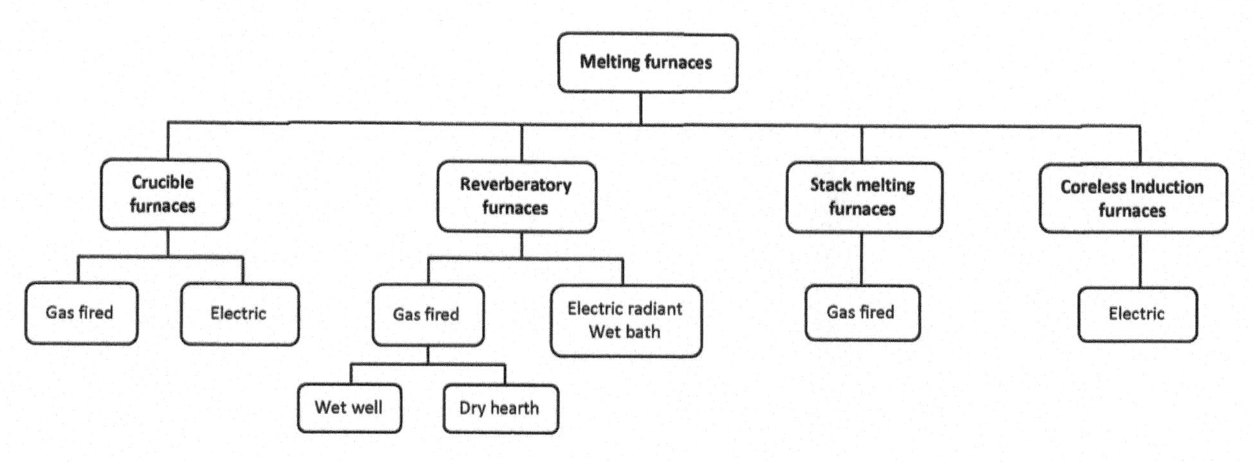

Fig. 7.1 Overview of aluminum melting furnace

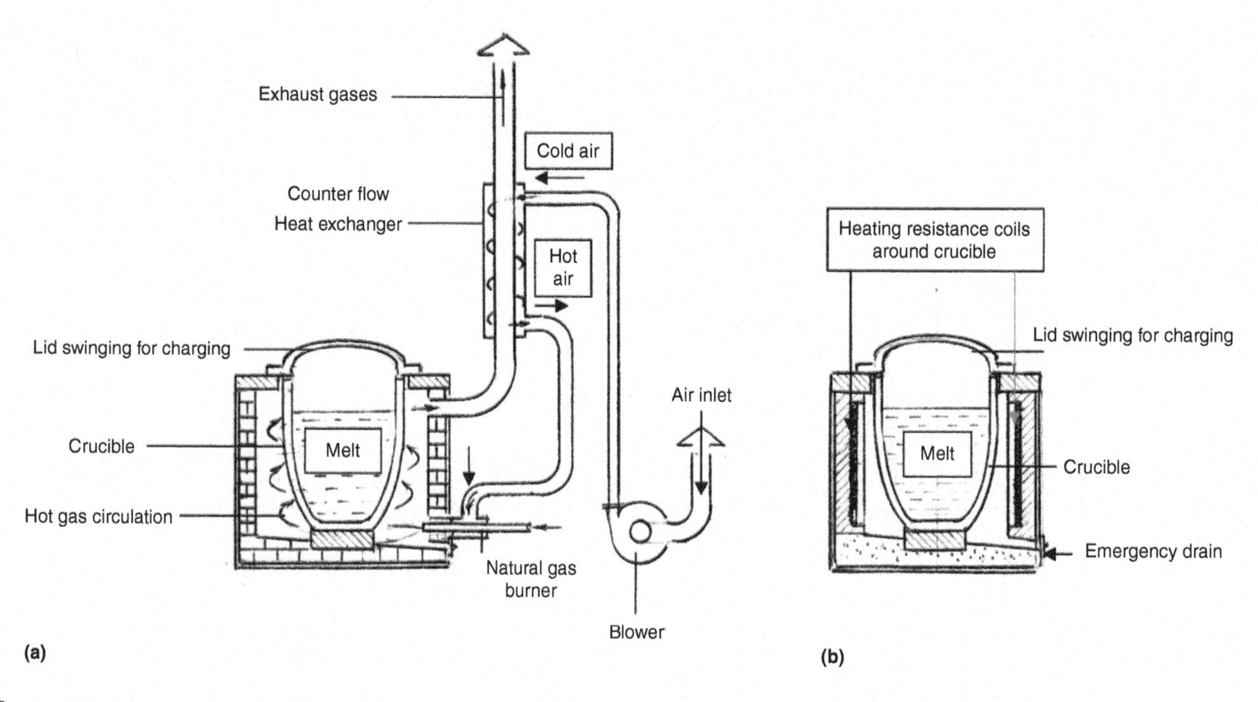

Fig. 7.2 Crucible furnaces, (a) gas-fired, (b) electric resistance

manually or by using a robotic ladler unit. Electric resistance furnaces are preferred, especially with manual ladling, because the ambient temperature in a resistance furnace tends to be lower than temperatures occurring in gas-fired crucible furnaces. Figure 7.3(c) shows a crucible furnace that is mounted on a tilting mechanism for pouring out, usually into another pouring ladle.

7.3 Reverberatory Furnaces

The reverberatory furnace is installed as a central melting unit for delivery of large volumes of metal. Coupled with degassing and filtration furnaces, this furnace is used for die casting (or pressure die casting), gravity and low-pressure permanent mold casting, counter-pressure casting, and squeeze casting operations. Primary melt producers also use this furnace with degassing and filtration units attached. These furnaces are designed with low hearth depths; they are large in length and width, and the shallow roof generates a significant amount of heat due to radiation from the burners located on the low roofs.

Table 7.1 Comparison of gas-fired and electric crucible furnaces

Attribute	Gas fired	Electric resistance
Melt loss, %	3 to 5	<1
Fuel/power consumed, kcals/kg (BTU/lb)	167 to 3900 (300 to 7000)	452 to 512 (854 to 92)
Energy consumed, kWh/kg (kWh/lb)	1.98 to 4.4 (0.9 to 2.0)	0.5 to 0.6 (0.23 to 0.27)
Thermal efficiency, %	7 to 17	57 to 61
Metal quality	Acceptable	Superior
Emissions	High, needs a stack	Low, no need for stack
Operator environment	Hot	Less hot
Meltdown time	Short	Longer
Downtime	Little	Potential for element breakage

Source: Ref 1, 2, 3

Reverbs are classified into gas-fired or electric with radiant electrodes. Gas-fired reverbs are classified as wet-well (also called wet-bath) or dry-hearth. The bottom of the wet-well furnace is shaped to partially submerge the charged ingots, sows, or T-bars in the liquid metal bath. As the charge melts, ingots, sows, or T-bars move down, allowing more of the charge to be submerged. Figure 7.4(a) is a schematic illustration of this process. Wet-well reverbs have the advantage of quick melting for shot biscuits, gates, runners, feeders, and flash without the excessive oxidation that can occur in a dry hearth furnace. Care should be taken in wet-well reverbs to prevent any moisture from contaminating the melt. The potential for hydrogen gas inclusions and the chance of an explosion due to moisture can be averted by installing a preheating section for the sows and T-bars before they are loaded into the wet-well baths. Shot biscuits and gating systems in die castings, gates, runners, and feeders from gravity permanent mold castings are liable to retain moisture when the castings are submerged in water for cooling. Some furnace designers combine the front-end section of a wet-well furnace with a charge-preheating section, either by adding a separate burner or by engineering to use the heat of the exhaust gases to expel any moisture before the charge is immersed into the liquid metal.

Dry-hearth furnaces have two sets of burners, as shown in Fig. 7.4(b). The cold charge is placed over the sloping bottom below the burner. Any water that may have settled on the charge material is evaporated, averting the potential for explosion (which might occur if the material were dunked directly into the melt). It also prevents any contamination of the melt by dissociating hydrogen from the water. As the metal melts, it flows down the slope to the main bath, where it is heated to the required pouring temperature.

While dry-hearth furnaces offer the advantage of expelling any moisture that might contaminate the melt and avoid the potential for explosion (as might occur with immersing damp charges into the liquid metal), they have the disadvantage of higher melt loss compared to wet-well furnaces. Wet-bath reverbs with a preheating section offer the advantage of quick melting of shot biscuits, gates, runners, feeders, and flash without the excessive

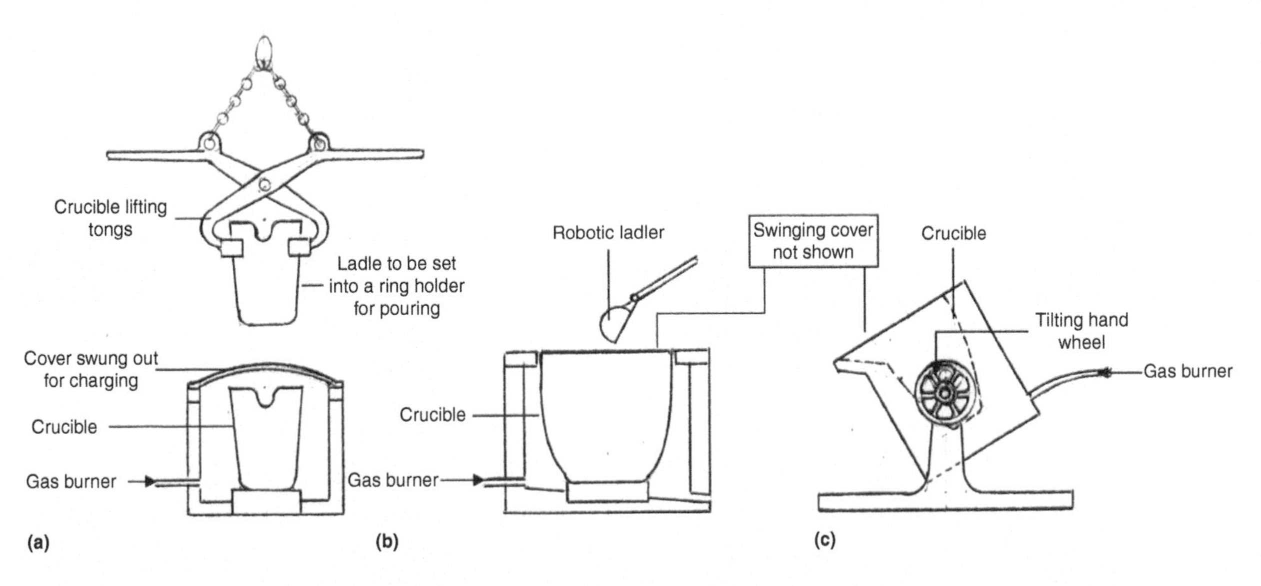

Fig. 7.3 Crucible furnaces, (a) lift out, (b) bale out, (c) tilting

oxidation that could occur in the dry hearth. The thermal efficiency of reverbs can be increased by nearly 50% by installing a regenerator, which utilizes the temperature of the exhaust gases.

The ratio of the amount of metal in the furnace to the amount of each charge influences the fluctuation of the bath temperature. This fluctuation is more pronounced in the case of wet-well reverbs. The ratio maintained is about 10 to 1 for the furnace bath to charge, to limit wide swings in bath temperatures.

Electric reverbs are similar in construction to wet-well reverbs, except that they are designed with electric resistance heating elements on the roof instead of gas burners. Electric reverbs are preferred for holding the metal instead of melting. They offer many advantages, which include low emissions and low melt loss due to oxidation. There are, however, a few disadvantages, such as higher capital and operating costs and a lower rate of throughput.

7.3.1 Comparison between Gas-Fired (Wet-well and Dry-hearth) and Electric Reverb Furnaces

Reverb furnaces differ in capital and operating costs, power consumed, and melt losses. Table 7.2 summarizes some of the salient differences. The data in Table 7.2 is compiled from multiple sources (Ref 4, 5, 6, 7) which reflect different furnace sizes. Therefore, these figures are useful only for general guidance.

Electric reverbs have significant advantages, including lower melt loss, higher thermal efficiency, no emissions, and a more comfortable environment. But the recovery of temperature is low after each charge, and the capital costs are slightly higher.

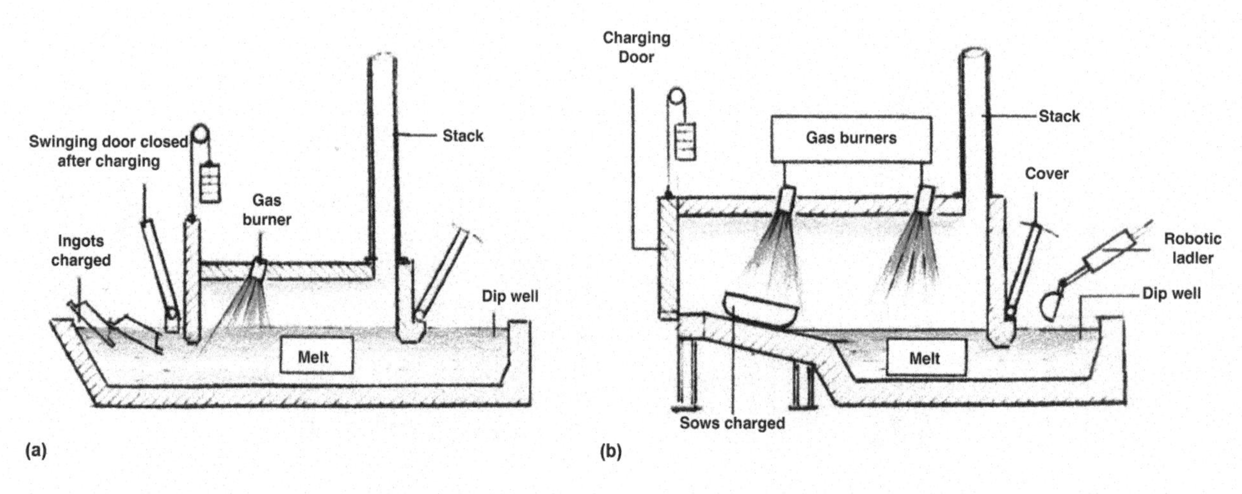

Fig. 7.4 Reverberatory furnaces, (a) wet-well, (b) dry hearth

Table 7.2 Comparison of gas fired and electric reverberatory furnaces

Attributes	Gas fired: Wet well	Gas fired: Dry hearth	Electric resistance
Melt loss, estimated, %	3 to 7	3 to 9	About 1
Fuel/power consumed, kcals/kg (BTU/lb)	About 1112 (2000)	1028 to 1112 With improvements, 695 to 834 (1850 to 2000) With improvements, (1250 to 1500)	426 (767)
Energy consumed, kclas/h (BTU/h)	**Melting** 504 (2000)	**Melting** 453.8 (1800)	**Melting** 0.5 to 0.55 kW/kg (0.23 to 0.25 kW/lb) **Holding** 0.0132 kW/kg/hr (0.005968 kW/lb/hr)
Thermal efficiency, %	About 22	15 to 40, without recuperation 18 to 45, with recuperation	About 67
Metal quality	Acceptable	Acceptable	Very good
Emissions, kg/hr (lb/hr)	Similar to gas fired dry hearth	67 (148) of CO_2 53 (118) of H_2O 321 (708) of NO_x	0 of CO_2 0 of H_2O 0 of NO_x
Operation environment	Hot	Very hot	Less hot
Melting time	Very fast	Fast	Slow
Down time	Low	Low	Medium
Capacities	Similar to gas fired dry hearth	Up to 136 metric tons (Up to 150 tons)	545 to 9066 kg (1200 to 20,000 lb)
Recovery rates	High	High	Slow

Source: Ref 2 to 7

7.4 Stack Melting Furnaces

Stack melting furnaces, or stack melters, are also known as tower or jet melters.

The stack melting furnace offers higher thermal efficiencies (40 to 50% over gas-fired reverbs) and a lower melt loss of 1 to 3%. These attractive attributes are due to the heat exchange between the charge material as it descends the stack and the combustion gases as they ascend the stack, like the cupolas addressed in Chapter 6, "Iron and Steel Melting Furnaces," in this book. Any moisture in the flash or gating systems which may become trapped during the quenching for casting-cooling is evaporated during the descent of the charge.

In addition to the heat exchange in the counter-flow, the furnace features another burner system for holding metal in the forehearth, which is independently controlled. These furnaces are designed for continuous operation by balancing the melt rate to match the casting cell needs. The thermal efficiencies range from 40 to 77%, depending on the design and operation (Ref 2). The energy consumption is around 556 to 778 kcals per kg (or 1000 to1400 BTUs per pound).

Figure 7.5 shows two different configurations of stack melting furnaces. Some installations have bottom bucket discharge systems (shown in Fig. 6.15 in Chapter 6) instead of the *skip* system (shown in Fig. 7.5) for metal charging. Figure 7.5(a) shows a forehearth connected to the stack melter with a tap-out system for metal transfer to either a degassing and filtration furnace (used for structural castings in gravity and low-pressure permanent molding) or to a ladle (used to transfer the metal to holders in die casting machines). Figure 7.5(b) illustrates a top-charging system with a forehearth engineered for robotic dipping for pouring (used in die casting). Different configurations are possible, depending on the type of castings produced and the process used.

Stack melting furnaces offer many advantages, but the charging system cannot handle sows of secondary metal or T-bars of primary metal because the large pieces of metal can damage the lining with its impact. Also, the mass of the sows and T-bars is not conducive to efficient heat exchange from the ascending hot gases.

Stack melting furnaces need to be run continuously with a full stack of charge materials to maximize the heat exchange and assure high thermal efficiency. This requires a good matching of the melting rate with the metal requirements of the cell or the group of machines. A stack melter that is too large for a group of small and lightweight parts run on a cell is neither energy efficient nor economical. Table 7.3 is a comparison of stack melters and gas-fired dry-hearth reverbs (Ref 5, 7).

7.5 Coreless Induction Furnaces

Chapter 6 in this book addresses the features of coreless induction furnaces for iron and steel melting. Coreless induction furnaces, which are specially designed with the best frequency and power for controlled stirring and limited humping, (bulged metal meniscus as shown in Fig. 7.6), offer several advantages, including:

- High efficiency of 60 to 90%
- Low melt loss due to oxidation
- Low emissions
- Ability to start and stop on short notice
- Efficient alloying distribution by mixing control
- High-quality melt is achieved because the power source is not in contact with the melt
- Low operating costs
- Ability to melt aluminum chips by pulling them into the melt by adjusting the stirring (this operation is detailed later in this chapter)

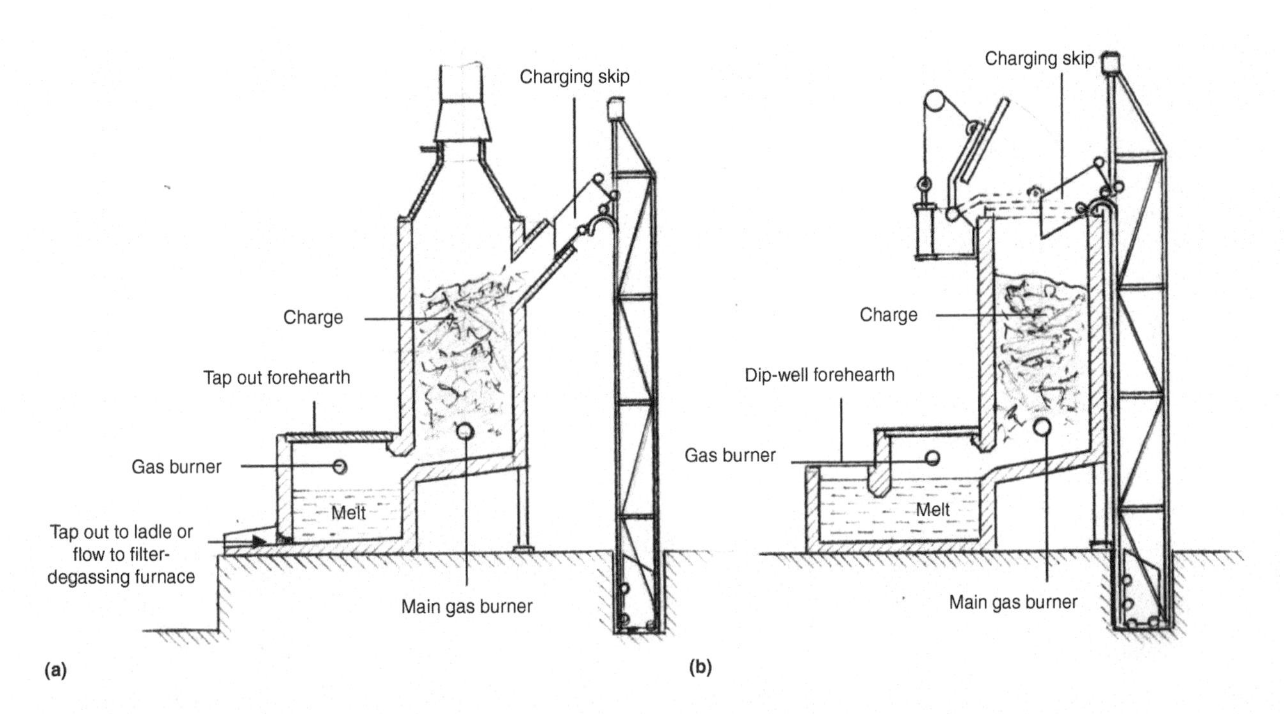

Fig. 7.5 Schematics of stack melting furnaces, (a) with tapout forehearth, (b) with dip well forehearth

Table 7.3 Comparison of stack and gas-fired reverberatory furnaces

Attributes	Stack melting furnaces	Gas fired reverberatory furnaces
Melt loss, %	1 to 3	3 to 9
Fuel/power consumed, kcal/kg (Btu/lb)	556 to 778 (1000 to 1400)	1028 to 1112 (1850 to 2000) With improvement, 695 to 834 (1250 to 1500)
Thermal efficiency, %	40 to 77	15 to 40, without recuperation 18 to 45, with recuperation
Metal quality	Good	Good
Emissions	Low	High
Operation environment	Hot	Very hot
Down time	Low	Low
Type of charge materials	Ingots, gates, feeders and scrapped castings	Ingots, sows, t-bars, gates, feeders, and scrapped castings
Alloying	Not feasible in stack furnace, done in holder attached	Feasible, but losses occur
Operation	Continuous operation needed for keeping stack full for high thermal efficiency	Flexible operation, but melt losses occur during long holding times
Space needed	Smaller footprint, but higher roof structure needed	Larger footprint for furnace and emission control but lower roof structure

Sources: Ref 5, 7

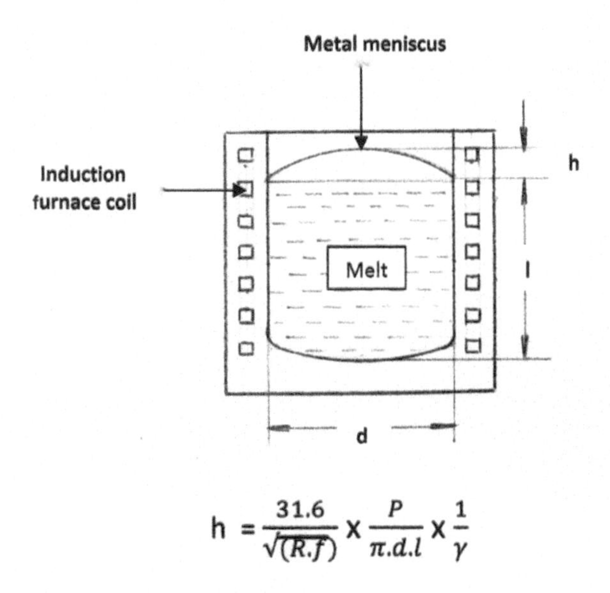

$$h = \frac{31.6}{\sqrt{(R.f)}} \times \frac{P}{\pi.d.l} \times \frac{1}{\gamma}$$

where,

h = height of the hump in dm
R = resistivity of the charge in Ω sq.mm per meter
f = power supply frequency in Hz
P = power supply in kW
d = furnace diameter in dm
l = height of the charge in furnace in dm
γ = Liquid metal density kg/cu.dm

Fig. 7.6 Metal meniscus computation

The one drawback to this method is the high initial capital outlay. Currently, only operations such as aluminum wheel producers (using the low-pressure permanent mold process) which generate large volumes of aluminum chips that need to be melted can justify the high capital investment in electric induction furnaces. The emphasis on reducing global warming and the pursuit of low emissions, along with the push for effective material recyclability, make the induction furnace a strong candidate for aluminum melting.

The extent of melt oxidation in induction furnaces is influenced by the height of the metal meniscus (or hump, as shown in Fig. 7.6), which exposes the metal layer to air (or oxygen in the air). The intent is to keep the hump at a minimum while ensuring adequate stirring. The equation in Fig. 7.6 (Ref 8) shows the relationship between the height of the metal meniscus, frequency, and power. The height of the hump decreases as frequency increases and power input decreases.

7.6 Casting Process and Furnace Options

Aluminum alloys are expensive. Aluminum recyclability is an important consideration for cost competitiveness and emission reduction compared to smelting operations that process the ore to obtain primary metal. Every casting producer strives for100% recyclability of gates, runners, feeders, and flash. The amount of new charge material needed in a melt depends on the percentage of the returns. (Returns are the gates, feeders, and flash that are *returned* to the furnace for remelting.)

Returns for remelting include:

- Die casting—gates, shot-biscuits, vents, flash, and rejections.
- Gravity permanent, and semipermanent casting—feeders, gates, runners, machined chips, and scrapped castings
- Low-pressure permanent mold and semipermanent mold casting—flash, machined chips (in wheel casting plants), gating system and scrapped castings

The casting yield, the ratio of the weight of the raw casting divided by the weight of the metal poured, varies depending on the casting process. The furnace charge consists of a remelt of returns combined with the addition of new charge material. The new charge material added comes in different forms:

- **Secondary metal**—the remelted metal by secondary smelters to produce sows and ingots. Die casting and some gravity permanent operations use secondary metal, where the iron content in the remelted alloys is higher (compared to primary metal).
- **Primary metal**—produced with low iron content and low impurities and produced as T-bars or ingots by primary smelters.

Gravity permanent and semipermanent mold castings, low-pressure permanent and semipermanent mold casting processes, and counter-pressure casting processes that are used to produce high-integrity structural castings for automotive and other industries need a high percentage of primary metal to keep the iron content and tramp elements low. Lower iron content is required to achieve high mechanical properties and ensure good feeding ability to eliminate shrink porosity.

The prices of secondary ingots and sows for the same chemistry vary; sows are slightly less expensive than ingots. But this difference varies when prices are compared between different suppliers. Similarly, the price of primary ingots and T-bars varies among different sources or suppliers.

Crucible furnaces can melt ingots but not sows or T-bars because of the limitations of the crucible size. Sows or T-bars cannot be charged in stack furnaces. Reverbs can handle diverse materials including ingots, sows, T-bars, gates, and feeders. Figure 7.7 illustrates ingots, sows, and T-bars.

The fluctuating price differential between secondary metal ingots, sows, primary ingots, and T-bars has encouraged manufacturers to offer melting furnaces that combine a stack melter with a reverberatory furnace. This combination handles ingots and returns in the stack melter and sows and T-bars in the reverberatory furnace section. The downside of this flexible arrangement is the increased capital cost.

Figure 7.8 (Ref 4) illustrates one such combination furnace that can melt ingots, gates, runners, and biscuits, along with sows or T-bars.

7.6.1 Casting Process Yield Improvement and Impact on Charge Makeup

Casting yield depends on the process used and the type of casting. This section is intended to show how the charge makeup is determined by the process. The figures quoted illustrate the effect of process differences on the charge makeup. These can vary depending on the product and process engineering. Figure 7.9 presents an overview.

The casting yield is the ratio of the weight of the rough casting divided by the weight of the metal poured into the mold. Among the three common processes listed, the yield is highest for the low-pressure permanent and semipermanent mold processes and the counter-pressure casting process, followed by die casting.

The gravity permanent or semipermanent mold process has the lowest yield of the three processes. The amount of gating, biscuits, sprues, and feeders to be recycled varies accordingly. After recycling all the returns (gating, biscuits, sprues, and feeders), the balance of the charge materials needs to be made up by new material purchased as either ingots, sows, or T-bars.

Hot castings after ejection from the die or the mold are cooled by quenching in water. The gatings and flash that are separated from the castings still contain moisture, which needs to be removed as a part of the furnace design. A preheater can be installed before charging into a wet-well reverb, or the moisture can be evaporated automatically in a stack melter as the charge descends the stack.

The percentage of secondary ingots, sows, T-bars, or ingots in the primary metal charge depends on the percentage of gates, feeders, and sprues to be recycled (based on the process being used). The flash generated depends on the casting injection pressure. The amount of flash is likely to be higher in high-pressure die casting, followed by counter-pressure casting and low-pressure permanent mold casting. The least amount of flash is generated by the gravity permanent mold process.

The amount of chips generated depends upon the process and the product. Die castings have high dimensional capability, and the machine stock for finishing is minimum. Therefore, the amount of chips generated is low. Usually, the chips are accumulated for a few weeks to a month and then sent to secondary melting houses which are equipped to clean the coolant from the chips, and then dry, degas, and filter before casting them into ingots or sows. Die-casting gates are usually sheared and therefore do not generate chips during the degating process.

The feeders and gating from gravity permanent molding are cut by saws, generating chips. Depending on the casting design, some excess stock may have to be added in some product designs for feeding. The premachining of gravity permanent molding generates chips, but its amount may not be large enough to process for proper recycling without affecting the melt quality.

The low-pressure and counter-pressure processes use small sprues (about 250 gm or 0.5 pounds). But the steel screens used to filter the oxides are embedded in the sprues. Sprues are either drilled out of the cast aluminum wheels or cut by saws.

When sprues are drilled out in wheels, the gate stubs are contaminated with iron in the steel screens. Such metal is used for preparing secondary ingots or sows. When the sprues are cut using a saw, the steel screens need to be separated by heating

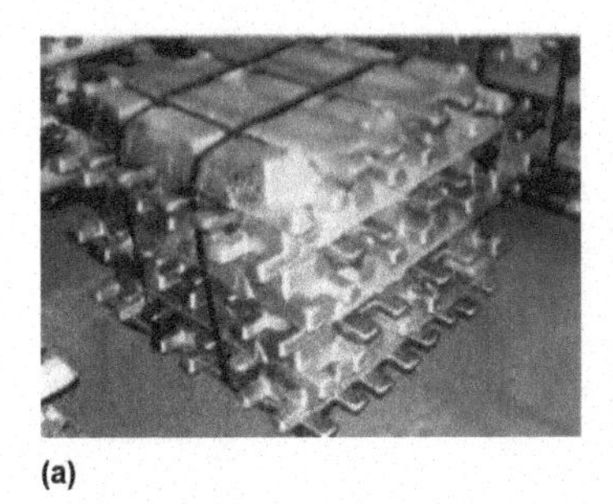

(a)

(b)

(c)

Fig. 7.7 Melt materials, (a) ingots, (b) sows, (c) T-bars

them over a grid where the aluminum melt drains down, leaving the screens to be discarded. The aluminum recovered can be reused as primary metal.

Fig. 7.8 Furnace combining features of reverberatory and stack melting. Reprinted with permission from Ref 4

Cast aluminum wheels (produced out of primary metal) represent the most popular application of the low-pressure permanent mold process. A high percentage of the wheel surface is machined for dynamic balancing as well as for aesthetics. This machining generates a significant amount of chips (of primary metal) that need to be recycled by feeding them back into the furnace instead of shipping them to a remelting facility. This process is detailed later in this chapter.

7.6.2 Effect of Multiple Material Grades and Melting Unit Selection

Casting facilities differ in the amount of metal throughput and the number of different metal grades they handle. This influences the number of melting units and whether a central melting unit is preferred or individual melting units for each cell are needed. There are many alloys suitable for each process, but a few of them are more popular than others. Captive casting facilities that specialize in one or two products use one or two alloys. The minimum number of melting units is limited to the number of alloys. For example: a captive facility producing automotive wheels (using the low-pressure permanent mold process) will use one primary alloy, ASTM A356. A manufacturing unit producing a group of suspension components either using the low-pressure process, the gravity permanent mold process, or the counter-pressure process uses one common alloy, A356.

A gravity permanent molding facility producing brake calipers uses one primary alloy, ASTM 357, along with a secondary alloy, ASTM 356, for producing brake master cylinders. Figure 7.10 is an overview of the different alloys used for the three common casting processes: die casting or high pressure die casting, gravity permanent molding, and low-pressure permanent molding. The more frequently used alloys are shown in bold

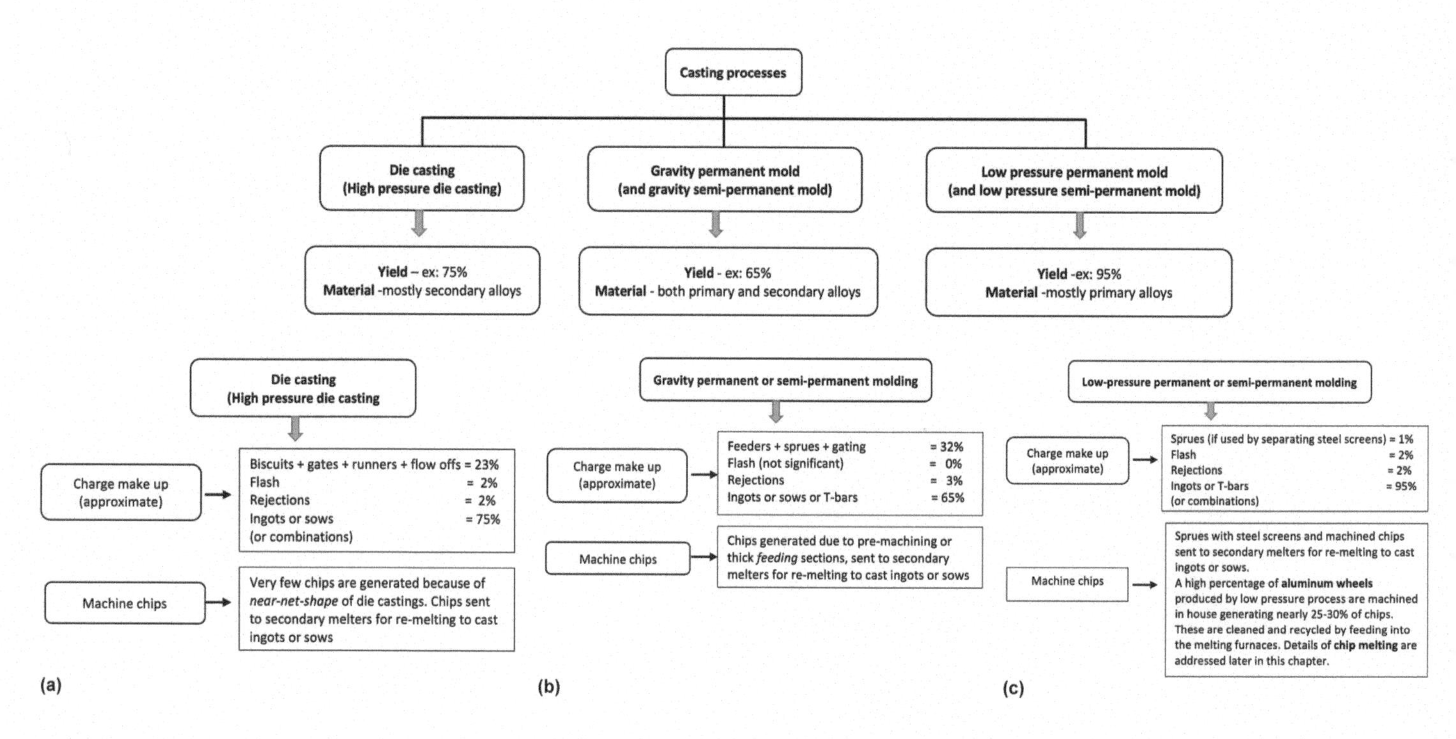

Fig. 7.9 Differences in charge makeup based on the casting process

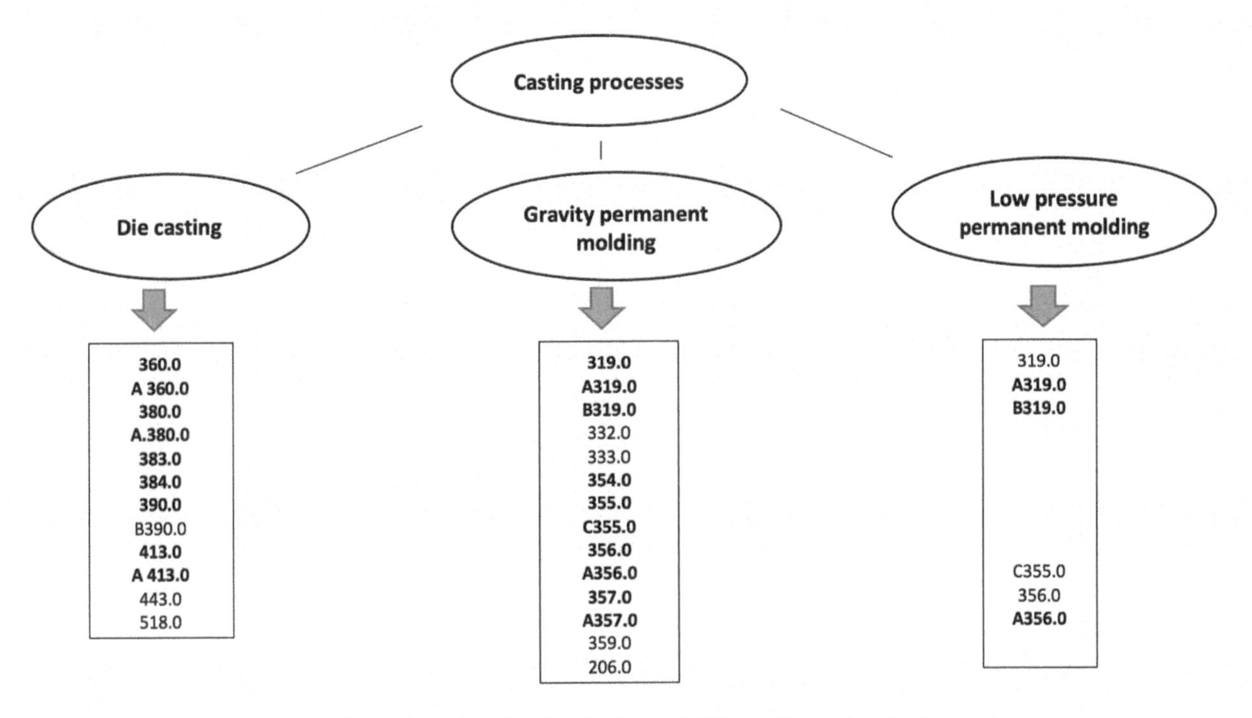

Fig. 7.10 Processes and alloys

type. The number of alloys used, and their volumes, are good indicators of the number and type of melting units that are required.

Jobbing operations producing a smaller number of components per week, but using different alloys, may prefer multiple crucible furnaces that are suitable for batch operations and the changing of grades. Medium-volume operations may consider electric furnaces that are suitable for emptying and changing grades; these have the additional advantages of high efficiency and zero emissions. Medium- and large-volume operations using one or two alloy grades have the option of choosing central melting systems such as stack melting or reverberatory furnaces. The low melt loss of stack furnaces is a significant cost factor. Captive foundries that use just one alloy may find a reverberatory furnace suitable to meet their metal needs continuously for several weeks at a time, despite the higher melt losses and the cost of emission control.

7.6.3 Cost and Other Factors Influencing Melting Unit Selection

The selection of the right kind and the suitable size of furnace depends upon several factors including:

- Cost factors
- Metallurgical factors
- Environmental factors
- Size factors

Each of these factors involves several considerations. Figure 7.11 summarizes the several factors to be considered in the selection of a suitable furnace.

7.6.3.1 Cost Factors

The cost factors include:

- **Capital costs:** These are the expenses for melting units, holder furnaces, metal distribution equipment from the main melting furnace to individual smaller holding furnaces at each cell or machine, and equipment to satisfy environmental regulations.
- **Process costs:** These expenses include energy for melting and holding, along with material costs. The material costs are comprised of charge materials and other process materials.
- **Maintenance costs:** Costs for preventive and breakdown maintenance need to be estimated from similar equipment in-house or other comparable industries.
- **Labor costs:** These expenses need to be accounted for after a critical assessment of the manpower needed, depending on the extent of mechanization and operator skills.

7.6.3.2 Metallurgical Factors

Metallurgical factors are process-oriented factors that have a significant impact on the selection of the most suitable furnace:

- **Charge materials**: The ingots, sows, and T-bars differ in costs and the suitability of different furnaces to handle them (see Table 7.3).
- **Melt Loss**: This is one of the most critical factors influencing furnace selection. Aluminum alloys are expensive compared to ferrous materials. The melt loss is a significant cost factor impacting the choice of the furnace. Tables 7.2 and 7.3 quantify the percentage of melt loss for different types of melting units.
- **Melt quality:** Section 7.6.2 in this chapter addresses the comparison between primary alloys required for structural

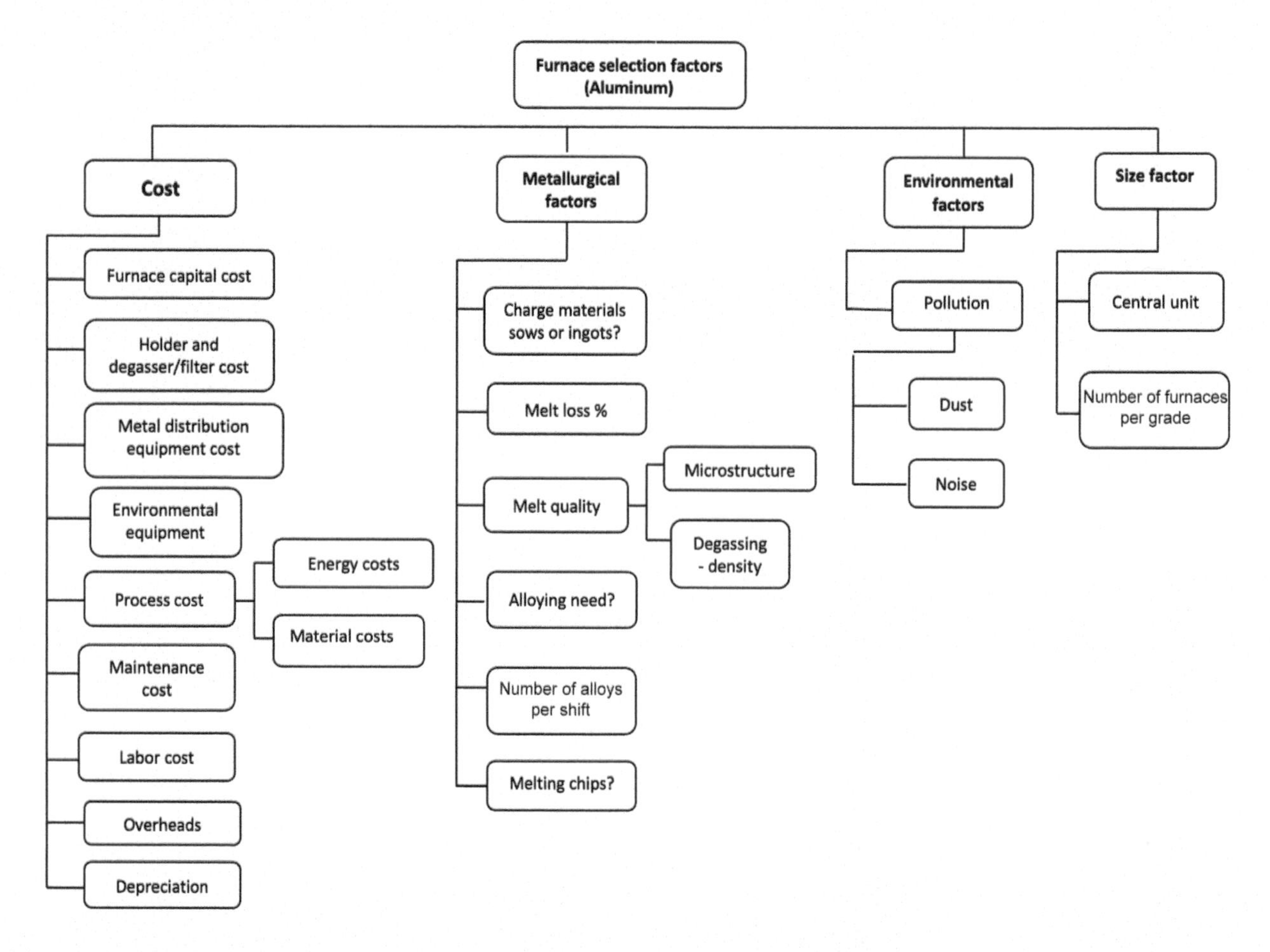

Fig. 7.11 Factors in furnace selection

applications and secondary alloys for a commercial grade of castings. The microstructure requirements and the metal treatment such as degassing, filtration, grain refinement, and silicon modification are process elements whose costs need to be accounted for.

- **Alloying needs:** The furnace selection is influenced by the need for alloying in the furnace.
- **Number of alloys per shift:** If several alloys are needed in each shift, smaller furnaces that can be started and stopped are required. If there are only one or two alloys needed in capital casting operations, a central melting unit is more economical (see Fig. 7.12).
- **Melting chips:** Large casting operations such as those for the manufacture of automotive wheels generate large volumes of chips as the wheels are machined. It is economical to plan on cleaning and recycling them in house. The selection of the furnace is also influenced by the need to melt the chips generated in house.

7.6.3.3 Environmental Factors

Furnaces differ in the amount of smoke and dust created and the noise in the vicinity. Federal and local regulations mandate emission and pollution controls, which have a bearing on the selection of furnaces.

7.6.3.4 Size Factor

A central melting unit producing one grade of alloy would be large, compared to several smaller units that provide the flexibility of different grades or alloys.

7.6.3.5 Other Factors

Tables 7.1, 7.2, and 7.3 provide comparisons of furnace options. The previous sections in this chapter highlight other considerations that influence the selection of suitable furnaces. The choice of the right kind of furnace depends upon several factors. Installing individual melting units for each high-productivity cell provides the flexibility of using multiple alloy grades, and this involves a higher capital outlay. Centralized melting units lower the capital investment but limit the alloy flexibility and add the cost of distribution to individual holding and dosing furnaces of the manufacturing cells. In addition, each transfer of metal from the centralized unit to the cell results in metal cascading and the creation of oxides which may affect high-quality casings produced by the gravity permanent molding process. Low-pressure permanent mold operations using a centralized melting unit or a common melting unit to serve four to five low-pressure machines use steel screens at the metal entry to the mold cavity to offset the effect of oxide creation due to the metal transfer.

Figure 7.12(a) to (d) illustrates alternative options for melting unit locations. These are examples of jobbing, dedicated cell melting furnaces, group melting furnaces, and central melting furnaces. Figure 7.12(a) is a schematic of a jobbing foundry where low-volume parts are manufactured using different alloys. The furnace should be capable of completely emptying and restarting a new heat (melt) with a new alloy. The molds are ready for the metal from the furnace. Crucible furnaces (gas-fired

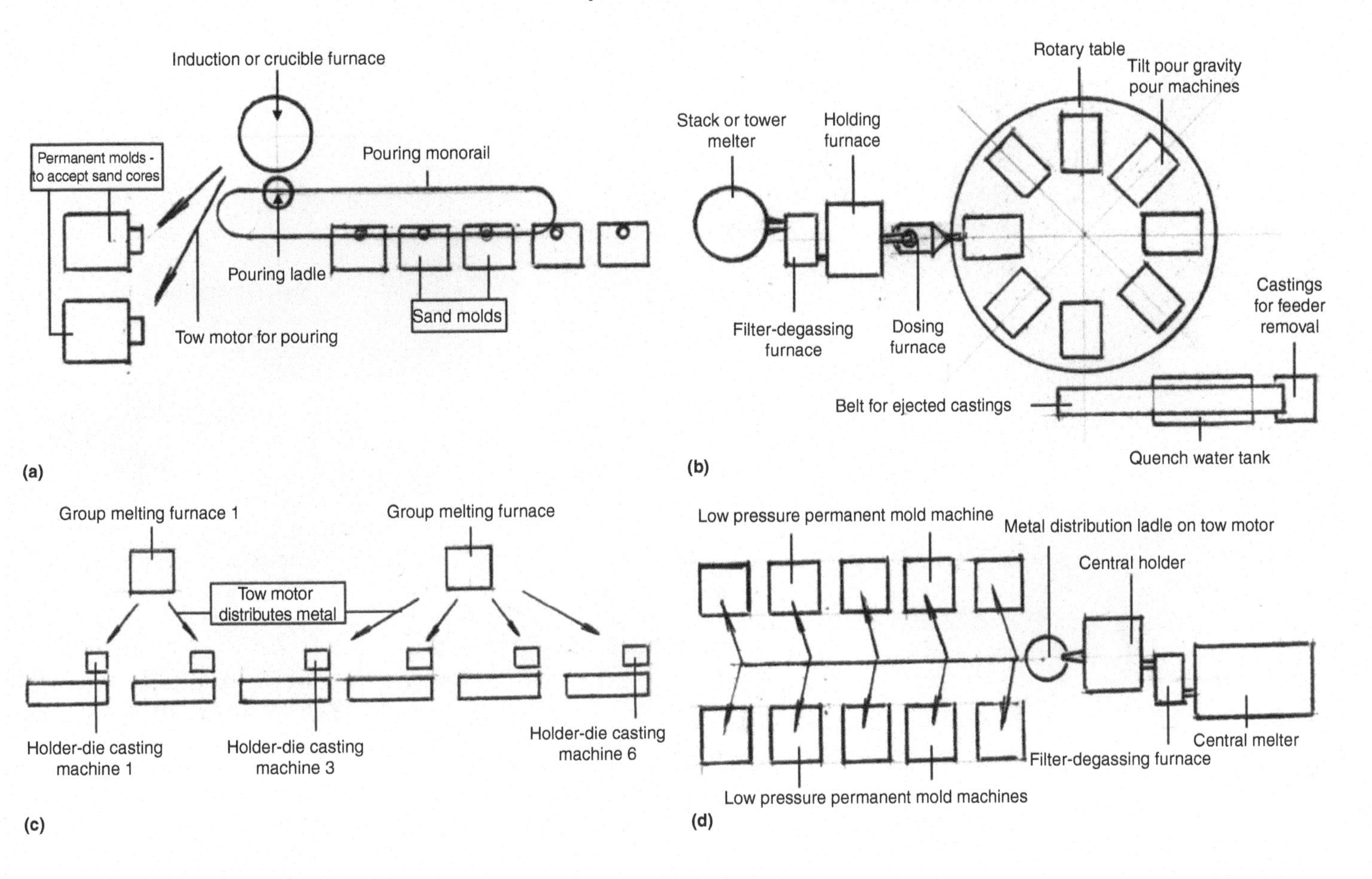

Fig. 7.12 Melting system concepts for, (a) jobbing facility, (b) large manufacturing GPM operation, (c) die casting facility melting two alloys, (d) low-pressure permanent mold manufacturing operation melting one alloy

or electric) or electric induction furnaces offer the flexibility that is needed. Figure 7.12(b) illustrates a rotary manufacturing cell using the gravity permanent mold (GPM) process. The tilt-pouring casting machines are mounted on a carousel that brings the machines in alignment with the doser spout. The melting unit serves the needs of multiple machines on the rotary table.

A stack or tower melter discharges metal into a degassing and filtration furnace. This type of furnace may be preferred because of the low melt loss and the ability to melt the heavy feeders that are typical of GPM. Degassing is carried out using a nitrogen lance with a graphite impeller that disperses the gas into finer bubbles. Any hydrogen dissolved in the melt due to the dissociation of moisture contamination is liberated along with the nitrogen bubbles. The melt flows across a foam filter to arrest any oxides. The filtered metal flows into a holder on a doser, where the metered amount of metal is discharged into the pouring cup of the casting machine.

After pouring, the table indexes align the next casting machine with the doser. Heat is extracted from the poured mold, and the casting solidifies as the table indexes. The ejection time is designed for the feeder to solidify partially or almost completely, ensuring that the feeder does not break prematurely, as its removal from the casting machines if breakage should occur would interrupt the cycle.

The casting with the feeder attached is ejected onto a belt with the aid of a robot. The belt dips the castings into a water-cooling tank, and the cooled castings collect into a basket ready for feeder removal and grinding.

Figure 7.12(c) is an example of a die-casting facility producing two grades of alloys on two groups of die-casting machines. Each group of machines is served by two independent melting units. Stack melters or gas-fired reverbs are good candidates for consideration. The stack melters are suitable for handling the wet gates from the quench tank. The advantage of gas-fired reverbs is that they can handle sows.

Metal from the melting furnace is transferred to the individual holders at each die-casting machine using a distribution ladle carried on a tow motor or forklift. A few furnaces have installed heated launders, instead of using a distribution ladle.

Figure 7.12(d) illustrates a central melter which meets the needs of 6 to 10 low-pressure casting machines. Gas-fired reverberatory furnaces are popular for such an application because a huge hearth can supply metal continuously to the casting machines. Such a layout is typical of low-pressure wheel casting plants. The metal is usually distributed using a ladle mounted on a forklift. The degassing and filtration furnace as well as the holder furnace deliver high-quality metal for wheels and other structural castings.

In making a decision about the best choice of melting furnace, a couple of alternatives need to be analyzed and brainstormed with the team, involving both the operation and furnace personnel. This is essential in order to select the best options.

Compromises will need be made, and investment costs should be estimated to be recovered in a span of 7 to 8 years.

Emission controls to mitigate global warming compel technological developments to reduce emissions to zero, reduce melt loss significantly, improve energy efficiency, and drastically improve the working environment for the operators.

Possible or likely future developments include:

- Improvement of stack or tower melting furnaces with a plasma torch instead of the natural gas burner.
- Modifications of induction furnaces to improve charging recovery rates (i.e., the extent to which the melt temperature decreases after each charge and how soon the melt temperature increases back to the initial value before charging)
- Improvements in reverb and forehearth (and holding furnace) heating using immersion heating

7.7 Holding Furnaces and Dosing Equipment

Holding and dosing furnaces are critical elements of the metal handling system that influence the casting quality through temperature control of the melt and consistency of the amount of metal poured into a mold.

7.7.1 Holding Furnaces

Group melting and central melting furnaces feed metal to the holding furnaces next to the casting machines to ensure a continuous and homogeneous metal supply to the machines at the targeted temperatures and with specified chemistry. Gravity and low-pressure casting processes producing high-integrity castings require the metal to be grain-refined and the silicon microstructure to be modified by alloy additions for enhancement of the mechanical properties. The effect of these additions, especially the alloy addition, fades over time. Therefore, these elements are best added to the holding furnaces in real time for maximum effectiveness.

The bath temperature drops each time the melting furnace is charged. The furnace energy source tries to recover the drop in the melt temperature, but this takes time. Therefore, temperature swings in furnace baths where the charge is introduced into the melt, such as in reverberatory furnaces, are inevitable. These swings can be minimized by maintaining the charge-to-bath ratio at 1:10. Stack melting furnaces do not have these large swings, and the forehearth experiences only minor temperature swings. The forehearth attached to a stack meter is sized to have a capacity ratio of the volume of charge to forehearth capacity of 1:2.5 or more.

Casting processes require close temperature control, within a range of about 5 °C, for consistent casting quality. This may be achieved by installing a holding furnace that is fed by the melting furnace because it has its own energy source to maintain the temperature. In addition, it serves to offset minor chemistry fluctuations from charge to charge. The reservoir of metal also serves to take care of slight imbalances of melting rates that are needed by the casting machines or cells periodically.

Reverberatory furnaces can also hold metal for long durations. Recent advances in burner technology have reduced energy consumption and emissions. Other holding systems have been popular for die casting and permanent mold process applications.

Holding furnaces for die castings have three sections: filling, heating, and dip-out (Ref 9). The filling section is the receiving end of the metal from the main melter, either by a distribution ladle or through a heated launder. The heating section, which is covered with a lid, houses a bank of either electrical heating elements or glow-bars which radiate heat to the surface of the metal. The dipping section is the open section where a ladle dipper or a robot scoops a designated amount of metal and pours it into the shot sleeve of a die casting machine. These ladle dippers or robots are also used in high-production rotary table gravity permanent mold cells where the robot pours into the pouring cups attached to the molds. Figure 7.13 is a schematic showing the three basic sections of a holding furnace.

Resistance elements or radiant tubes, which are used extensively, may occasionally be damaged by metal splashing, especially during the transfer of molten metal using a distribution ladle. Heating from the top of the metal produces a temperature gradient over the depth of the bath. The radiating energy is affected by the distance of the radiant tubes from the surface of the metal. Recent innovations have found that immersion heating coils or ceramic heating tubes submerged in the metal are effective in maintaining bath temperature within a range of about 2 °C. These electrical systems eliminate emissions, having a very low energy consumption rate of 5 kW per hour (Ref 9).

Coreless induction furnaces, which are specially designed for holding, are also energy efficient, although they are capital intensive. Channel holding furnaces shown in Fig. 6.25 and 6.26 in Chapter 6 in this book, are also energy efficient choices for aluminum holding.

7.7.2 Dosing Equipment

The consistency of the amount of metal poured into a die-casting or gravity permanent mold significantly influences casting

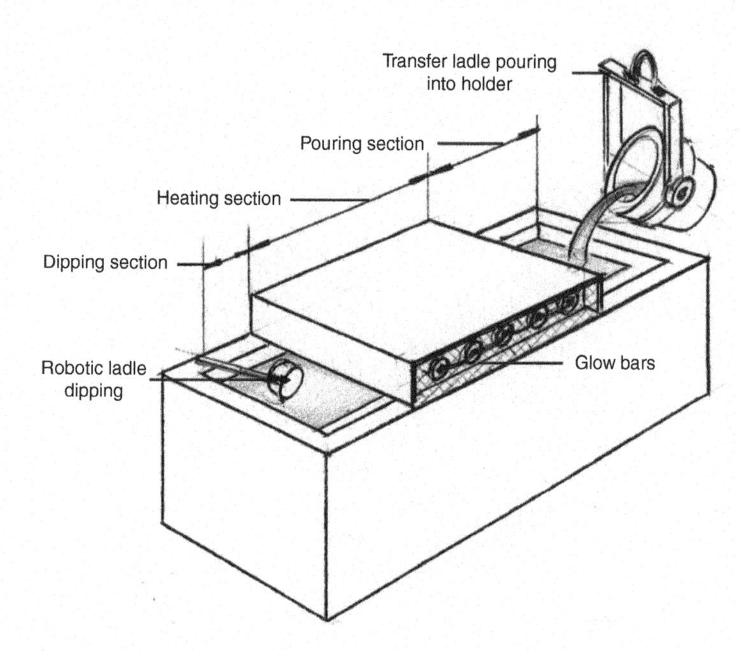

Fig. 7.13 Schematic of a holding furnace

quality. The consistency of automated ladle dippers in die casting operations is controllable and good. But the exposure of the metal in the pouring cup forms an oxide skin which should be prevented from getting into the die cavity. Oxide traps are designed in the tooling, just at the bottom of the shot biscuit, to trap the oxide. But prevention of the oxide film is preferable to methods to trap it. This is achieved by avoiding the exposure of the liquid metal to the air, using a doser or a metallic pump.

Figure 7.14 (Ref 10) illustrates a Westomat doser used for gravity permanent (and semipermanent) molding or die casting. Figure 7.14(a) is an external view, and (b) is a cross sectional view. The doser has a hot face castable refractory over a castable insulating layer, backed by layers of insulation next to the shell. The fill funnel is made up of two parts. The upper part, above the liquid metal, is made of alloy cast iron for good heat resistance. The lower portion of the fill funnel, which dips into the aluminum, is made of a refractory. The riser tube is made of aluminum titanate ceramic. As the meniscus of the metal is pressurized, the metal flows out of the riser tube (alternatively called a stalk tube). A metal sensor at the exit side of the riser tube registers the start of the metal flow, and the programmable controller maintains pressure until the targeted amount of metal has flowed out. The accuracy of this process ranges from 1 to 2%.

Figure 7.15(a) illustrates the application of the doser to a die cast machine. Figure 7.15(b) details the interface between the die casting machine and the doser.

Metallic pumps for metal dosing for die casting, gravity permanent molding, and squeeze casting processes are less capital intensive and are highly reliable. Figure 7.16 (Ref 11) is a cross section showing the pump and the ladle pouring metal into the pump sump. These pumps are highly accurate and reliable.

7.8 Fluxing, Degassing, and Metal Treatment

The mechanical properties of castings are influenced by the microstructure of the alloy, porosity, and oxides. The microstructure is influenced by the cooling rate within the mold and the metal treatment to refine the grain and silicon modification refinement. Oxides are removed from the melt by introducing fluxes.

Figure 7.17 is a schematic illustrating the factors that influence the mechanical properties and the process improvements to enhance the mechanical properties of the alloy. The three main determinants for high-quality castings with superior mechanical properties are:

- Freedom from oxide inclusions (aluminum oxide inclusions)
- Freedom from gas porosity (hydrogen gas porosity)
- Freedom from shrinkage porosity (macro and microshrinkage)

Figure 7.17 illustrates these three determinants along with resolutions to address them. Fluxes play a major role in the elimination of oxide inclusions and gas porosity.

7.8.1 Fluxes and Their Role

Fluxes are salts or compounds in granular form which are added to the melt in the furnaces for refining or modifying the metal with the objectives to:

- Reducing melt oxidation by minimizing contact of melt surface with the atmosphere, reducing the potential for hydrogen formation and gas porosity
- Separating metal from oxides by converting the dross or oxides into powdery form
- Removing nonmetallic inclusions that impair soundness and strength
- Modifying the melt by refining grain and modifying silicon

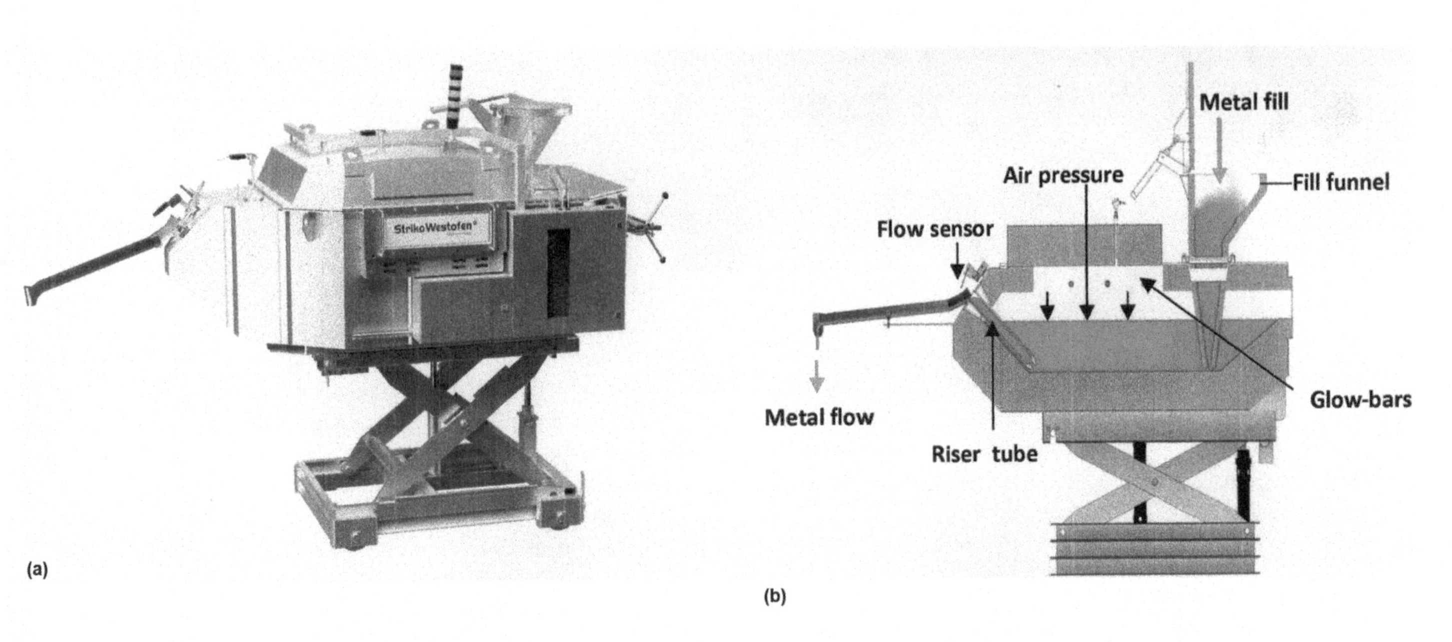

Fig. 7.14 Doser used for molding or die casting, (a) external view, (b) cross-sectional view. Reprinted with permission from Ref 10

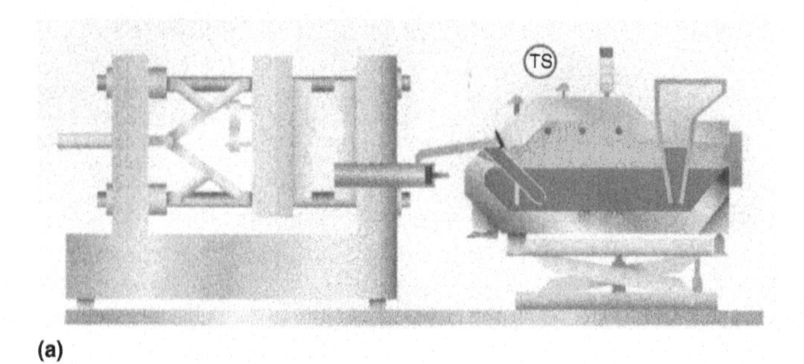

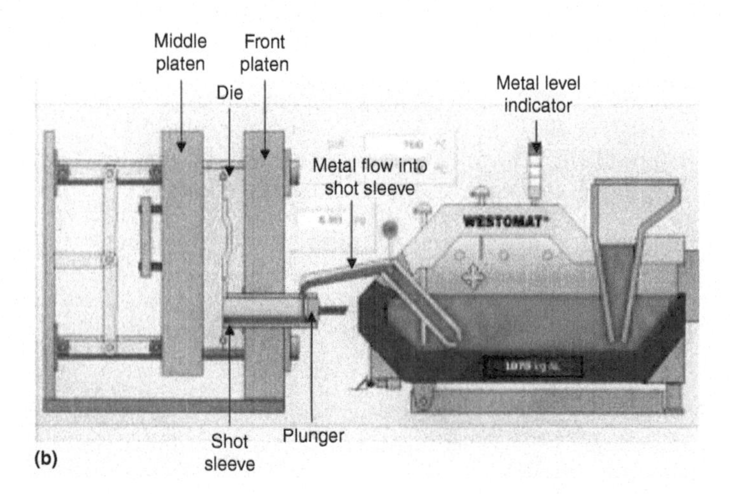

Fig. 7.15 Dosing equipment (a) applied to die casting machine, (b) filling shot sleeve. Reprinted with permission from Ref 10

Fig. 7.16 A metallic pump for die casting and gravity permanent mold applications. Reprinted with permission from Ref 11

- Maintaining furnace walls and ladles free from oxide buildup
- Reducing magnesium in the melt, as needed

Figure 7.18 represents a broad grouping of the fluxes. Fluxes can be grouped based on:

- Oxide reduction—prevention, removal, and separation
- Gas porosity reduction—hydrogen removal in conjunction with a rotary impeller degasser (RID)

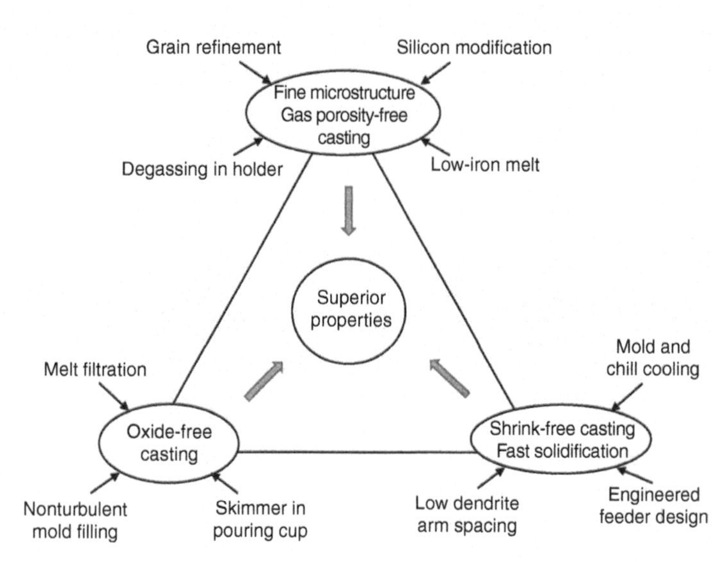

Fig. 7.17 Factors for achieving superior properties

- Metal modification—grain refinement, silicon modification, and magnesium removal
- Furnace and ladle cleaning

7.8.2 Chemical Elements in Fluxes and their Function

Fluxes are mixtures of salts. The chemical elements chosen for fluxes target seven characteristics:

- **Melting point**: The melting point of the salt combinations, which form binary or ternary eutectics, is lower than that of the aluminum alloy. The low melting point increases their fluidity for maximum effectiveness.
- **Cover fluxes**: The typical composition of the fluxes is NaCl (45%) + KCl (45%) + CaF_2 or KF_2 (5%), which forms a ternary eutectic with a melting point of 607 °C.
- **Cleaning fluxes**: These have a higher percentage of alkali fluorides CaF_2 of KF_2 along with NaCl and KCl. These lower the surface tension between the metal and the flux and between the flux and the oxides. The reduction in surface tension increases the wettability, resulting in the separation of oxide inclusions from the metal.
- **Drossing fluxes**: This group of fluxes has an oxidizing agent such as potassium nitrate, KNO_3, added to the combination of the NaCl + KCl + fluoride CaF_2 or KF_2. The oxidizing agent causes the exothermic salt to release heat. The oxygen liberated from the nitrate salt reacts with aluminum to form aluminum oxide (Al_2O_3), releasing a significant amount of heat which results in the separation of aluminum melt entrainment in oxides. These salts are also useful in cleaning furnace walls and ladles, as the exothermic reaction enables penetration into the walls.
- **Degassing fluxes**: These are salts that liberate gases such as chlorine or carbon dioxide. The hydrogen bubbles in the melt coalesce with bubbles of chlorine or carbon dioxide and rise to the surface and exit. Compounds such as hexachloroethane (C_2Cl_6) for chlorine liberation are no longer used due to emission regulations. The usage of aluminum chloride or

aluminum fluoride requires an effective exhaust system for operator protection. Other methods of bubbling nitrogen or argon have replaced the use of chlorides for degassing.

- **Grain refinement and silicon modification:** These are addressed in Table 7.4 along with other fluxes and their compositions.

Table 7.5 (Ref12) lists the physical properties of some of the fluxing materials used by several manufacturers. The density of liquid aluminum alloys is about 2.67 gm/cm^3. The densities of fluxing materials listed indicate that most of the chloride-based fluxes tend to float on the surface of the melt. Only fluoride-based fluxes have a higher density than that of liquid aluminum.

Cover fluxes are designed to float on the melt surface. Drossing and degassing fluxes must be submerged and stirred gently and thoroughly to ensure intimate contact with the melt for maximum effectiveness.

Chlorides have a lower melting point than fluorides. The fluxes are formulated to match the liquidus temperature of the aluminum alloys for which they are intended.

After the complete reaction of the flux, the dross is skimmed off to ensure that it does not get entrained in the casting.

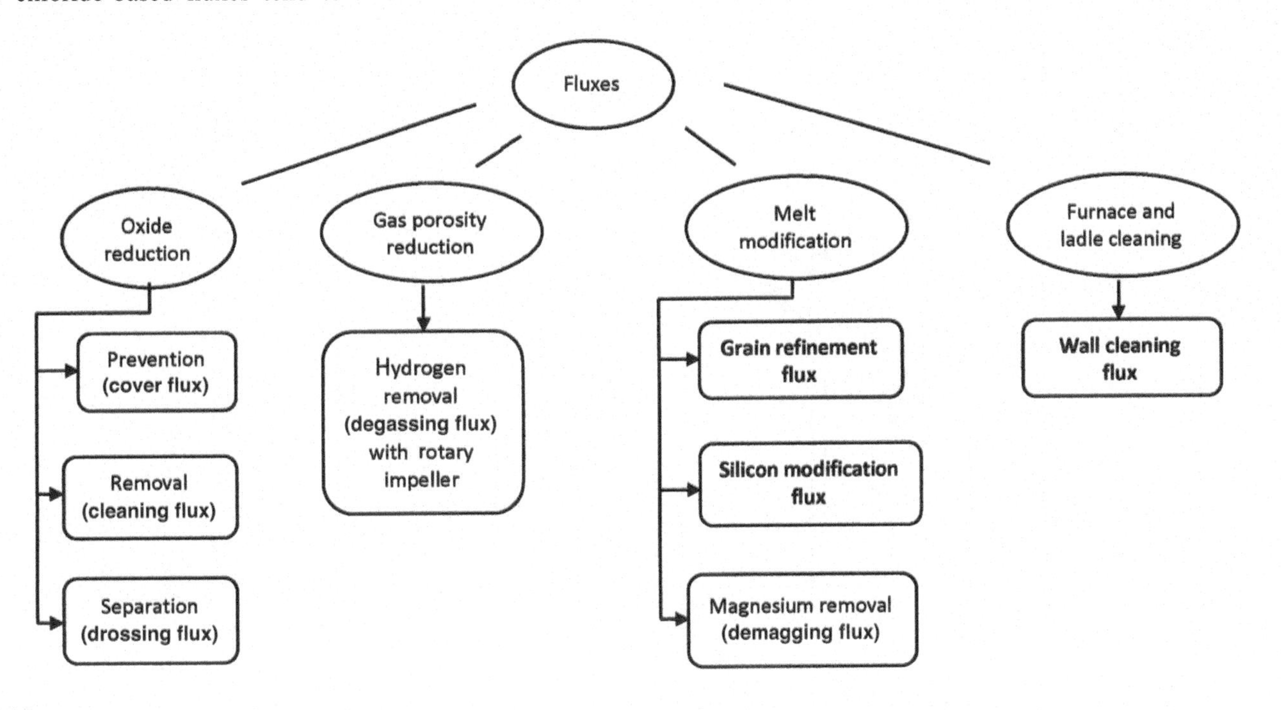

Fig. 7.18 Grouping and role of fluxes

Table 7.4 Fluxes, compositions, and applications

Group	Flux type	Composition	Comments
Oxide reduction	Cover	Mixture of NaCl and KCl with $CaCl_2$, CaF_2 or KF_2	Liquidus temperature of cover flux is lower than that of aluminum. The liquid layer floats on melt surface, preventing exposure to air, reducing oxide formation. Fumes containing fluoride are harmful and should be properly evacuated away from operators
	Cleaning	Mixture of NaCl and KCl, CaF_2 or KF_2 + oxidizing agent	These absorb oxide inclusions from the melt.
	Drossing	NaCl+KCl+CaF_2 or KF_2+ oxidizing agent KNO_3	The exothermic reaction improves the wettability and increases fluidity of entrapped aluminum. The separated aluminum mixes with the melt. The powdery and dry drossing flux is easily skimmed out. Drossing fluxes are useful for melting returns and chips. Dross not treated with flux may contain 60 to 65% of entrained aluminum; after treatment it reduces to 25 to 30%.
Gas porosity reduction	Degassing	Tablets of salts containing chlorine and fluorine	Tablets plunged into the melt using a preheated perforated bell. The aluminum chloride and aluminum fluoride gas bubbles rise through the melt. Hydrogen bubbles from the melt diffuse into the bubbles formed by aluminum chloride and fluoride. The exhaust system evacuates the fumes. Alternatively an inert gas such as argon or nitrogen can be used to carry the granulated degassing flux through an injector lance.
Melt modification	Grain refiner	Salts containing titanium and boron: K_2TiF_2 and KBF_4	The flux forms stable nuclei of $TiAl_2$ and TiB_2 which form as nucleation sites for the fine grains formed during solidification. Alternatively, grain refining can be done using a continuously fed rod of a material alloy containing titanium and boron (ex. Al-5%Ti+1%B).
	Silicon modifier	Salt combinations containing NaF to introduce Na or Sr	The shape of eutectic silicon crystalized during solidification is modified and refined.
Magnesium reduction	Demagging	Flux reduces magnesium by burning or oxidizing	Used when the magnesium level needs to be reduced.
Furnace cleaning	Wall-cleaning	Double fluorides Na_2SiF_6 + $NaAlF_2$ + oxidizing agent	Oxidizing ingredient reacts with aluminum to generate heat, promoting penetration of fluorides into the oxide buildup on the walls, softening them. The operator can then scrape the softened oxides from the walls.

7.8.3 Alloy Compositions and Types of Fluxes

Different alloys have diverse tendencies for oxidation, depending on the magnesium and silicon levels. The fluxes for alloys with high levels of magnesium do not contain oxidizing agents such as KNO_3. Alloys for sand casting and gravity permanent molds, such as 319.0, 355.0, and 356.0, do not contain high levels of magnesium. Drossing fluxes containing oxidizing agents can be used. Alloys such as 357.0 and 359.0, which contain higher levels of magnesium, use cleaning fluxes that contain low levels of oxidizing agents. Die-casting alloys such as 518.0, which contain very high levels of magnesium, do not use fluxes containing oxidizing agents.

7.8.4 Flux Storage and Flux Addition Methods

7.8.4.1 Flux Storage

Care is needed in storing the fluxes because these salts are hygroscopic. The moisture introduced through the fluxes can result in hydrogen gas porosity in the castings. Higher moisture contamination can also result in an explosion when the fluxes are plunged into the melt.

Fluxes are stored in sealed containers indoors close to the melt shop. Tools used for introducing the fluxes into the metal should be dry and coated with a refractory clay to prevent rusting and pick-up of iron when dipped into the melt. These fluxes need to be preheated to 250 to 300 °C (500 to 575 °F) to ensure that they are free from any moisture contamination.

7.8.4.2 Flux Addition Methods

The flux addition methods depend upon the type of melting operation and size of the furnaces.

Table 7.5 Properties of fluxing materials

Material	Chemical formula	Density g/m³		Melting point		Boiling point	
		Solid	Liquid	°C	°F	°C	°F
Aluminum chloride	$AlCl_3$	2.440	1.31	190	374	182.7	360.9
Aluminum fluoride	AlF_3	3.070	...	1040	1904	...	...
Borax	$Na_2B_4O_7$	2.367	...	741	1366	1575	2867
Calcium chloride	$CaCl_2$	2.512	2.06	772	1422	1600	2912
Calcium fluoride	CaF_2	3.180	...	1360	2480	...	...
Carnallite	$MgCl_2KCl$	1.600	1.50	487	909	...	...
Zinc chloride	$ZnCl_2$	2.910	...	262	504	732	1349.6
Zinc fluoride	ZnF_2	4.840	...	872	1602	...	...
Cryolite	$3NaFAlF_3$	2.970	...	1000	1832	...	...
Lithium chloride	LiCl	2.068	1.50	613	1135	1363	2467.4
Lithium fluoride	LiF	2.295	1.80	870	1598	1676	3048.8
Magnesium chloride	$MgCl_2$	2.325	...	712	1314	1412	2573.6
Magnesium fluoride	MgF_2	3.000	...	1396	2545	2239	4062.2
Potassium chloride	KCl	1.984	1.53	776	1711	1500	2732
Potassium fluoride	KF	2.480	...	880	1616	1500	2732
Potassium borate	$K_2B_2O_4$	...	...	947	1737	...	...
Potassium sulfate	K_2SO_4	2.662	...	1076	1969	...	...
Potassium carbonate	K_2CO_3	2.290	...	891	1636	...	...
Sodium chloride	NaCl	2.165	1.55	801	1474	1413	2575.4
Sodium fluoride	NaF	2.790	1.91	980	1796	1700	3092

Source: Ref 12

- **Crucible furnaces**: Crucible furnaces, which are limited in size and suitable for batch operations, are normally tended by operators. Cover fluxes are sprinkled over the melt surface using a shovel. Cleaning, drossing, and degassing fluxes are plunged into the melt using a refractory coated perforated bell. Figure 7.19(a) illustrates a crucible furnace with a perforated plunger. Some crucible furnaces use a perforated plunger basket in which the fluxes are placed and plunged, as illustrated in Fig. 7.19(b)
- **Reverberatory and Holding Furnaces**: These furnaces need equipment and processes for long duration operations with minimum operator intervention. In larger furnaces such as reverberatory furnaces and holding furnaces (holders), the cover flux is blown over the surface using a pneumatic gun where the air carries the flux. Cleaning and drossing fluxes are fed below the metal surface through a refractory-coated steel lance, graphite, or silicon nitride lance using an inert carrier gas such as nitrogen. Figure 7.20 illustrates the equipment used for blowing the flux using nitrogen as a carrier gas.

Figure 7.21 illustrates the mechanism of removal of inclusions using flux and nitrogen.

Oxides and other inclusions in the melt are suspended, depending on their relative density. As the flux is carried through by the nitrogen, the inclusions attach themselves to the nitrogen bubbles. As the bubbles rise to the surface of the metal, they take with them the inclusions, which float on the metal surface. These are skimmed out manually or by using a sweeping skimmer actuated by two pneumatic cylinders on each side of the furnace.

7.8.5 Degassing Methods and Degassing-Filtration Combination

7.8.5.1 Hydrogen Gas Inclusions

Aluminum alloys are susceptible to the absorption of hydrogen mainly from the furnace atmosphere and other moist sources such as gates and feeders that have been water-quenched. Water

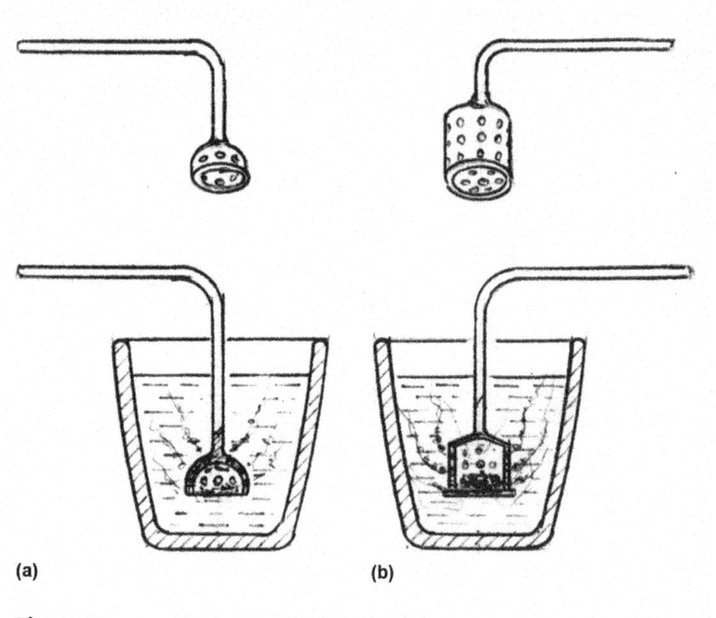

Fig. 7.19 Crucible furnaces with (a) bell, and (b) basket plungers

vapor reacts with aluminum, releasing hydrogen and creating aluminum oxide. Equation 7.1 describes the chemical reaction:

$$2\,Al + 3\,H_2O \rightarrow 6\,H + Al_2O_3$$
$$(Aluminum) + (Moisture) \rightarrow (Atomic\ hydrogen) + Aluminum\ Oxide \quad (Eq\ 7.1)$$

The active atomic or nascent hydrogen is quickly absorbed into liquid aluminum, which creates gas porosity.

It is necessary to take action to prevent contamination by moisture. There are several potential sources of moisture contamination:

- Furnace atmosphere (which contains 18% moisture)
- Uncured refractories and furnace operation tools
- Charge materials exposed to rain or water due to casting quenching with gates and feeders

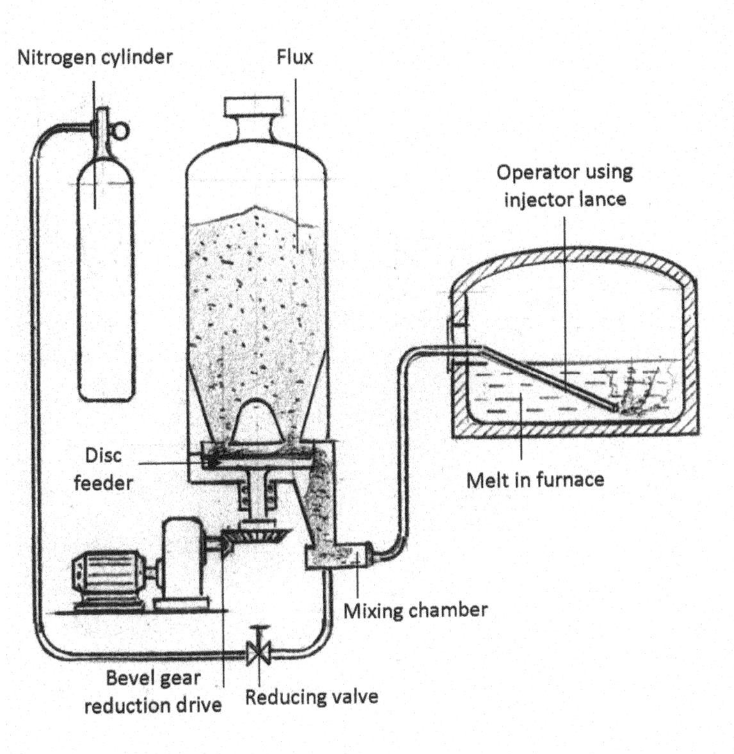

Fig. 7.20 Flux blowing using nitrogen as a carrier gas

- Turnings and borings contaminated by oil
- Improperly stored fluxes

Nitrogen, which is a nonreactive gas, has been effective in purging hydrogen by various methods. Argon is also nonreactive but is more expensive than nitrogen.

A small addition of chlorine (10 to 20%) to nitrogen makes the purging gas reactive. But environmental regulations discourage the use of reactive purging gases unless there is an effective exhaust system.

Bubbling nitrogen through the melt for the hydrogen bubbles to coalesce and float to the surface, or out of the melt, has been an effective and environmentally acceptable method of degassing the melt. Figure 7.22 illustrates the principle schematically.

7.8.5.2 Degassing Effectiveness Measurement

It is essential to confirm the extent of degassing of the melt before pouring the casting or before transferring the metal from the main melting furnace to the holding furnace. This test is called the reduced pressure test or the Straube-Pfeiffer test (Ref 12). A sample of melt is poured into a cup, and the melt is subjected to a vacuum pressure of 26 inches of mercury. The density of the test coupon is measured, and the coupon is sectioned for a visual check. The grades are rated into 12 categories. Figure 7.23 illustrates the 12 grades and the corresponding density (Ref 12).

7.8.5.3 Degassing in Medium-Volume Operations

The size of the nitrogen bubble is important for effective degassing because the finer bubbles can carry finely dispersed hydrogen in the melt. The two popular methods of fine nitrogen gas bubble dispersion are:

- Porous plug ladles
- Rotary impeller degassing

Medium-volume casting operations use distribution ladles fitted with a porous plug at the bottom to bubble the nitrogen for degassing. The porous plugs mounted at the bottom of the ladles allow the nitrogen to flow through without being clogged by the melt. Figure 7.24 (Ref 13) illustrates a standard ladle fitted with a porous plug at the bottom. Some casting operations use drum ladles with porous plugs for metal degassing and distribution.

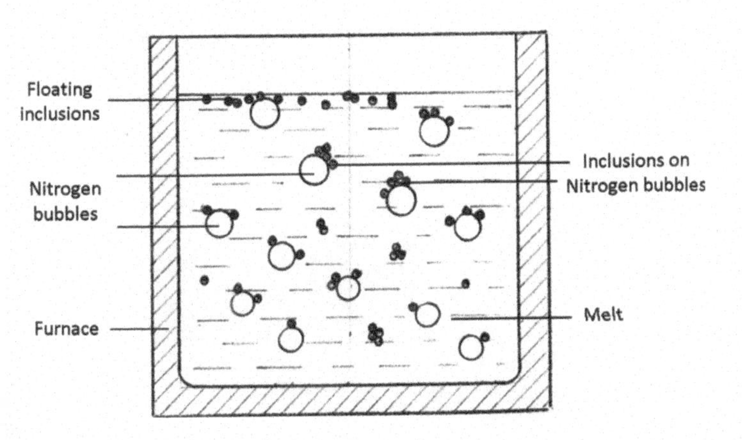

Fig. 7.21 Nitrogen gas bubbles floating inclusions

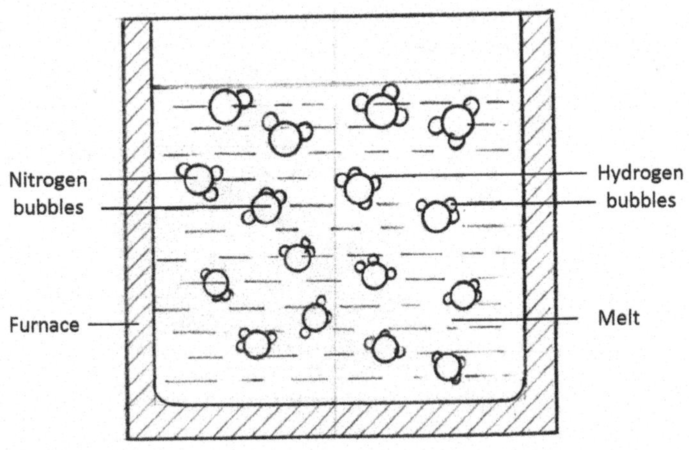

Fig. 7.22 Nitrogen gas bubbles floating hydrogen bubbles

Porous plugs are manufactured (Ref 14) with a refractory mix of Al_2O_3,(89%), SiO_2(6%), ZrO_2(3%), and Cr_2O_3(2%) with a porosity of 23.8%, which allows a flow rate of 330 to 430 l/min at a pressure of 1 atm. gauge. The height of the plug measures 121 to 135 mm, with a diameter at the melt side of 80 to 90 mm.

Fig. 7.23 Reduced pressure test (Straube-Pfeiffer test) coupons and densities. Reprinted with permission from Ref 12

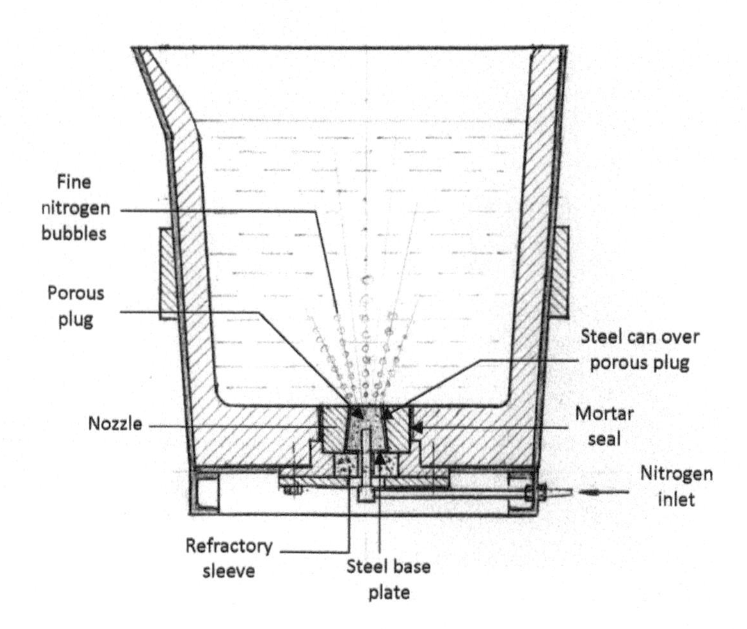

Fig. 7.24 Ladle with porous plug for degassing

Porous plugs are fitted snugly in the bottom of the ladles with a nozzle set into the refractory lining using a sealing mortar. This construction ensures an easy change and a metal-tight fit. The fine pores allow dispersion of small bubbles of nitrogen throughout the melt for degassing. The porous plugs are also used for desulfurization before magnesium treatment to obtain ductile iron.

7.8.5.4 Rotary Impeller Degassing

Large volume and captive casting operations require continuous degassing of the melt. A rotary impeller made of graphite or silicon nitride dipped into the melt with a hollow shaft through which nitrogen can be bubbled is effective for fine dispersion. Several configurations of the impeller are available for efficient dispersion of nitrogen.

Some casting producers employ a system of flux injection through nitrogen gas dispersed through a rotary impeller unit. Figure 7.25 illustrates such an arrangement.

Standalone units are available for batch operations where a distribution ladle needs to be degassed before it is delivered to the individual holders. Low-pressure permanent mold automotive aluminum wheel producers typically use such an arrangement. Figure 7.26 illustrates such a mobile arrangement (Ref 15).

High-volume gravity permanent mold casting facilities have these units engineered in-line. The Rotary Impeller Degassing units (RID units) generate oxides, which need to be removed with minimum manual intervention.

Figure 7.27 illustrates an impeller degassing unit for continuous operation. The melt to be degassed flows into a furnace which is equipped with a rotary impeller degasser driven by an overhead motor. Two immersion heaters keep the metal hot. Nitrogen is passed through a manifold into a hollow shaft and onto an impeller. The drive shaft and the impeller are made of graphite or silicon nitride, which has nonwetting characteristics, minimizing any buildup. A skimmer dam allows the degassed

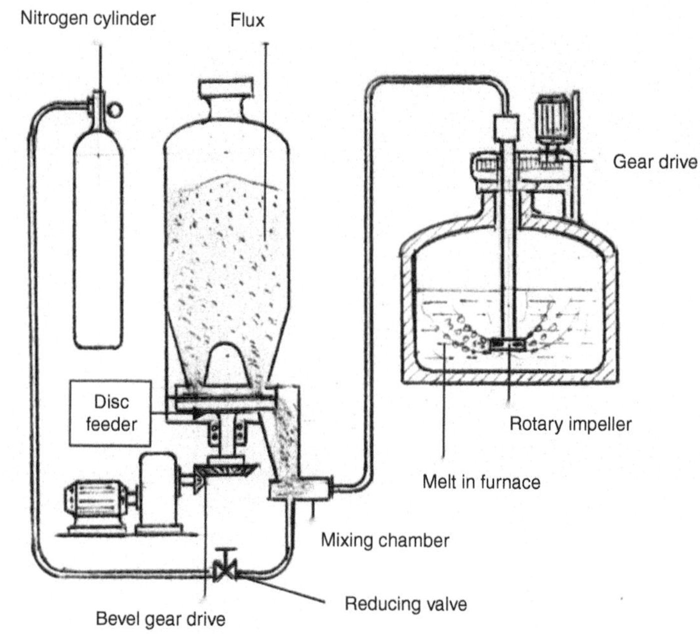

Fig. 7.25 Flux injection with nitrogen using a rotary impeller

metal into the melt outflow section. The furnace lid is attached to a lifting mechanism for periodic servicing.

The shapes of the impeller will vary. They disperse nitrogen gas into fine bubbles for maximum effectiveness in degassing.

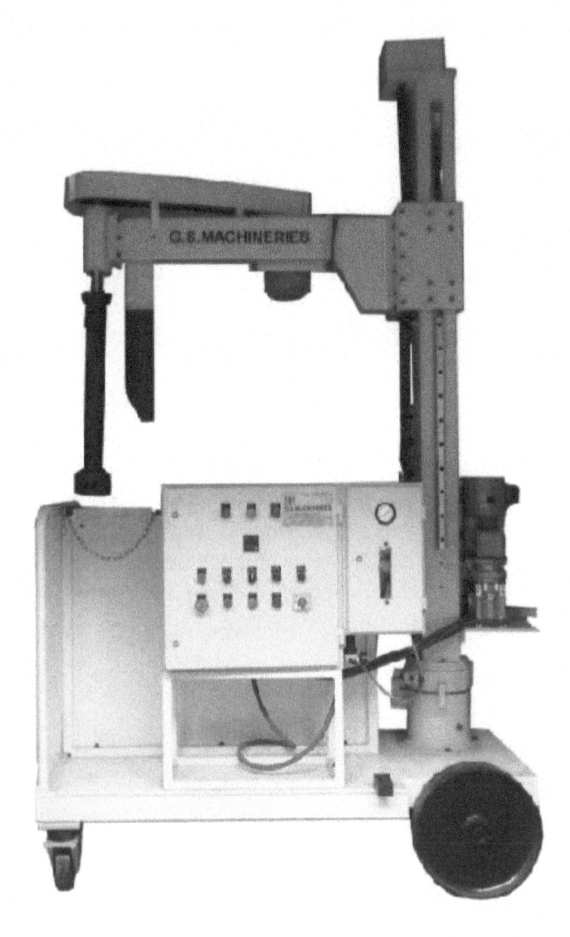

Fig. 7.26 Mobile degassing unit. Reprinted with permission from Ref 15

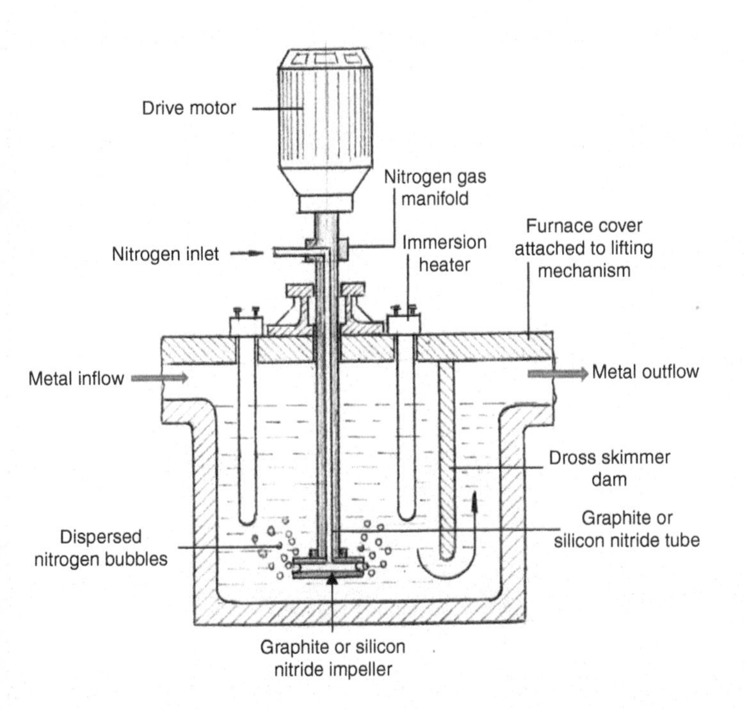

Fig. 7.27 Continuous degassing operation using an RID unit

Figure 7.28 illustrates some of the shapes available for good dispersion of fine bubbles of nitrogen.

The churning of metal for nitrogen gas dispersion generates aluminum oxides. The oxide inclusions in the melt are removed by filtration. Degassing and filtration furnaces combining an RID unit with a replaceable, perforated ceramic filter are popular for gravity permanent mold installations producing high-quality aluminum structural castings. Figure 7.29 represents a schematic layout of such an installation.

Molten metal from a stack melter flows into a degasser-filter furnace that has an RID unit used for the dispersion of nitrogen. Degassed metal passes through a filter section with vertical ceramic foam filters. Metal passes through a 10 to 20 ppi (pores per inch) filter to filter out larger particles. The metal then passes through a 30 ppi filter that filters smaller particles. Using this combination of two filters prolongs the life of the fine filter

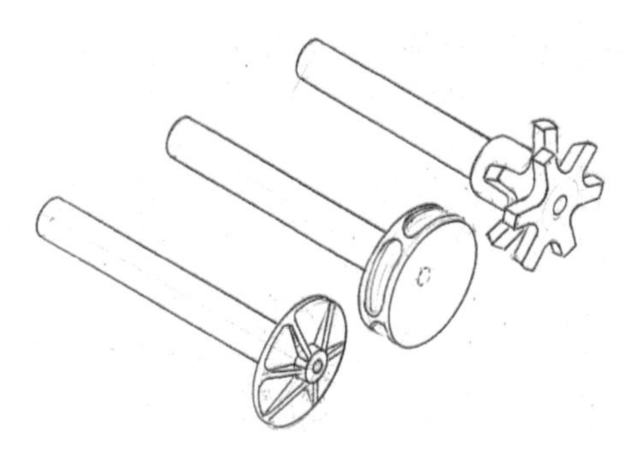

Fig. 7.28 Different shapes of impellers for efficient nitrogen dispersion

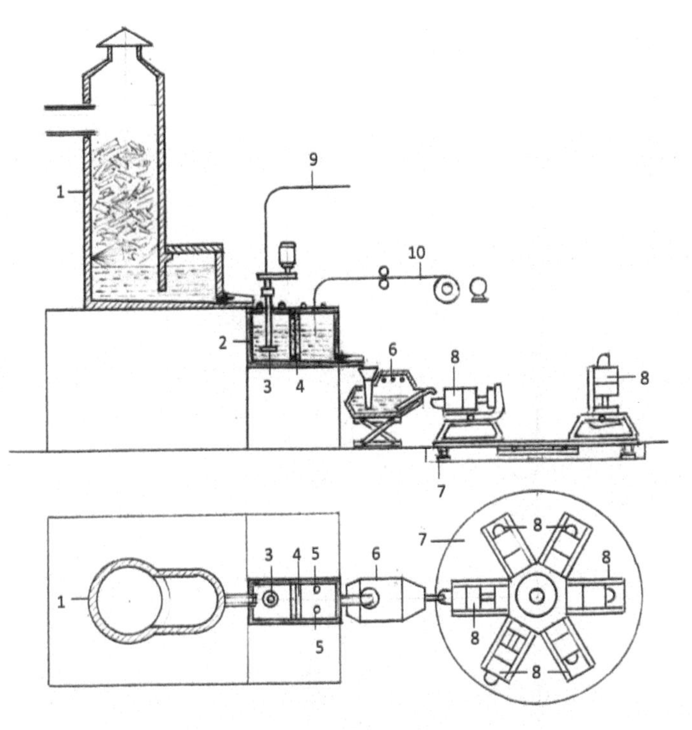

Notes: 1. Stack melter 2. Degassing-filtration furnace 3. Rotary impeller degassing unit (RID unit) 4. Filter assembly 5. Immersion heaters 6. Dosing furnace 7. Rotary table 8. Tippers or tilt pour machines 9. Nitrogen gas supply 10. Aluminum-titanium boride grain refining feeder

Fig. 7.29 Rotary manufacturing cell with a degassing-filtration furnace

(30 ppi filter). The ability to change one filter at a time ensures that no unfiltered metal flows downstream while the filters are being changed.

Metal from the filter furnace flows into a dosing furnace which delivers metered metal to the individual molds mounted on the tilting machines, or tippers. Immersion heaters in the degassed-melt section enable improved temperature control of the melt.

7.8.5.5 Grain Refinement and Modification

Many casting producers prefer to use master alloys for grain refinement and silicon modification, over the use of fluxes, to achieve a refined structure. Master alloys offer the advantage of easier storage, handling, and control.

The fluxes for grain refinement are typically combinations of salts such as potassium fluoro-titanate and potassium fluoroborate, which are used to introduce titanium and boron. The fluxes for modification are combinations of salts to introduce sodium or strontium into the melt for silicon modification.

7.8.5.5.1 Grain Refinement

The refinement of aluminum grain offers several advantages: improved mechanical properties, improved ability to feed the shrinkage, increased pressure tightness, and reduced tendency to hot tear.

The master alloys are obtained in rod form, either sectioned to pieces for batch additions or as wire for continuous additions. Grain refiners are:

- Aluminum titanium boron (AlTiB)
- Aluminum titanium (AlTi)
- Aluminum boron (AlB)

Aluminum titanium boron is the most effective as a grain refiner. It is available in different titanium-to-boron ratios. The most popular are AlTiB5/1 with Ti to B = 5 : 1, or AlTiB3/1with Ti to B = 3:1.

Figure 7.29 illustrates the addition of AlTiB rod to the section of the furnace that has the degassed and filtered metal. The addition is metered as (x) inches or (x) centimeters advancement of the rod into the melt every 15 minutes or 30 minutes.

7.8.5.5.2 Silicon Modification

The modification of silicon morphology improves mechanical properties such as tensile strength and elongation, the ability to feed, and machinability.

The master alloy is added to the melt to refine the morphology of the silicon from a coarse needle-like structure to fine fibrous particles, enhancing the mechanical properties. The master alloys used are:

- Aluminum strontium (AlSr)
- Aluminum strontium silicon (AlSrSi)

Strontium additions became more popular than sodium additions years ago because they offer slower fade rates. However, since the effect of strontium modification fades over time, casting operations prefer that the furnace operator should add the master alloy in the shape of cut rods or pellets to each of the distribution ladles or to the doser, instead of automating the process as in the aluminum titanium addition.

7.9 Aluminum Chip Melting Equipment

The efficient recycling of aluminum chips in-house is important for cost competitiveness in a wheel manufacturing facility where the machining of rims generates a high percentage of chips. Sometimes the wheel face is also machined for aesthetics, which generates additional chips. The original equipment manufacturers (OEM) produce wheels from a primary aluminum alloy, A356. Before aluminum wheels became popular, casting facilities shipped chips to a secondary melter where they would be melted and sold as a secondary metal. Wheel manufacturers are now equipping themselves with the ability to accomplish their own efficient recycling of the primary alloy.

Decorative wheels are cosmetic components and should not show any blemishes (resulting from pinholes or microporosity). These small blemishes are revealed during the clear-coat process, which is one of the final operations. Scrapping wheels at a near-final stage affects profitability significantly.

It is essential to submerge chips into the charge well as quickly as possible, without creating any dross due to oxidation. This is achieved by introducing well prepared chips into a metal vortex created either by a ceramic pump or by an induction coil (Ref 6, 16).

The preparation of chips for melting is important to ensure freedom from coolant, moisture, and any contamination of ferrous tramp elements. Figure 7.30 is a schematic showing the sequence of steps for chip preparation.

The collected chips are fed into a crusher to create pieces of uniform size (10 to 30 mm) (Ref 16) and to increase the density

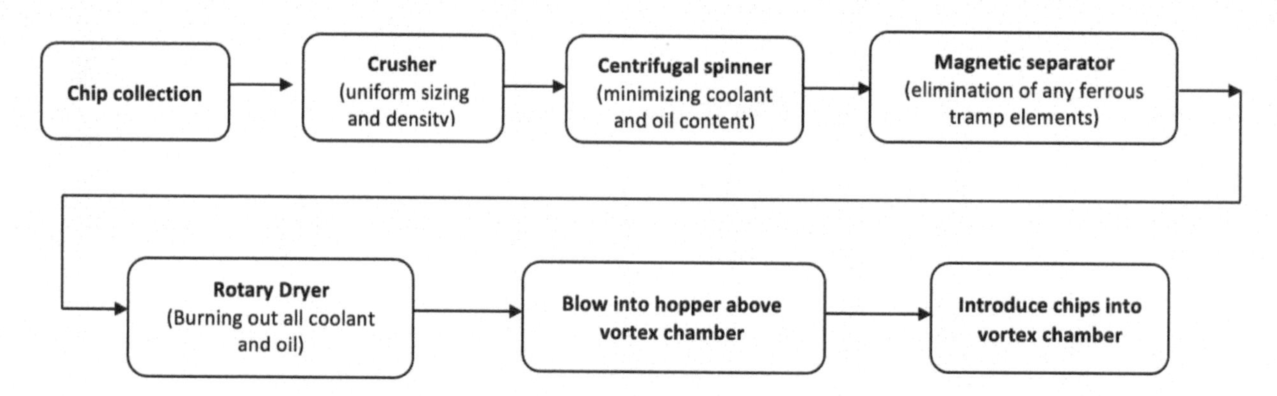

Fig. 7.30 Schematic for aluminum chip melting preparation

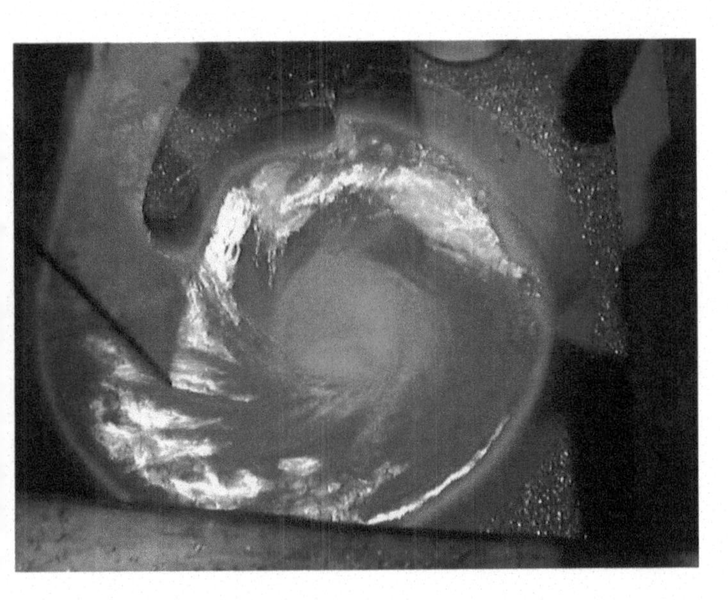

Fig. 7.31 Vortex created for rapid suction of chips. Reprinted with permission from Ref 16

of the scrap pieces. At this stage the chips are damp with the coolant and with oil contamination. They are spun at high speed through a centrifugal spinner to expel most of the coolant and oil, up to 3 to 5%. The chips are then processed through a magnetic separator to isolate any magnetic material that might have been mixed in inadvertently. The chips are then dried by roasting them in a rotary dryer to expel any remnant of coolant or oil. The hot chips are blown into a hopper and introduced at a controlled rate over the vortex chamber. The chips are quickly drawn into the melt, and the dissolved melt flows into the forehearth and then into the degassing and filter furnace. From there, the liquid metal flows into the holder. Figure 7.31 shows the vortex created by the metal pump that draws the aluminum chips in and submerges them without oxidation.

Coreless induction furnaces with properly designed frequency and power for vortex creation and chip suction are also used for aluminum chip melting.

REFERENCES

1. M. Franz-Huster, Selecting the Right Unit for Efficient Induction Melting, *Foundry Management & Technology*, Jan. 2021
2. R.J. Schmitt and J. Kolar, Electric Resistance Melting Center for Metals Production, Mellon Institute, Pittsburgh, PA
3. W.A. Butler, Melting and Holding Furnaces for Die Casting, *Die Casting Engineer Magazine*, March 2006
4. E. Lange, How to Select the Right Aluminum Furnace for Your Casting Operation, Dynamo Furnaces, Oct. 21, 2020
5. D.W. White, *Foundry Management & Technology Magazine*, Jan. 29, 2015
6. Inductotherm Corporation brochure
7. Electroheat Induction brochure
8. Superb Electromachinery Co. Ltd.
9. Lindberg/MPH brochure
10. StrikoWestofen brochure
11. Stotek Inc.
12. *Aluminum Permanent Mold Handbook*, American Foundry Society, 2015
13. Miller and Company, Rosemont, IL
14. Pyrotec brochure
15. G.S. Machineries, India
16. Insertec Furnaces & Refractories brochure

Casting Equipment Engineering Guide
Jagan Nath
https://doi.org/10.31399/asm.tb.ceeg.t59370167

Copyright © 2023 ASM International®
All rights reserved
www.asminternational.org

CHAPTER 8

Aluminum Die Casting Equipment

THE ALUMINUM DIE CASTING process is the most widely used process for high-volume aluminum castings for a variety of markets, including the automotive field. The markets served by the process are:

- Automotive components
- Transportation, including trucking, railroad, and aircraft interiors
- Home appliances
- Agricultural and farm equipment
- Boats and aquatic sport vehicles
- Off-road vehicles, all-terrain vehicles, and snowmobiles
- Industrial machinery and tools
- Office equipment and furniture
- Electrical and electronic equipment

Figure 8.1 offers illustrations of some typical applications.

Figure 8.1(a) illustrates steering gear housings, which need to withstand steering fluid pressure. Figure 8.1(b) (Ref 1) displays the intricate designs of transmission valve bodies, which must be leakproof. Figure 8.1(c) (Ref 2) shows a 6-cylinder engine block with cooling water jackets. Engine cylinder blocks need to have sufficient strength and stiffness to withstand mechanical and thermal stresses.

8.1 Advantages and Limitations

The die casting process is suitable for fast rates of production using high automation that result in excellent consistency in product quality. The advantages of die casting include:

- Excellent surface finish (about 63 μ in. or 1.6 microns of the root-mean square value or RMS)
- Close tolerances
- Near net shape, needing minimum drafts and machine stock, with the ability to cast coarse threads in some applications, avoiding the need for machining
- Thin walls (as little as 2 mm or 0.08 in.), which allow for lightweight applications
- High surface integrity
- Higher mechanical properties compared to sand castings
- Leakproof casting capability to handle fluids (engine blocks, crankcases, front engine covers, oil pans, transmission housings, differential carriers, and steering gear housings are only some of the typical leakproof casting applications)
- Ability to chrome-plate for styling (such as parts for motorcycles and lamp housings)
- Minimum need for manual intervention in operations
- Competitive costs for high-volume production (about 200,000 pieces per year)

The limitations of this process include:

- The inability to use profiled sand cores, which limits the scope of die casting for hollow configurations. Straight hollow passages can be formed using a metal pin insert that can be pulled out before the ejection of the casting.
- Undercut configurations are difficult and would require complex tooling.
- Castings produced from conventional die casting processes cannot usually be heat-treated without causing blisters. The trapped

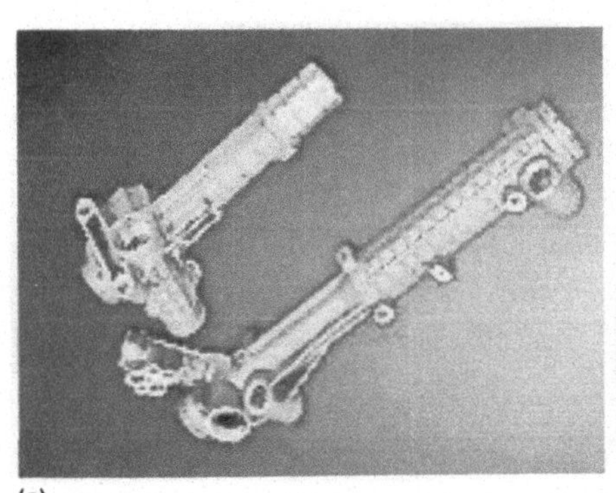
(a)

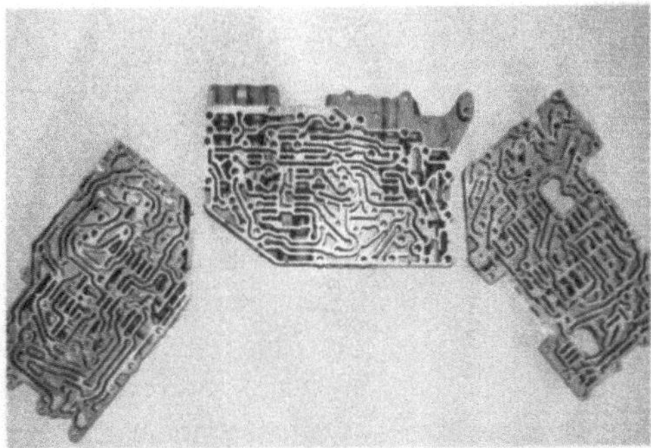
(b)

(c)

Fig. 8.1 Typical die castings (a) steering gear, (b) transmission valves (Ref 1), (c) engine cylinder block (Ref 2)

air due to high shot velocities and the combustible gases from the die lubricant expand during heat treatment, causing blisters which impair surface integrity. Processes, such as welding or clear coating, which require high temperatures also cause blisters. The application of high vacuum during the casting process is required to prevent blisters, but this requires equipment that is expensive to implement and difficult to maintain. Some products can only be produced using the vacuum application.

- Conventional die castings (produced without vacuum) cannot be easily welded, due to the formation of blisters (the die lube penetrates the porosity of the aluminum and releases gases due to combustion on heating) as a result of the heat generated during welding. Only castings produced using the high-vacuum process can be welded (for example, A and B pillars in automotive body structures). Metal Inert Gas (MIG) and Tungsten Inert Gas (TIG) welding processes are used for welding in such applications.
- The tendency toward die-soldering (aluminum sticking to the die, which damages the casting surface) requires a higher percentage of iron (min. 0.6%), which reduces the elongation, impact, and fatigue properties of the alloy. Suitable applications are those where these properties are not functionally required. Special alloys are used with high vacuum for applications where welding is required (Ref 3).

Table 8.1 lists typical applications of the die casting process.

8.2 Die Casting Machine Features

The die casting machine was invented in 1838 to produce movable type for printing, which used zinc. The Soss die casting machine, which was manufactured before 1890, was a breakthrough that set in motion the process development and extension to aluminum and other alloys used for consumer goods. There are two versions of die castings:

- Hot-chamber
- Cold-chamber

Table 8.1 Die Casting Applications

Market	Applications
Automotive	Cylinder blocks, crank cases, engine front covers, thermostat housings, pulleys, alternator housings, transmission housings and valves, differential carriers, cross members, dashboard frames, steering gear housings, air conditioner housings, brackets battery housings, seat supports, shock towers, A, B and C pillar supports
Motor bike parts	Cylinder blocks, gear boxes, crank cases
Outboard marine parts	Blocks, brackets, gear box housings, marine hardware
Agricultural machinery	Crank cases, gear boxes
Aircraft parts	Seat supports, luggage compartment brackets, fittings
Household appliances	Clothes washer parts, dish washer parts, fan housings, room air conditioner parts
Food processing and diary	Hardware, cookware, food processing machinery, pipe fittings
Public utilities	Street light housings

In hot-chamber machines (also called gooseneck machines), the pressurizing chamber is submerged in the molten bath. Pressure applied over the liquid metal in the pressurizing chamber forces the metal into a die through a gooseneck that aligns with the die inlet. This process is suited mainly to low-melting alloys, such as zinc.

Cold-chamber die casting machines are suitable for aluminum alloys. The cold-chamber die cast machine consists of a shot sleeve fitted into a fixed half of the die. A movable die half is mounted on a sliding system for opening and closing the dies. Metal is transferred from a holding furnace into the shot sleeve through a fill port, using a pouring device. A plunger moving at high speed injects the metal into the cavity formed by the dies under high pressure, to form the casting shape. The term "cold-chamber" refers to the shot sleeve, which is not externally heated. The fast-cooling casting is ejected out of the dies using an ejector system. Post-ejection cleaning of the dies is accomplished via air blowing, and lubrication of the dies via spray application of lubricant, which prepare the dies for the next casting cycle.

The casting machine is the heart of the die casting process. It has to be sturdy and rigid to withstand the clamping forces and the casting shot pressure. The dies are hotter at the parting surface than their back faces, which are bolted to the machine platens. This temperature differential separates the die halves at the mating surface, forming the shape of a bow. The clamping pressure should overcome the tendency to warp or bow and ensure a good mating surface at the contact face.

Figure 8.2 illustrates the features of a cold-chamber die casting machine. The machine consists of a stationary platen (2) and a rear platen (4) connected by four tie rods (6) which are bolted together. Sliding in between is a moving platen connected to a hydraulic cylinder called the clamping cylinder (1). The fixed half of the die (11) is bolted to the fixed platen (2), and the half of the die with an ejector mechanism is bolted to the moving platen. The shot sleeve (8), which fits into the fixed half of the die (11), features a slot through which metal can be poured using an automated ladler, a robot, or a dozer. A plunger (10) advances to push the metal into the die cavity at a high velocity and high pressure. A hydraulic cylinder (7) advances and retracts the plunger. The solidifying casting is ejected out, preparing both halves of the die for the next shot. The ejection of the casting is accomplished using an ejection cylinder (13). A robot grabs the casting as it is being ejected and places it on a belt for air cooling or on a tray for water quenching. Simultaneously, the die-cleaning and lubricating mechanism swings in between the two die halves, blows out any flash, and sprays a lubricant on the die cavity for easier release of the casting and also to cool the die down to the initial temperature for the next shot.

Figure 8.2(a) shows the die open and the toggle mechanism in a retracted position. Figure 8.2(b) illustrates the toggle mechanism extended along with the moving platen and the movable die half closed against the fixed half. Figure 8.3 shows a view of a die casting machine (Ref 4).

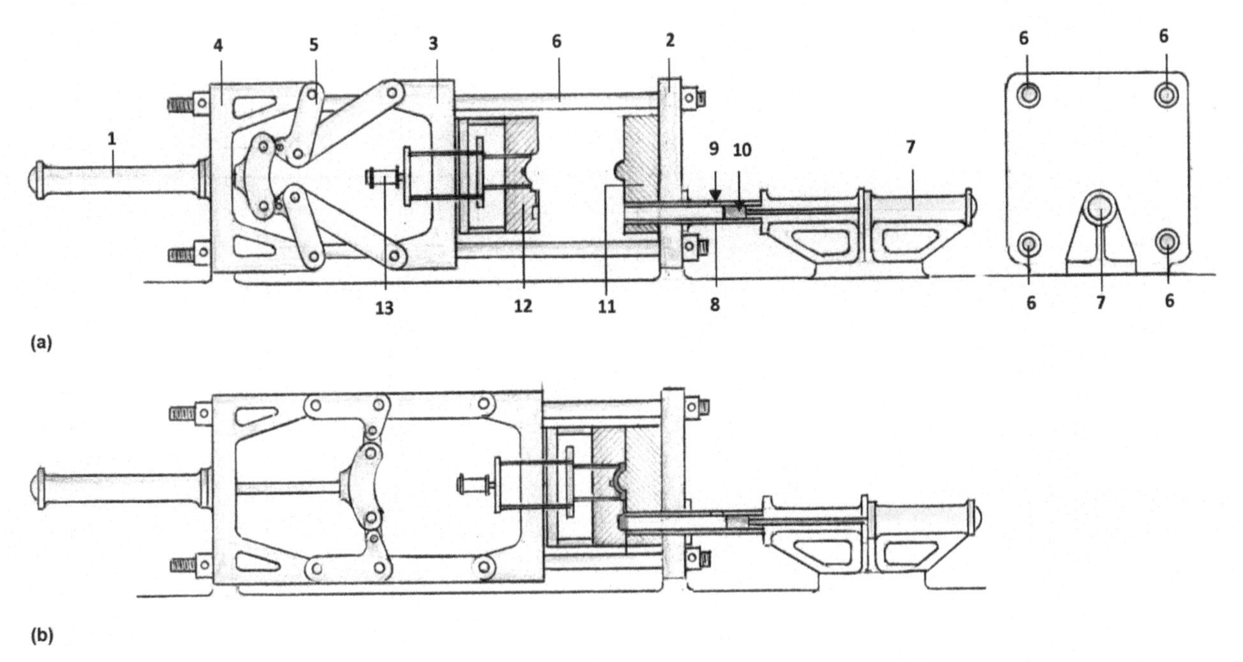

Note: 1. Clamping cylinder 2. Stationary platen 3. Moving platen 4. Rear platen 5. Toggle mechanism 6. Tie rods 7. Shot cylinder 8. Shot sleeve 9. Metal fill port 10. Plunger 11. Fixed die 12. Ejector die 13. Ejection cylinder

Fig. 8.2 Main features of a die casting machine, (a) open, (b) closed

Fig. 8.3 View of a die casting machine with operator safety enclosures. Reprinted with permission from Ref 4

8.2.1 Die Casting Process and Cycle Timings

The die casting process is a high-volume production operation with short cycle times; it is amenable to a high degree of automation. The overall sequence of operations is:

1. The movable platen carrying half of the ejector die moves forward to close against the fixed die mounted on the stationary platen. Any metallic core pin (slides or inserts) advances on a slide, ready for the casting.
2. The ladler or the robot scoops the metal from the holding furnace and pours it into the shot sleeve. The plunger advances at a high velocity to inject the metal under pressure.
3. The casting solidifies, and the core pin slides out.
4. The ejector half of the die moves away, along with the moving platen. The casting is ejected with the help of the ejection cylinder.
5. The robot grabs the casting and places it onto the water immersion tray. The tray moves downward, immersing the casting in water to cool it. Depending on the part, the robot may pick up the cooled casting and deliver it to the gate-shearing press.
6. The blow and spray robot advances in between the two die halves from the top and blows out the die halves and sprays the die lube.
7. The gate-shearing press shears the gates and a pick-and-place unit picks up the casting and delivers it to the deburring station. The pick-and-place unit picks up the gating and dumps it into the remelting hopper. Subsequently, it blows out the shearing dies to clean them of any flash.
8. Castings are deburred, and the pick-and-place unit moves them to the final inspection station. Any gauging that may be required is carried out at this stage.
9. Visual inspection is performed, along with any fine, touch-up grinding that may be required, and the inspection sign-off is completed. The castings are stacked in the dunnage and are ready for shipment.

Figure 8.4 (Ref 3) is an example of the individual operations and their cycle times.

8.2.2 Machine Clamping Systems

The function of the clamping system is to open and close the dies and hold them tightly together against the shot injection load to ensure that there is no metal leakage between the two halves. There are two types of die-clamping systems (Ref 5):

- Direct-pressure clamping system or positive mode locking system
- Toggle locking system

The clamping load should be higher than the opening force due to the injection pressure over the cavity area to ensure:

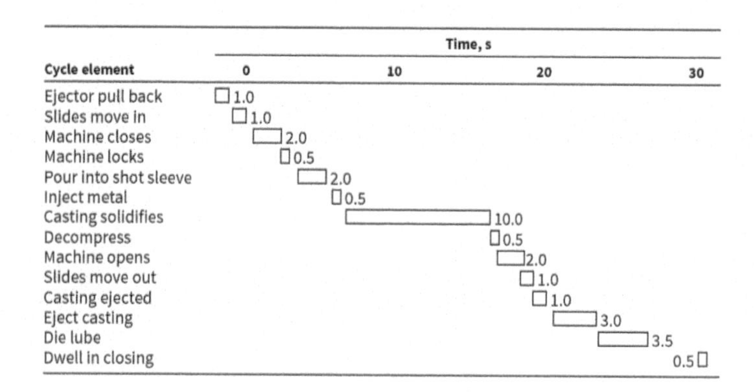

Fig. 8.4 Example of die casting cycle times. Source: Ref 3

- No metal leakage at the parting plane
- No unintended flash thickness
- No undue increase in wall thickness due to the die opening or buckling

8.2.2.1 Direct-Pressure Clamping System

The direct-pressure clamping system is also called the positive-mode locking system. In this system, the closing cylinder is designed to provide extra locking force to prevent any gaps from forming between the two die halves; gaps are undesirable, as they lead to metal leakage or undue flash thickness. An alternative system is to design a separate, shorter locking cylinder in addition to the closing cylinder.

8.2.2.2 Toggle Locking System

The positive-fit toggle locking system is popular over a wide range of locking forces or closing tonnage. It consists of a toggle mechanism to preload the assembly and provide a locking system, as illustrated in Fig. 8.2 (a) and (b). The toggle mechanism provides a constant and consistent locking force, reducing the need for higher exterior force.

The clamping force needed depends upon several factors such as:

- Tie bar load
- Distribution of cavity areas with four tie bars
- Die growth due to the temperature
- The load exerted due to cross core-pin pulls (perpendicular or angled to closing cylinder axis)
- Areas of the shot biscuit, runner, and flow-off vents
- Machine dynamics – injection velocity, intensification ratio, and accumulator pressure drop

Because variables are difficult to measure, some assumptions need to be made to arrive at the tonnage needed for a targeted new part.

Assumptions include:

- The clamping force should be at least 10% greater than the opening force
- The runner area can be 20% of the cavity area, when the gating system is designed
- The flow-off vent area can be 12% of the cavity area, when the venting system is designed
- The load that is due to the core equals the cavity pressure times the sliding core area times tan θ, where θ Is the angle of the core pull (if there is a core pull. Otherwise, it is 0).

Assuming a simple case of a die cavity layout which is symmetric with four tie bars in both X and Y axes and without a core pull, the clamping force C per Eq 8.1 (Ref 6) is:

$$C = P \times A \quad \text{(Eq 8.1)}$$

where C is the clamping force in Newtons (or pounds), P is the maximum metal pressure in the die cavity in MPa (or pounds per square inch), and A is the projected area of the cavity and shot biscuit, runner, flowoff in cm^2 (or in.^{2}).

Converting to different units:

$$\text{Clamping force } C \text{ (in kN)} = \frac{P\text{(in Mpa) x } A \text{ (sq.cm)}}{10} \quad \text{(Eq 8.2)}$$

or

$$\text{Clamping force } C \text{ (in tons)} = \frac{P \text{ (in psi) x } A \text{ (sq.in)}}{2000} \quad \text{(Eq 8.3)}$$

Most often, the shot biscuit is designed at the bottom of the die, and the cavities are laid out for the metal to flow upward, as shown in Fig. 8.5. The cavity force distribution is asymmetric, or unequal, between the top and the bottom tie bars. The centroid of the cavity distribution is computed as shown in Fig. 8.5, and the force on each of the four tie bars is calculated. The maximum load on one tie bar is multiplied by four, and the safety multiplying factor of 1.1 is used to obtain the tonnage of the machine.

Figure 8.5 is a schematic of two cavity layouts that are symmetric across the Y axis. Tie bars A and B experience the same load, and tie bars C and D are subjected to the same load. Because the cavities are not symmetric across the X axis, tie bars A and B experience a different load compared to tie bars D or C. If the cavity is not symmetric in both X and Y axes, all four tie bars will experience different loads.

The percentage load distribution is proportional to the cavity area shared by tie bars A and D. The individual areas are numbered from 1 to 6. The effective load center is determined by multiplying the product of each area from 1 to 6 times the distance from the tie bar A and dividing this product by the sum of the areas from 1 to 6.

Table 8.2 lists the areas of the elements 1 to 6, the distance of the centroid of each area to the tie Bar A, and the product of the area times the distance of centroid of each area. Since the layout is symmetric around the Y-axis, only one cavity and two tie bars, A and D, are used in the calculation. If the die cavity does not have symmetry, the load centers in both the X and Y axes need to be computed on the same lines, as illustrated in this example:

> Because the two-cavity layout is symmetric around the Y-axis, only one cavity and two tie bars, A and D, are considered for calculations.

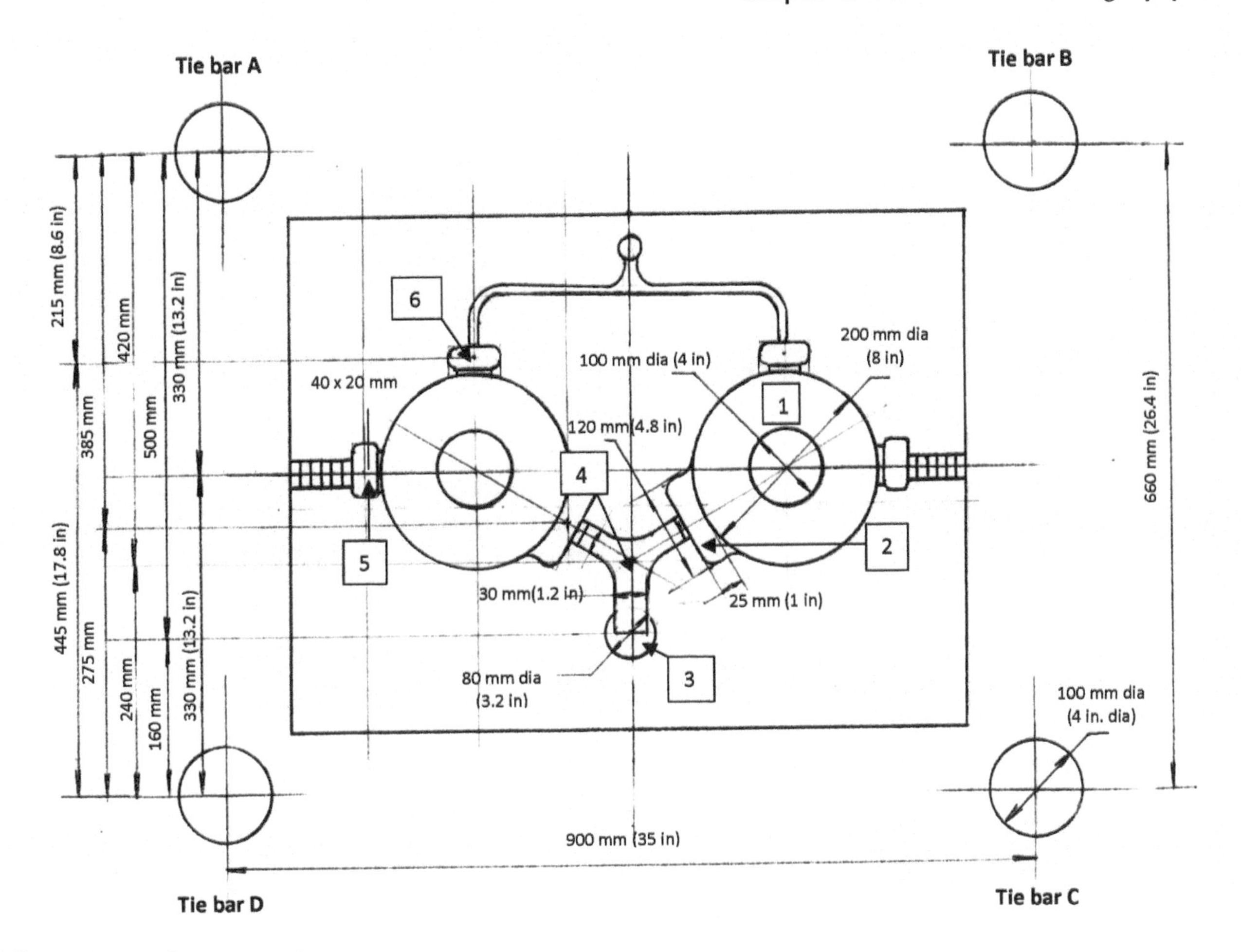

Fig. 8.5 Die clamping force computation

Table 8.2 Computation of load center in symmetric layout of cavities

Element number	Area cm²	Area in²	Vertical distance from tie bar A cm	Vertical distance from tie bar A in.	Area X distance cm³	Area X distance in³
1	235	36.4	33.0	13.0	7755	473.20
2	30	4.65	38.5	20.2	1155	93.93
3	50	7.75	50.0	19.68	2500	152.52
4	45	6.98	42.0	16.53	1890	115.37
5	6	0.93	33.0	12.99	198	12.08
6	6	0.93	21.5	8.46	129	7.87
Totals	**372**	**57.64**	**218**	**90.86**	**13,627**	**831.24**

Using the dimensions as in Fig. 8.5, the load center $= \frac{136.27}{372} = 36.63$ cm

or $\frac{831.24}{57.64} = 14.42$ in

Using Eq 8.2 and 8.3 and maximum metal pressure in the die cavity at 27.6 MPa or 4000 psi,

$$\text{Total opening force} = \frac{27.6 \text{ MPa x } 372 \text{ sq.cm}}{10} = 1026 \text{ kN} \quad \text{or}$$

$$= \frac{4000 \text{ psi x } 57.64 \text{ sq.in}}{2000} = 115.28 \text{ tons}$$

$$\text{Load shared by tie bar D} = \frac{1026 \, x \, 36}{66} = 559.64 \text{ kN} \quad \text{or}$$

$$= \frac{115.28 \, x \, 14.42}{25.98} = 63.9 \text{ tons}$$

$$\text{Load shared by tie Bar A} = \frac{1026 \, x \, 30}{66} = 466.36 \text{ kN} \quad \text{or}$$

$$= \frac{115.28 \, x \, 11.56}{25.98} = 51.2 \text{ tons}$$

Tie bar D shares the higher load of 559.64 kN, or 63.9 tons. The machine should be designed for 559.64 kN x 4 or 63.9 tons x 4. A machine rated at approximately 3000 kN or 300 tons would be a safe choice for this application.

8.2.2.2.1 Toggle Mechanism

The toggle mechanism enables a lock or preload before the first shot is made. Figure 8.2(a) illustrates the toggle mechanism retracted, and Fig. 8.2(b) shows the toggle mechanism in a fully extended position. Preloading requires that the toggles extend to a dimension slightly more than the die in closed position, creating a tight fit, and setting the tie bars in tension.

The extended dimension of the toggle mechanism is established by calculating the tie bar strain or elongation percentage. This is dependent on the force applied on the tie bars, cross-sectional area of the tie bar, and the modulus of elasticity of the tie-bar steel. Equation 8.4 (Ref 6) provides the relationships to establish the percent of elongation of the tie bars:

$$e = 10 \text{ x } \frac{p}{a \, x \, E} \qquad \text{(Eq 8.4)}$$

where e is the unit of elongation mm/mm, p is the force applied on one tie bar in kN, a is the cross-sectional area of one tie bar in cm^2, and E is the modulus of elasticity of steel – 207,000 MPa.

$$\text{Or} \quad e = 2000 \text{ x } \frac{p}{a \, x \, E} \qquad \text{(Eq 8.5)}$$

where e is the unit of elongation in/in, p is the force applied on one tie bar in tons, a is the cross-sectional area of one tie bar in in^{-2}, and E is the modulus of elasticity of steel – 30,000,000 psi.

Substituting the dimensions from Fig. 8.5 in Eq 8.3, and rounding off the maximum force in tie bar D from 559.64 kN to 600 kN, and cross-sectional area of $\pi/4$ x (10) x (10) = 78.54 cm^2:

$$e = 10 \text{ x } \frac{600}{78.54 \, x \, 207{,}000} = 0.000369 \text{ mm/mm}$$

or substituting in Eq 8.4 and rounding off the maximum force in tie bar D from 63.9 to 70 tons, and cross-sectional area of $\pi/4$ x ((4) x (4) = 12.566 in^2:

$$e = 2000 \text{ x } \frac{70}{12.566 \, x \, 30{,}000{,}000}$$

$$= 0.000371 \text{ in/in (close to 0.000369 mm/mm)}$$

If the length of the tie bar is 150cm (60 in.), the toggle extends it to 150 plus 150 times 0.00036, which equals 150.05535 cm. Or, if the length of the tie bar is 60 in., the toggle extends it to 6 plus 6 times 0.000371, which equals 6.002226 in.

8.2.2.2.2 Effect of Die Expansion

The strain on the tie bar is also influenced by expansion due to the die temperature increase. The temperature variation during a shift could be as much as ±30 ° C (± 55° F). The coefficient of expansion for H 13 steel between 25 and 205 °C (80 and 400 °F) is 0.0000115 mm/mm °C, or 0.0000063 in/in ° F.

The die thickness changes per Eq 8.6 or 8.7:

$$\Delta t = (t)\,(\Delta C) \text{ x } (0.0000115) \text{ mm} \qquad \text{(Eq 8.6)}$$

$$\text{or} \quad \Delta t = (t)\,(\Delta F) \text{ x } (0.0000063) \text{ in.} \qquad \text{(Eq 8.7)}$$

where Δt is the change in die thickness in mm or in., t is the die thickness at room temperature, ΔC is the change in die temperature in °C, and ΔF is the change in die temperature in °F.

A die that is 150 mm (6 in) thick with a temperature variation of 50 °C (or 90 °F) grows by Δt = 150 x 50 x 0.0000115 = 0.8625 mm, or Δt = 6 x 90 x 0.0000063 = 0.003 in.

8.2.2.2.3 Effect of Die Warpage

Dies can warp during service due to the temperature differences between the cavity and the back of the dies. The die temperature at the cavity is approximately 250 °C (480 °F), and the back (clamping) faces are approximately 90 °C (200 °F). The die expands more at the cavity face compared to the back face, resulting in warpage. Both halves of the die with warpage (bowing) at the mating face need to be clamped together to overcome warpage and to prevent any metal leakage. There is also a potential for the casting thickness to increase if the clamping force is inadequate.

8.2.2.2.4 Effect of Plunger Diameter

The required tonnage of the machine depends upon the plunger diameter. Equation 8.1 shows clamping pressure proportional to metal pressure in the cavity. This pressure is inversely proportional to the square of the plunger diameter. The cavity pressure can be reduced by increasing the shot sleeve diameter. The machine tonnage should be verified if the shot sleeve diameter has been changed.

8.2.2.2.5 Clamping Force Measurement

Load cells mounted on the tie bars, linear variable differential transformers (LVDTs), or linear optical encoders are used for measuring the tie bar strain, which is the means of measuring clamping force. Tie bar load control is achieved by continuous feedback control systems.

8.2.3 Die Materials

The selection of the most suitable die materials is critical to ensure that the expensive dies have a service life that justifies their cost. The typical die service life for aluminum castings is around 100,000 cycles. Die life depends upon many factors, such as the complexity and design of the die. Wear or erosion are the main causes of failure, followed by heat checking due to fatigue because of rapid thermal cycling. Table 8.3 lists the common materials used in die design, along with their hardness values.

8.2.4 Metal Injection System

The system of metal injection in die casting is unique among metal casting methods. The metal injection system is engineered to inject metal at very high pressures (maximum 275 MPa or 40,000 psi) in milliseconds (0.01 to 0.1 secs). The high velocities at the gate result in dispersed droplets or an atomized jet metal stream that rapidly fills the thin walls.

Hot metal is poured into a cylinder called the shot cylinder (or shot sleeve), and the plunger advances rapidly to shoot the metal into the die cavity. The high energy is stored in the form of

Table 8.3 Die component materials and hardness values for aluminum die castings

Die component	Material	Hardness, HRC(a)
Cavity inserts	H 13	42–48
	H 11	42–48
Cores	H 13	44–48
Core pins	H 13	37–40
Sprue inserts	H 13	46–48
Nozzles	H 13	42–48
Ejector pins	H 13	46–50
Plunger shot sleeves	H 13	42–48
Holder blocks	4140 prehardened	about 300 HB(b)

(a) HRC is Rockwell hardness to C scale. (b) HB is Brinell Hardness. Note: H11 and H13 are high alloyed chromium-molybdenum heat-resistant steels for die cavities and other parts. 4140 is a low alloyed chromium-molybdenum-manganese steel prehardened to 300 HB for holder blocks and general-purpose applications.

compressed nitrogen gas in an accumulator and then is released quickly to the hydraulic fluid in the hydraulic cylinder and the plunger through fast-acting control valves. The hydraulic system consists of an intensifier that boosts the hydraulic pressure in the system.

The sudden burst of energy drives the plunger to shoot the metal into the die cavity at high velocities under very high pressures. Figure 8.3 shows the nitrogen tanks on the left side. Figure 8.6 is a schematic of the shot sleeve and plunger system.

The system involving the shot sleeve and plunger consists of a thick-walled, high-temperature-resistant steel sleeve (1) to hold the metal being transferred from a holding furnace, using a dedicated ladler or a programmable robot. The hot metal is poured into the shot sleeve through the port (2). The plunger (3), with an internal water cooling circuit, is connected to the cooling passage (4). The plunger pushes the metal in the shot sleeve into the die cavity at high pressure and velocity. The plunger rod is connected to the shot cylinder (6) through a coupling (5). An intensifier (7) helps to boost the pressure. Nitrogen contained in the cylinder (8) is compressed by the hydraulic fluid and is ready to unload into the shot cylinder. As soon as the timed valve opens, the hydraulic fluid bursts into the shot cylinder (6), advancing the plunger to shoot the metal into the die cavity.

8.2.5 Die Lubrication

Dies are sprayed with a release agent to prevent the castings from sticking to the die while they are being ejected. This release agent is known by various names, such as die lube, die spray, or parting spray. In addition to ensuring the ease of release, die lubrication helps to improve the surface finish. Die lube helps to cool the dies nearly back to the temperature they had at the start of the cycle. Die lubrication also improves uptime due to reduced interruptions, because castings do not stick to the die. Also, casting warpage is reduced due to the ease of release without sticking.

Complex dies are sprayed each time before the die is closed. Simpler dies are sprayed once every three or four shots. The die spray unit, reaching down from the top of the open dies, combines a blowing operation to dislodge any scale or flash before spraying the die lube. Sometimes the die spray unit reaches out from the side, if the die design requires any cores to be pulled from the top. Programmable robots or units suspended from a gantry crane above the machine are more common.

8.2.5.1 Performance Factors of Lube

The four performance factors of the die lube are:

- Ability to release the casting
- Prevention of any soldering
- Lubrication of moving parts, such as devices that pull the core pins
- Environmentally safe for operators

The thin film that is sprayed prevents close contact between the metal and the die surface, ensuring easy release. Due to environmental requirements, lubes are water-based emulsions formulated with oil. The water in the lube evaporates as it is sprayed onto the hot dies, leaving an oil film that helps to release the shot or the casting.

Emulsions of water with oil are typically formulated with a ratio of water to oil as 30 to 1. Formulations vary with combinations and different proportions of heavy residual oil (HRO), animal fat, vegetable fat, or synthetic oil. The emulsification process is assisted by the addition of emulsifiers such as soap, alcohol esters, or ethylene oxides (Ref 7).

The chemical affinity of molten aluminum to iron in the die steel can result in reactions at the interface causing soldering. The interfacial thermal properties are influenced by additions

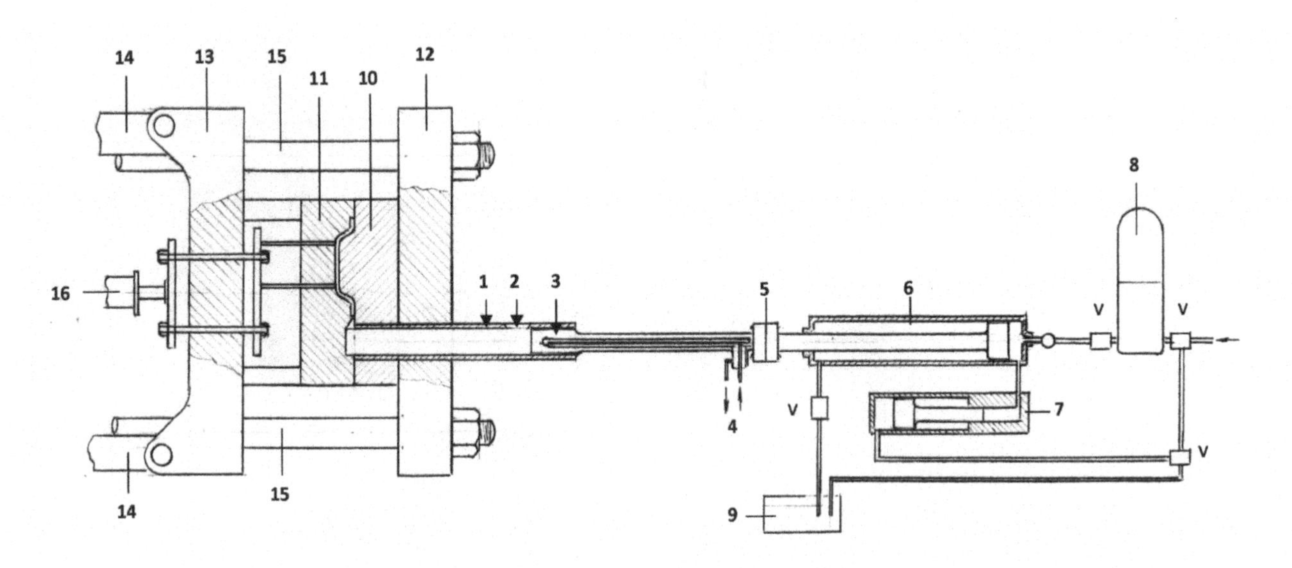

Fig. 8.6 Shot sleeve and plunger schematic

such as mica and graphite. Rusting and oxidation at the die surface are prevented by other proprietary additions.

Water-based lubricants are most common because they are environmentally friendly and are required for the health and safety of operators.

High-heat zones such as the biscuit and runner need a heavier spray compared to the die cavity and core faces. Complex and deep areas also need a heavier spray. The flow-offs only require a light spray. There is no dependable method of measuring the spray thickness, but the frequency of die-solder occurring on a particular zone is a good indication that the spray should be heavier in a particular zone. Normally, the lubrication time programmed for any zone is an indication of the spray thickness and the local die temperature.

Casting process engineers prepare and document coating maps to guide the robotic programmer. The programmer is responsible for adjusting the number of lubrication cycles repeated at the particular zone and the total time allocated for each zone. Figure 8.7 (Ref 7) is an illustration of a die lube sprayer for the two die halves. Figure 8.8 (Ref 8) shows a bank of spray nozzles.

8.3 Die Casting Cell Layout

Most die casting operations have multiple die casting machines served by one or two central melting units. Typically, one central melting furnace serves about four to six machines.

Metal is distributed to the individual casting machine holders using a distribution ladle mounted on a tow motor. Installations with heated launders are also common. Some die cast operations use a monorail with a ladle for metal distribution. The metal distribution method influences the layout of the cells.

Figure 8.9 is a schematic of a die cast machine layout. Although only two cells are illustrated, the concept applies to more machines. More than one central melting unit is installed if more than one metal grade is used simultaneously.

The die cast alloys used are listed in Fig. 7.10 in Chapter 7, "Aluminum Melting, Holding, and Dosing Furnaces," in this book. The common material grades used per ASTM specifications are: 380.0, A380.0, 383.0, 384.0, 390.0, A390.0, B390.0, 360.0, A360.0, 413.0, and A413.0. Alternate grades used are: 413, A413.0, 443.0, and 518.0. The alloys used in high-vacuum cast structural applications are: Silafont 36, Castasil 37, Aural 2/3/5, and Mercalloy 362, 367, and 368 (Ref 3).

Figure 8.9 illustrates a central melting unit (14) capable of charging sows (15), biscuits, and runners at the other end using a stack melter (16) attached. A skip (17) hauls the remelt up the top of the stack for charging. A distribution ladle (18) carried on a tow motor provides the flexibility to distribute the metal to the individual machines, depending on the amount of metal poured into each cell.

Cell 1 is a die cast machine equipped with a doser, as described in section 7.6.2 in Chapter 7.

Cell 2 shows a die casting machine with a holding furnace or a holder (4), as detailed in section 7.6.1. A launder delivers the metal from the dozer to the port in the shot sleeve in cell 1. A ladler (5) serves to scoop the metal from the holder and pour it into the port in the shot sleeve in cell 2.

The castings are unloaded by the robot (6) and placed on the cooling belt, dipping into the cooling water tank (8). Castings move on the conveyor (9) under a station for air cooling and blowing off the water. The castings are picked off by an operator who inspects the castings visually and feeds the castings into a trim press (10) to remove the biscuit, runner, and flash. The castings are transported by conveyor (12) for deburring and machining operations. The remelt collected in totes (13) moves to the skip (17) for charging into the stack meter.

Fig. 8.7 Die lube in progress. Reprinted with permission from Ref 7

Fig. 8.8. Bank of spray nozzles. Reprinted with permission from Ref 8

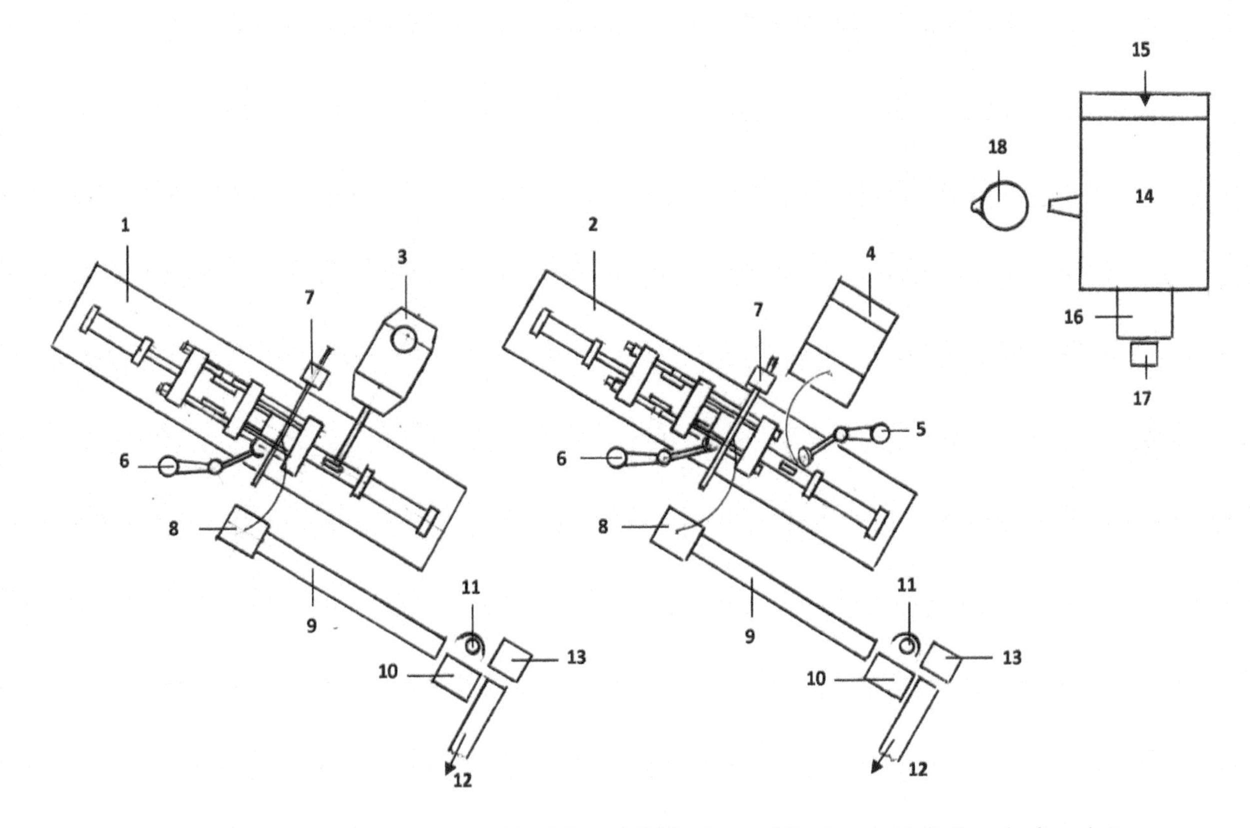

Note: 1. Cell 1 Die casting machine 2. Cell 2 Die casting machine 3. Dozer 4. Holding furnace 5. Pouring robot 6. Casting unloading robot 7. Die lube gantry robot 8. Quench tank 9. Air drying conveyor 10. Trim press 11. Operator 12. Castings for deburring and machining 13. Biscuits and runners for re-melting 14. Central melting furnace 15. Well for charging sows 16. Stack melter for remelt 17. skip hoist 18. Metal distribution ladle on tow motor

Fig. 8.9 Die casting machine cell layout

The trim operation can be automated by using another robot to pick up the castings and feed them to the trim press. But the manual operation provides an opportunity for hot inspection to find drag marks due to solder buildup on the dies, which allows refinements to be made without causing much scrap.

REFERENCES

1. CMT Imports, Cincinnati, Ohio
2. Wikipedia: Cylinder block picture Fig. 8.1(c)
3. J. Nath, *Aluminum Castings Engineering Guide*, ASM International, 2018
4. Buhler Die Casting Machines, Inc.
5. Giesserei Lexicon from Internet
6. Die Casting Process, Engineering and Control NADCA Publication # E – 410, 1991
7. Hill & Griffith, Cincinnati, Ohio
8. HERCO Llc, Rochester Hills, Michigan

Casting Equipment Engineering Guide
Jagan Nath
https://doi.org/10.31399/asm.tb.ceeg.t59370177

Copyright © 2023 ASM International®
All rights reserved
www.asminternational.org

CHAPTER 9

Gravity Permanent and Semipermanent Mold Equipment

THE GRAVITY PERMANENT MOLD (GPM) process produces castings of high strength and integrity with good surface finish and consistent dimensions. Recent advances in metal handling and treatment as well as the capability for computerized solidification and flow simulation techniques have positioned the GPM process as a very cost competitive method to meet the most stringent quality and property requirements of critical safety components.

Critical safety components are those in which deficiency in quality or an anomaly may cause an accident resulting in injury or death to the occupants of a vehicle. In the gravity permanent mold (GPM) Process, both the method of pouring the molds and feeding of the casting shrinkage use *gravity* as the driving force. The metal molds that are reused for every pour cycle have a long lifespan of nearly 100,000 pours. This process is known as the gravity permanent mold process, although the molds do have a finite life. When expendable sand cores are used for each pour in a metal mold, the process is known as the semipermanent mold process (SPM). The industries served by these processes are:

- Automotive
- Truck
- Farm equipment
- Air compressors
- Marine hardware
- Appliances and household goods
- Electrical components

Figure 9.1 illustrates some of the high-integrity castings such as knuckles, control arms, and brake calipers which have been successfully converted to aluminum alloys from ductile iron and stamped steel, resulting in weight savings of nearly 35%. Other applications illustrated include brake master cylinders, antibrake skidding (ABS) housings, and air compressor housings. Over more than two decades, millions of these parts have been cast, assembled and installed in passenger vehicles without any quality issues related to the casting process. Table 9.1 lists castings produced using GPM and SPM.

When molten aluminum reacts with oxygen, it oxidizes to form aluminum oxide. Any entrainment of aluminum oxide in the casting impairs its strength. Turbulence due to the high velocity of metal flow during pouring generates aluminum oxide. One effective method of eliminating or reducing turbulence is to *tilt pour*. The tilt pouring process consists of titling the mold from a horizontal to a vertical position to aid filling and feeding, using a tilt pouring machine or a tipper.

Fig. 9.1 GPM applications

Table 9.1 Gravity permanent and semipermanent molded parts

Markets	Gravity permanent molding (GPM) Metal cores and pull backs	Semipermanent molds (SPM) Sand cores
Automotive	Pistons, **knuckles, control arms, brake master cylinders, brake wheel cylinders, brake calipers, ABS housings,** engine brackets, cam housings and covers, air conditioning housings, engine front covers, super charger lobes and housings	Cylinder heads, twin piston calipers, engine crank cases, oiler adapters, differential carriers, intake manifolds, water pump bodies
Truck and diesel	Transmission housings, engine front covers, gear boxes	
Farm equipment	Transmission housings, oil pans, gear boxes, lawn mower parts	Oil pans, water pump bodies
Air compressors	Compressor bodies	
Marine	Propellers (for soft water lake boats)	
Hardware	Brackets	
Appliance and household goods	Vacuum cleaner housings, washing machine agitators and gear boxes, electric griddles	
Electrical	Motor housings, lamp post parts, junction boxes	

Note: Products shown in **bold** are **safety-critical** components. A safety-critical component is one where the quality requirement is the most critical because a failure of any of these components due to any deficiency of integrity may result in an accident, injury, or death to the occupants of the vehicle.

Figure 9.2 illustrates the reduction of turbulence by velocity reduction in tilt pouring compared to static pouring:

$$V = \sqrt{2gH} \quad \text{(Eq 9.1)}$$

where V is the velocity of the metal entering the mold cavity, H is the metal head above the metal inlet (gate) into the mold, and g is the acceleration due to gravity.

The metal head H in tilt pouring (Fig. 9.2b) is much lower compared to static pouring (Fig. 9.2a). The liquid metal experiences a much lower velocity in tilt pouring, as in Eq 9.1. The tilt pouring process reduces metal velocity, turbulence, and oxide formation, resulting in improved melt quality and casting integrity.

Depending upon the casting configuration and the design for feeding the shrinkage, tilt pouring equipment is designed either for:

- Parting plane pouring, or
- Cover or reverse pouring

9.1 Parting Plane Pouring

Figure 9.3 illustrates the concept of parting plane pouring. The part configuration allows the parting plane to split, as shown with the pouring cup attached to the bottom mold (Fig. 9.3a). The refractory pouring cup reduces the loss of metal temperature. As the mold and pouring cup gradually tilt, the metal in the cup is slowly transferred to the mold cavity (Fig. 9.3b). The mold tilts 90° (Fig. 9.3c). In this orientation, the top feeder starts to feed the solidifying casting. As soon as the solidification is complete and the casting is solid enough for ejection, the mold rotates back 90° to the horizontal position (Fig. 9.3d). The mold now starts to open, and the catcher (catch pan) slides in, ready to receive the casting. The top mold moves farther for the ejector pin rack to bump against a bump plate, which causes the casting to be ejected onto the catch pan. (Alternatively, a cylinder attached to the ejector rack is actuated to eject the casting from the mold.) The operator removes the foil in the pouring cup and blows out the mold of any debris in preparation for the next pour. The machine is designed for the operator to actuate two buttons, using each hand, to close the mold (designed as a safety feature) and prepare it for the next pour.

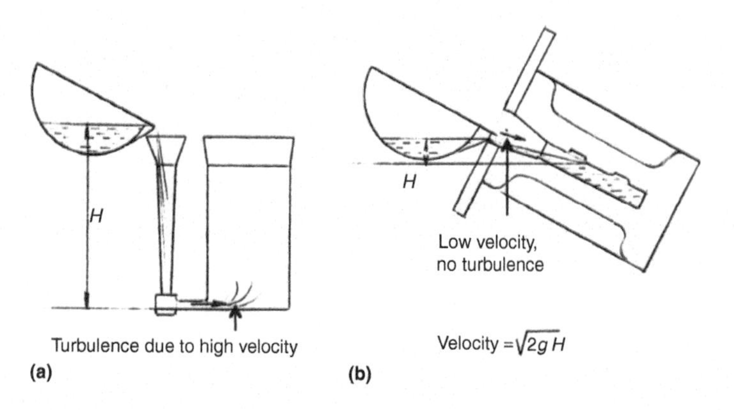

Fig. 9.2 Reduction of metal velocity from (a) static to (b) tilt pouring

9.2 Cover or Reverse Pouring

Some casting configurations have an uneven parting plane, which makes them more suitable for cover (or reverse) pouring, as shown in Fig. 9.4. In such cases, the pouring cup is attached to the cover mold along with the fixed platen. Metal is poured into the pouring cup, as shown in Fig. 9.4(a). As the mold assembly tilts (Fig. 9.4b), the metal flows into the mold cavity at a slow and controlled rate. The metal in the pouring cup empties into the mold cavity as it turns 90° (Fig. 9.4c). The top feeder starts to feed the shrinkage in the casting as the casting solidifies. After solidification is complete, the moving half of the mold retracts and the ejector pins lift the casting off the mold, as shown in Fig. 9.4(d). The casting is picked up by a retrieving device and moved to the cooling station. The operator removes the foil from the pouring cup, blows out the mold, and closes it in preparation for the next pour.

9.3 Tilt Pouring Machine Designs

Gravity tilt pour machines are based on the principles illustrated in Fig. 9.3 and 9.4. Most of the machines are designed for a 90° tilt, although machines that are capable of tilting 180° are available for special applications. Machines that use hydraulics for clamping and tilting are more popular than those that use electric power.

Figure 9.5 illustrates the basic elements of a conventional tilt pour machine. The machine consists of a C -frame (2) hinged on two trunnions or a fulcrum (3), which is mounted on a base frame (1). Two hydraulic tilting cylinders (4) mounted on the

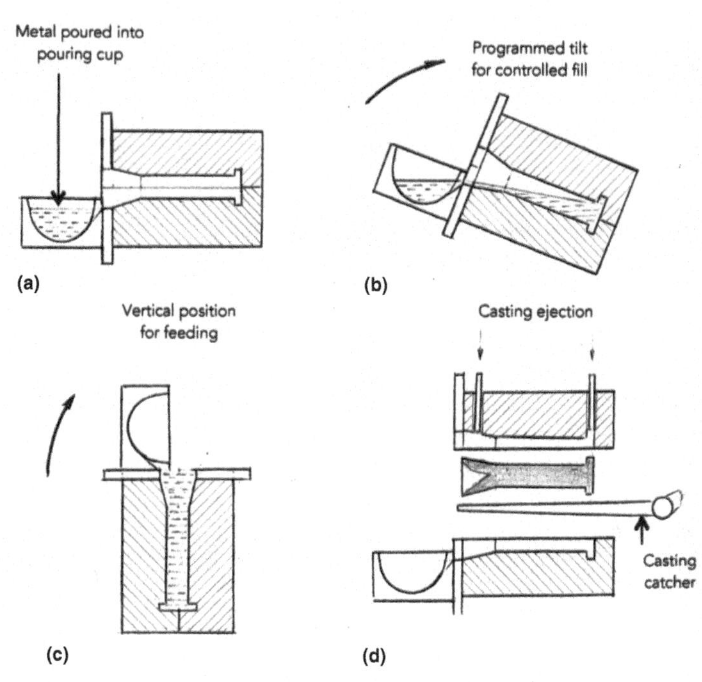

Fig. 9.3 Concept of parting plane pour, (a) parting plane splits, (b) metal transfers to mold cavity, (c) mold tilts 90°, (d) solidification is complete and the mold rotates back 90° to horizontal position

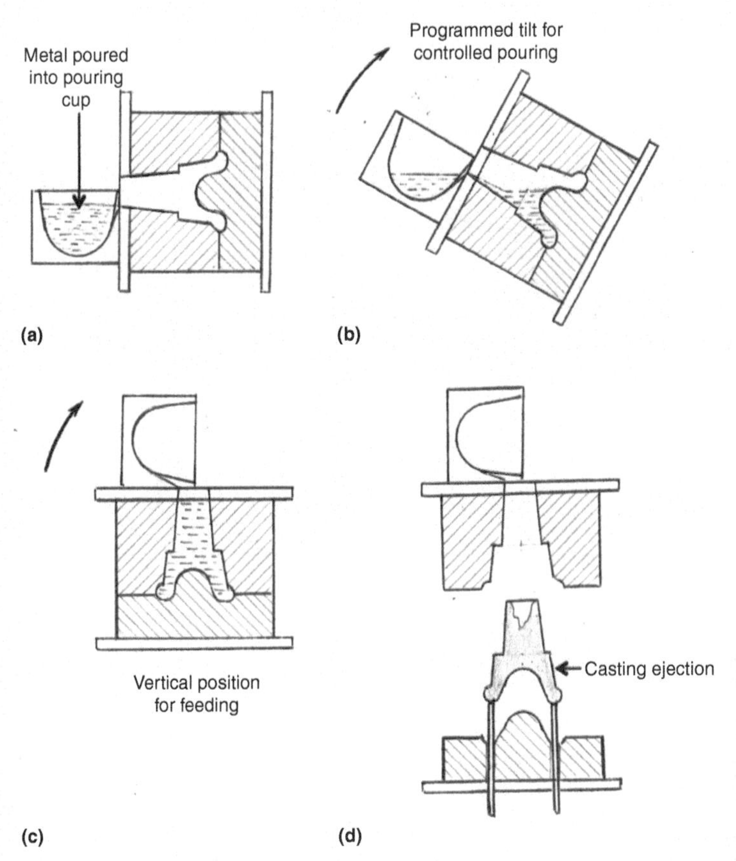

Fig. 9.4 Cover or reverse pouring

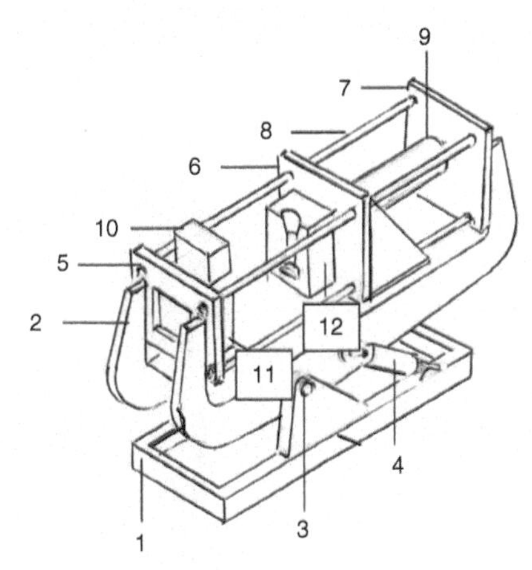

Note: 1. Base frame 2. C-frame 3. Fulcrum 4. Tilting cylinder 5. Front fixed platen 6. Middle moving platen. 7. Rear fixed platen 8. Tie rods 9. Closing cylinder 10. Pouring cup 11. Fixed mold half 12. Moving mold half with ejector at the back

Fig. 9.5 Tilt pouring machine or tipper schematic

base frame tilt the C-frame and the mold assembly by 90°. Four tie rods (8) connect the front platen (5), the moving platen (6), and the rear platen (7). The cylinder that closes the mold (9) is mounted on the rear platen. The pouring cup (10) and the stationary half of the mold (11) are mounted on the front platen, and the moving half of the mold (12) is mounted on the moving platen (6). The closing cylinder (9) closes the mold halves together and the tilting cylinders (4) are actuated to tilt the C-frame in a vertical position, orienting the molds in a horizontal position for pouring. The sequence of operations follows as described in section 9.1.

Figure 9.6 (Ref 1) is an illustration of a tilt pouring machine (also called a tipper) which can interface with a robot for casting extraction. Figure 9.7 (Ref 2) shows a tilt pouring machine capable of tilting 180°. One or two machines can be installed for low-volume work with a common automated pouring system or with manual pouring, depending on the casting and size of the tilt pouring machine.

9.4 High-Volume Production Rotary Table

Multiple tilt pouring machines (6, 8, or 12) can be mounted on a rotary table or turntable for high-volume work. They share common melting and metal treatment furnaces as well as shared dosing equipment. Such cells use a robot for extraction of the castings.

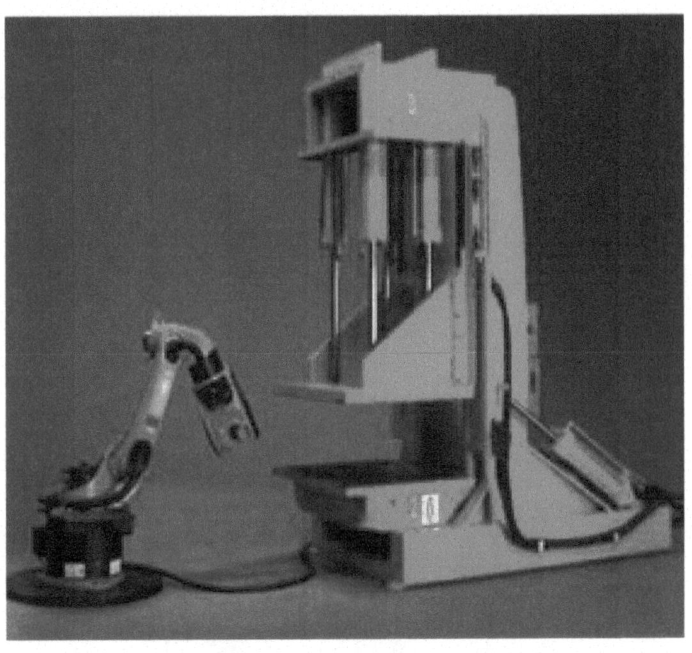

Fig. 9.6 90° tilt pouring machine. Reprinted with permission from Ref 1

Figure 9.8 illustrates a rotary table with eight tilting machines mounted on it. Multiple tilt pouring machines are mounted radially and symmetrically over a turntable supported by heavy-duty central bearings and a series of supporting rollers on the periphery. The electrical and cooling water connections for each tilt pouring machine are located in the center. The electrical connections are mounted on a rotary encoder. The incoming and outgoing cooling water connections are connected to a central manifold and a trough. The table rotation is controlled, based on the encoder and proximity limit switches. Each tilt pouring machine has local programmable logic controllers (PLCs) that interface with a central PLC, mounted toward the center. The titling machines align with the dosing unit and swing away as

Fig. 9.7 180° tilt pouring machine. Reprinted with permission from Ref 2

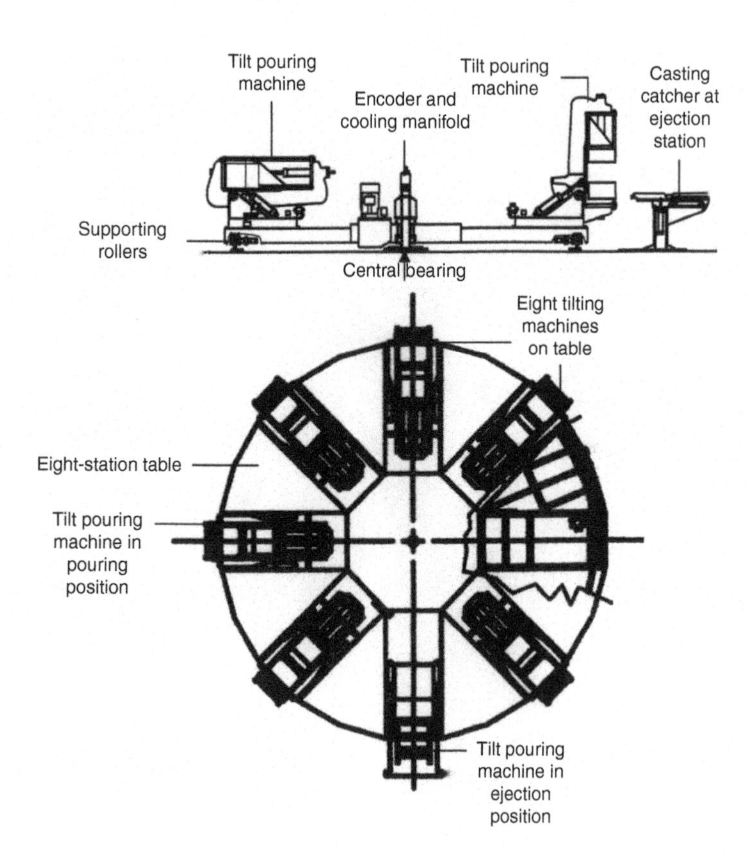

Fig. 9.8 GPM rotary manufacturing table with tilt pouring machines. Source: Ref 1

the mold tilts to pour. The molds open after a designated time by which the feeder would have solidified to have adequate strength to allow for ejection of the casting. A robot or casting catcher receives the casting as the table is realigned. Figure 9.9 is a three-dimensional view of the turntable arrangement.

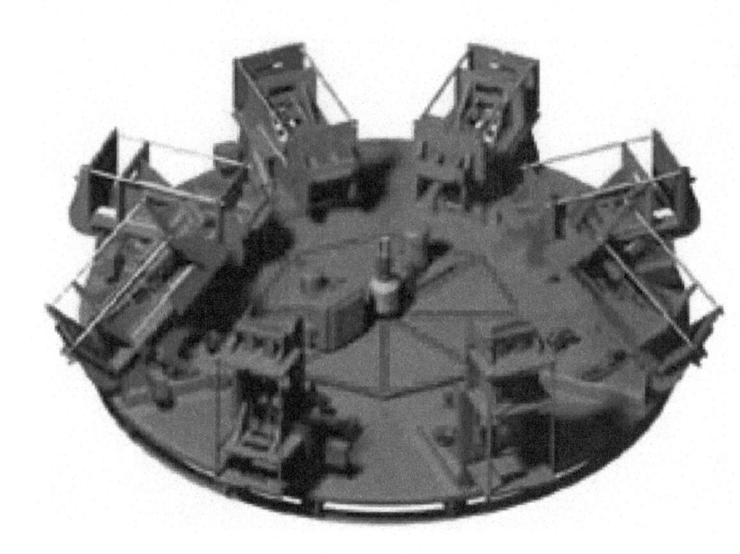

Fig. 9.9 Three-dimensional unit of the turntable. Reprinted with permission from Ref 1

9.5 High-Volume Production Rotary Manufacturing Cell

Many capital and high-volume operations install highly productive automated manufacturing cells using the turntables illustrated in Fig. 9.10. The cell consists of a melting unit such as a stack melter (1), a reverberatory furnace (gas or electric), or an electric induction furnace (for low emissions). A skip unit (2) meets the charging needs of the stack melter. Metal from the melting furnace flows into a degassing and filter furnace (3), which is equipped with a rotary impeller degassing (RID) unit (4) to remove any dissolved hydrogen. Any oxides that are generated due to the rotary motion of the impeller are removed by the filter (5). Metal treatment such as grain refinement is carried out in the furnace section past the filter. Degassed, filtered, and grain refined metal flows into a doser (7) which dispenses a consistent amount of metal, as described in Chapter 7, "Aluminum Melting, Holding, and Dosing Furnaces," in this book.

The rotary table (8) carries six to twelve tilt pouring machines. The position in which the table stops to align the tilt pouring machines in front of the doser is accurately controlled by the rotary encoder. Proximity switches can be mounted to improve the accuracy of the stopping position of the table. After the metal has been dispensed into the pouring cup, the table rotates to the next position , where the machine starts tilting the mold assembly, emptying the metal into the mold. During the next three to four stations, the casting continues to cool and solidify. By the time the tilting machine reaches the five o'clock position, the casting is solidified completely, and the feeder is solid enough for the casting to be ejected from the mold. The ejected casting (or castings) is gripped by the unloading robot (10) or by a robotic retrieving device and placed on the conveyor. The casting moves to station 13, where the operator (12) checks for any drag marks or other defects. The operator acts quickly to fix mold coatings or other malfunctions of the water-cooling system. The castings then move on the conveyor, dipping into a tank with

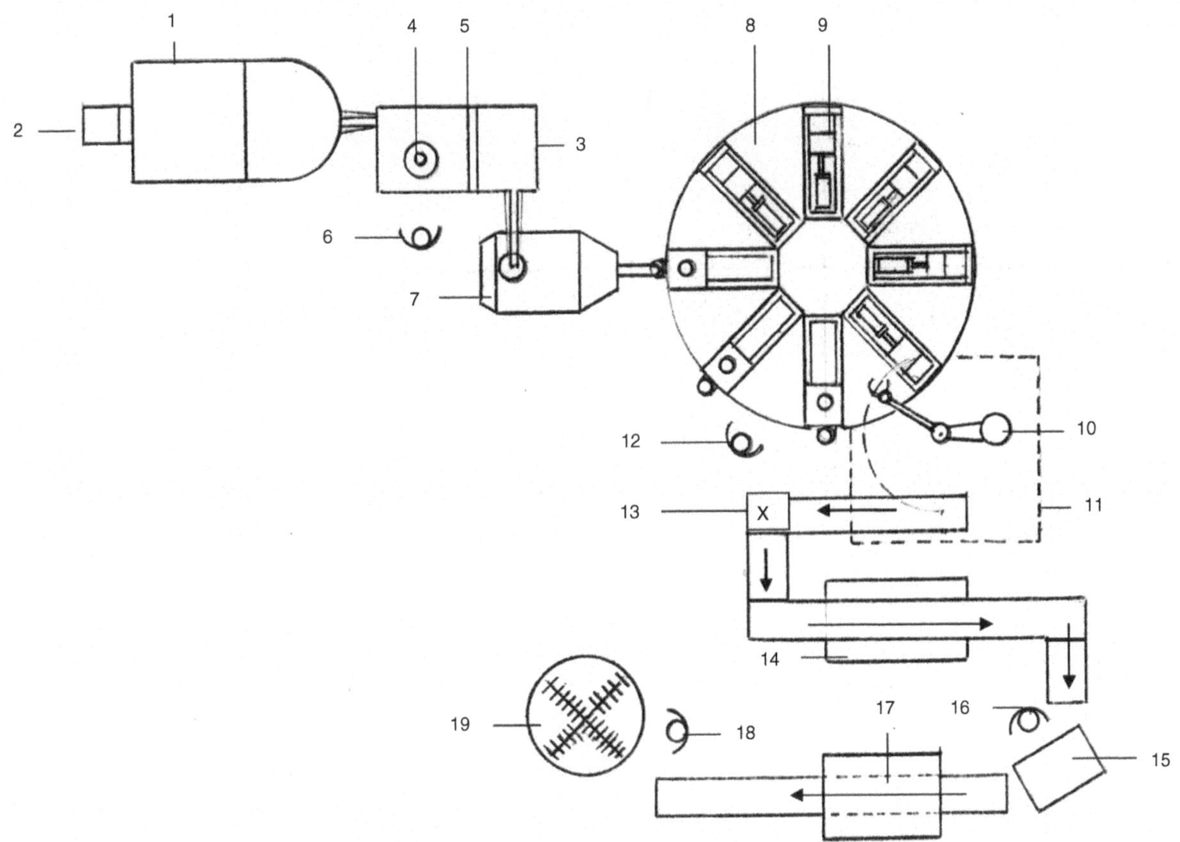

Note: 1. Stack melter 2. Charging skip 3. Degassing and filtration furnace 4. RID unit, rotary impeller degassing unit 5. Filter 6. Melt operator 7. Dosing furnace 8. Turntable 9. Tilt pouring machines (8 total) 10. Casting unloading robot 11. Safety enclosure 12. Turntable operator 13. Hot check station 14. Cooling dip tank 15. Trim station 16. Trim operator 17. X -ray ADT, automatic defect recognition 18. Final inspector 19. Heat treat rack

Fig. 9.10 GPM rotary manufacturing cell

water for cooling (14). The trim machine operator picks up the cooled casting, visually inspects it for any surface defects, and coordinates with the table operator (12) for any immediate corrective action. The operator feeds the casting to the trim saw (also called chop saw) (15), where the feeder and gates are machined off using circular saws. The operator (16) also deburrs the casting if there are any remaining burrs. The castings now move to the x-ray unit, where a pick-and-place unit or a robot manipulates the casting in different orientations to check for any internal defects. The computer matches the level of porosity against the master template previously set up to direct an "accept or reject" decision. The acceptable castings move on the conveyor to the operator (18), who visually inspects the castings before stacking them on a rack for heat treatment (19). A tow motor moves the rack to another location for the heat treatment process. *Aluminum Castings Engineering Guide* (Ref 4) provides more details of the process.

9.6 Cycle Time Planning

The cycle time for each casting is driven by the nature of the product and equipment. The pouring time, tipping speed, and solidification times are influenced by the casting weight, configuration, wall thickness, and feeder sizing. The table movement (indexing) time and the mold closing time are dependent on the equipment design. The time necessary for servicing the mold is dependent on the training and skill of the operator. Experienced operators maintain a rhythm in servicing the mold without interrupting the timing of the cycle.

Planning for the cycle time of a twin piston front brake caliper is provided as an example. The caliper is illustrated in Fig. 9.11. The casting details are:

- Weight of each casting is 2.40 kg (5.3 lb.)
- Number of cavities per mold is 2 (1 each left-hand and right-hand)
- Weight of total metal poured is 7 kg (15.4 lb.)
- Material is aluminum alloy ASTM 357
- Heat treatment is T6

The total cycle time is 240 seconds; solidification time is 96 seconds; ejection time is 144 seconds after the end of tipping; and the next pour starts 240 seconds after the first pour. Table 9.2 summarizes the different events and cycle times. Table 9.3 lists typical values of weights and process cycle times.

9.7 Planning the Number of Tilt Pour Machines on the Table

The number of tilt pour machines needed on the table depends on:

- Annual volumes of the castings
- Number of shifts per day
- Number of hours per shift
- Number of working days in a year

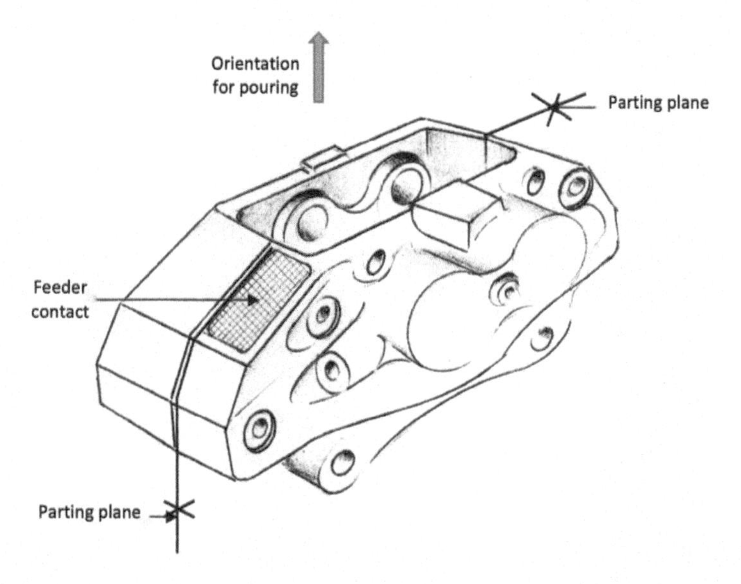

Fig. 9.11 Twin bore front brake caliper

- Uptime of the table
- Number of cavities per mold
- Cycle time
- Scrap

Examples of assumptions taken into consideration may include:

- Annual volume is 300,000 vehicles; 600,000 castings at two front calipers per vehicle
- Number of shifts per day is 2
- Number of hours per shift is 8
- Number of working days in a year is 300
- Uptime of table is 75%
- Number of cavities per mold is 2 (1 each left-hand and right-hand)
- Cycle time is 240 seconds
- Scrap is 4 percent (conservatively)

Calculations:
Number of castings per hour per machine is:

$$\frac{3600}{\text{Cycle time}} \times 2$$

Number of castings per mold times uptime percent is:

$$\frac{3600}{240} \times 2 \times 0.75 = 22.5 \text{ castings}$$

Number of good castings per hour per machine, considering:

$$\text{scrap} \times \frac{100 - 4}{100} = 22.5 \times 0.96 = 21.6$$

Table 9.2 Summary of front brake caliper cycle timing

Activity	Cumulative time, s	0	20	40	60	80	100	120	140	160	180	200	220	240
Pour in pour cup	8	x	...	...	...	...	...	...	...	...	...	...	...	...
Mold tip	28	...	x	...	...	...	...	...	...	...	...	...	...	...
Solidification	132	...	...	...	...	...	...	...	...	...	...	...	...	...
Tip return	160	...	...	...	...	...	...	...	...	x	...	...	...	...
Mold open	163	...	...	...	...	...	...	...	...	x	...	...	...	...
Core pull	168	...	...	...	...	...	...	...	...	x	...	...	...	...
Eject	180	...	...	...	...	...	...	...	...	...	x	...	...	...
Mold servicing	40	...	...	...	...	...	...	...	...	...	...	...	...	...
Mold close	220	...	...	...	...	...	...	...	...	...	...	...	x	...
Index to pouring	230	...	...	...	...	...	...	...	...	...	...	...	...	x
Next pour starts	240	...	...	...	...	...	...	...	...	...	...	...	...	x

Table 9.3 Cycle times and yields of some typical GPM castings (average values)

Part name	Gross weight with feeders, kg (lb)	Rough weight without feeders, kg (lb)	Machined weight, kg (lb)	Yield= $\frac{\text{Rough weight}}{\text{Gross weight}}$, percent	Time, seconds					
					Table index	Pour	Tip	Core pull	Die open	Cycle
Passenger car rear knuckle	3.45 (7.6)	2.40 (5.3)	2.2 (4.9)	69	8	6+/- 1	20+/-2	50	120	200+/-5
Truck rear knuckle	4.3 (9.4)	3.1 (6.8)	2.9 (6.4)	72	8	6+/-1	20+/-2	60	120	220+/-5
SUV rear knuckle	5.8 (12.7)	4.3 (9.4)	3.0 (6.6)	74	8	7 +/-1	20+/-2	70	120	240+/-5
SUV upper control arm	5.3 (11.7)	3.4 (7.50)	3.0 (6.6)	64	8	7+/-1	23+/-3	80	130	220+/-5
SUV differential carrier	13 (29)	8.3 (18.4)	8.0 (17.6)	64	8	9+/-2	16+/-2	...	120	270+/-5

Number of castings per day per machine with 16 hours of work is $21.6 \times 16 = 345.6$

Number of castings per machine in one year of 300 days working is $345.6 \times 300 = 103{,}680$

Number of machines needed on the table is:

$$\frac{\text{Annual volume of castings}}{\text{No.of casting per machine}} = \frac{600,000}{103,680} = 5.78$$

or approximately 6 machines.

A rotary table with six tilt pour machines will meet the required annual volume.

9.8 Sizes of Tilt Pour Machine and Table

The machine size is fixed based on the platen dimensions and the nearest suitable standard machine. If there is no suitable configuration available, a customized machine with a suitable platen size needs to be designed.

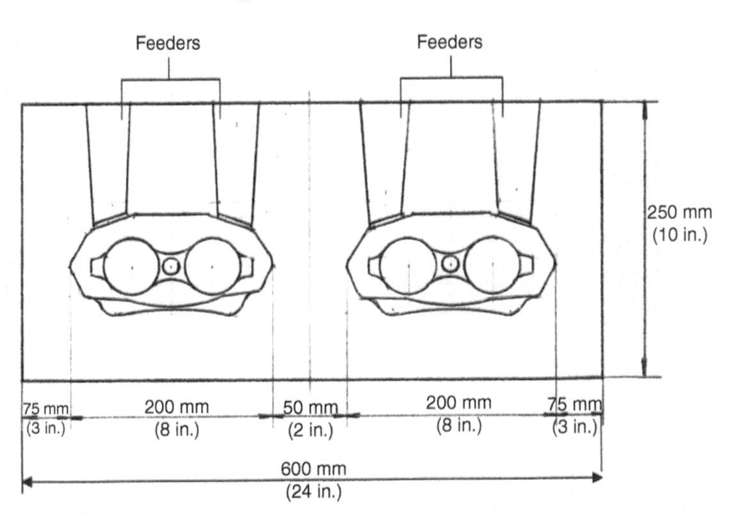

Fig. 9.12 Tooling layout for platen size confirmation

The platen size is fixed based on the cavity layout. Figure 9.12 illustrates a layout with a platen size of 600 by 250 mm (24 by 10 in.). Figure 9.13 provides the sizes of the tilt pour machines suitable for a width of 600 mm (24 in.). The figure shows the length of the machine as 2695.58 mm (106.13 in.). With a central pedestal for the cooling manifolds and other connections of 200 cm square (or 200 by 200 cm) for the electrical box, the diameter of the table is 7.4 meters (24 feet).

9.9 Improvement of Turntable Uptime

The uptime of the turntables influences the output and return on investment. Ready-to-use tilt pour machines with spare molds mounted and being heated facilitate the ability to switch the tilt pour machines with minimum downtime and loss of productivity. Otherwise, if the table must be stopped to switch molds, the quality of castings could be affected because the molds on the table will cool. Frequent interruptions have a negative impact on productivity and quality.

It is essential to ensure that the limit switches on the tilt pour machines are set and programmed properly, in addition to ensuring that there are no water leaks in the cooling circuits. These requirements are addressed by a well-organized mold reconditioning shop located near the turntables. Figure 9.14 shows the tasks carried out by a mold reconditioning shop.

Figure 9.15 is a schematic of a mold reconditioning shop. The tilt pour machine, along with the mold, is brought to the mold reconditioning shop using a tow motor. The mold is dismantled from the machine using the bridge crane (1) and delivered to station 4. The tilt pour machine (or tipper) is moved to station 11, awaiting the mold. The mold halves are sand blasted or blasted with dry ice at station 5 to remove the old coating. The molds are then preheated to 230 °C (450 °F) in the oven (6). The mold halves are then moved to station 7, where a manipulator or a tilt table orients the mold halves at an angle suitable for spray coating. They are spray coated at station 8 by an operator (using

SPECIFICATIONS

Clamping Pressure	28,875 lbs.	(13,125 kg.)
Max. Mold Size	24 x 26 in.	(61 x 66 cm.)
Max. Mold Weight	900 lbs.	(410 kg.)
Ram Stroke	22 in.	(55.8 cm.)
Daylight Opening	12 in.	(30.5 cm.)
Pressure Height Center	8.5 in.	(21.6 cm.)
Machine Weight	3,500 lbs.	(1,588 kg.)
Tilt Speed (Min.)	7 sec.	

Tie Bars and Extensions Available on Request Only.

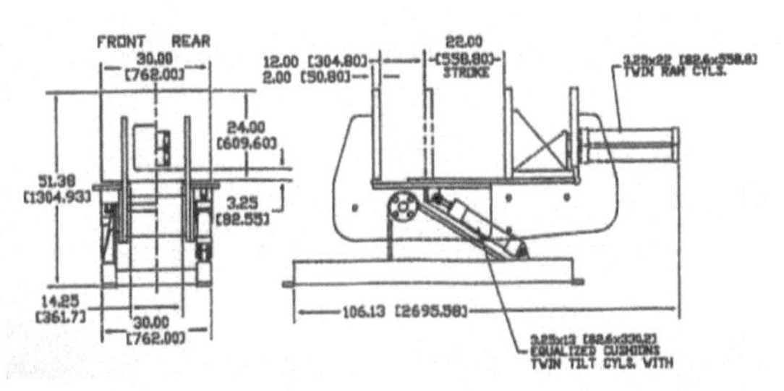

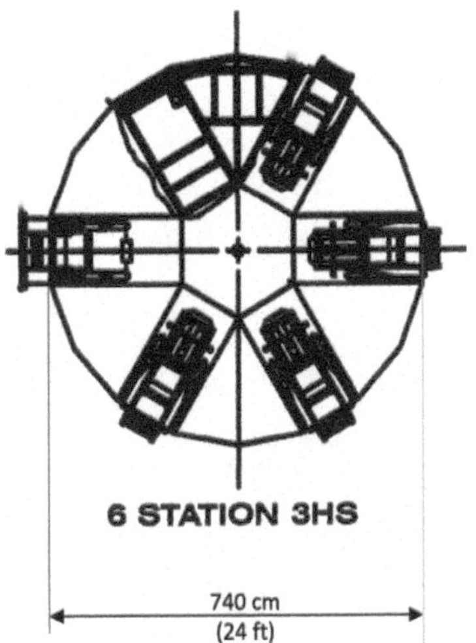

Fig. 9.13 Tilt pour machine and turntable sizing. Reprinted with permission from Ref 1

mold conditioner). The molds are moved to station 10 by another manipulator (or tilt table) to orient them for mounting back on the machine. At station 11, another operator mounts the molds on the machine, connects the water lines, and mounts the electrical limit switches.

The molds are moved to the test stand at station 12, where they undergo dry cycling with the water lines turned on. The molds are tested for proper process sequencing and the absence of any water leakage. At station 13 they are approved to be ready for movement back to the turntable. Burners are introduced (sandwiched) in between the mold halves to keep them hot and ready for use at any time they are needed on the table to replace any malfunctioning equipment.

9.10 Semipermanent Gravity Molding

Gravity permanent molded castings produced in a metal (permanent) mold with expendable sand cores set in each mold are called semipermanent molded castings. Castings that feature passages for handling fluids need sand cores that disintegrate at the casting temperature, facilitating easy cleaning. These castings are usually thin-walled and are required to be leakproof for oil or water passages in most cases. Table 9.4 lists some of these common semipermanent molded castings.

Figure 9.16 illustrates castings produced by the semipermanent mold process. Figure 9.16 (a) is the front cover of an engine; Fig. 9.16 (b) is a differential carrier; and Fig. 9.16(c) is an intake manifold. The type of sand cores used range from shell-phenolic resin-coated sand to furan resin, no-bake and furan resin, and sulfur dioxide gas-cured.

The tilt pour machines used are similar to those illustrated in Fig. 9.6, with rotary manufacturing cells as shown in Fig. 9.8 and 9.9.

9.10.1 Postcasting Operations in Semipermanent Molding

Post casting operations essentially consist of decoring, degating, deheading, flash grinding, deburring, x-ray inspection, premachining, testing for leaks, and final heat treatment. Figure 9.17 illustrates the postcasting operations for decoring, feeder removal, deburring, and testing for leaks.

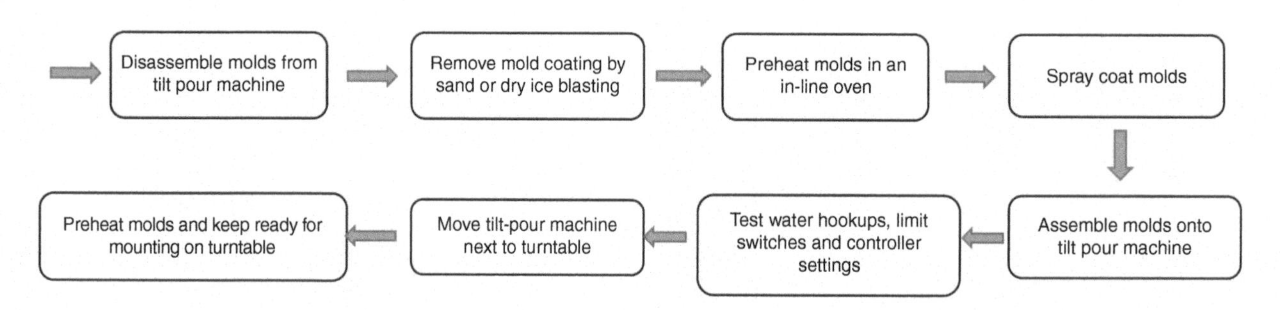

Fig. 9.14 Mold reconditioning activities

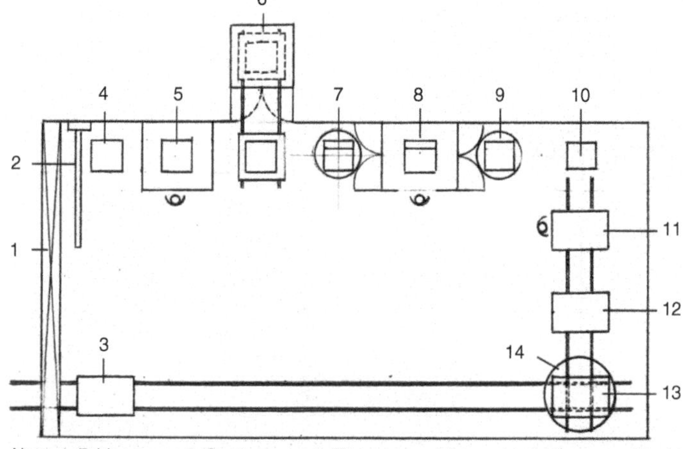

Note: 1. Bridge crane 2. Gantry crane 3. Tilt-pour machine and mold for reconditioning 4. Disassembled mold 5. Mold in sand or dry-blast cleaning 6. Mold in preheating oven 7. Mold manipulator of orientation 8. Mold being spray coated 9. Mold manipulator for re-orientation 10. Mold ready for mounting 11. Mold being assembled onto tilt-pour machine 12. Mold and tilt-pour machine under testing for connections and settings 13. Certified mold on tilt-pour machine. 14. Turntable for orientation

Fig. 9.15 Mold reconditioning shop schematic

Table 9.4 Common castings produced by semipermanent mold casting process

Casting	Fluid in passage	Comments
Intake manifold	Intake air	Also produced in greensand molds with cores and core-molds
Cylinder head	Cooling water	Produced in sand molds with cores and core molds, and also in low pressure permanent molding
Engine block	Cooling water	Also produced in core molds and in low pressure permanent molding, as well as in die casting with design modifications
Crank case	Oil	Also produced in die casting with design modifications to suit
Engine front covers	Water	Produced in no-bake sand molds also
Differential carrier	Oil	Some designs produced in gravity permanent molding and die castings with design modifications to suit
Transmission pump housings	Oil	...
Oilers	Oil	...
Water pump	Water	Produced in green sand also
Hydraulic pump bodies	Oil	...
Hydraulic valve bodies	Oil	Some designs produced in die castings
Oil pans	Oil	...
Control arms	None	Hollow design for increased stiffness

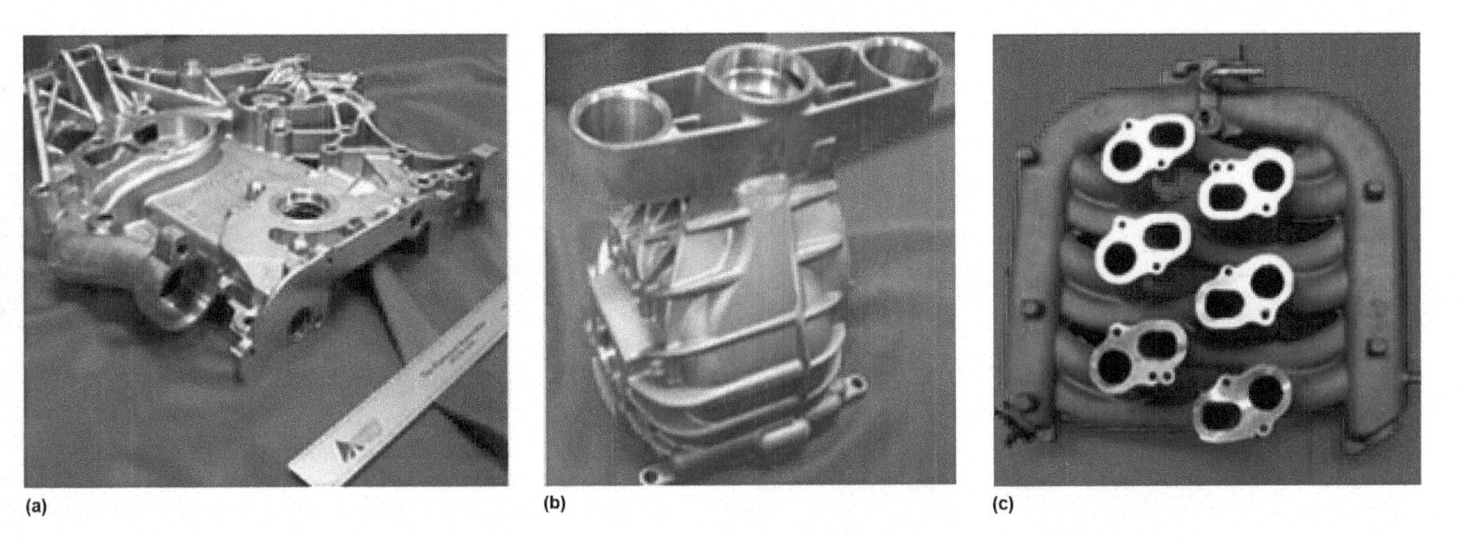

Fig. 9.16 Examples of semipermanent molded castings, (a) front cover of an engine, (b) differential carrier, (c) intake manifold. Reprinted with permission from Ref 3

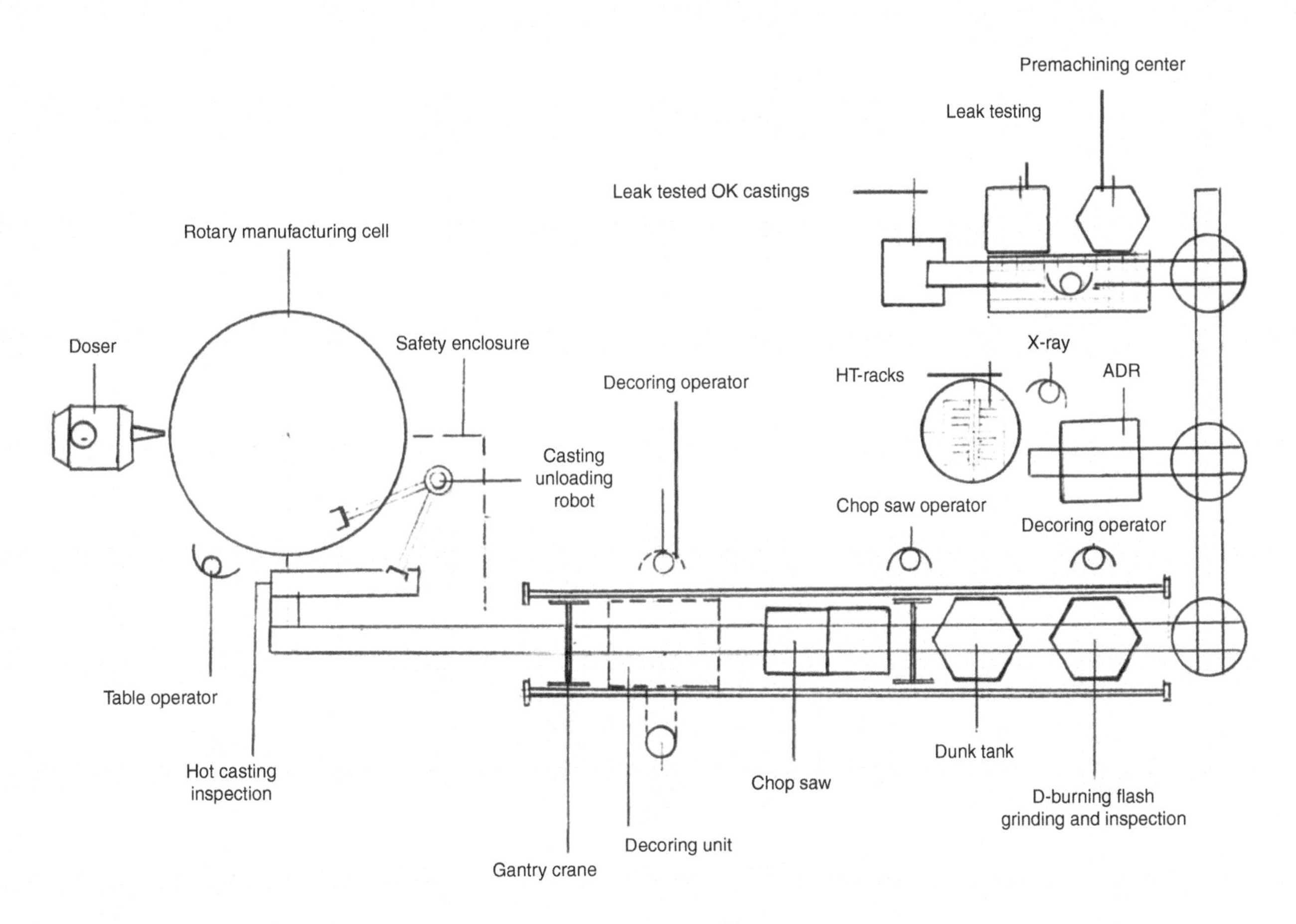

Fig. 9.17 Postcasting operations for castings produced using the semipermanent mold process

Castings unloaded from the turntable pass through a hot-check station to inspect for any major defects or indications that would require a mold coating touch-up. A gantry crane picks up the casting and moves it to a decoring station. Decored castings move to a chop saw station where the feeders are removed using a saw. The castings are then moved to a dunk tank and submerged in water to cool. (This step is needed for operator safety if deflashing and deburring are done manually.) The castings move to the deburring station, where the burrs left from sawing off the feeders and other sharp edges remaining from the flash are removed.

The castings move through an x-ray unit with automatic defect recognition (ADR) capability, and they are then stacked on the racks for heat treatment. Castings that must be leakproof pass through a premachining center on to a station where they are tested for leaks and approved. Figure 9.18 illustrates a decoring machine for an aluminum cylinder head casting.

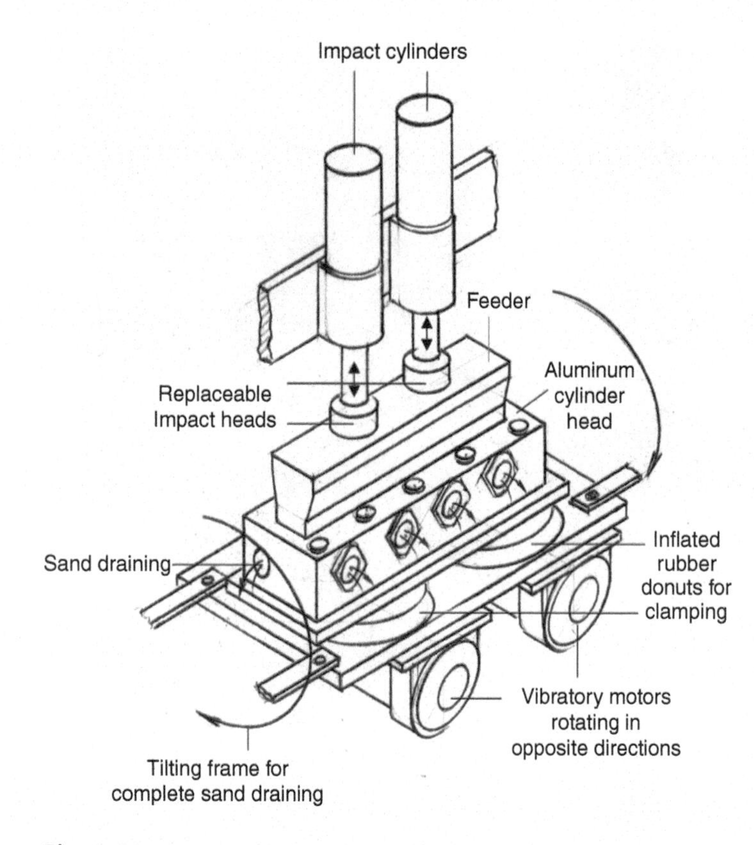

Fig. 9.18 Aluminum cylinder head decoring machine

9.10.2 Decoring Machines

The decoring machines have features such as:

- A soundproof cabinet for isolating the noise. The cabinet is connected to an exhaust unit for dust control.
- A sturdy frame to mount the vibrating motors to generate high acceleration to break the cores. The frame is designed for tilting to empty the core sand.
- A fixed frame that supports the high-frequency hammerheads with replaceable heads for separating the core from the casting. The hammers are capable of amplitudes up to 36 mm.
- A pair of air bags to hold the casting against the hammerheads
- Four isolators, from which the frame is suspended
- A trough to collect the core sand and a conveyor to haul the sand out of the cabinet
- A robot for automatic loading and unloading, for large volume castings of capital foundries

REFERENCES

1. C.M.H. Manufacturing, Hall Permanent Mold Machines, Lubbock, Texas
2. Reliable Castings, Cincinnati, Ohio
3. General Aluminum Manufacturing Company, Ravenna, Ohio
4. J. Nath, *Aluminum Castings Engineering Guide*, ASM International, 2018

Casting Equipment Engineering Guide
Jagan Nath
https://doi.org/10.31399/asm.tb.ceeg.t59370187

Copyright © 2023 ASM International®
All rights reserved
www.asminternational.org

CHAPTER 10

Low-Pressure Permanent and Semipermanent Mold Casting and Counter-Pressure Casting

10.1 Low-Pressure Permanent and Semipermanent Mold Casting

THE LOW-PRESSURE PERMANENT MOLD (LPPM) and semipermanent mold (LPSPM) casting processes are versatile, and they meet the quality requirements of a variety of high-integrity, large-sized, and thin-walled aluminum castings for various industries. A nonturbulent mold fill is critical for ensuring oxide-free metal in the casting. While this is achieved in GPM by tilt pouring, the nonturbulent mold fill in LPPM and LPSPM is achieved by filling the mold from the bottom and controlling the fill pressure to avoid turbulence.

The application of pressure during solidification achieves two objectives:

- Reduces the size of gas pores
- Improves the ability to feed the shrinkage through partially solidified sections (mushy zones)

The low-pressure permanent mold casting process uses steel molds (also known as low-pressure permanent molding) repetitively for each casting cycle, similar to the gravity permanent molding process. The low-pressure semipermanent molding process uses steel molds with expendable sand cores, like the gravity semipermanent molding process.

The process applications range from decorative automotive wheels to critical safety components for automotive chassis (substructure) and complex castings such as engine blocks and cylinder heads. Over 90% of automotive wheel styles manufactured by original equipment manufacturers (OEMs) are produced using low-pressure permanent molding.

Machine tool manufacturers use the LPPM process to produce bed frames and machine tables. Furniture manufacturers use this process for large tabletops and frames. Figure 10.1 illustrates a decorative automotive wheel and a rear suspension trailing arm produced using the low-pressure permanent mold process.

Geometric factors such as selection of parting planes, drafts, shrink factors, dimensional tolerances, and selection of locating and clamping pads are common between the GPM and LPPM processes.

Fig. 10.1 Decorative automotive wheel and a rear control arm produced by LPPM

10.1.1 Successful Low-Pressure Permanent and Semipermanent Molding Products

Many products have been developed and launched using both LPPM and LPSPM processes. Millions of these castings are working successfully on very many vehicles, proving the capability and versatility of these processes. Table 10.1 lists some of the successful products launched using LPPM and LPSPM processes.

10.1.2 Low-Pressure Permanent Molding Schematic

Low-pressure molding machines are less expensive than die casting machines because of the lower pressures and less expensive automation that is needed to keep up with the rate of production.

Figure 10.2 is a schematic of an LPPM machine, highlighting the major features (Ref 1). Figure 10.2 features a four-post machine

Table 10.1 Typical LPPM and LPSPM products launched

Product	Process LPPM	LPSPM	Comments
Wheels	X	...	Development work with hollow spoked LPSPM done. Not launched in production
Front knuckles for passenger cars and SUVs	X	X	Limited LPSPM launches for SUVs Squeeze cast and cast-forge knuckles in market
Upper and lower control arms	X	X	Upper control arms in LPPM. More hollow lower control arms in LPSPM production. Fabricated, squeeze cast and cast-forge control arms in use
Engine blocks	...	X	Many in production. Also produced in core-molds
Cylinder heads	...	X	Many in production. Also produced in core-molds
Crank cases	X	X	Crank cases also produced in GPM, SPM and die casting
Engine front covers	X	X	Castings produced in GPM, SPM and die casting also
Cross members and cradles	X	X	Cored cast end nodes with rolled mid-section also in use
Differential carriers	...	X	Differential carriers in GPM and (high pressure) die casting in use
Truck transmission housings	X	...	...
Intake manifolds	...	X	Also produced in GPM and core sand molds
Turbocharger pump bodies	...	X	
Brake calipers	X	X	Produced in GPM and SPM also
Valve bodies	X	X	Produced in GPM and SPM also
Transmission housing of farm equipment	X	...	...
Air compressor bodies	X	X	Also produced in GPM and SPM
Machine tool beds and tables	X	X	Produced in no-bake and vacuum molding also
Wind power equipment	X	X	...
Furniture tables and benches	X	...	Produced in no-bake sand molding also

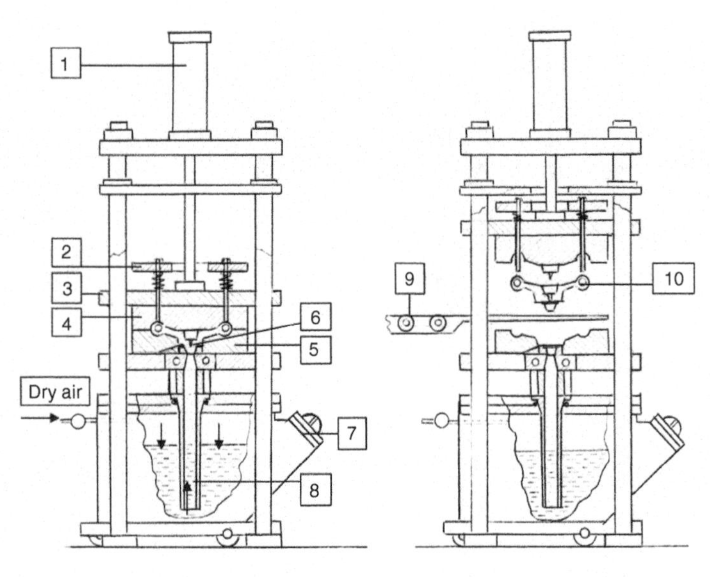

Note: 1. Clamping cylinder, 2. Bump ejector plate, 3. Top platen, 4. Top mold, 5. Bottom mold, 6. Screen hold down pin, 7. Holding furnace, 8. Stalk tube 9. Casting receiver, 10. Metal filling and slag skimming door, 11. Ejected casting

Fig. 10.2 Low-pressure permanent mold casting machine. Source: Ref 1

with a sealed holding furnace that can be moved between the posts for maintenance. The four posts are connected by three platens. A hydraulic cylinder (1) mounted on the top fixed platen serves to move the middle platen (3) carrying the top half of the mold (4). The bottom platen carries the bottom half of the mold (5). A stalk tube (8) with one end submerged in the metal in the holding furnace (7) is connected to the bottom mold, as shown in the figure. Dry air injected into the sealed furnace pressurizes the top of the metal meniscus. The metal rises in the stalk tube (as shown by the arrow marks) into the mold cavity. A steel screen introduced into the stalk tube top prevents oxides from entering the mold cavity. After the mold is filled, pressure is increased to feed the shrinkage, and this increased pressure is maintained until the solidification of the casting is complete. The hydraulic cylinder pulls the middle platen and the top of the mold to bump up against the bumper plate (2), resulting in the casting (11) being ejected onto a casting receiver (9). The operator hoists the casting off the casting receiver and onto the perforated platform of a water dunk tank. The casting cools in the air over the platform for a short duration, and during this time, the machine operator checks the casting for any surface blemishes. The operator records areas that need a paint touchup before closing the mold for the next shot. The casting is lowered and dipped into a water tank for further cooling. The cooled casting is hoisted out and hung on an overhead conveyor. A robot can be programmed to reach out and move each casting to the next operation. The cooled casting moves along to be desprued and deburred before moving to an in-line automated x -ray unit (automated defect recognition, ADR) to have its internal soundness checked. Surface defects are checked visually by an operator before the castings are stacked on racks for heat treatment.

10.1.3 Low-Pressure Permanent Molding Machines for Wheel Manufacturing

LPPM machines used for producing wheels have tooling in six pieces, including the top and bottom molds and four sides, as illustrated in Fig. 10.3 (Ref 1). In addition to the top

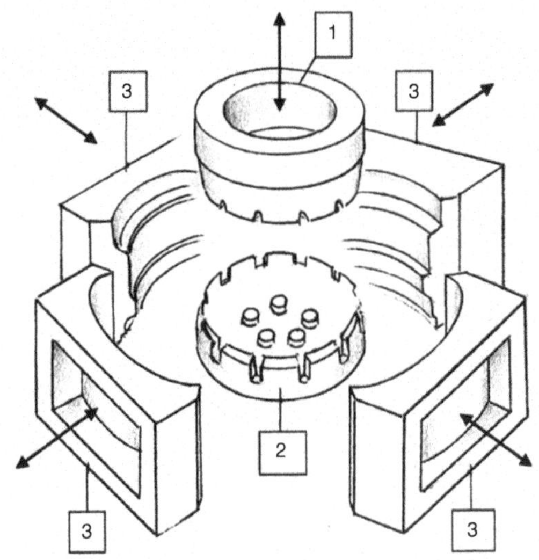

Note: 1. Top mold, 2. Bottom mold insert, 3. Side mold pull backs

Fig. 10.3 Wheel mold tooling slides. Source: Ref 1

hydraulic cylinder, the machine has four hydraulic cylinders which move the side molds out before the casting is ejected. Some facilities use two- or three-sided tooling pieces instead of the traditional four-sided pieces. Figure 10.4 illustrates this sequence of operations.

10.1.4 Low-Pressure Permanent Molding Machine Details

Figure 10.5 shows a line of LPPM machines (Ref 2). The furnaces with a fail-safe seal for the fill port can be seen, along with the mainframes of the machines. Figure 10.6

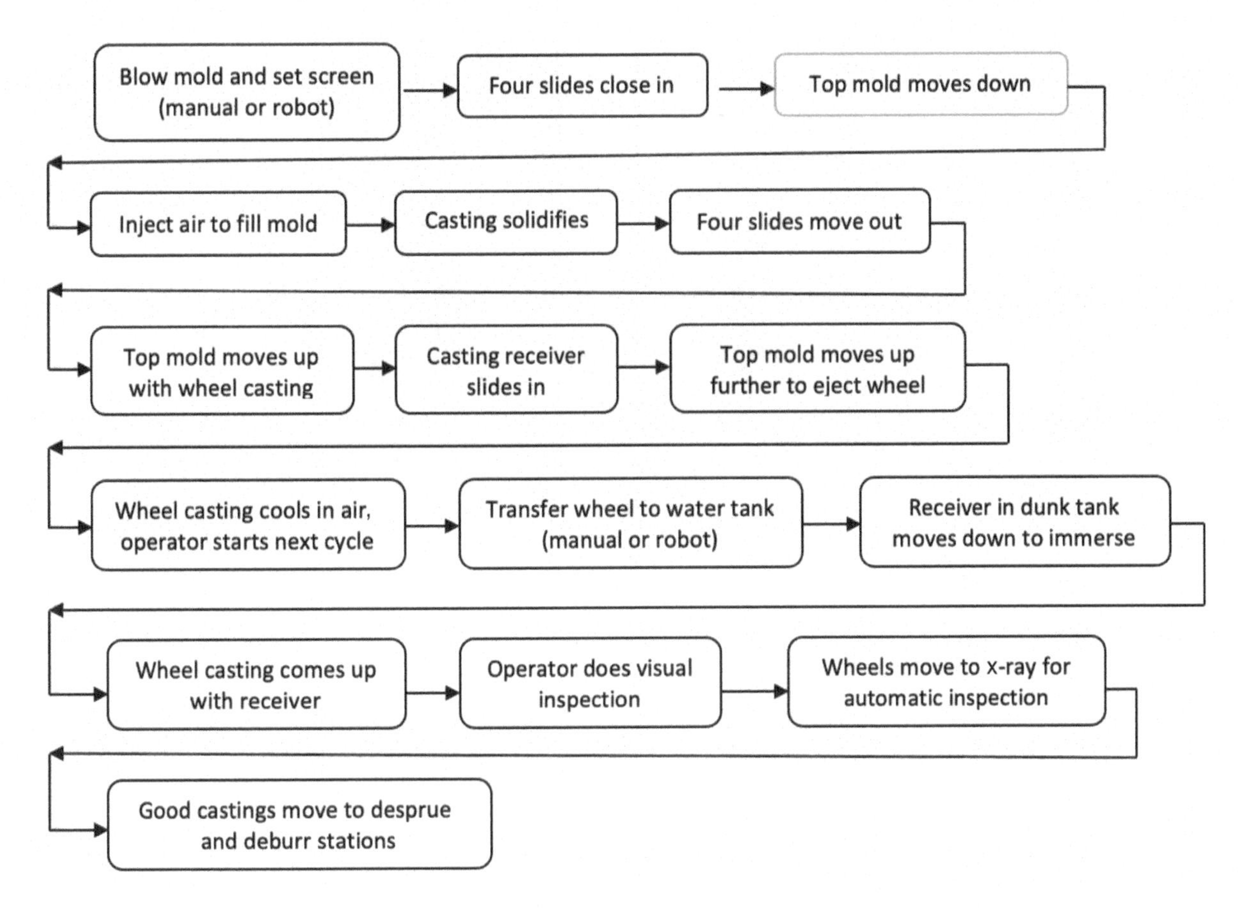

Fig. 10.4 Wheel casting process schematic

Fig. 10.5 A row of LPPM machines. Reprinted with permission from Ref 2

Fig. 10.6 LPPM machine closeup. Reprinted with permission from Ref 2

Fig. 10.7 LPPM wheel casting machines, with shared robot. Reprinted with permission from Ref 2

(Ref 2) shows the movable furnace at the bottom of the machine. The furnace is mounted on a scissor lift, which facilitates the emptying of the furnace when it moves out from the machine frame. Figure 10.7 (Ref 2) shows a layout where a robot serves two-wheel casting machines. Figure 10.8 (Ref 3) shows an LPPM machine designed to switch the crucible furnaces for continuous operation. When the first crucible furnace has been emptied, the second crucible furnace moves in, ready for production. This interchangeable furnace design reduces downtime.

10.1.5 Pressure Ramps

The design and control of the pressure ramp is a critical process element. The programmable controller (PLC) and the valves must react quickly. Pressurization is divided into three stages, as shown in Fig. 10.9 (Ref 1).

- Stage 1 — controls the rate at which the metal rises in the stalk tube
- Stage 2 — controls the rate for nonturbulent fill of the mold cavity
- Stage 3 — pressure maintained ensures the feeding of shrinkage

Stage 1: As dry air enters the sealed furnace, the top of the metal is depressed, and the metal rises in the stalk tube. The rate of pressurization in stage 1 controls the rise of the liquid metal in the stalk tube, up to the entry of metal into the mold cavity. An increase that is too slow inhibits the metal flow, causing the melt temperature to drop and extending the cycle time. An increase that is too fast causes the metal to surge into the mold cavity before the intended onset of the second stage. The stage 1 pressurization is influenced by the level of metal in the furnace. The PLC is designed to compensate for the change in the level of the metal approximately every five pours. Normally a value of 0.006895 Mpa/sec (1 psi/sec) is used for programming.

Stage 2: The second stage pressurization controls the filling of the mold cavity. The bottom mold cavity has a thermocouple embedded in it, as shown in Fig. 10.9. As the metal touches the thermocouple, the second stage of the mold cavity fill is initiated. The rate of pressurization in the second stage controls the tranquility or the nonturbulent flow into the mold cavity. Turbulent flow results in the creation of oxides, which must be avoided. A fill rate that is too slow may result in nonfills (or leakage in wheels). Normally, a slow increase of 0.0006895 MPa/sec (0.1 psi/sec) is a good start in process development. The PLC registers the last value and corrects for the drop in liquid level in the furnace based on the number of wheels cast or castings that are produced.

Stage 3: The third stage impacts the feeding of the shrinkage through the application of pressure from the metal inlet to the casting or sprue. The casting profile is engineered for the solidification to start from the area farthest from the sprue and to proceed progressively toward the sprue. The pressure applied in the third stage feeds the potential shrink-prone areas that are the thickest and closest to the sprue.

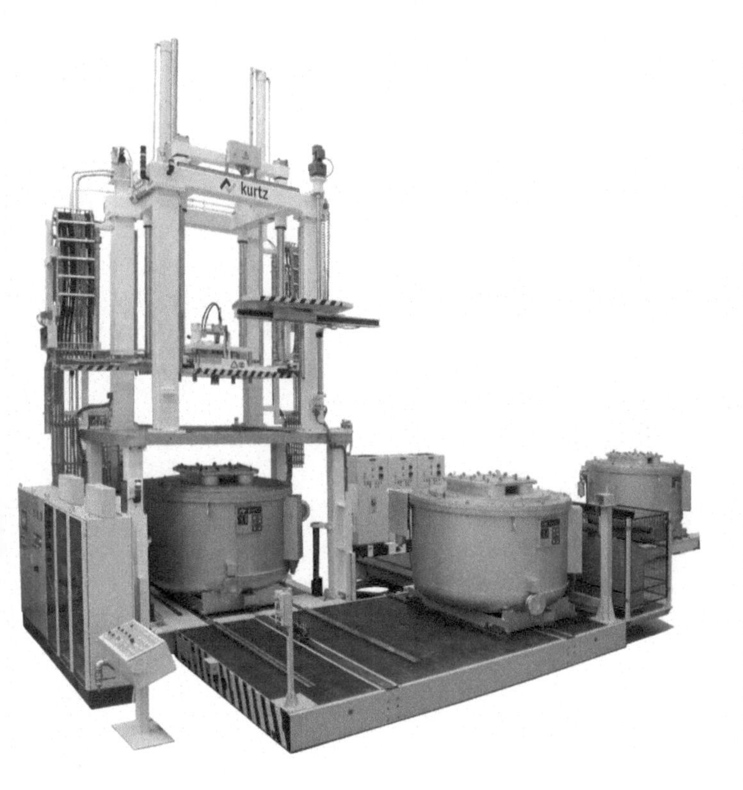

Fig. 10.8 LPPM machine with interchangeable furnaces. Reprinted with permission from Ref 3

The rate of pressurization is fast, at about 0.01379 MPa /sec (2 psi/sec) until a pressure value of 0.1034 MPa (15 psi) is reached. This pressure is maintained until solidification is complete. The air pressure is then dropped quickly as can be seen in Fig. 10.9. The casting is then ready to be ejected onto the receiver, and this process sequence is detailed in Fig. 10.4. Although Fig. 10.9 shows the start of the second cycle immediately after depressurization, in practice, there is a time lag of a couple of seconds for blowing out the mold and setting the screen before the mold is closed and made ready for the next cycle to begin.

10.1.6 Oxide Entrapment

The liquid metal moves up the stalk tube when the top of the metal in the furnace is pressurized. As the air exits the furnace at the end of the cycle, the metal in the stalk tube falls rapidly. The metal movement up and down and its exposure to air creates oxides (dross) that stick to the walls of the stalk tube. When the furnace is pressurized for the next cycle, these oxides would be carried into the mold cavity, without a screen or a filter set at the top of the stalk tube.

Any oxide entrainment in the casting is unacceptable. Decorative wheels and wheels that are chrome plated cannot have any blemishes in their finish. Wheels also need to be able to withstand air pressure during service, and oxides can cause leakage.

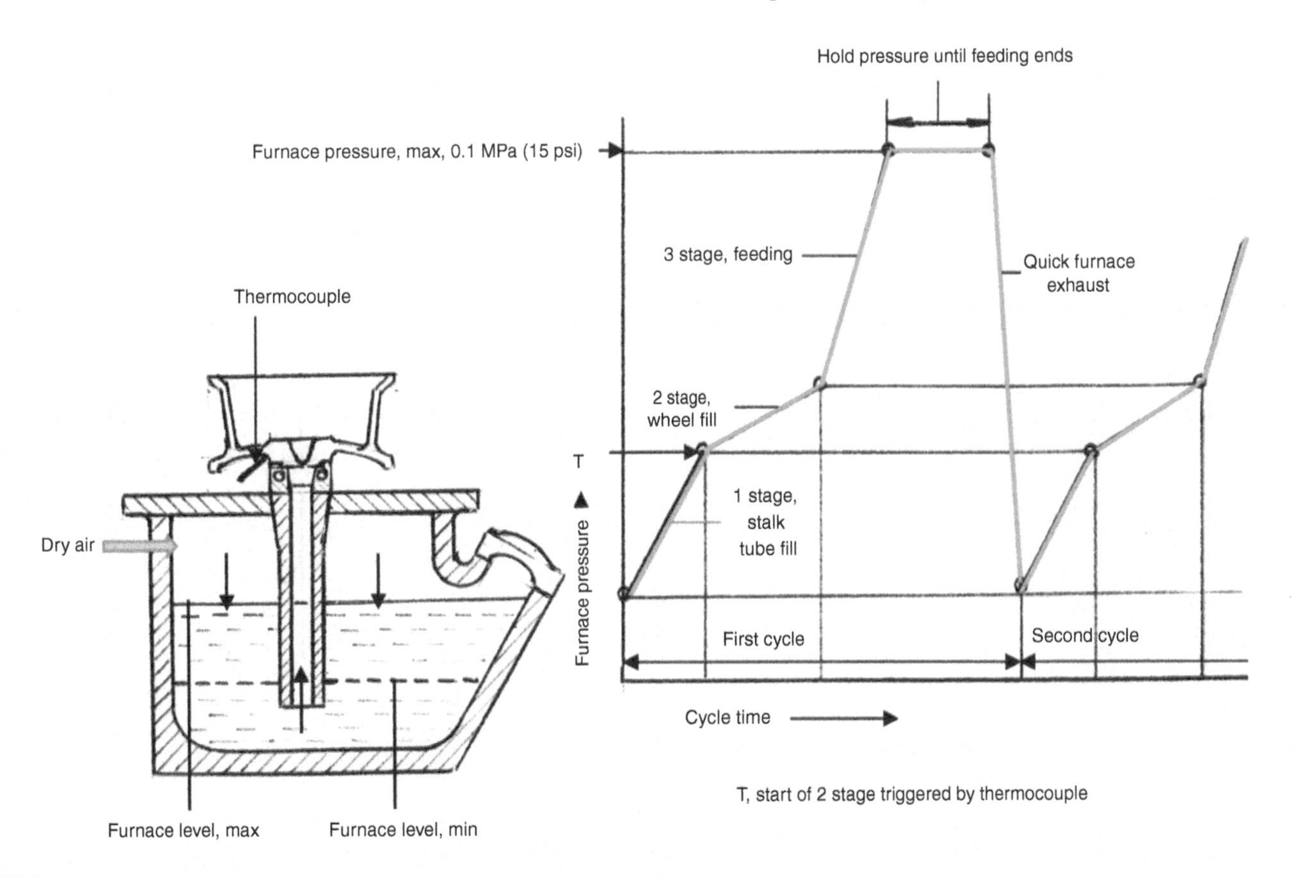

Fig. 10.9 Pressure ramp stages

Structural parts that are subjected to high stress, fatigue, and impact loads also cannot have any oxide inclusions.

Figure 10.10(a) illustrates the depression of the metal in the furnace as dry air enters the sealed furnace. The metal rises up the stalk tube into the mold cavity. After the stage 3 increase in pressure, when air exits the furnace, the metal level falls from the stalk tube, leaving potential oxides sticking to the sides of the stalk tube, as shown in Fig. 10.10(b). If there is no screen or filter at the entry to the mold cavity, there would be a possibility for the oxides to be carried through with the metal, into the mold cavity in the next cycle.

The potential for oxide entrainment may be prevented by including a perforated screen or a filter at the point where the metal enters the mold cavity. Figure 10.10(c) shows a thin steel disc with small, perforated holes of about 1.5 mm (0.06 in). These are coated with tin or zinc to prevent them from rusting. A pin is designed in the tooling to hold the screen in place so it does not lift under pressure from the metal. Figure 10.10(d) shows a conical cup made of steel wire mesh. These cups are pressed into the tapered sprue and held tight by friction. Fiberglass conical cups can also be used. It is essential that the fiberglass screens do not disintegrate and cause oxide and gas inclusions. This is especially critical for wheels that are chrome plated.

10.1.7 Multiple Sprues

Larger machines use multiple cavities of the same casting (for example, knuckles, brake calipers, or turbocharger compressor pump bodies), and they use most of the platen area to achieve higher productivity. Each cavity is designed with a sprue and stalk tube. Four to six stalk tubes that dip into one common holding furnace are designed in such cases. Figure 10.11 illustrates a multiple-cavity layout for six knuckles with six stalk tubes.

Large thin-walled castings such as crossmembers require multiple metal entry points to fill the casting. In these cases, multiple sprues (usually four) are used, or two to three sprues with a runner and gating system, as shown in Fig. 10.12 (Ref 1), are used. Automation in screen setting saves cycle time when multiple sprues are used.

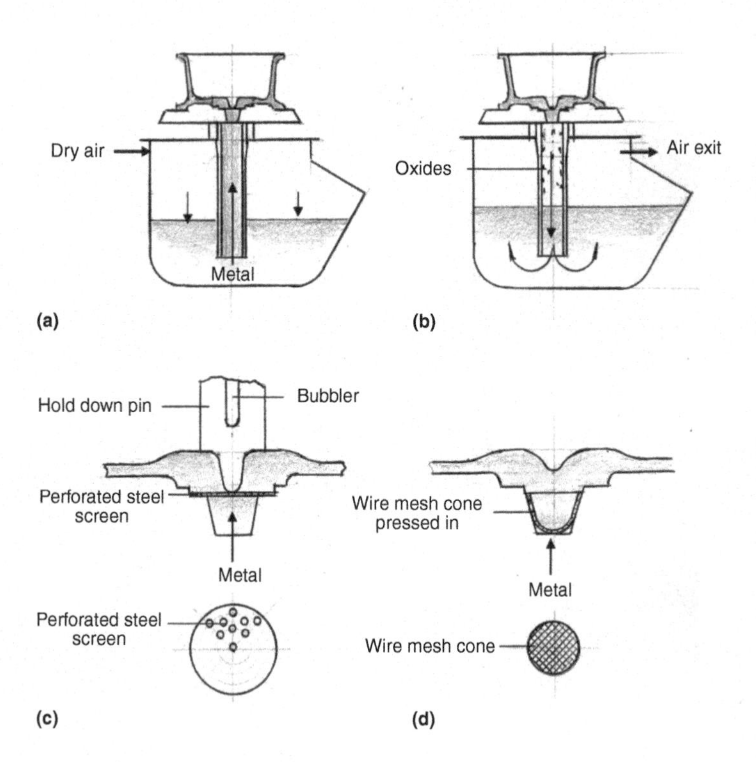

Fig 10.10 Oxide formation and prevention; (a) depression of metal as dry air enters sealed furnace, (b) potential oxides sticking to sides of stalk tube, (c) pin holds thin steel disc in place, (d) steel wire mesh conical cup

10.1.8 Additional Features of Low-Pressure Permanent Molding Machines

10.1.8.1 Quick Stalk Tube Changeover

Stalk tubes need to be replaced when they are cracked or when the oxide buildup is excessive. One method for reducing the

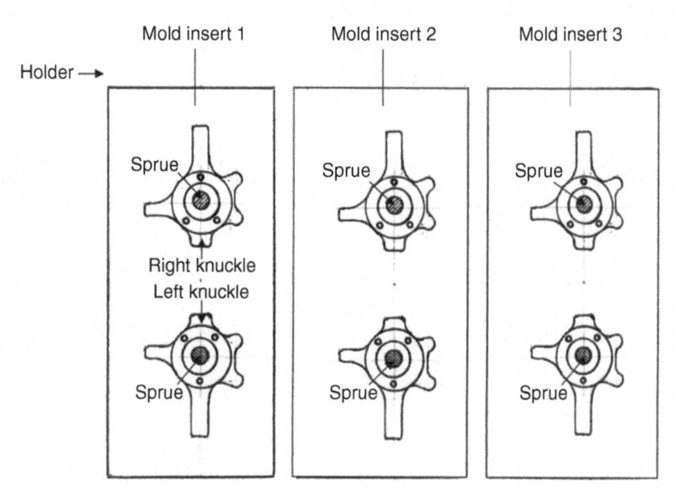

Fig. 10.11 Multistalk tube layout

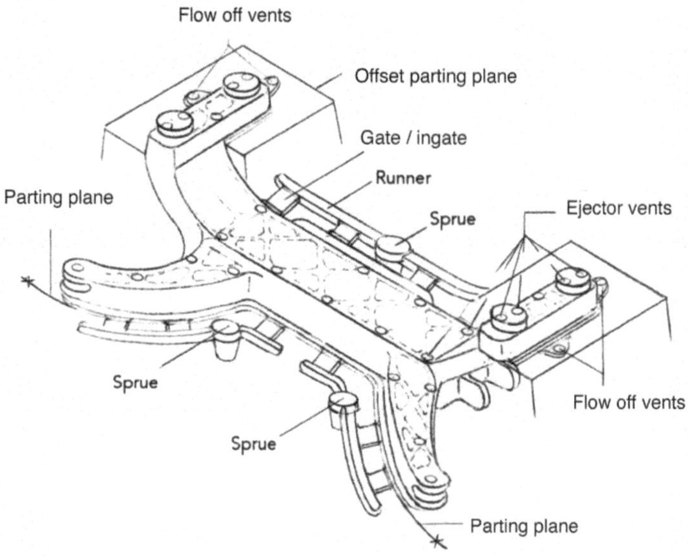

Fig. 10.12 Multiple sprues for a crossmember casting. Source: Ref 1

changeover time is to move the furnace out of the machine and lower it to remove the old stalk tube and replace it with a new one, using a steel guide tube. Proper sealing below the stalk tube flange is important to prevent excessive compression on the stalk tube. Sealing above the stalk tube is critical to prevent leakage of the air used to pressurize the furnace.

10.1.8.2 Automatic Opening of the Furnace Door

The furnace door must be airtight to prevent any leakage of the pressurizing air and to prevent any metal from getting out through improper sealing. Normally, a mechanical toggle mechanism is provided for operation of the door with a fail-safe locking mechanism (see Fig. 10.6). Some manufacturers offer an automatic feature for opening and closing the door that saves time and reduces the loss of mold temperature. Other manufacturers use two furnaces (Fig. 10.8), to be switched alternately.

10.1.8.3 Ability to Tilt the Upper Mold

Molds must be periodically inspected while the casting is unloaded onto the receiver. If any anomaly is observed, or any mold coating wear is noticed, the operator needs to spray the coating during the touch-up operation. The top of the mold, constrained by the four columns, does not permit a complete view for inspection or touch-up. The top mold tooling is specially designed with a provision for tilting. The ability to tilt the top of the mold is of special importance to medium- and large-sized castings such as control arms and crossmembers. Tilting is of less importance to wheel production, where the top mold is cylindrical and not profiled with deep ribs or deep pockets.

Some machine manufacturers offer the ability to tilt the top mold, as illustrated in Fig. 10.13. (Ref 2). Other manufacturers install a mirror that can be swung into position to provide the operator with a good look at the top mold.

10.1.9 Post-Casting Processes

There are several process steps that take place following casting and degating before a casting is ready to be machined. Operations such as deflashing and deburring are required for finishing, and operations such as x-ray inspection, straightening (as required only for cradles and cross members) and stacking control on heat treatment racks are required to ensure high casting integrity and close dimensional control. The straightening process is addressed later in section 10.1.10.4. Figure 10.14 summarizes the sequence of processes, which include:

- Desprueing, or sprue removal
- Deflashing, or flash removal
- Deburring, or burr removal
- Automated x-ray inspection to verify integrity
- Solution heat treatment
- Aging heat treatment
- Quality checks before final approval

Fig. 10.13 Top mold tilting feature. Reprinted with permission from Ref 2

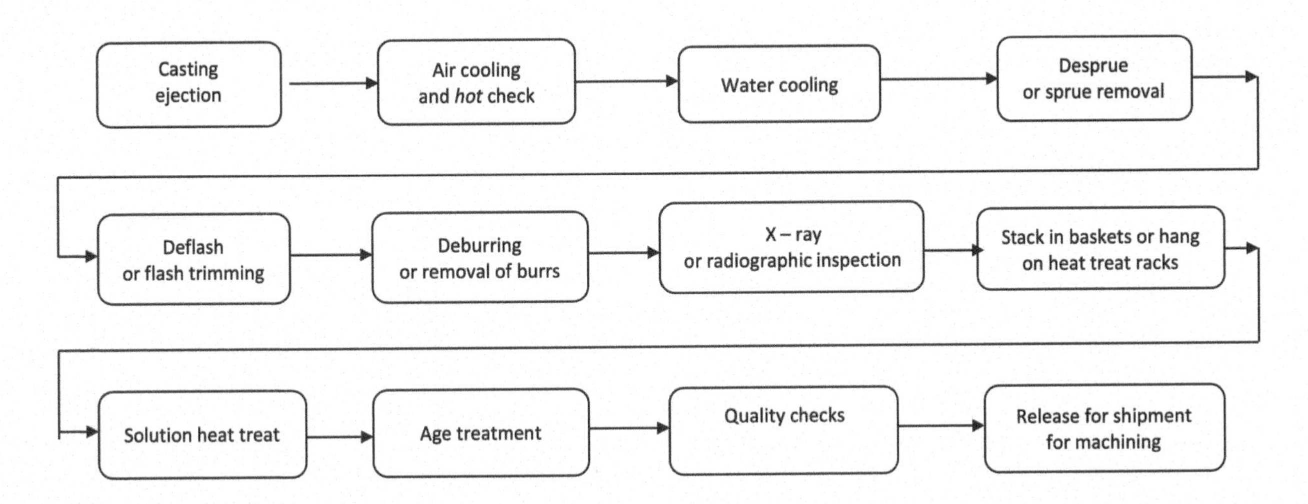

Fig. 10.14 Post-casting operations schematic

10.1.9.1 Desprueing, or Sprue Removal

Sprues are removed by:

- Punching out
- Drilling through
- Sawing or chopping off

In each case, the sprue with the steel screen embedded is removed. The screens are separated from the sprue by sweating them over a burner or in a furnace to melt out the aluminum. The sprues are placed over a grating with a gas burner at the bottom. A refractory-lined receiving pan is provided at the bottom to collect the melted aluminum. Since the melting temperature of carbon steel is 1425 to 1540 °C (2600 to 2800 °F) and that of aluminum is about 660 °C (1221 °F), the aluminum melts first, leaving the steel screens to be raked and removed. Punching the sprue out is suitable for components where the area around the punched hole is subsequently machined, for example for larger wheels for SUVs and trucks. Punching is liable to strain the sheared area with a potential for minor cracks. Machining around the punched hole removes the strained area from the casting.

Chopping off or drilling through the sprue may be alternative options, as illustrated in Fig. 10.15(a) and (b). Figure 10.15(a) illustrates the drill press for wheels and Fig. 10.15(b) illustrates a multidrill head machine for knuckle sprues (Ref 2).

If the sprue has a different shape, such as a truncated cone, the sprue and the transition conical section are both cut using a circular saw, as shown in Fig. 10.16. Figure 10.16(a) shows a front knuckle for an SUV produced by LPPM. It is gated into the bearing hub, which is a central location for metal flow and feeding. A conical gate connects the sprue and the hub. Figure 10.16(c) illustrates a knuckle turned with the sprue upward and a circular saw moving in to chop off the conical gate along with the sprue. A boss into the central hub in the fixture locates the casting. One or two stops in the fixture help to prevent the knuckle from displacement during the sawing operation, allowing the knuckle to be clamped rigidly to prevent any movement. Sprues, runners, and gates are removed from castings using cutoff saws. The castings then move to the next operation of flash removal, deflashing, or trimming by a conveyor or a robot.

(a)

(b)

Fig. 10.15 Drilling sprues; (a) wheel sprue drilling, (b) multidrill heads for knuckle sprues. Reprinted with permission from Ref 2

10.1.9.2 Deflashing, Flash Removal, or Trimming

Small and medium castings are deflashed during the trim press operation. Unlike the die-casting process, degating and deflashing are not done simultaneously in a trim press because the sprue and gates from LPPM are much thicker compared to the flash generated in the die-casting process.

Medium- and large-sized castings in LPPM may have flash around the parting plane due to the pressure during the third stage of the ramp and due to mold distortion at the operating temperatures. The flash needs to be removed by shearing in a trim press.

The orientation of the casting for trimming is maintained the same as it is for deflashing to avoid additional time for orientation manipulation. Figure 10.17(a) shows the front knuckle casting with a parting plane. Figure 10.17(b) illustrates the trim die concept for the front knuckle shown in Fig. 10.16 (Ref 4). The casting is supported by the bottom trim die and is held rigid by hydraulic clamps. The top trim die is actuated by a hydraulic cylinder, and it shears the flash off the knuckle as the top trim die lowers onto the bottom trim die. Four posts with bronze bearings ensure that the top and bottom dies are aligned to prevent any damage to the casting surface. One method sometimes used to prevent shaving of the casting surface during trimming is to provide a *trim bead* around the parting plane, as shown in Fig. 10.17(c). The trim bead ensures that even with a minor mismatch of less than 1 mm (0.04 in.), the casting surface will not be damaged if the top die shaves off any surface of the trim bead. The provision of the trim bead also helps in deburring the area of the flash that is shaved off during the trim operation.

10.1.9.3 Deburring, or Removal of Burrs

Burrs can result when the saw cuts the sprue and the gating, as well as during deflashing. The shearing of the flash (or deflashing) also results in burrs along the parting plane and the core pin prints. These burrs can cause injuries to the operators who

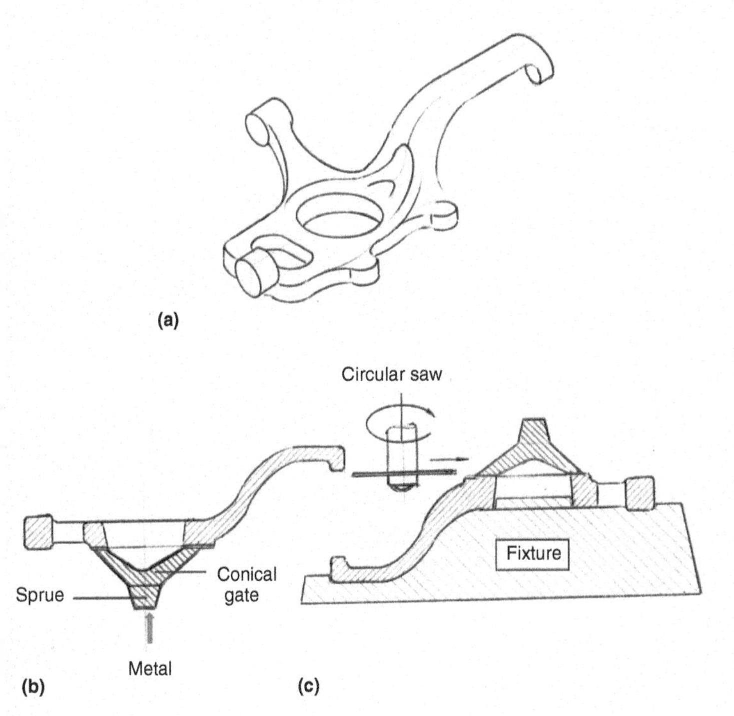

Fig. 10.16 Sprue removal by sawing or chopping; (a) front knuckle of SUV, (b) sprue and gate, (c) sprue and gate cut off

handle the castings. The burrs need to be removed manually for small and medium volumes or by automated devices such as computer numerical control (CNC) systems or robotics for larger volumes of production.

Figure 10.18 illustrates the saw-cut gate areas and the deflashed areas along the parting plane. Casting configurations and parting planes extend to the three dimensions. The carbide-tipped deburring tool (Ref 5) is guided as indicated. CNC programmers can use the 3D computer aided design (CAD) files to guide the tool in three dimensions.

10.1.9.4 Automated X-Ray

Automated x-ray or automatic defect recognition (ADR) is vital to high-volume aluminum wheel manufacturers and other structural casting producers. X-rays are produced by an x-ray tube mounted at the top of a lead-lined cabinet, which ensures containment of the radiation for operator safety. The x-rays produce an image that is captured by a detector. The detector converts the x-rays into visible light which is imaged by an optical camera. Alternatively, the detector reads the image using an x-ray sensor array. The differences in absorption of x-rays through the casting reveal any discontinuities or anomalies, such as shrink or gas porosity.

It is necessary to capture many views from different orientations, because complex castings are three-dimensional. The x-ray tube mounted at the top of the cabinet is designed to be manipulated in different orientations. The casting mounted in a fixture at the bottom of the tube is also designed to be oriented at different angles.

Figure 10.19 shows an image of an aluminum wheel with porosity defects (Ref 6). Selected views of the most critical areas

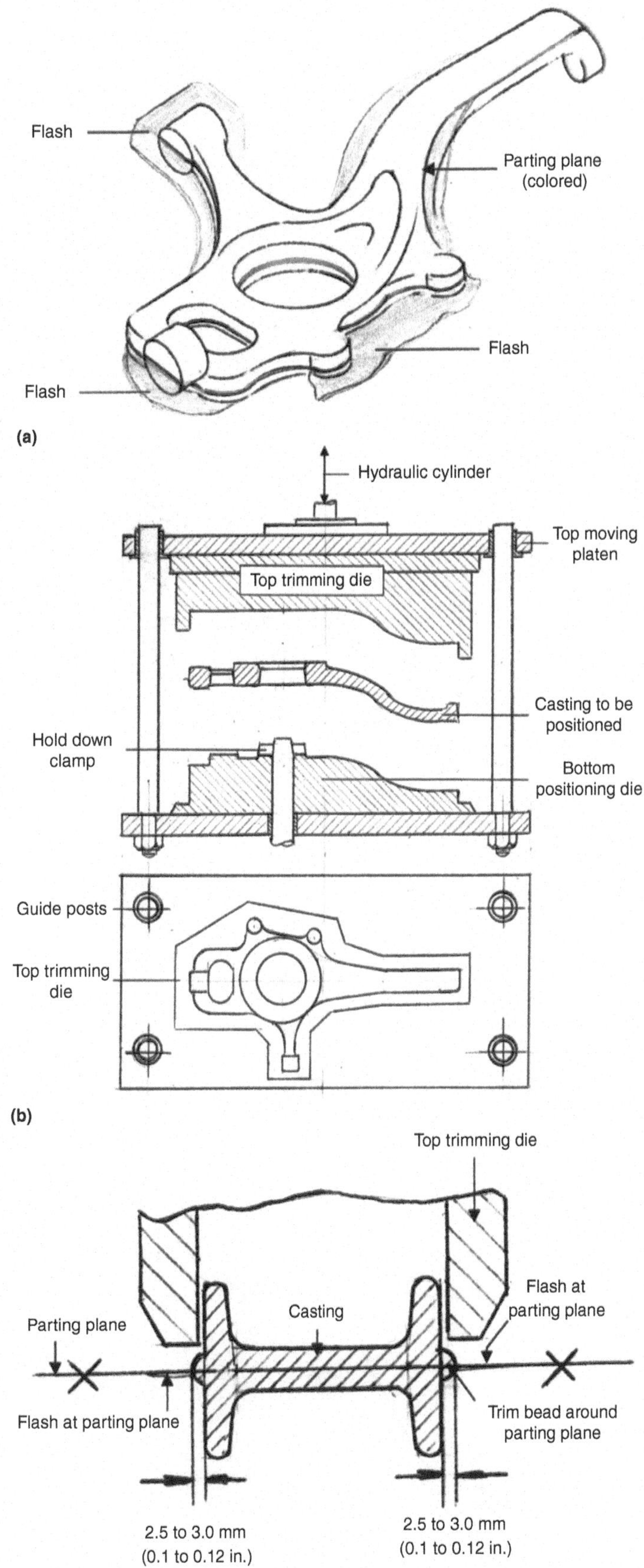

Fig. 10.17 Front knuckle casting; (a) parting plane and flash, (b) flash trimming die concept, (c) trim bead schematic

of the casting with acceptable and unacceptable levels of anomalies, as established by the American Society of Testing Materials (ASTM) Standard E155, are compiled and stored as computer database masters. The masters are saved for reference and verification of x-ray machine settings.

Each casting passes through the x-ray machine, and each view is compared to the master. Different filters are used for enhancement to see the differences in gray shades. Figure 10.20 illustrates this schematically (Ref 6).

Several views in position *p* are combined and coordinated. Different filters are used to accentuate the image resolution. The computer compares the read image against the master and decides whether to accept or reject the casting.

10.1.10 Heat Treatment

Aluminum castings (A356) are heat treated to improve their mechanical properties and machinability. The heat treatment, known as T6, is a two-step process, comprised of solutionzing and aging. The typical values applied to wheels produced in LPPM are provided.

10.1.10.1 Heat Treatment Cycles

Castings are heated to a solutionizing temperature of 527 °C, ± 5 °C (980 °F, ±10 °F) with the temperature gradually increasing to the set value in about 1.5 to 2 hours and held at the set temperature of 527 °C for a duration of 5 to 6 hours. The castings are quenched in water quickly, within 8 to 10 seconds, and maintained in water with agitation at 40 to 60 °C (105 to 140 °F). The castings are then removed from the water, and the water is allowed to drain for a maximum of 24 hours.

The second step consists of aging the castings by heating for 4 to 4.5 hours, typically at 155 °C, ± 2 °C (310 °F, ± 5 °F) to attain a minimum hardness of 90 Brinell hardness number (BHN). This aging cycle is chosen to create high strength, the required ductility, and good machinability. Figure 10.21 illustrates the process schematically.

10.1.10.2 Types of Heat Treatment Furnaces

The type of heat-treating equipment selected depends upon the production volume and the method of handling during deflashing and deburring. Smaller castings (up to 4.5 kg, or 10 lb. approximately) in medium volumes, produced either by GPM or LPPM, are deflashed and deburred manually and stacked in heat treatment

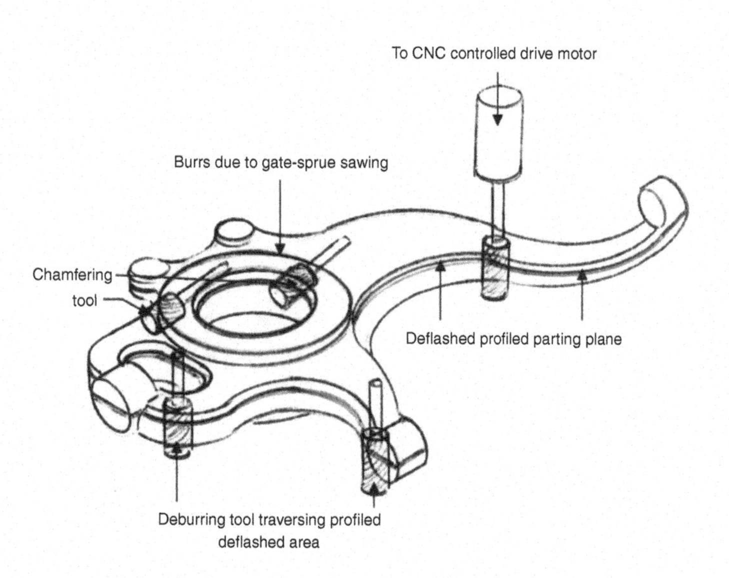

Fig. 10.18 CNC deburring schematic

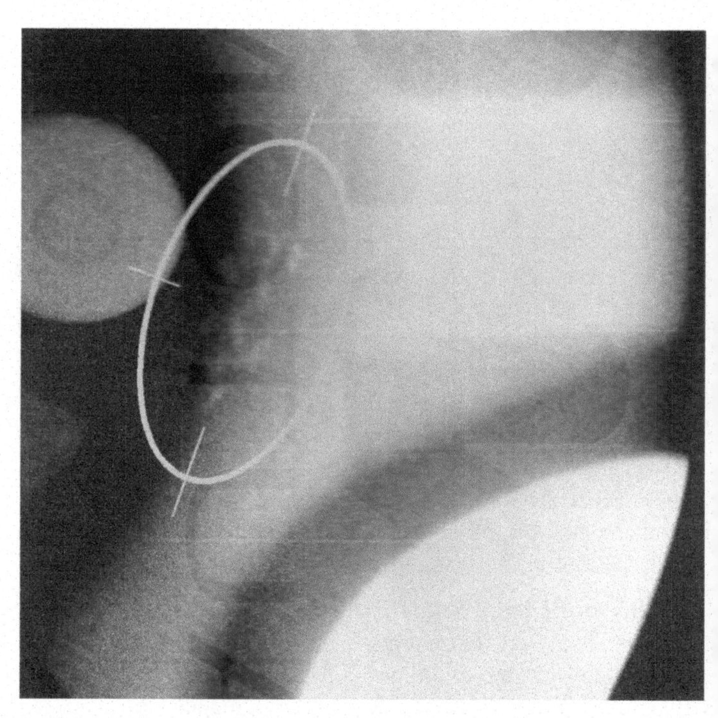

Fig. 10.19 X-ray showing porosity in an aluminum wheel. Reprinted with permission from Ref 6

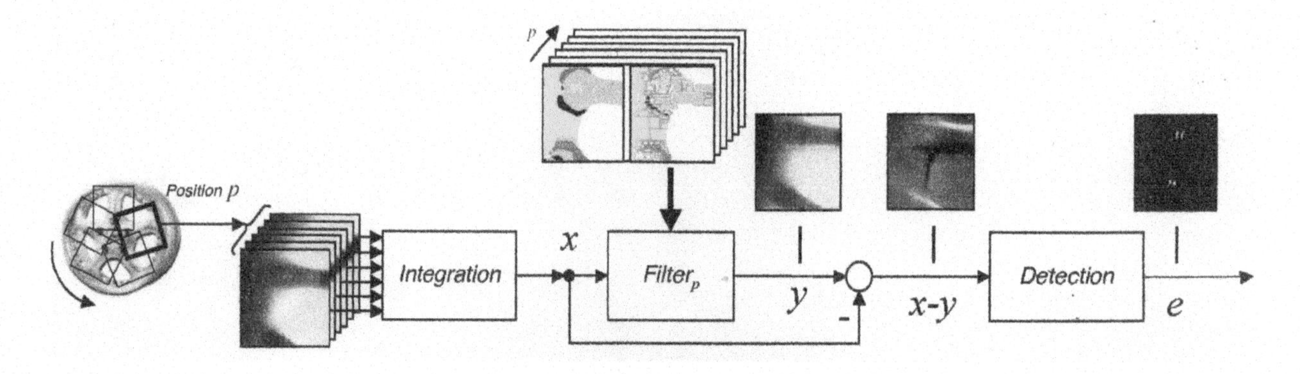

Fig. 10.20 Automatic x-ray inspection schematic. Source: Ref 6

baskets. Typical castings that fall in this category are brake master cylinders, calipers, knuckles, brackets, and passenger car wheels.

These castings are heat treated in solution heat treated furnaces, and the stack of baskets is handled by an overhead crane. The individual shallow baskets are convenient for the deburring operators. Figure 10.22 is a schematic of such an arrangement. If adequate head room is not available in the facility, the solutionizing furnaces can be located below ground level in concrete pits.

Medium-sized castings such as larger wheels, control arms, and large-sized knuckles for SUVs are suspended on racks, as shown in Fig. 10.23. Rack mounting is particularly well suited for drop-bottom solutionizing furnaces. The loaded racks are positioned below the drop-bottom furnace and hoisted up into the solution furnace, mounted at an elevation. After the load is hoisted, the bottom doors are closed and the furnace starts the solutionizing cycle, as illustrated in Fig. 10.21. The quench tank is positioned below the solutionizing furnace. The water is maintained at 40 to 60 °C (105 to 140 °F) by agitation using pumps. (The water tank is connected to a cooling tower for temperature regulation.)

Once the holding period of 5 to 6 hours is completed, the bottom doors are opened and the load is quickly lowered into the water in the tank for the quenching operation, as shown in Fig. 10.21. The load is removed after a couple of hours and naturally aged at room temperature before loading into the aging oven to carry out the artificial aging cycle shown in Fig. 10.21.

Figure 10.24 (Ref 7) illustrates a drop-bottom furnace showing an elevated solutionizing furnace with the quench tank below. This kind of furnace is versatile for jobbing as well as for mass production of a variety of products that are manufactured in one casting operation. Captive foundries engaged in producing one product (for example, wheel production or cylinder heads) use a continuous solutionizing furnace. The castings are stacked in baskets as shown in Fig. 10.23(a), and the quenching is done in baskets in succession, automatically. Figure 10.25 (Ref 8) illustrates this type of furnace.

10.1.10.3 Basketless Heat Treatment

The heat treatment baskets and racks are costly items when it comes to capital and maintenance. These items are made of

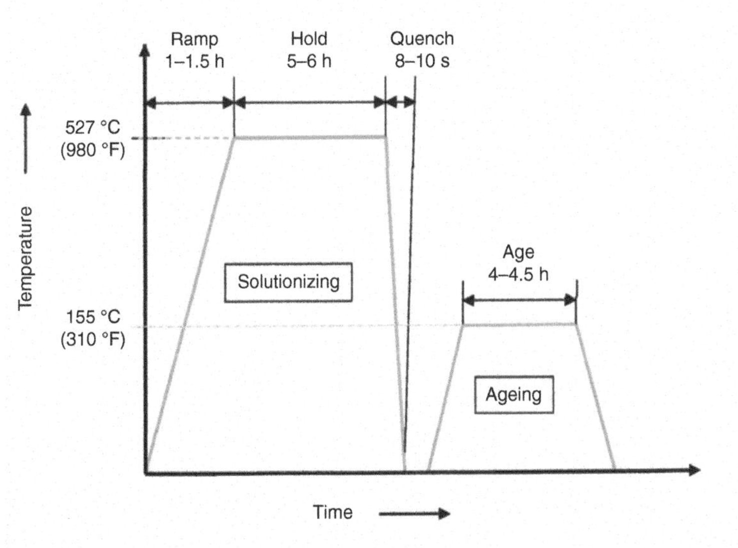

Fig. 10.21 Heat treatment cycles

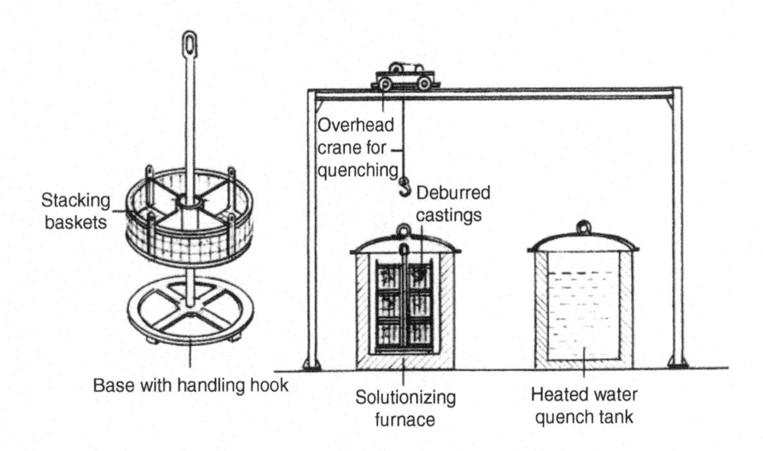

Fig. 10.22 Solutionizing heat treatment schematic

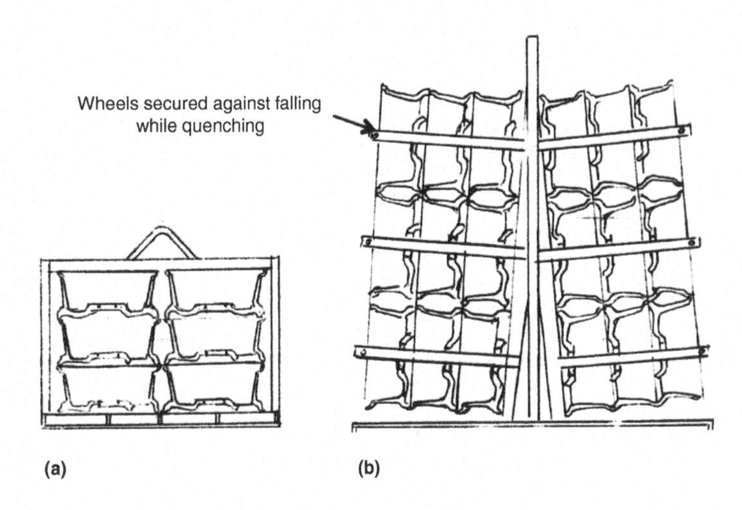

Fig. 10.23 Stacking methods, (a) basket, (b) rack mounting

Fig. 10.24 Drop-bottom solutionizing furnace. Reprinted with permission from Ref 7

Fig. 10.25 Continuous solutionizing furnace. Reprinted with permission from Ref 8

heat-resistant steel, and damaged baskets and racks need to be replaced periodically to prevent castings from slipping into the dip tank. Producers of medium- and large-sized castings such as wheels, cylinder heads, engine blocks, and crankcases use a basketless heat treatment system. Figure 10.26 (Ref 8) illustrates an installation for engine blocks, using a robot for handling.

The advantages of basketless heat treatment include:

- Reduced capital and maintenance costs
- Reduction in energy consumption as the mass of baskets or racks to heat is eliminated
- Cycle time reduction due to reduced mass to be heated
- Improved temperature distribution of casting and reduced property variability
- Increased productivity due to reduced cycle times
- Reduced need for handling equipment
- Lower space requirements

Fig. 10.26 Basketless solutionizing heat treatment. Reprinted with permission from Ref 8

10.1.10.4 Casting Quenching Distortion

Castings are prone to distortion due to stresses created during quenching, because different section thicknesses of the casting contract at different rates. Higher quench rates improve mechanical properties but also result in higher rates of distortion. The quench water temperature is selected as a compromise between creating high mechanical properties and limiting casting distortion.

Stacking shallow castings in baskets increases the tendency for distortion due to the weight of the castings bearing on one another. Castings such as lower and upper control arms are preferably hung on racks instead of stacking them in baskets. Knuckles for SUVs are also suspended on racks. The quenching orientation ensures that the heavier sections enter the water first to equalize the quench rates between the thick and thin sections. Figure 10.27 shows lower control arms suspended on a rack. Large castings such as the front cradle or the front crossmembers (see Fig. 10.12) are subject to distortion because the designed combinations of wall sections restrict contraction. The middle and the rear crossmembers are less complex and do not pose distortion issues. Some tooling designs are built with a camber (or a bow) in the opposite direction to compensate for the distortion (Ref 9).

Straightening of bent or distorted castings is done after solutionizing and before natural aging to bring the casting dimensions within the tolerance limits. This is to ensure that the casting fits

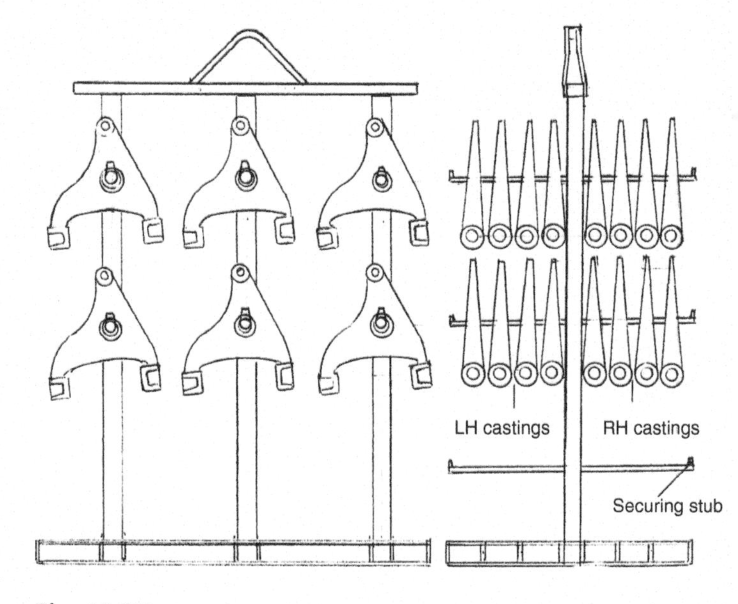

Fig. 10.27 Control arms hanging on a rack

in the assembly of mating parts and that the machine stock provided is adequate. Aerospace casting producers store the castings at freezing temperatures using dry ice, which slows down the natural aging process. But storing large castings such as front crossmembers in such an environment is too expensive. So, casting producers use straightening machines designed to bring the casting size within acceptable limits, allowing for a spring-back of nearly 33% during straightening. Some designers have opted to eliminate solutionizing and the aging treatment by using crossmembers in an as-cast condition. Simpler designs of crossmembers do not encounter distortion issues. Figure 10.28 (Ref 10) illustrates a schematic of such a crossmember casting.

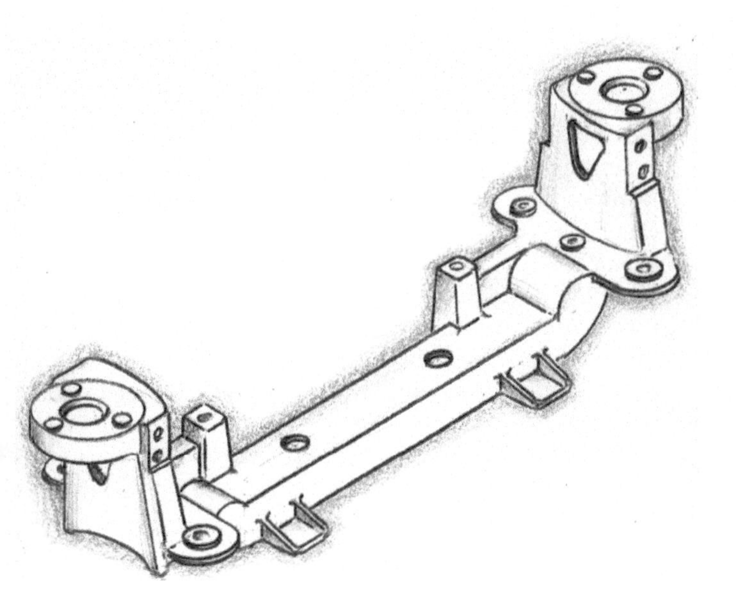

Fig. 10.28 Crossmember casting schematic. Source: Ref 10

10.1.10.5 Quality Checks

Product specifications are established to ensure fitness for function when the design is developed. Quality checks are performed on castings before machining to confirm that the material and the product meet the specifications.

Quality checks are grouped into three categories:

- Material checks
- Product checks
- Functional checks

Figure 10.29 provides an overview of the product quality checks for critical safety structural castings produced by gravity permanent or semipermanent molding, as well as low-pressure permanent and semipermanent molding. The quality checks for counter-pressure castings are identical to those produced by the low-pressure permanent mold process.

10.1.10.5.1 Material Checks

Material checks are divided into chemistry, microstructure, and mechanical property checks.

- **Chemistry checks:** The chemical composition of the melt (after grain refinement and silicon modification) is checked before the casting is poured. Records are maintained to document the castings poured, furnace used, ladle numbers, and time of pouring. Casting facilities use a spectrograph for quick feedback about the chemistry and to confirm the process capability. Trends in composition are tracked, and actions are taken to ensure that the chemistry is corrected to be within the specification limits.
- **Microstructure verification:** The specifications establish the acceptable microstructure from a part of the casting that experiences high stress. Because microstructure verification is a destructive test, it is conducted on a representative sample of the production run. Normally, for an aluminum silicon

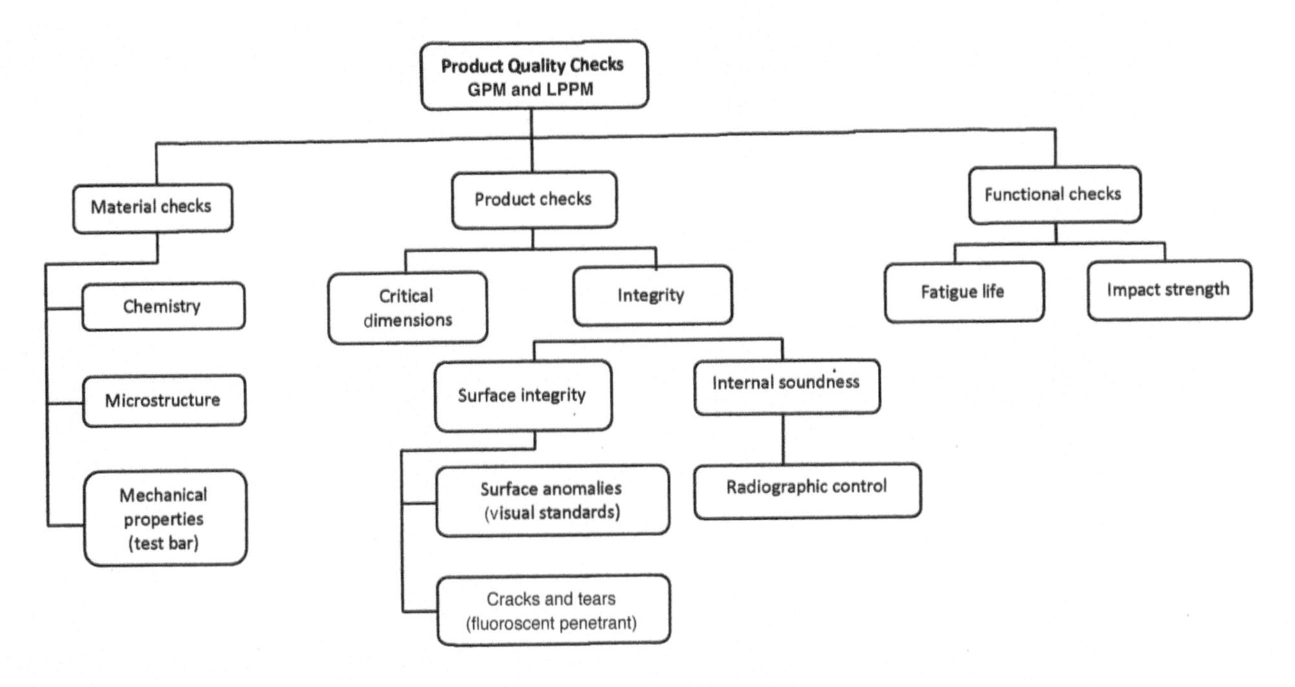

Fig. 10.29 Product quality check overview

alloy (A356-T6) used for structural applications, the grain size, degree of silicon modification, and the dendrite arm spacing (DAS) are specified and confirmed at regular and periodic intervals per the specifications.

- **Mechanical properties:** The designer specifies the locations for the test bars, the hardness (BHN), and the microstructure checks based on high-stress areas.

Figure 10.30 (Ref 1) illustrates the identified locations for a front knuckle used on an SUV for product validation (PV) and production process approval (PPAP). Only one location of the test bar (test bar1), microstructure (area1), and BHN (area1) are chosen as a basis for approval or for release of production shipments. Because these tests need a sectioned part, only one to three castings per heat treatment batch are tested. These are randomly selected from the top of the heat treatment basket or rack because these locations, due to delayed or insufficient quenching, may reflect the worst-case events.

10.1.10.5.2 Product Checks

- **Critical dimensions:** Castings for validation and pre-production approval are checked for designated dimensions. Critical dimensions are identified by the designer for fitness for function and assembly. The process capability is established by statistical analysis on a batch of 20 to 25 castings from the preproduction run. Dimensions that may change over time, due to variability such as pattern wear, are measured by gauging.
- **Checks for integrity:**
  Surface integrity is checked manually against visual standards established and agreed upon before the product launch (Ref 1). Castings are hot checked for surface anomalies as soon as the castings are ejected and moved to the cooling belt; this allows anomalies to be caught *just in time* and identifies the locations where mold coatings may need to be touched up.

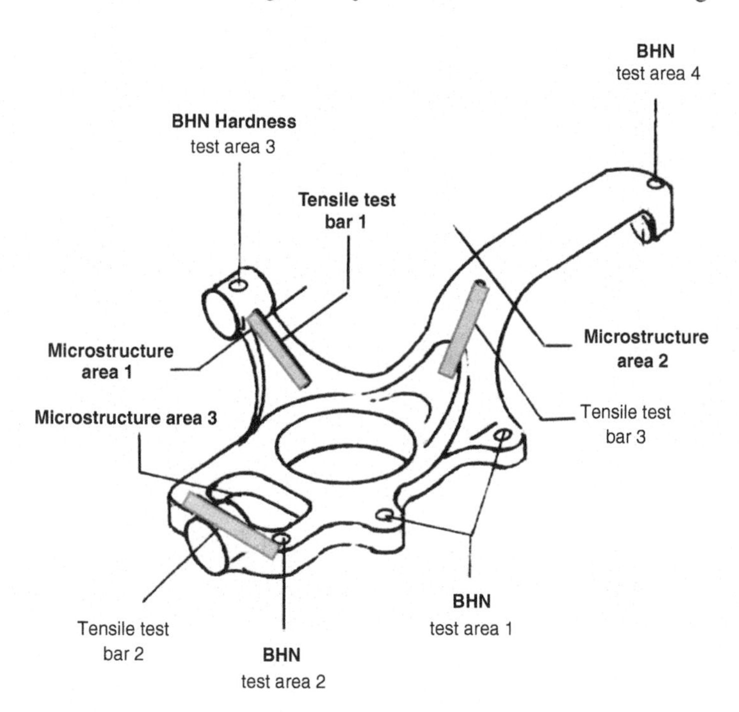

Fig. 10.30 Front steering knuckle showing locations for test bars, BHN measurement, and microstructure confirmation for product validation

Detailed visual inspection requires the part to be cold enough to handle if the inspection is performed manually. Automated dimensional inspection is performed using cameras set up to view the part in all three dimensions, and the computer inspects the castings against previous standard images of acceptable standards and alerts.

Critical safety structural castings used for suspension (knuckles and control arms) and brakes (calipers and master cylinders) are inspected for surface cracks or tears using fluoroscopic inspection or dye penetrant inspection (standard ASTM E165). Heat-treated castings are dunked automatically in a dye, allowing time for the dye to penetrate any cracks. The castings are then washed to clean the surface of any dye, and they are inspected manually using an ultraviolet light that highlights the cracks through fluorescence.

Internal integrity or soundness for any shrinkage, oxide inclusion, or gas porosity is checked using radiographic or x-ray units (Fig. 10.19 and 10.20) located in lead-lined cabinets. The automation includes a conveyor that moves the casting into the cabinet using an access door. Alternatively, a robot can be designed to grasp the casting and orient it in a designated position under the x-ray tube. The images are compared against the master images (or standard) or templates of acceptable and unacceptable levels of anomalies. The PLC sends a signal to the robot either to accept for further processing or isolate the suspect casting in a scrap review bin for the quality team to review and take corrective action.

10.1.10.5.3 Functional Checks

- **Fatigue life:** Test fixtures and loading devices are set up to simulate the varying loads experienced by the component in service (for example, a front steering knuckle in a passenger car). During the product development and validation stages, the castings with acceptable levels of anomalies are subjected to simulated loads of designed magnitude and designated directions. Three castings are tested to ensure that the fatigue life exceeds the specified number of cycles equivalent to the vehicle running for 100,000 miles.

  One or two of the castings are fatigue tested until the casting fails. The location and mode of failure are confirmed against the highest stress location, per the finite element analysis (FEA) used in design for fatigue life. This is a validation of the design analysis. During production, one or two products are annually bench-tested to ensure that there have been no unacceptable changes in the quality or process, over time.
- **Impact strength:** Products such as aluminum wheels and knuckles are subjected to impact loads when the wheel hits a curb or goes over a pothole in the road. The stresses created by these events are excessive. Tests are set up to simulate these shock loads by dropping a weight to confirm if the part can withstand these excessive stresses. Wheels or steering knuckle designs will be modified as needed for conformance with standards.

10.1.11 Low-Pressure Semipermanent Molding Examples

Castings that require fluid containment or that need to be designed for high stiffness require hollow or sand cored casting designs. Low-pressure semipermanent molding (LPSPM) examples include:

- Air passages: intake manifolds, turbocharger pump housings
- Water cooling passages: engine blocks, cylinder heads, front engine covers
- Oil passages: crankcases, differential carriers
- Structural stiffness: control arms, crossmembers, steering knuckles for SUVs.

10.1.11.1 Post-Casting Processes for Low-Pressure Semipermanent Molding

The post-casting processes of LPSPM castings are similar to those of LPPM, with minor modifications due to the sand cores. Castings of LPSPM ejected from molds need to be decored before being submerged in the water for cooling. Automation in handling ejection to decoring is desired. Also, because many of the cored castings contain fluid (water or oil), they must be tested for leaks or for residue after washing.

Figure 10.31 outlines the process sequence for cored components after casting. Castings, after they are ejected, are hot checked for any major issues such as non-fills or core floats. Because the burning of the bond in the cores releases environmentally unacceptable gases, castings with cores move to an air-cooling chamber where the effluent gases are drawn out and the incoming fresh air cools the castings before decoring. Castings need to be decored before they are quenched in water.

10.1.11.2 Decoring, or Core Removal, for Low-Pressure Semipermanent Molding

Because LPSPM does not have large feeders, the areas for hammering need to be chosen carefully so as not to damage the casting. Sometimes special impact pads need to be designed in consultation with the product designer.

Impact hammers vibrate or shake the casting to decore them. Figure 9.18 in Chapter 9, "Gravity Permanent and Semipermanent Mold Equipment" of this book, illustrates the decoring stand for a cylinder head. Figure 10.32 illustrates a hollow-cored lower control arm and holes designed for decoring.

Structural castings such as hollow lower control arms do not require the same degree of core passage cleanliness as crankcases, which feature oil passages. Metal penetration into the sand grains of such castings is decreased by reducing the third stage pressure to 0.055 to 0.069 MPa (8 to 10 psi) for no-bake cores.

The water-cooling passages of cylinder heads and engine blocks, and the oil passages of crankcases, need to be checked for any remnants of the sand cores using a video borescope. The rest of the processes, such as desprueing, deflashing, deburring, x-ray inspection, heat treatment, and quality checks are the same between LPPM and LPSPM processes.

Castings are premachined to ensure that there are no leak-prone areas exposed during machining. Also, the premachined surface provides a better sealing surface during the leak testing process. Industrial leak testers are based on pressure decay or by using helium gas (Ref 10). Castings such as crankcases which feature an oil passage are washed after machining, and the residue is measured in a sample. The cooling passages of cylinder heads are also washed. The amount of residue should be less than the specified maximum limit.

Other quality checks such as hardness, mechanical properties, microstructure, and critical dimensions are common between LPPM and LPSPM processes.

10.1.12 Cycle Time and Number of Machines

Unlike GPM castings, LPPM and LPSPM castings do not have large feeders. The cycle times are not affected significantly by the time required for the sprue and the runners to solidify.

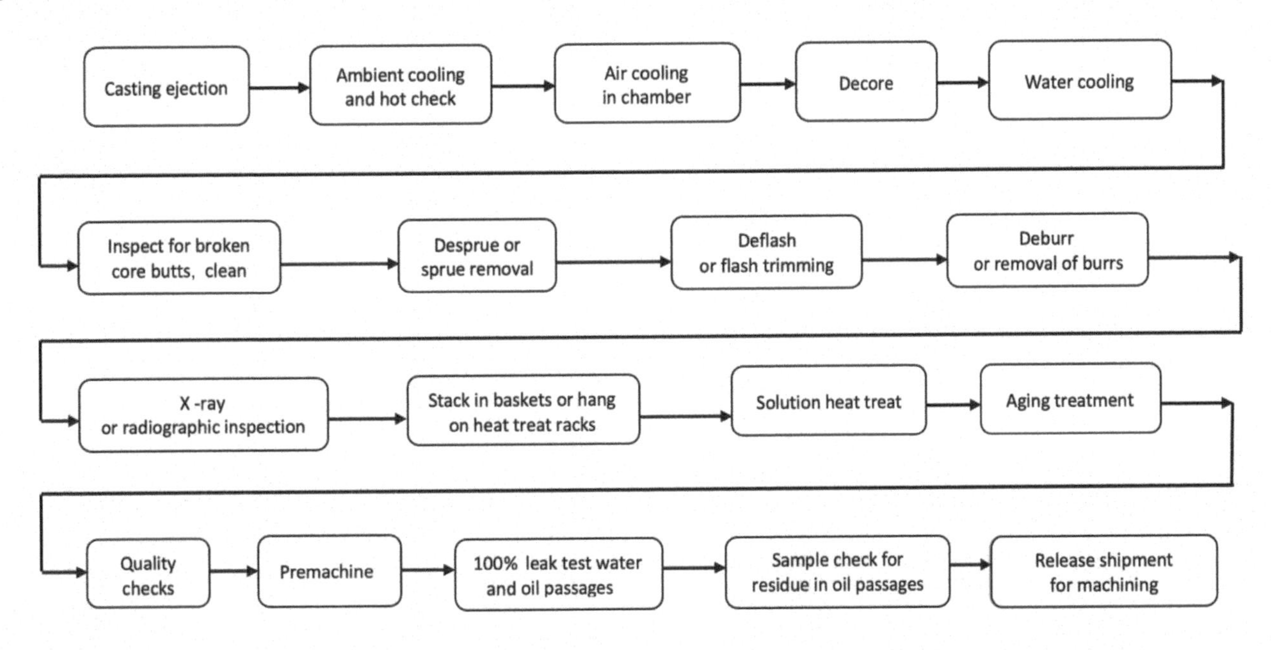

Fig. 10.31 Processes of LPSPM following casting

10.1.12.1 Cycle Times

The cycle times are established initially using computer simulation and finalized by actual trials. The variables are the poured weight, section thickness, and configuration. Table 10.2 lists the cycle times of castings produced by LPPM.

LPSPM cycle times are assumed to be 10 to 15% more than the LPPM process for general planning purposes. LPSPM castings are usually thin-walled, but the heat transfer from the metal to the cored surfaces is also significantly less.

10.1.12.2 Calculating the Number of Machines

Planning the number of machines in an LPPM casting facility requires certain assumptions to be made:

- Size of the machine
- Number of cavities in the tooling
- Cycle times
- Machine uptime
- Number of production shifts
- Number of working days in a year
- Scrap rates

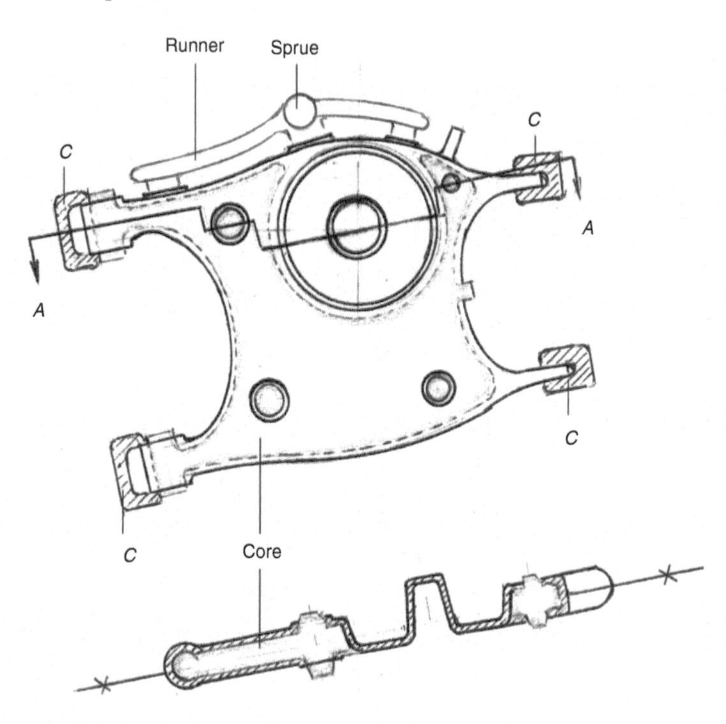

Fig. 10.32 Decoring a lower control arm casting

When planning for a captive foundry, it is preferable to use computer simulation to establish the cycle times. But even a casting facility for wheels produces different styles and sizes that affect cycle time assumptions. Table 10.2 can be used for suspension casting cycle assessment.

10.1.12.3 Calculating the Number of Machines Example

An LPPM facility is being planned for 500,000 SUV/truck rear lower control arms (RLCA) per year with a weight of 8.6 kg (19 lb.) each.

Assumptions:

- Size of machine: The machine can accommodate 1 cavity for a single control arm
- Cycle time: Table 10.2 shows a control arm with a weight of 8.8 kg (19.6 lb.) with a cycle time of 270 seconds
- The machine uptime is 75%
- The casting facility works 24 hours a day in 3 shifts
- The number of working days in a year is 300
- Scrap rate is 6% (conservative assumption)

In one hour, each machine produces in 3600 divided by 270, which equals 13.33 shots per hour.
The number of castings per day is 13.33 times 24, which equals 320.
The number of productive days in a year is 300 times the percent of uptime, which is 300 times 0.75, or 225 days.
The number of castings produced per machine per year is 320 times 225, which equals 72,000 RLCAs.
The number of good RLCAs produced per machine per year is (100 scrap rate) divided by 100 times the number of RLCA castings, which is 0.94 times 72,000, or 67,680.
The number of machines needed for 500,000 RLCAs per year is 500,000 divided by 67,680, which is 7.4 machines.

Therefore, the number of machines planned is 8.

10.1.13 Low-Pressure Permanent Molding and Low-Pressure Semipermanent Molding Plant Layouts

Chapter 1 of this book ("Casting Manufacturing Layout: Principles and Guidelines") highlights the importance of engineering a suitable layout for productivity and competitiveness. Figure 10.33 represents a layout of eight machines with a central

Table 10.2 Cycle times of castings produced in LPPM

Part	Weight, kg (lb) Cast	Weight, kg (lb) After desprue	Number per mold	Total cycle time, s	Wall thickness, max, mm (in.)
16 by 6 in. wheel	8.34 (18.4)	8.25 (18.2)	1	260	15 (0.6)
17 by 7 in. wheel	9.79 (21.6)	9.66 (21.3)	1	285	17 (0.67)
SUV rear lower control arm	8.14 (17.96)	8.05 (17.76)	1	266	20 (0.79)
SUV/truck rear lower control arm	8.97 (19.8)	8.88 (19.6)	1	270	20 (0.79)
Sedan, 6-passenger, rear trailing arm	7.66 (16.9)	7.48 (16.5	2	280–305	26 (1.02)
Crossover small SUV, front lower control arm	4.90 (10.8)	4.62 (10.2)	2	230	23 (0.91)
Luxury passenger vehicle, rear lower control arm	3.90 (8.6)	3.81 (8.4)	2	240	40 (1.57)
Sedan, 6-passenger, front control arm	2.9 (6.4)	2.8 (6.2)	4	240	23 (0.91)
Luxury passenger vehicle, rear upper control arm	1.18 (2.6)	1.09 (2.4)	4	150	20 (0.79)

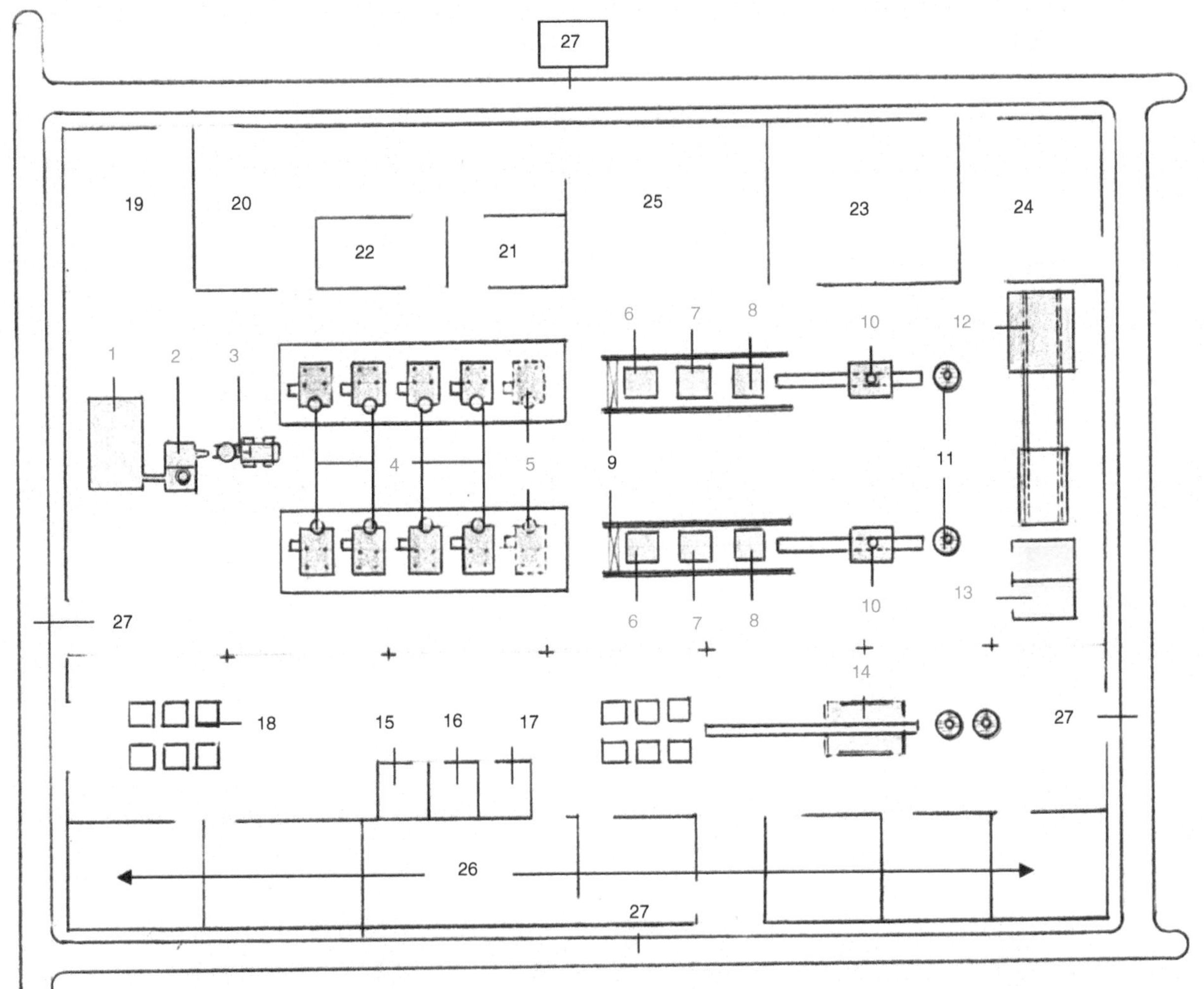

Note: 1. Central melter, 2. Degassing and filter furnace 3. Tow motor with distribution ladle 4. LPPM machines 5. Future expansion provision 6. De-flashing 7. Desprueing 8. Deburring 9. Gantry crane robot 10. X-ray unit 11. Heat treatment racks 12. Solutionizing furnace 13. Ageing ovens 14. Dye penetrant inspection 15. Spectrograph 16. Dimensional inspection 17. Metallography 18. Released shipments 19. Charge material storage. 20. Tooling storage 21. Tooling reconditioning 22. Mold coating 23. Maintenance 24. Utilities 25. Provision for core storage 26. Office space 27. Access roads

Fig. 10.33 Layout of an LPPM manufacturing unit

melting and metal treatment furnace. The machines are aligned in a row with four machines on either side; the machines receive metal periodically using a tow motor through a central aisle. There is also an aisle at the back of the machines to facilitate delivery of the tooling and unloading without interfering with the metal distribution. Many LPPM facilities operate with one operator who operates two machines. A casting cycle time of 270 seconds allows enough time for the operator to operate two machines alternately. Some use a robotic screen setter that operates from the back of the machine. In such cases, one operator can operate three machines. The screen-setting robot can also be engineered for core setting.

Overhead conveyors enable the transportation of the castings to common deflashing and deburring areas. A robot picks up the castings after deburring and delivers them to the x-ray unit with automatic defect recognition (ADR) capability. An operator can supervise the x-ray machine, perform visual inspection, and stack the castings in the heat treatment rack.

Heat-treated castings pass through a dye-penetrant inspection to look for any anomalies such as cracks and folds. The quality department oversees the quality check functions such as checking hardness, tensile testing, chemistry certification, layout on the coordinate measuring machines (CMMs), and checking by gauging. The quality department has the authority for final product release.

The layout provides for future expansion within limits. The offices are also laid out for ready access to all the activities, and the approach roads are planned for material movement in and out.

10.1.14 Vacuum Application to Low-Pressure Semipermanent Molding

Manufacturers can apply vacuum to draw the metal into the mold cavity. A vacuum manifold is built over the top mold, and vacuum is applied through sinter vents. The ejector pins need to be sealed against vacuum leaks using high-temperature-resistant gaskets. Their provision adds to the tooling costs, and their proper maintenance is critical for product quality. Thicker sections that need feeding to offset shrinkage porosity require low-pressure application once the cavity is filled. Cost competitiveness requires that the application of vacuum should be considered only if a well- engineered LPPM process is unable to meet the mold filling needs.

10.2 Counter-Pressure Casting Process

In both LPPM and LPSPM processes, the sealed furnace is pressurized with low pressure for both filling and feeding while the mold tooling is maintained at atmospheric pressure. In the counter-pressure process, both the furnace and the tooling are sealed and pressurized using medium pressure. The pressure in the furnace is countered by the pressure in the tooling, and the pressure differential between them is altered for controlled filling and feeding.

The counter-pressure process targets casting solidification at medium pressures that are higher than those used in low-pressure permanent mold process. The key features of this process are:

- The mold cavity fills due to the difference in pressures between the furnace and the mold.
- Both the furnace and the mold are subjected to pressure.
- Casting solidifies under medium pressures of 0.40 to 0.6 MPa (58 to 87 psi or 4 to 6 bars).

The higher pressure used in the process during solidification suppresses the precipitation of hydrogen, resulting in significantly less gas porosity. These medium pressure levels substantially improve the ability to feed the shrinkage cavities through the partially solidified mushy zone.

Figure 10.34 is a schematic of a four-post machine with an electric resistance heated crucible encased in a pressure chamber capable of withstanding up to 1 MPa (145 psi), although the maximum working pressure is less than 0.6 Mpa (87 psi). The metal mold is also encased in a pressure chamber, as illustrated. The mold pressurization chamber is split and is provided with a gasket seal in between the two halves; the lower half protects the operator against any accidental metal leaks from the mating surfaces. The sides of the mold are pulled out using the angled pins, as shown. Hydraulic pull-back cylinders are not feasible because of the circular pressure chamber that surrounds the tooling. The stalk tubes are made of aluminum titanate, which is nonwetting and heat-resistant. A hydraulic cylinder is provided at the top for ejection of the casting.

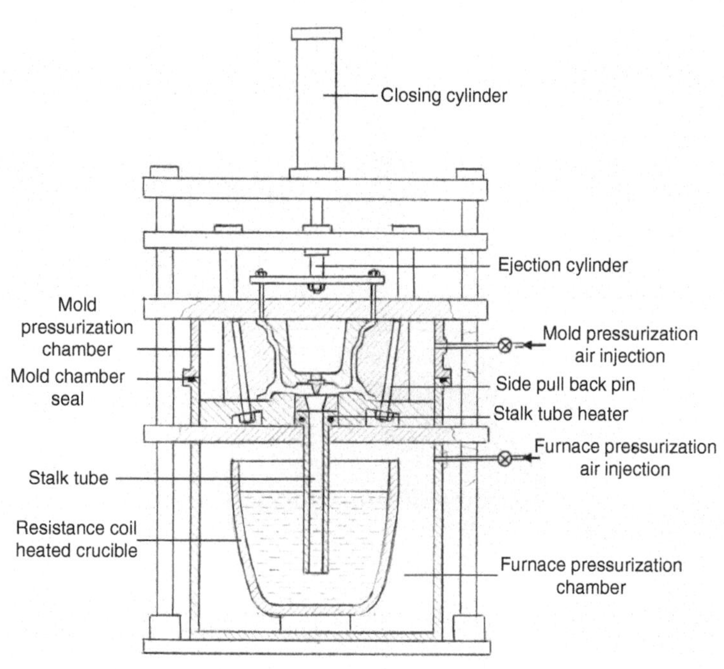

Fig. 10.34 Counter-pressure machine schematic

Figure 10.35 (Ref 11) illustrates the sequence of process steps, and Fig. 10.36 (Ref 11) shows the pressurization increases of the process.

- **Step 1** —The mold is blown out to clean any flash. The steel screen (or screens in the case of multicavities) is set manually or by a robot. The mold is closed, and the cycle is initiated.
- **Step 2** — The furnace and the mold chamber are pressurized equally. There is no counterpressure, and there is no metal flow.
- **Step 3** — The pressure in the furnace is gradually increased (0.03 to 0.1 MPa) to build the differential pressure for controlled filling without turbulence.
- **Step 4** —After the cavity is filled, the pressure in the furnace is increased to build the differential pressure, ranging from 0.4 to 0.6 MPa, for feeding the shrink porosity as the casting solidifies. Water- and air-cooling circuits built into the tooling promote directional solidification. Pressure is maintained until the feed cut-off point, where the liquid fraction in the mushy zone is too low for the pressurization to be effective. This can be established by solidification modeling.
- **Step 5** — Solidification continues, and the casting reaches a temperature at which it can be ejected without distortion or damage.
- **Step 6** — The top mold is pulled up by the hydraulic cylinder at the top of the machine. The receiver table swings (or slides) in between the bottom and the top mold. The ejector cylinder ejects the castings onto the receiving table. The table swings out (or slides out) to air-cool the castings before dunking them into the water tank beneath. The mold is blown out to clean it, and the screens are set to get the mold ready for the next cycle.

The counter-pressure casting process (CPC) has been successfully used for producing critical safety suspension components such as front steering knuckles, rear knuckles, front lower control arms, and rear lower control arms. It has been used for large components such as cored and noncored crossmembers, cradles, and crossmember nodes to be welded to rolled aluminum sections.

Figure 10.37 (Ref 11) illustrates a front knuckle, rear knuckle, front lower control arm, and a rear lower control arm. The rear knuckle shown in Fig. 10.37(b) weighs 4.5 kg. These knuckles are produced with a cycle time of 175 seconds. The mechanical properties taken from a test bar are: tensile strength, 345 MPa; yield strength, 291 MPa; and elongation, 10%. The microstructure showed an SDAS of 20 μm.

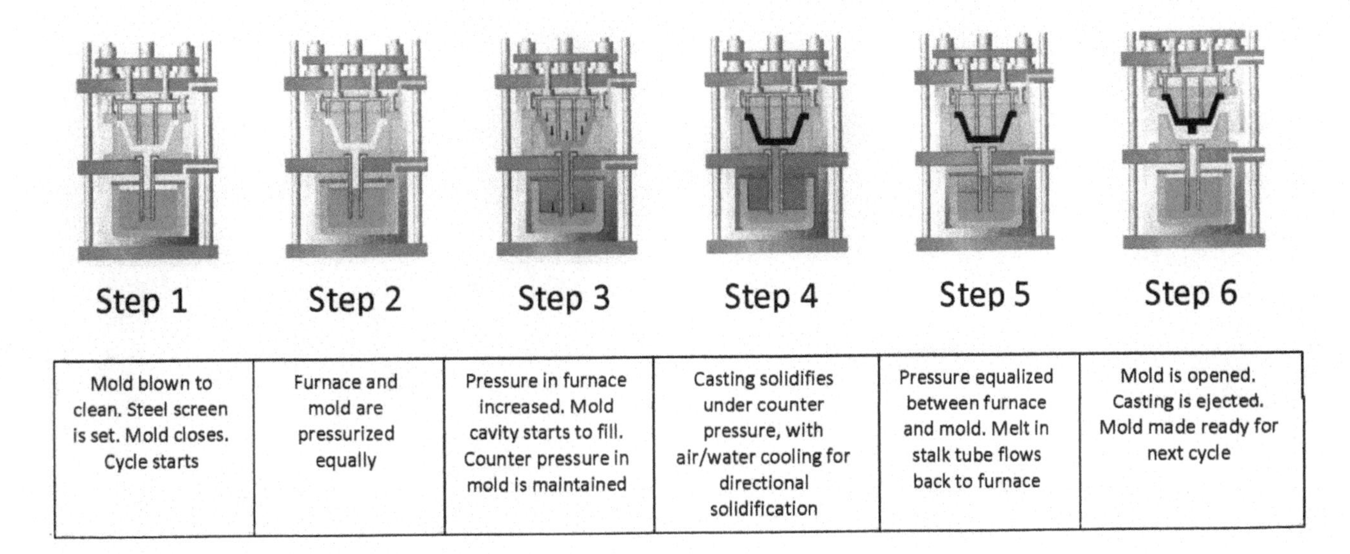

Fig. 10.35 Counter-pressure casting process schematic. Source: Ref 11

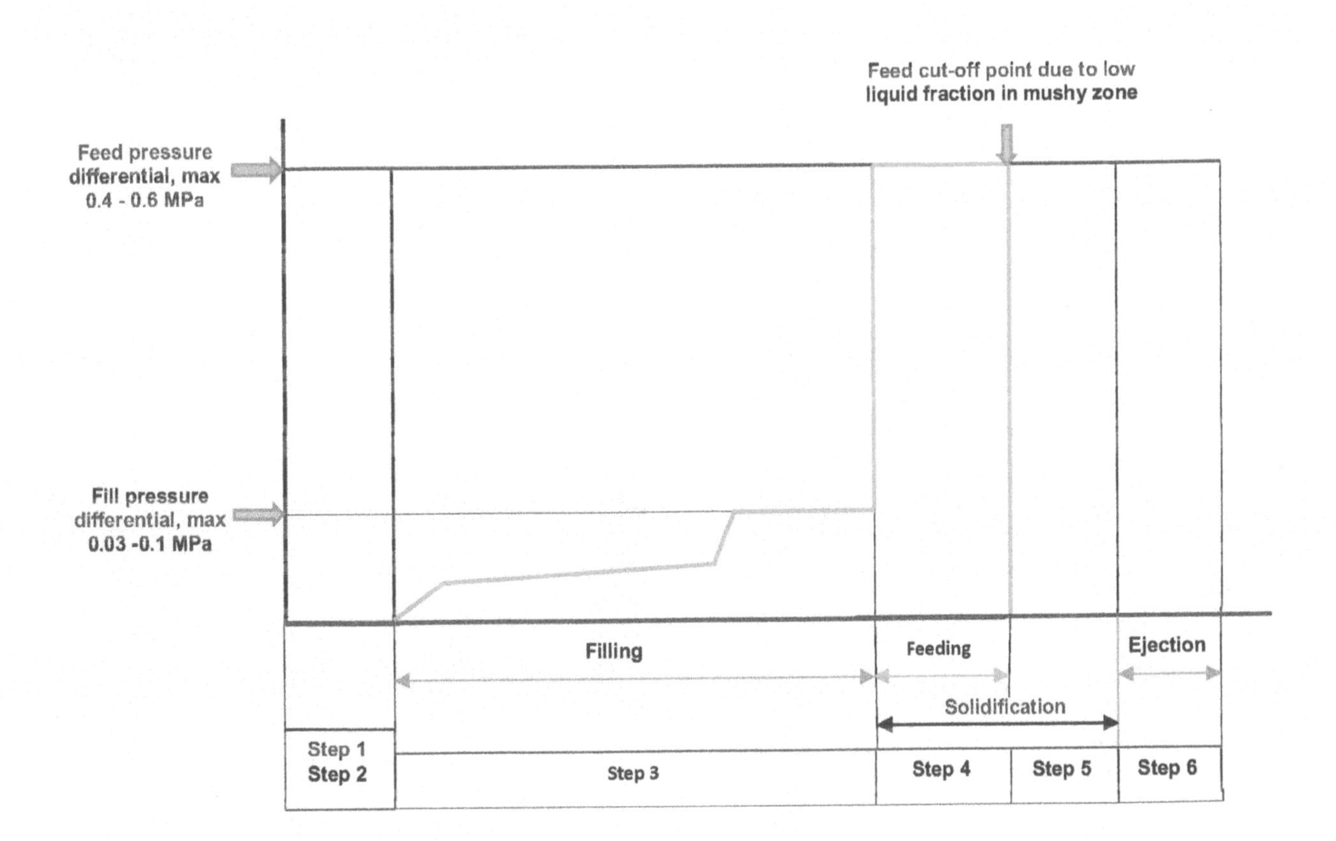

Fig. 10.36 Pressurization process schematic. Source: Ref 11

Figure 10.38 (Ref 11) illustrates a cored cradle casting with a centrally located sprue and side feeders to feed the thick sections on the periphery.

Figure 10.39 (Ref11) illustrates one of the larger machines suitable for producing castings such as cradles, subframes, and crossmembers. The large machines are used to produce eight knuckles, as can be seen in the figure. Figure 10.40 (Ref 11) provides a closer view of the knuckle castings on the receiving tray ready to be water-quenched. Table 10.3 (Ref 11) provides an overview of the machines and their applications.

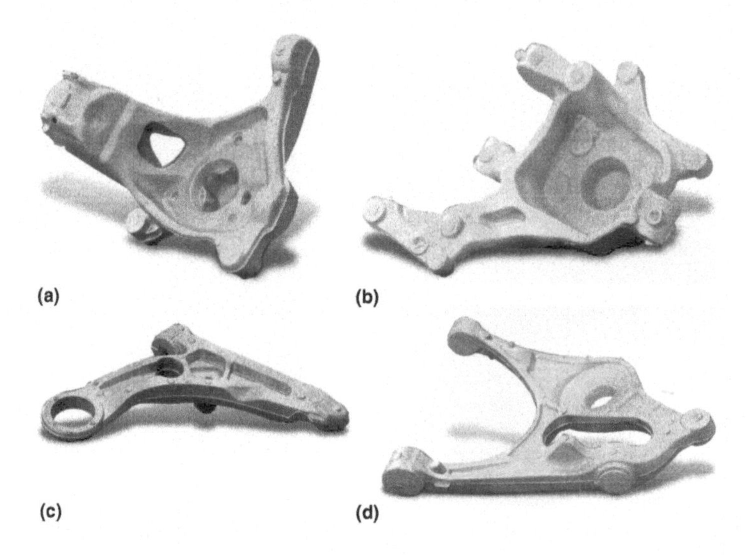

Fig. 10.37 Knuckles and control arms produced by CPC counter-pressure machines; (a) front knuckle, (b) rear knuckle, (c) front lower control arm, (d) rear lower control arm. Reprinted with permission from Ref 11

Fig. 10.38 Cored cradle casting with gating system. Source: Ref 11

Fig. 10.39 Large counter-pressure casting machine. Reprinted with permission from Ref 11

Fig. 10.40 Closeup view of the eight knuckles on the receiving tray over the quench tank. Reprinted with permission from Ref 11

Table 10.3 Counter-pressure machines and applications

Machine	Applications
CPC 1300 C 12	CPC machine that can be used as low-pressure machine
CPC 1500 C 64	...
CPC 1600 C 96	Can produce 6 knuckles, most popular size
CPC 1800 C 128	Can produce 8 knuckles or 1 cradle or subframe
CPC 2000 C 128	Can produce 2 cradles or subframes

Source: Ref 11

REFERENCES

1. J. Nath, *Aluminum Castings Engineering Guide*, ASM International, 2018
2. LPM Group, Italy
3. KURTZ ERSA Inc., Plymouth, Wisconsin
4. Lion Tool & Die, Algonquin, Illinois
5. Carbide Burr Inc., Morrow, Ohio
6. Automated Radioscopic Inspection of Aluminum Die Castings By Domingo Mery Departmento de Ciencia de la Computacion, Pontifica Universidad, Catolica de Chile
7. SECO/WARWICK Corp.
8. CAN-ENG
9. S. Sakamoto et al., Toyota Motor Co., The Casting Technology for the Aluminum Suspension Cross Member, SAE International Paper No. 901730
10. InterTech Development Co., Skokie, Illinois
11. CPC-USA Ltd, Lake Forest, California

Copyright © 2023 ASM International®
All rights reserved
www.asminternational.org

Index

C

D

E

F

G

H

I

L

M

T

V

W

Y

Z